Computational Biology

The *Computational Biology* series publishes the very latest, high-quality research devoted to specific issues in computer-assisted analysis of biological data. The main emphasis is on current scientific developments and innovative techniques in computational biology (bioinformatics), bringing to light methods from mathematics, statistics and computer science that directly address biological problems currently under investigation.

The series offers publications that present the state-of-the-art regarding the problems in question; show computational biology/bioinformatics methods at work; and finally discuss anticipated demands regarding developments in future methodology. Titles can range from focused monographs, to undergraduate and graduate textbooks, and professional text/reference works.

Author guidelines: springer.com > Authors > Author Guidelines

For further volumes:
http://www.springer.com/series/5769

Naruya Saitou

Introduction
to Evolutionary Genomics

 Springer

Naruya Saitou
Division of Population Genetics
National Institute of Genetics (NIG)
Mishima, Shizuoka, Japan

ISSN 1568-2684
ISBN 978-1-4471-7159-1   ISBN 978-1-4471-5304-7 (eBook)
DOI 10.1007/978-1-4471-5304-7
Springer London Heidelberg New York Dordrecht

Printed on acid-free paper

Springer is part of Springer Science+Business Media (www.springer.com)

I would like to dedicate this book to Dr. Nei, my academic supervisor at Houston.

Preface

Necessity of Evolutionary Studies

Organisms on the earth are rich in diversity. Each organism also contains its own genome with many genes. This complex genetic system has been generated and constantly modified through eons of evolution since the origin of life. Evolutionary study is thus indispensable for gaining the unified view of life. Because even a single-cell bacterium is so complex, we have to study its genetic entity, that is, its genome, to acquire a comprehensive view of the organism. I will discuss evolution of organisms from the viewpoint of temporal changes of genomes and methods for their study in this book, titled *Introduction to Evolutionary Genomics*.

Evolutionary changes already started even before the origin of life, known as chemical evolution. Therefore, we need to dig down to the molecular level, starting from nucleotides and amino acids. In this sense, it is logically straightforward to study evolution of life at the molecular level, that is, molecular evolution. Molecular evolutionary study as a discipline was established only after biochemistry and genetics became the center of biology in the middle of the twentieth century. Because of this late start, there are still some molecular biologists who consider the study of evolution as carried out by specialized researchers, while there are some old-fashioned evolutionists who do not appreciate molecular-level studies. It would be my great pleasure if such people change their minds after reading this book. But, of course, I hope that the majority of readers of this book are young students and researchers.

What Is Evolution?

Evolution is the temporal process of life. Originally, the word "evolution" meant the development of embryo from egg. Charles Lyell was probably the first person to use this word in the modern meaning in his *Principles of Geology* published in the 1830s. Thanks to the pioneering works of Lamarck, Wallace, and Darwin, evolution as a biological phenomenon was gradually accepted during the last 200 years (e.g., [14]). Evolution, however, does not contain a predetermined pathway, unlike developmental processes. As the time arrow moves from past to present, life forms change. Therefore, any temporal change of organisms is evolution. Nowadays, the concept of evolution is sometimes extended to nonlife, such as evolution of the universe or evolution of human society.

What Is a Genome?

The word "genome" was coined by Hans Winkler, botanist, in 1920 [15]. Genes were already localized in chromosomes in the cell nucleus at that time, and Winkler joined two words, "gene" and "chromosome," to produce a new word "genome." Plants are often polyploid, and there was a need to designate a certain unit of chromosome sets. Later, Hitoshi Kihara defined genome as a minimum set of genes that are necessary for that organism [6]. This function-oriented definition is still used today, but a structure-oriented definition needs to be invented. Thus, I redefined genome as "a maximum unit of self-replicating body" [12]. A "self-replicating body" includes not only usual organisms but also organella and virus that need some help from organisms for their replication.

Kihara and his group conducted genome analysis on various wheat species in the early twentieth century, and he coined this famous couplet:
The history of the earth is recorded in the layers of its crust;
The history of all organisms is inscribed in the chromosomes.

This couplet was originally mentioned in his book written in Japanese [7] and is increasingly become evident as we now study evolution at the nucleotide sequence level.

The word genome also implies completeness. It is important to grasp all genetic information contained in a single genome, because this gene set mostly determines life patterns of its organism. However, all genes in one genome are not sufficient for that organism to exist. This insufficiency is clear for parasites. For example, leprosy-causing bacteria, *Mycobacterium leprae*, has many pseudogenes [1]. These bacteria probably lost their functional genes through a long parasite life due to dependence on host genomes. We should remember that all organisms on earth are interacting with each other. These kind of known host-parasite relationships are the only prominent examples. Even our own human genome gives a good example of dependency on nonhuman genomes.

Vitamins, by definition, cannot be synthesized inside human body, and we need to obtain them through various foods. For example, deficiency of vitamin C, or ascorbic acid, causes scurvy. Many nonhuman organisms do produce ascorbic acid, as they have its chemical pathway. A gene for enzyme L-gulonolactone oxidase (E.C. no. 1.1.3.8) became a pseudogene (nonfunctional) in the common ancestor of human and Old World monkeys, and we are no longer producing ascorbic acid [9]. In any case, an organism cannot survive alone. We have to always consider the environment surrounding an organism.

Vitalism Versus Mechanism

If we consider the history of biology, one viewpoint presents a controversy between vitalism and mechanism. Vitalism maintains that life has a unique law that does not exist in nonlife forms; thus, it is dualistic. Mechanism is monistic, for it states that life only follows physicochemical laws that govern inorganic matters. In other words, there is no specific difference between organism and inorganic matters according to the mechanistic view. The long history of biology may be considered a series of victories of mechanism against vitalism (e.g., [11]). For example, we easily recall the theory of heart as a pump for blood circulation by William Harvey in the seventeenth century and the discovery of enzyme function in a cell-free system by Eduard Buchner in the nineteenth century. Biochemistry and genetics are two main fields of biology where the mechanistic viewpoint is always emphasized. Molecular biology inherited this aspect from these two disciplines.

Some biologists, however, were strong proponents of vitalism. Hans Driesch, developmental biologist in the early twentieth century, examined the development of sea urchin and discovered that two- or four-cell-stage embryos can develop adult individuals even after they were separated. Because of this utterly mysterious process, he proposed the existence of "entelechy" only in organism [4]. It is true that the animal development is still not completely known. Yet, the modern developmental biology is clearly on the side of Wilhelm Roux, a contemporary of Driesch. Roux strictly followed the mechanistic view in his study.

There is still the remnant of a dualistic view similar to vitalism, to consider mind and body as totally separate entities. As vitalism tried to demarcate organism from inorganic matters, this dualistic view tries to demarcate mind from body. However, a logical consequence of the mechanistic view of life is of course to explain mind as some special organismic process, most probably a neuronal one, that is, mind exists in a body, and these two are inseparable. The mind-body dualism is illogical to begin with and scientifically wrong (e.g., [2]). It is unfortunate that some eminent neurologists such as Wilder Penfield [10] and John Eccles [5] maintained the dualistic view.

Everything Is History

Mechanism is about to declare its victory over vitalism, including the mind-body dualism. Yet, it still remains whether we can explain the whole life phenomenon completely only through mechanism. Life, with its eons of history, is a product of evolution, and there are so many chance effects. For example, spontaneous mutations appear randomly. Most of the mutations that last a long time in the history of life are selectively neutral, and they were chosen through the random genetic drift [8]. Furthermore, there are so many inorganic factors that drastically change the environment of the earth. Examples are volcanic activities, ice ages, continental drifts, and asteroid impacts. These seem to be all random from the organismic world. These historical processes where chance dominates are out of control of mechanism.

As mutations arise, some disappear while the others remain. This process is impossible to be fully explained through the logical cause and effect style of mechanism. This is not restricted to life. *Shinra bansho* (Japanese; "all matters and events in this universe," in English) themselves are transient and there always exists a history, as Hitoshi Kihara pointed out in his couplet. After all, everything is history. Therefore, the essence of natural science is to describe the history of the universe at various levels. Often it is claimed that the ultimate goal of natural science is to discover the laws of nature, and the description of nature is only one process to the eventual finding of laws. It fails to put first things first. So-called natural laws are mere tools for an effective description of natural phenomena. A phenomenon that can be described succinctly is relatively simple, while from a complex phenomenon, it is difficult to extract some laws. However, such difference comes from the phenomena themselves, and the objective of natural science should not be restricted to phenomena from which it is easy to find some laws. It should be noted, however, that giving a mere description of everything is not enough. Human ability to recognize the outer world is physically limited, and a structured description of the historical process is definitely necessary, depending on the content of each phenomenon. In this sense, the time axis, which is most important for organismal evolution, is obligatory for the description of nature itself. With the above argument in mind, I am quite confident that the very historical nature of genes with its self-replication mechanism has the key to overcome the mechanistic view of this universe.

Genome as a Republic of Genes

Another important feature of the genome is its completeness in the finite world. As biology experienced the mechanism versus vitalism controversy, it suffers another controversy on methodology, that is, reductionism versus holism. Organismal evolution is the summation of small genetic changes, and the whole process can be understood through the divide-and-conquer strategy. This reductionistic approach is very useful for the evolution of genes. Some people stress the importance of the holistic approach. However, I do not know of any profound discovery in natural science where a holistic approach was truly effective. Of course, I am not trying to say that the whole is a simple sum of parts. Reductionism is the only approach to understand any phenomenon. Then the "genome" comes in. Because of its completeness, the reductionistic approach in genome studies naturally brings us the whole world. In this sense, evolutionary genomics plays an important role as a unique test case of methodology in biology.

The genome of one organism usually contains many genes, and they are interacting with each other in a complex way. In this sense, Saitou [12] characterized a genome as a "republic of genes." This implies that one particular gene or gene group is not controlling the other genes as the "master control gene." In other words, there is no fixed role for one gene as the master or slave. Any gene has the potential to control or influence functions of other genes.

It is true that in some genomes there exist systems in which a single "master control" gene plays a crucial role and only after this gene is expressed, other downstream genes are expressed. This kind of master control gene, however, has also the possibility of receiving some influence from other genes. Therefore, a top-down system, or the "empire of genes" viewpoint, may not apply to genes in every genome [13].

The "republic of genes" also implies another assertion that locations of genes in chromosomes are not so important. Since RNA and protein molecules are transcribed and translated, respectively, they are expected to show a "trans effect" – to influence genes on other chromosomes or those at a remotely related location of the same chromosome. In contrast, the "cis effect" is to influence the genes on the same chromosome at a close location. In fact, phenomena such as enhancers and chromatin remodeling recently received much attention. However, their effects may be restricted.

Synteny, or the gene order conservation between species, may also play only the passive role. If the gene order is important for the coordination of gene expression, as in bacterial operons, this order may be selectively conserved. However, most of the syntenic regions seem to be only results of chance effects, that is, a gene order happened to be void of disruption.

Genome sequence reporting papers often try to stress the importance of the cis effect, for a considerable part of any genome sequence information is the gene order. Yet, we have to first consider the trans effect when proteins or small RNA molecules are in question. In contrast, the cis effect should first be considered when a DNA sequence itself is influencing other genome regions, because this may involve neither transcription nor translation.

Structure of This Book

This book consists of three parts: Part I – Basic Processes of Genome Evolution; Part II – Evolving Genes and Genomes; and Part III – Methods for Evolutionary Genomics. Part I includes five chapters (Chaps. 1–5). I explain the basics of molecular biology in Chap. 1, while the following four chapters are more specific to evolution. Charles Darwin defined evolution as "descent with modification" in the *Origin of Species* [3]. DNA replication, explained in Chap. 1, is fundamental to "descent," or the connection from parents to offsprings, while "modification" in modern terms is mutation. Mutation is thus covered in Chap. 2, and phylogeny is described in Chap. 3 as the descriptor of evolution. Mutation is already a random process, but randomness also dominates throughout the default process of genome evolution, namely, neutral evolution, covered in Chap. 4. The description of natural selection follows in Chap. 5.

There are five chapters (Chaps. 6–10) in Part II. A brief history of life starting from the origin of life is discussed in Chap. 6, followed by an explanation of genomes of various organism groups; prokaryotes in Chap. 7; eukaryotes in Chap. 8; vertebrates in Chap. 9; and humans in Chap. 10. The evolution of many genomes is amply discussed in this part. Lineage-specific evolutionary problems are also covered in appropriate chapters.

We then move to methods in Part III (Chaps. 11–17). Genome sequencing, phenotype data collection, and databases are explained in Chaps. 11–13, followed by homology searches, multiple alignments, and evolutionary distances in Chaps. 14 and 15. Methods of tree and network building are discussed rather in detail in Chap. 16. Chapter 17 is devoted to population genomics.

I created a page on evolutionary genomics under my laboratory website (http://www.saitou-naruya-laboratory.org/Evolutionary_Genomics/), which contains a web appendix, detailed references, and a detailed index. The web appendix consists of four parts: basic statistics, worked-out examples of evolutionary genomics, and updates of each chapter.

It should be noted that I published a textbook on evolutionary genomics, written in Japanese, in 2007 [13]. This book is an extension of that book. If you have any questions, please contact me at saitou.naruya@gmail.com. I hope you will enjoy reading this book.

Following Walter Pater's maxim – "all art constantly aspires towards the condition of music" – I would like to conclude this preface with my epigram:

All biology aspires to evolution.

Acknowledgments

I appreciate the kindness of many people who helped me to complete this book: Mrs. Mizuguchi Masako for her programming and drawing figures, Mrs. Kawamoto Tatsuko for her excellent secretarial work, and Mrs. Noaki Yoshimi for providing me some gel photos and chromatograms. I thank current Saitou Lab members (Dr. Sumiyama Kenta, Dr. Kirill Kryukov, Mr. Kanzawa Hideaki, Ms. Nilmini Hettiatachchi, and Mr. Isaac Babarinde), my former students (Dr. Matsunami Masatoshi, Dr. Timothy Jinam, Dr. Takahashi Mahoko, Dr. Suzuki Rumiko, Dr. Masuyama Waka, Dr. Kim Hyungcheol, Dr. Tomiki Takeshi, Dr. Kaneko Mika, Dr. Noda Reiko, Dr. Kitano Takashi, and Dr. OOta Satoshi), and my former postdocs (Dr. Sato Yukuto, Dr. Kawai Yosuke, Dr. Shimada Makoto, Dr. Takahashi Aya, Dr. Kim Chungon, and Dr. Liu Yuhua) for their various comments and for providing some parts of their works to the manuscripts of this book. I also thank Dr. Xie Lu for providing me some materials for Chap. 12. I thank Dr. Andreas Dress and Dr. Martin Vingnon, series editors of Computational Biology, to include this book into their series. Finally, I thank people of my good old Houston days (1982–1986), the most exciting period of my academic life: Dr. Lydia Aguilar-Bryan, Dr. Ranajit Chakraborty, Dr. Stephen P. Daiger, Dr. Robert Ferrell, Dr. Takashi Gojobori, Dr. Dan Graur, Dr. Craig Hanis, Dr. Wen-Hsiung Li, Dr. Chi-Cheng Luo, Dr. Partha Majumder, Dr. Pekka Pamilo, Dr. William Jack Schull, Dr. Bob Schwartz, Dr. Claiborne Stephens, Dr. Fumio Tajima, Dr. Naoyuki Takahata, Dr. Ken Weiss, Dr. Chung-I Wu, and, of course, Dr. Masatoshi Nei.

September, A.S. 0013 (2013 A.D.)

References

1. Cole, S. T., et al. (2001). Massive gene decay in the leprosy bacillus. *Nature, 409*, 1007–1011.
2. Damasio, A. (1994). *Descartes' error*. New York: Putnam.
3. Darwin, C. (1859). *On the origin of species*. London: John Murray.
4. Driesch, H. (1914). *The history and theory of vitalism*. London: MaCmillan.
5. Eccles, J. C. (1989). *Evolution of the brain: Creation of the self*. London/New York: Routledge.
6. Kihara, H. (1930). Genomanalyse bei *Triticum* und *Aegilops. Cytologia, 1*, 263–284.
7. Kihara, H. (1946). *Story on wheets (written in Japanese)*. Tokyo: Sogen-sha.
8. Kimura, M. (1983). *The neutral theory of molecular evolution*. Cambridge: Cambridge University Press.
9. Nishikimi, M., Fukuyama, R., Minoshima, S., Shimizu, N., & Yagi, K. (1994). Cloning and chromosomal mapping of the human nonfunctional gene for L-gulono-gamma-lactone oxidase, the enzyme for L-ascorbic acid biosynthesis missing in man. *Journal of Biological Chemistry, 269*, 13685–13688.
10. Penfield, W. (1975) The mystery of the mind: A critical study of consciousness and the human brain. Princeton: Princeton University Press.
11. Saitou, N. (1997). *Genes dream for 3.5 billion years (written in Japanese)*. Tokyo: Daiwa Shobo.
12. Saitou, N. (2004). *Genome and evolution (written in Japanese)*. Tokyo: Shinyo-sha.
13. Saitou, N. (2007). *Introduction to evolutionary genomics (written in Japanese)*. Tokyo: Kyoritsu Shuppan.
14. Saitou, N. (2011). *Darwin: An introduction (written in Japanese)*. Tokyo: Chikuma Shobo.
15. Winkler, H. (1920). *Verbreitung und Ursache der Parthenogenesis im Pflanzen- und Tierreiche*. Jena: Fischer.

Contents

Part I

Basic Processes of Genome Evolution

Basic Metabolism Surrounding DNAs

1

Chapter Summary

Basic molecular processes of living beings with special reference to DNA are discussed in this chapter, including replication, transcription, and translation. Molecular natures of DNAs, RNAs, and proteins are described, as well as their informational sides such as genetic codes and protein diversity.

1.1 Central Dogma of Molecular Biology

The most basic characteristic of life is self-replication. Although "self" is not easy to define, one individual may correspond to self in a multicellular organism, while one cell corresponds to self for a single-cell organism. The center of self-replication in both cases is replication of DNA molecules. We therefore first consider replication of DNA in this chapter.

Another important characteristic of life is metabolism. RNAs and proteins are major players in metabolism. Nucleotide sequence information is transcribed from DNA to RNA, and then this information is translated into protein or polypeptide. DNA and RNA molecules are linear connection of four kinds of nucleotides, while protein molecules are linear connection of 20 kinds of amino acids. This colinearity is the basis of translation mechanism.

Figure 1.1 is the summary of information flow among DNA, RNA, and protein. This is often called "central dogma" of molecular biology. It should be noted that RNA may be transcribed into DNA by reverse transcription. DNA and RNA are both nucleic acids, and interchangeability may be understandable. In contrast, protein molecules are quite different from nucleic acids, and so far "reverse-translation" (information flow from protein to RNA or DNA) was never observed. Discussion given in this chapter is rather brief. For detailed description on molecular biology, readers may refer to Alberts et al. [1] and Brown [2].

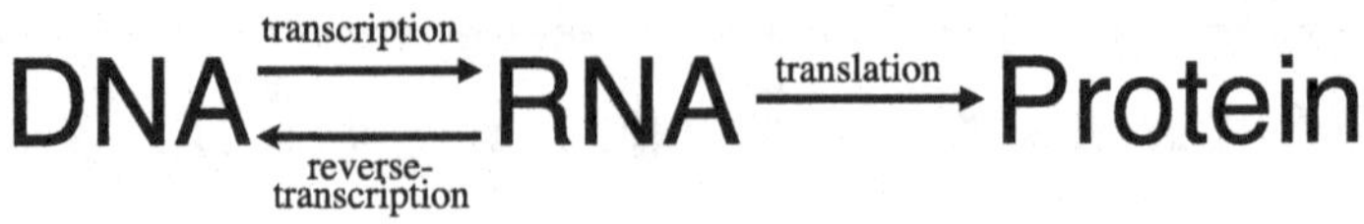
transcription
DNA
RNA
translation
Protein
reverse-
transcription

Fig. 1.1 Direction of genetic information

Fig. 1.2 Three molecules for nucleotide: base, phosphate, and deoxyribose

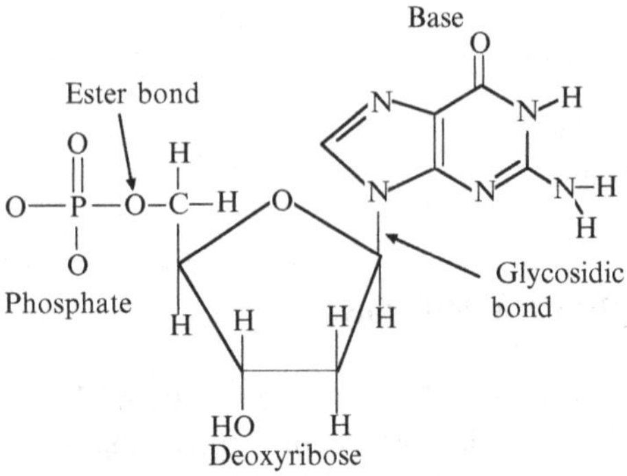
Base
Ester bond
Phosphate
Glycosidic
bond
Deoxyribose

1.2 DNA Replication

Major characteristics of life can be summarized as self-replication and metabolism. Because DNA, material basis of genes, is in the center of self-replication, often self-replication alone is considered as the basic feature of life. However, metabolism can include any physical and chemical reactions of organism, and self-replication can be considered as one of metabolism. We thus need more reasonable definition of "life" without using self-replication. This question is, however, not yet completely solved.

1.2.1 Chemical Nature of DNA Molecules

DNA is the abbreviation of "deoxyribonucleic acid," and its basic building block is a nucleotide. One nucleotide molecule consists of base, phosphate, and deoxyribose (see Fig. 1.2). Backbone structure of the DNA molecule consists of phosphate and deoxyribose. There are four major kinds of base; adenine, cytosine, guanine, and thymine (Fig. 1.3). They are abbreviated as A, C, G, and T, after their acronyms. Adenine and guanine are chemically similar and belong to purine with 5 carbon and 5 nitrogen atoms, while cytosine and thymine are somewhat smaller pyrimidine with 4 carbon and 3 nitrogen atoms.

Because of molecular size difference, the purine and pyrimidine pair can make similar-sized composite forms. There are four possibilities: A and C, A and T, G and C, and G and T (Fig. 1.4). Which pairs are really used for DNA double helix?

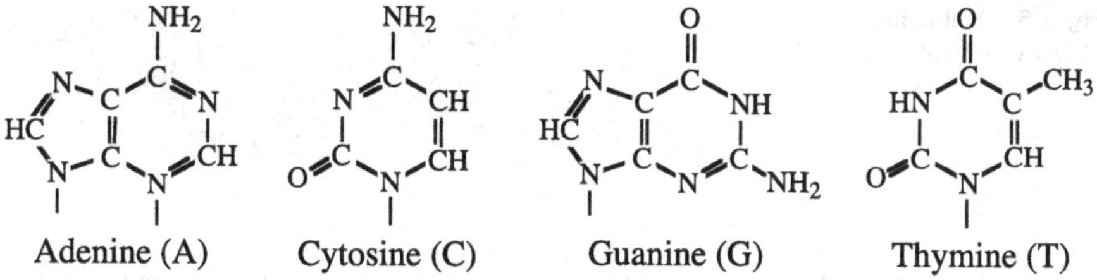

Fig. 1.3 Four kinds of bases: adenine, cytosine, guanine, and thymine

Fig. 1.4 Two types of chemical bonding of purine and pyrimidine

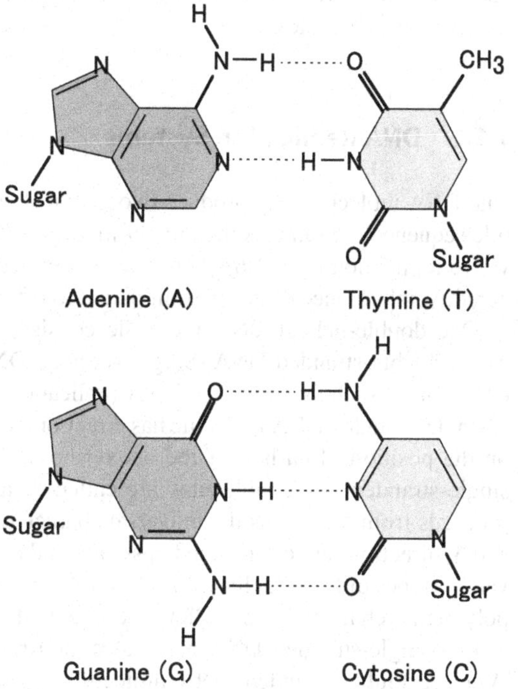

Chargaff et al. [3] found that an amount of adenine and thymine and amount of guanine and cytosine are similar with each other, so-called Chargaff's rule. This suggests that A and T and G and C pairs are really used in nature. Watson and Crick [4] in fact showed that these two pairs are most reasonable, and these pairs are nowadays called Watson–Crick bonding or pairing. Please note that there are two and three hydrogen bonding for A and T and G and C pairs, respectively. Therefore, nucleotide sequence with higher GC content should be chemically more stable.

DNA molecules usually exist in double-helix form, as shown in Fig. 1.5. DNA is acyclic crystal, for there are four kinds of bases, and the order of these bases is

Fig. 1.5 Double-helix form
of DNA molecule

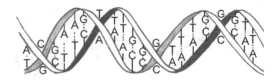

essentially arbitrary. If we note that the atomic compositions of the GC pair and AT pair nucleotides are 2P + 14O + 8N + 19C + 24H and 2P + 14O + 7N + 20C + 25H, respectively, the molecular weight of one nucleotide pair is 650.4 and 649.4 for the GC pair and AT pair, respectively. Therefore, one nucleotide pair is approximately 650 atomic weight, or 1.08×10^{-21} g, if we consider Avogadro's number ($\sim 6.02 \times 10^{23}$) for 1 mol of molecules.

1.2.2 DNA Replication Systems

One DNA molecule may produce two identical child DNA molecules with nucleotide sequence the same as the mother molecule. This process is called semiconservative replication (Fig. 1.6). This semiconservative replication is the very basis of heredity, inheritance of genes from parent to offspring.

One double-helical DNA molecule consists of two "strands," and often it is called double-stranded DNA. Single-stranded DNA molecules also exist, but only temporarily, such as during the DNA replication (with some exception in viruses). A single-stranded DNA molecule has directionality, as shown in Fig. 1.7. Depending on the position of carbon of the deoxyribose, the two terminal positions of one single-stranded DNA molecules are called 5′ and 3′. Because DNA replication proceeds from 5′ to 3′ as the universal characteristics of DNA polymerases, the 5′ and 3′ directions are often called upstream and downstream, respectively. There are various types of DNA replications. A simple one, though an artificial one, is used for polymerase chain reaction (PCR) experiments (Fig. 1.8a). PCR primers are attached to two single-stranded DNA molecules, and replication proceeds symmetrically. When a double-stranded DNA molecule is circular and small, such as animal mitochondrial DNA genomes, displacement replication is used (Fig. 1.8b). Two single-stranded circular DNA molecules (first strand and displaced strand) are independently replicated. Another replication mechanism used in small and circular bacteriophages is rolling circle replication (Fig. 1.8c). This type of replication starts from a nick or a break of one DNA strand, and as 5′ to 3′ replication proceeds, the 5′ end region rolls off. The most common type of DNA replication used either in linear or in circular long DNA molecules is Okazaki fragment involved in leading/lagging strand-type replication (Fig. 1.8d). As the semiconservative replication proceeds, we have the replication fork, in which two new strands of incipient daughter DNA molecules are elongated. The strand which elongates from 5′ to 3′ is called the leading strand, and the remaining one which elongates from 3′ to 5′ is called the lagging strand. Because DNA polymerases polymerize DNAs only from the 5′ to 3′

Fig. 1.6 Semiconservative replication of DNA double helix

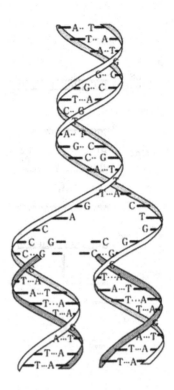

Fig. 1.7 Directionality of a single-stranded DNA molecule

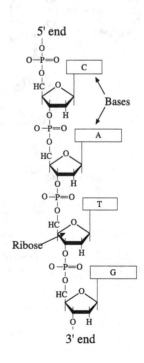

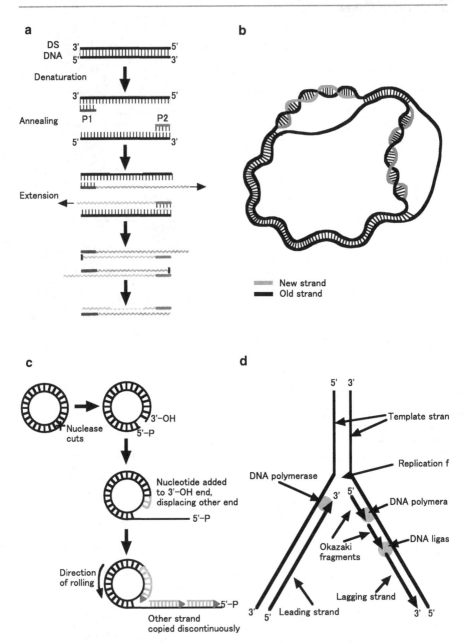

Fig. 1.8 Four types of DNA replications. (**a**) PCR type replication. (**b**) Displacement replication. (**c**) Rolling circle replication. (**d**) Leading/lagging strand type replication

direction, we face a dilemma for replicating the lagging strand. The solution is involvement of Okazaki fragments, named after Reiji Okazaki who discovered these fragments in 1967. Each Okazaki fragment is replicated from 5′ to 3′, but they are short, ~200 nucleotides in eukaryotes and up to 1,000–2,000 in prokaryotes. These short fragments are then ligated to form a long DNA strand.

DNA is a very long molecule, and often it is fragmented, and various DNA repairing systems evolved. If these DNA repairs do not function well, the mutation rate may be increased. There are four major mechanisms for DNA repair: direct repair, excision repair, mismatch repair, and nonhomologous end-joining. Interested readers may refer to Brown [2].

If one cell divides into two daughter cells, DNA replication should precede the cell division. Therefore, DNA replication is usually coupled with the cell cycle. In eukaryotes, a typical cell cycle involves the following four phases: G1, S, G2, and M. G, S, and M stand for gap, synthesis, and mitosis, respectively. DNA replication occurs at S phase. If we grow mammalian cells in tissue culture, one cell cycle is ~24 h, and G1, S, G2, and M phases are ~10, ~9, ~4, and ~1 h, respectively [5].

1.3 Transcription

DNA nucleotide sequence is not directly transferred to protein amino acid sequence, as we mentioned in Sect. 1.1. Sequence information of DNA is first transcribed to RNA (ribonucleic acid). This RNA is called messenger RNA (mRNA). Messengers in human society may be considered to have a relatively minor role. However, mRNA molecules can be considered to be located at the center of the information flow from DNA to proteins, and they may be the most ancient molecules, as we will discuss in Chap. 6.

1.3.1 RNA

The chemical nature of RNA is similar to DNA. Ribose is used instead of deoxyribose in RNA, and uracil is used instead of thymine (Fig. 1.9). An RNA variety of prokaryotes are essentially mRNA, tRNA, and rRNA (see Chap. 7), while eukaryotes have many more RNA species. This difference is the basis of different patterns of transcriptions between prokaryotes and eukaryotes. We only discuss the diversity of RNA molecules in eukaryotes in this section. Other RNA molecules are more or

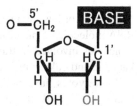

Fig. 1.9 Molecular structure of RNA. Light-gray OH of ribose is H in deoxyribose of DNA

less short, and depending on locations inside the cell, they are called small nuclear RNA (snRNA), small nucleolar RNA (snoRNA), and small cytoplasmic RNA (scRNA). The snRNA is involved in mRNA splicing, and snoRNAs are mostly responsible for modification of rRNAs. scRNA includes microRNA (miRNA) that is only 22-nucleotide short molecule. miRNAs are regulating posttranscriptional gene expression. An enzyme dicer cuts a long precursor RNA to produce an miRNA in animals, and these miRNAs bind to their complementary sequence of mRNA so as to suppress its expression.

Xist RNA is a long 17-kb RNA and is related to X chromosome inactivation in mammals. Mammalian sex is determined by two kinds of sex chromosomes: X and Y. Mammalian males are XY heterozygotes, and mammalian females are XX homozygotes. Because chromosome Y is much smaller than X and it has only a small number of functional genes, mammals need gene expression control of X chromosomes. This is fulfilled by inactivation of one of the X chromosomes in female cells. Once inactivated, this condition is kept after mitosis. Xist RNA is playing an important role for this inactivation [6].

1.3.2 Messenger RNA (mRNA)

RNA molecules that transmit nucleotide sequence information of DNA to amino acid sequence information of protein are called messenger RNA, or mRNA. In prokaryotes, mRNA is directly transcribed from DNA. In eukaryotes, splicing is often necessary to produce mRNA. The final form of mRNA is called "mature mRNA," and this molecule experienced splicing. Splicing is to cut precursor RNA corresponding to introns and connect remaining parts corresponding to exons. Precursor RNA is first transcribed and often quite long (can be more than 10,000 bp), while mature mRNA may be ~1,000 bp. Long precursor RNA is sometimes detected experimentally and was claimed to be functional even if there was no clear sequence conservation [7, 8]. However, van Bakel et al. [9] showed that most of these RNAs are infrequent. Because the majority of nonexonic regions in eukaryotes are considered to be junk DNA [10] and evolve in purely neutral fashion [11], we should be careful for claiming any function simply because they are transcribed. There may exist many "junk RNAs" in eukaryotic cells. We will return to this question in Chap. 4.

1.3.3 Transcription Machinery and Splicing

Nucleotide or base sequences of DNA molecules are used as templates to generate RNA molecules. This process is called "transcription." RNA polymerases are main molecules involved in transcription. Because there are various differences in prokaryotic and eukaryotic transcription machineries, we discuss them separately.

Prokaryotes have only one type of RNA polymerase, which consists of five subunits; α, α, β, β', and σ. The RNA polymerase first attaches to the target sequence, called "promoter," just upstream of the protein coding region. Promoter sequences vary from gene to gene, but there are two consensus promoter sequences: TTGACA and TATAAT at ~35 bp and ~10 bp upstream of the initiation codon in *Escherichia coli*. Independent genes or operons have slightly different promoters, such as TTTACA and TATGTT for lactose operon. This sequence specificity is achieved by the σ subunit of the RNA polymerase. After the contact with the promoter region, the DNA double strand is opened, and the σ subunit is detached from the remaining four subunits, called RNA polymerase core enzyme. After the RNA polymerization starts, it should be terminated. However, compared to the initiation process, the termination process is still not well known. This may be because translations are coupled with transcription in bacteria, and there are termination signals in translation.

Eukaryotes have three types of RNA polymerases, I, II, and III. RNA polymerase I transcribes genes for three ribosomal RNA genes (28S, 5.8S, and 18S), while RNA polymerase II transcribes all protein coding genes, most small nuclear RNA genes, and microRNA genes. RNA polymerase III is specialized in transcribing genes for tRNAs, 5S ribosomal RNA, U6 snRNA, and small nucleolar RNAs. Control of transcriptions in eukaryotes is much more complex than that of prokaryotes. Not only promoters but also enhancers are involved in many genes. Some enhancers are involved in tissue-specific transcription in multicellular organisms.

Most of the bacterial genes lack introns, but some have retrointrons which can do self-splicing. It is known that tRNA genes of Archaea have introns. Eukaryotes have the splicing system, and many eukaryotic genes have introns. Introns are sometimes also called as "intervening sequences." We will discuss about the evolution of introns in Chap. 8.

Spliceosomes are formed in eukaryotes to eliminate intron parts of the premature mRNA molecules. First, an RNA–protein complex called U1-snRNP (U1-snRNA and 10 proteins) binds to the 5′ splice site, followed by binding a series of proteins (SF1, U2AF^{35}, and U2AF^{65}) to the RNA molecule corresponding to the branch site, polypyrimidine tract, and the 3′ splice site, respectively. Another RNA–protein complex U2-snRNP then binds to the branch site, and it and U1-snRNP together bring the 3′ and 5′ splice sites into proximity. Finally, U4/U6-snRNP and U5-snRNP also come to the branch site and form the spliceosome, which eventually cut the premature mRNA molecule to complete the splicing. The content of this paragraph is after Brown [2].

Splicing is not always precise. If aberrant splicing products are formed, such incorrect mRNAs may not produce chemically stable peptides, and they will be decomposed. In some cases, however, a protein slightly different from the expected protein may be formed. If this new protein happens to be useful for the organism, this is the starting point of the formation of alternative splicing. Although alternative splicing is often considered to be a "good" mechanism to enlarge the diversity of peptides from only scores of 1,000 protein coding genes, it should be noted that most of alternative splicing frequently occurring in cells produces aberrant proteins.

1.3.4 RNA Editing

RNA molecules may be modified after splicing. It is well known that several specific nucleotides of tRNA are modified, some nucleotides of rRNA molecules are also modified. These modifications are considered to facilitate protein synthesis, while amino acid sequences themselves will be affected if mRNAs are modified. Such modification on mRNA is called "RNA editing." There is no specific pattern in RNA editing, and edited genes may vary from tissue to tissue in mammals. Because there are many RNA-binding proteins, it is possible that some of them acquired enzymatic activity to modify some bases of bound RNA, and these modifications of RNA sequences were evolutionarily accepted. If RNA editing occurred in some mRNA, resulting protein will have the amino acid sequence different from that inferred from genetic code table of that organism. This is a clear caveat to pure computational analyses of "protein" sequences solely based on genomic nucleotide sequences. We will discuss more on RNA editing in Chap. 8.

1.4 Translation and Genetic Codes

Translation, or conversion of nucleotide sequence information on mRNA molecule to protein amino acid information, is very complex, and mRNAs contain not only the coding information for amino acid sequence (protein coding region) but also non-translated region on both 5′ (upstream) and 3′ (downstream) regions. These non-translated regions of mRNA play important roles during translation.

The linear order of amino acid sequences in proteins is corresponding to the base or nucleotide sequence of mRNA, and these two chemically quite different molecules are connected via transfer RNA.

1.4.1 Transfer RNA (tRNA)

Translation is conducted at ribosome, and each amino acid is transported via special short-sequence RNA molecule called transfer RNA, or tRNA. Each tRNA molecule has its specific triplet nucleotide region called "anticodon" and amino acid anchoring region (see Fig. 1.10). One protein usually consists of up to 20 kinds of amino acids (see Fig. 1.11), and we need at least 20 kinds of tRNA. Animal mitochondrial DNA genomes may be the smallest ones with tRNA genes, and they have only 22 tRNA genes. Goodenbour and Pan [12] examined 11 eukaryote species genomes and found that the total number of tRNA genes per genome was between 170 and 570, and the number of tRNA isoacceptors ranged from 41 to 55.

One particular amino acid is attached at the opposite side of the anticodon at the tRNA molecule. This chemical reaction is catalyzed by a particular aminoacyl-tRNA synthetase. We will discuss evolution of aminoacyl-tRNA synthetases in Chap. 6.

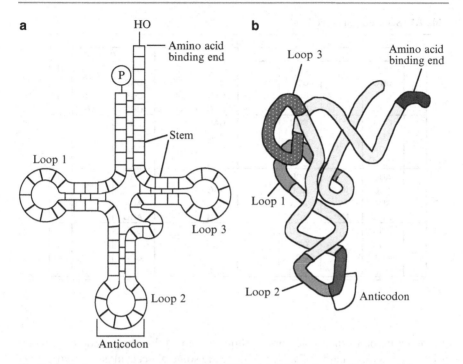

Fig. 1.10 Basic structure of transfer RNA (tRNA). (**a**) Schematic two-dimensional cloverleaf view. (**b**) Three-dimensional view

1.4.2 Genetic Codes

There are four types of nucleotides, the building unit of DNA. As described in Sect. 1.2.1, the difference among four nucleotides is caused by four kinds of bases: adenine (A), cytosine (C), guanine (G), and thymine (T). Proteins, however, usually consist of 20 kinds of amino acids (see Sect. 1.5). Therefore, through simple application of combinatorial mathematics, at least three nucleotides or bases should correspond to one amino acid. Two nucleotides can correspond to only 16 ($= 4 \times 4$) different possibilities, while three or four bases or nucleotides can correspond to 64 ($= 4 \times 4 \times 4$) or 256 ($= 4 \times 4 \times 4 \times 4$) possibilities, respectively. It turned out that three consecutive nucleotides in fact correspond to one amino acid, and this trio of nucleotides is called "triplet codon." "Triplet" means three small nucleotide, and "codon" means material unit of genetic code that ties nucleotide sequence information carried in DNA with amino acid sequence information carried in protein.

Because there are in average about three codons for one amino acid, often different codons may correspond to one amino acid. This relationship is usually represented in table form, so it is called "genetic code table," as you can see in Table 1.1. Decipherment of the genetic code is one of the greatest achievements

Table 1.1 Standard genetic code table

		Second Letter							
		U	C	A	G	
First Letter	U	UUU | Phe UUC | UUA | Leu UUG |	UCU | UCC | Ser UCA | UCG |	UAU | Tyr UAC | UAA | Stop UAG | Stop	UGU | Cys UGC | UGA | Stop UGG | Trp	U C A G	Third Letter
	C	CUU | CUC | CUA | Leu CUG |	CCU | CCC | CCA | Pro CCG |	CAU | His CAC | CAA | Gln CAG |	CGU | CGC | CGA | Arg CGG |	U C A G	
	A	AUU | AUC | Ile AUA | AUG | Met	ACU | ACC | ACA | Thr ACG |	AAU | Asn AAC | AAA | Lys AAG |	AGU | Ser AGC | AGA | Arg AGG |	U C A G	
	G	GUU | GUC | GUA | Val GUG |	GCU | GCC | GCA | Ala GCG |	GAU | Asp GAC | GAA | Glu GAG |	GGU | GGC | GGA | Gly GGG |	U C A G	

of biology in the twentieth century. It should be noted that the genetic code table in Table 1.1 is the "standard" code, for there are some dialects in genetic codes, as we will see later in this section. Traditionally, the standard genetic code is called "universal" code; however, because of the existence of many dialects, that one is no longer universal; hence, it should be called "standard" code table (see below for a historical perspective).

In any case, information for protein with 100 amino acids is coded in 300 nucleotide sequence, and this kind of nucleotide region is called "protein coding region." A protein coding region is often abbreviated as "coding region." However, "code" can apply to any DNA region that has genetic information, not only protein coding regions but also RNA coding regions. Therefore, I would like to reserve "coding region" for this broader meaning.

Three out of 64 codons in the standard genetic code have a special role called "termination codon" or simply "stop codon." They are TAA, TAG, and TGA. If we consider amino acid sequence as a sentence, these special codons correspond to a period (.). In a living cell, there is a special tRNA molecule that corresponds to these three stop codons, and this tRNA does not carry amino acid. Therefore, translation terminates. How about initiation of translation? As there is no special character for starting a sentence in most written languages, there is no special codon for starting protein translation. Single codon AUG corresponding to methionine (Met or M) plays as initiation or start codon in eukaryotes. Prokaryotes use GUG and UUG also for starting the translation, though less frequently compared to AUG.

When the connection between triplet nucleotides and amino acids was determined in 1960s, both *Escherichia coli* and we humans happened to have the identical codon table. This table was considered to be literally "universal," without no

Table 1.2 Dialects of the standard genetic code

<Mitochondrial genome>
AAA: Lys ==> Asn (Platyhelminths, Echinoderms)
AGA/G: Arg ==> Gly (tunicates)
AGA/G: Arg ==> Ser (invertebrates)
AGA/G: Arg ==> Stop (vertebrates)
AUA: Ile ==> Met (yeast, animals)
CGG: Arg ==> Nonsense (green alga)
CGN: Arg ==> Nonsense (*Torulopsis glabrata*)
CUN: Leu ==> Thr (yeast)
UAA: stop ==> Tyr (*Planaria*)
UAG: stop ==> Nonsense (green alga)
UGA: stop ==> Trp (All mitochondria except green plants)
<Nuclear genome>
AGA: Arg ==> Nonsense (*Micrococcus*)
AUA: Ile ==> Nonsense (*Micrococcus*)
CGG: Arg ==> Nonsense (Mycoplasma)
CUG: Leu ==> Ser (*Candida*)
UAA/G: stop ==> Gln (ciliates, *Acetabularia*)
UGA: stop ==> Trp (*Mycoplasma, Spiroplasma*)
UGA: stop ==> Cys (Euplotes)

Based on Fig. 6.1 at page 73 of Osawa (1995: [36])

change from the last common ancestor. Crick [13] called this as "frozen-accident theory." However, a somewhat different codon usage was found when the human mitochondrial DNA genome sequence was started to be determined by Barrell, Bankier, and Drouin in [14]; codon for methionine (Met) was not AUG but AUA, and codon for tryptophan (Trp) was not UGG but UGA. Later, mycoplasma nuclear genome was also shown to use UGA for Trp by Yamao et al. [15]. Nowadays, a significant number of organisms are known to use "dialects" of the standard genetic code, as shown in Table 1.2. NCBI presents 17 alternative genetic code tables [16]. Osawa and Jukes [17] proposed the codon-capture theory by applying the neutral evolutionary process.

1.4.3 Ribosome and Translation

Ribosome is a complex of protein and a special class of RNA molecules called ribosomal RNA, or rRNAs are the major component of ribosomes. There are three kinds of rRNA molecules, and they are called 5S rRNA, 16S rRNA, and 23S rRNA for prokaryotes, while eukaryote ribosomes consist of 5.8S, 18S, and 28S rRNA. Here, "S" is a unit of molecules according to ultracentrifuging process, after Theodor Svedberg, a pioneer of ultracentrifuge analysis. Because of the abundance of ribosomes in each cell, rRNA molecules are probably in the highest copy number in all RNA species. One ribosome is formed by small and large subunits. The bacterial small ribosome subunit is composed of 16S rRNA (about 1.5 kb) and 21 proteins, while large ribosome subunit contains 5S and 23S RNAs and 31 proteins. Translation takes place at ribosomes, and this is one of the most basic processes for producing proteins. Readers should refer to textbooks on molecular biology (e.g., [1, 2]) for a detailed description of the translational machinery.

1.5   Proteins

Jöns J. Berzelius coined the word "protein" in the early nineteenth century, meaning the "first" or "prime" nutrient. It was known for a long time that the basic component of proteins are amino acids, but how they are connected and how they are stable were not known until the middle of the twentieth century.

1.5.1   Amino Acids

Proteins are essentially linear polymers of amino acids. Any amino acid molecule used for the building block of proteins should have one amino radical and one carboxyl radical as well as a specific residue. Typical proteins consist of 20 amino acids (Fig. 1.11), and they can be classified into several categories depending on various criteria (Table 1.3). Interestingly, classification based on chemical property shown in Table 1.3A is quite similar to that based on the frequency of amino acid substitution shown in Table 1.3B. This indicates that chemically similar amino acids tend to substitute more often with each other. Selenocysteine (Sec or U) may be used in some dialect of universal genetic code; the stop codon UGA in the standard genetic code table is assigned to this amino acid here, where sulfur (S) atom was substituted to selenium (Se) in cysteine amino acid with the help of selenocysteine synthetase while selenocysteine tRNA was attaching cysteine. Because of this, selenocysteine is sometimes considered as the 21st amino acid for forming proteins. The 22nd amino acid in proteins found in Archaea is pyrrolysine (Pyl or O). This amino acid is connected to tRNA corresponding to UAG with the help of pyrrolysine synthetase and is introduced to proteins.

1.5.2   Four Strata of the Protein Structure

There are four strata for the protein structure. The linear array of amino acid sequence is called the primary structure of proteins. However, real proteins are very rare to be in a linear form. Adjacent amino acids will be connected through hydrogen bonding to form either alpha helix or beta sheet (see Fig. 1.12). These are called secondary structure. When some region of a protein is neither alpha helix nor beta sheet, it is called random coil. These three kinds of secondary structures finally build up the three-dimensional (3D) or tertiary structure of proteins. Figure 1.13

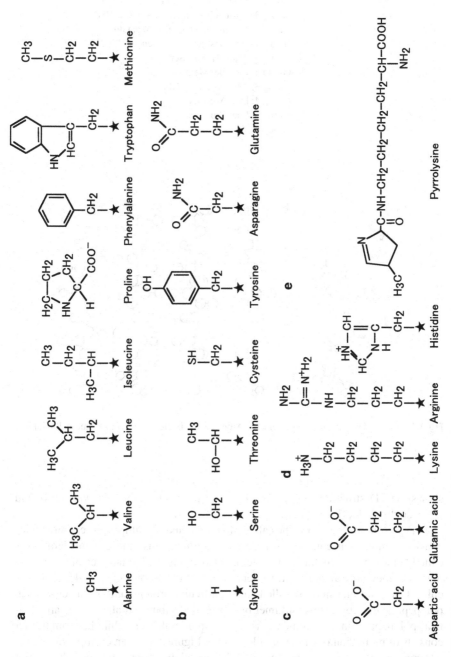

Fig. 1.11 List of amino acids used for protein

Table 1.3 Classification of amino acids

(A) Based on chemical property of residues (Brown 2007; [6])
Group 1 (Nonpolar): Ala, Leu, Ile, Met, Pro, Thr, Trp, Val
Group 2 (Polar): Asn, Cys, Gln, Gly, Ser, Tyr
Group 3 (Positively charged): Lys, Arg, His
Group 4 (Negatively charged): Asp, Glu
(B) Based on substitution frequency (Toh 2004; [37])
Group 1: Ala, Gly, Pro, Ser, Thr
Group 2: Leu, Ile, Met, Val
Group 3: Asp, Asn, Glu, Gln
Group 4: Lys, Arg, His
Group 5: Phe, Trp, Tyr
Group 6: Cys

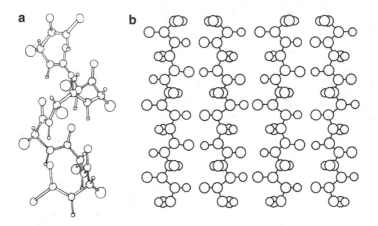

Fig. 1.12 Secondary structure of protein. (**a**) Alpha helix. (**b**) Beta sheet (From Saitou 2007; [34])

shows the 3D structure of the GAL4 transcription factor protein with its bound DNA double helix (From [18]).

One protein molecule may be called monomer, and if multiple protein molecules consist of molecular complex, it is called multimer. This kind of consideration is called "quaternary structure" of proteins. Depending on the number of molecules involved, they are called dimer, trimer, tetramer, and so on, for 2, 3, 4, etc. molecules. One multimer may be called some particular protein, and in this case, each polypeptide that forms this multimer may be called subunit. If all subunits are identical polypeptide in a multimer, it is called homomultimer, while heteromultimer consists of more than one type of polypeptide. Figure 1.14 is an example of heterotetramer where there are two types of subunits: alpha globin and beta globin. For more details, please refer to Branden and Tooze [19].

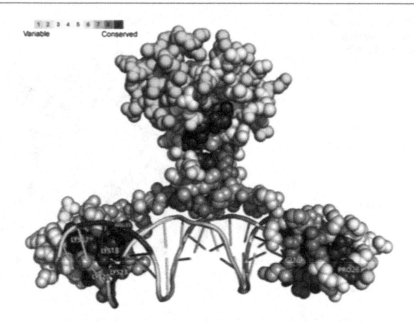

Fig. 1.13 3D structure of GAL4 transcription factor (From Ashkenazy et al. 2010; [18])

1.5.3 Protein Diversity

Metabolisms are mainly orchestrated by proteins. This is possible because proteins are quite diverse. If we consider a typical protein with 300 amino acids, there are 20^{300} or 2.04×10^{390} possible sequences. The number of amino acids may vary depending on proteins, and the sequence possibility further increases. Let us compare this combinatorial explosion with an English text. A typical one page of a paperback contains about 2,000 characters. If we consider both lowercase and capital letter alphabets and eight special characters such as punctuation, there are a total of 60 possible characters. Then the total possible text for one page becomes 60^{2000} possible texts. This number is much larger than the possible protein sequences, showing the immense space of writing. This comparison may raise the possibility of artificial life based on one relatively long computer program. However, amino acid sequence is just one kind of information to produce a full 3D structure of protein. It is still almost impossible to construct the 3D structure of one protein merely from its amino acid sequence. This shows the existence of many more parameters to form the 3D protein structure. In any case, a huge diversity of one protein leads to much more possible sets of thousands of proteins in one organism. This is one basis of the diversity of life and their emergence through evolution.

How to classify the existing diverse proteins is another matter. It is ideal to classify protein from the functional viewpoint. However, the majority of protein sequences registered to various databases are inferred from genomic nucleotide

Fig. 1.14 Example of
heterotetramer (hemoglobin)
(From Budd 2012; [35])

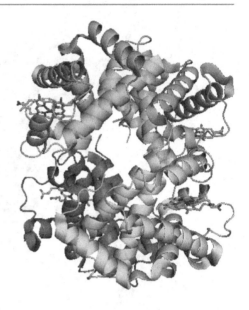

sequences using genetic code tables, and most of their functions are not yet examined. Therefore, amino acid sequence homology of a new putative protein is searched (see Chap. 14 for homology search), and if function of some homologous protein(s) is known in some species, such as *Escherichia coli* or *Saccharomyces cerevisiae*, this new protein is classified into this known protein group. Although homology search is usually purely examining sequence similarity from statistical point of view, often homologous sequences share function. This kind of homologous protein or gene group is called a "gene family." Different gene families may be further grouped to a broader gene family. For example, there are "alpha globin" and "beta globin" families in vertebrates, and because of sequence homology found between these two family proteins, we may call all of these homologous genes "globin gene family" generically. When sequence homology is found only in part of different protein gene families, they may be grouped into a supergene family. However, the boundary between gene family and supergene family is not clear.

Figure 1.15a, b shows the globin gene family and the immunoglobulin supergene family, respectively. There are three subfamilies (alpha globin, beta globin, and myoglobin) for the globin gene family, and their sequence homology spans the whole amino acid sequence (see their multiple alignment in website). Open diamond and full circles designate gene duplication and speciation events, respectively. The Pfam database (http://pfam.sanger.ac.uk/) and InterPro database (http://www.ebi.ac.uk/interpro/) contain 13,672 and 16,152 gene families, respectively, as of December 2012. The immunoglobulin supergene family, shown in Fig. 1.15b, is characterized by immunoglobulin domain (see next Sect. 1.5.4). Diverse protein molecules, such as immunoglobulin, T-cell receptor, and class 1 and 2 MHC proteins, are included in this supergene family. The SCOP (Structural Classification of Proteins) database contains 1,962 superfamilies as of December 2012.

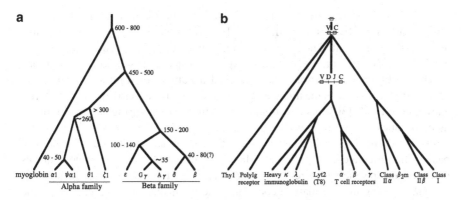

Fig. 1.15 Example of gene family and supergene family. (**a**) Globin gene family. (**b**) Immunoglobulin supergene family

1.5.4 Protein Domains and Motifs

Proteins often consist of one or a few domains. Domains typically form functional units. Each domain forms a compact 3D structure and often can be independently stable and folded. One domain may appear in a variety of evolutionarily related proteins, and multidomain protein may yield a novel function through combination of individual domain functions. Multidomain proteins were produced more frequently in eukaryotes than in prokaryotes, suggesting a correlation between the abundance of multidomain proteins and biological complexity [20].

A protein with many numbers of domains may appear at some evolutionary period, but later some domains may disappear in the course of time. Several notable examples of lineage-specific gene expansions illuminate the physiological changes of the organism. For example, the emergence of the mammal-specific vomeronasal organ receptor family correlates with the development of the vomeronasal organ in mammals [21]. Dorit et al. [22] postulated based on protein homology analysis that the whole human genes could arise from a limited set of 1,000–7,000 exon subunits. This estimate was further refined using the whole-genome sequence to the 1,865 distinct domain families in human, 1,218 in *Drosophila*, 1,183 in *Caenorhabditis elegans*, and 973 in *Saccharomyces cerevisiae* [23]. If we compare these with those of other organisms, 61 %, 43 %, and 46 % of the fly, worm, and yeast proteomes, respectively, are retained in the human genome [24].

Combining existing domains in novel gene architectures, also known as domain shuffling, is an important mechanism for creating protein complexity. Ikeo et al. [25] showed that Kunitz-type trypsin inhibitor domain was inserted in the amyloid beta precursor protein. Domain shuffling was estimated to be involved up to 20 % of eukaryotic exons [26]. The signs of ancient domain shuffling can be detected in current-day eukaryotes as the predominance of intron boundaries in linker regions connecting the domains [27] and the correlation of age-prevalent symmetrical intron phases and the age of their protein domains [28]. There are many cases of chimeric proteins thought to have arisen by domain shuffling; most notable examples include blood coagulation factors V, VIII, and XIII b and protein S [29]. Thus, extensive domain shuffling resulted in a great boost in complexity of domain architectures and combined with differential gene family expansions and their subsequent evolution.

The evolution of vertebrates has included a number of important events: the development of cartilage, the immune system, and complicated craniofacial structures. Kawashima et al. [30] mapped domain-shuffling events during the evolution of deuterostomes to examine how domain shuffling contributed to the evolution of verte-brate- and chordate-specific characteristics. They identified about 1,000 new domain pairs in the vertebrate lineage. Some of these pairs occurred in the protein components of vertebrate-specific structures, such as cartilage and the inner ear, suggesting that domain shuffling made a marked contribution to the evolution of vertebrate-specific characteristics. They also identified genes that were created as a result of domain shuf-fling in ancestral chordates. Some of these are involved in the functions of chordate structures, such as the endostyle, Reissner's fiber of the neural tube, and the notochord. It should be noted that a significant portion of this section relied on Masuyama [31].

1.6 Genes

Mendel [32] imagined a microscopic matter, which is transmitted from parents to offsprings. After the rediscovery of Mendel's laws in 1900, Wilhelm Johannsen coined the word "gene" in 1909. It has been known for a long time that a gene may code nucleotide sequence of an RNA molecule which is used in organism. Such genes include those for tRNAs and rRNAs. We now know that RNA molecules are much more diverse, as explained in Sect. 1.2.0. Enhancers are often quite distantly located from the protein coding genes, and we can consider them as part of genes. In this sense, it is unfortunate that still "genes" are often considered to be only coding for protein amino acid sequences.

1.6.1 Protein Coding Genes

Prokaryotes (Eubacteria and Archaea) do not have nuclear membrane in their cell, though genomic DNAs are localized in a particular area of a cell, called nucleus. Eukaryotic cells have nuclear membrane, and genomic DNA molecules are inside that membrane, called nucleus. Figure 1.16 shows the typical structure of protein coding gene in prokaryotes (From [1]).

As for eukaryotic protein coding gene, ABCC11 gene, responsible for human earwax dry/wet polymorphism [33], is used as an example. Figure 1.17a shows a schematic view of human chromosome 16. One chromosome has one centromere, which separates the short arm (designated as "p") and long arm (designated as "q") of the chromosome. When a particular chemical such as Giemsa is used to dye chromosomes, particular regions of chromosome are more strongly dyed than other regions. This results in band-like structure formation in each chromosome. These structures are called chromosomal bands and are inherited for a long time.

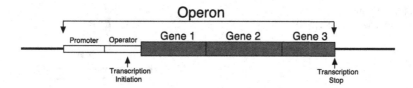

Fig. 1.16 Basic structure of protein coding gene in prokaryotes

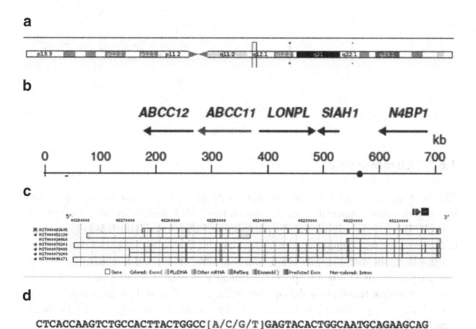

Fig. 1.17 Basic structure of protein coding gene in eukaryotes

The ABCC11 gene is located in 16q12.1 region of human chromosome 16, and it is surrounded by several genes such as ABCC12 and LONPL (Fig. 1.17b). Each gene is shown with one arrow to designate its direction of transcription. Figure 1.17c shows the exon–intron structure of ABCC11 gene. This gene has more than 20 exons. Finally we reach the nucleotide sequence itself, shown in Fig. 1.17d. This short sequence is the flanking region of the nonsynonymous SNP (A or G) responsible for enzymatic activity of this gene product.

If we include both exons and introns, the average size of protein coding gene in the human genome becomes about one quarter. However, most of intron sequences probably do not have functions. Therefore, excluding introns, but including promoters and enhancers that are affecting gene expression, we can consider "the effective number of nucleotides" [34]. Its average for the human genome is probably less than 5,000 bp.

Fig. 1.18 From nucleus to
DNA (From Saitou 2007; [34])

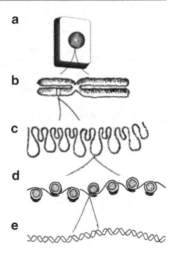

1.6.2 Chromosomes

We saw genes from informational aspects in the previous section. We now proceed to see genes from material aspects. That is, how DNA molecules are contained in the cell. There are five levels for this perspective. Level 1 is cell nucleus (Fig. 1.18a). Chromosomes that contain genomic DNA are inside the nucleus, and they do not get out from the nucleus. The word "chromosome" was from a cellular body that was easy to be dyed when a special chemical was used. Chromosomes at level 2 pursue complex reassembly during the cell cycle and may become thick or thin (Fig. 1.18b). In particular, systematic changes occur during mitosis and meiosis. Level 3 is chromatin fiber (Fig. 1.18c). Level 4 is nucleosome (Fig. 1.18d), in which DNA molecules twine around the special protein complex mainly composed of histones. Level 5, the last level, is DNA double helix (Fig. 1.18e).

References

1. Alberts, B., et al. (2008). *Molecular biology of the cell* (5th ed.). New York: Garland.
2. Brown, T. A. (2006). *Genomes 3*. New York: Garland.
3. Chargaff, E., Lipshitz, R., Green, C., & Hodes, M. E. (1951). The composition of the deoxyribonucleic acid of salmon sperm. *Journal of Biological Chemistry, 192*, 223–230.
4. Watson, J., & Crick, F. (1953). A structure for deoxyribose nucleic acid. *Nature, 171*, 737–738.
5. Hartl, D. L., & Ruvolo, M. (2012). *Genetics – Analysis of genes and genomes* (8th ed.). Burlington: Jones & Bartlett Learning.
6. Brockdorff, N. (2002). X-chromosome inactivation: Closing in on proteins that bind Xist RNA. *Trends in Genetics, 18*, 352–358.
7. Birney, E., et al. (2007). Identification and analysis of functional elements in 1% of the human genome by the ENCODE pilot project. *Nature, 447*, 799–816.

8. Kaspanov, P., et al. (2010). The majority of total nuclear-encoded non-ribosomal RNA in a human cell is 'dark matter' un-annotated RNA. *BMC Biology, 8*, 149.

9. van Bakel, H., Nislow, C., Blencowe, B. J., & Hughes, T. R. (2010). Most "dark matter" transcripts are associated with known genes. *Plos Biology, 8*, e1000371.

10. Ohno, S. (1972). So much "junk" DNA in our genome. *Brookhaven Symposia in Biology, 23*, 366–370.

11. Kimura, M. (1983). *The neutral theory of molecular evolution.* Cambridge: Cambridge University Press.

12. Goodenbour, J. M., & Pan, T. (2006). Diversity of tRNA genes in eukaryotes. *Nucleic Acids Research, 34*, 6137–6146.

13. Crick, F. (1968). The origin of the genetic code. *Journal of Molecular Biology, 38*, 367–379.

14. Barrell, B. G., Bankier, A. T., & Drouin, J. (1979). A different genetic code in human mito-chondria. *Nature, 282*, 189–194.

15. Yamao, F., Muto, A., Kawauchi, Y., Iwami, M., Iwagami, S., Azumi, Y., & Osawa, S. (1985). UGA is read as tryptophan in *Mycoplasma capricolum. Proceedings of the National Academy of Sciences of the United States of America, 82*, 2306–2309.

16. Elzanowski, A., & Ostell, J. (2010). *The genetic codes.* http://www.ncbi.nlm.nih.gov/Taxonomy/taxonomyhome.html/index.cgi?chapter=cgencodes

17. Osawa, S., & Jukes, T. H. (1989). Codon reassignment (codon capture) in evolution. *Journal of Molecular Evolution, 28*, 271–278.

18. Ashkenazy, H., Erez, E., Martz, E., Pupko, T., & Ben-Tal, N. (2010). ConSurf 2010: Calculating evolutionary conservation in sequence and structure of proteins and nucleic acids. *Nucleic Acids Research, 38*, W529–W533.

19. Branden, C., & Tooze, J. (1991). *Introduction to protein structure.* New York: Garland.

20. Koonin, E. V., Wolf, Y. I., & Karev, G. P. (2002). The structure of the protein universe and genome evolution. *Nature, 420*, 218–223.

21. Hillier, L., Miller, W., Birney, E., Warren, W., Hardison, R. C., Ponting, C. P., Bork, P., Burt, D. W., Groenen, M. A., Delany, M. E., Dodgson, J. B., & Chinwalla, A. T. (2004). Sequence and comparative analysis of the chicken genome provide unique perspectives on vertebrate evolution. *Nature, 432*, 695–716.

22. Dorit, R. L., Schoenbach, L., & Gilbert, W. (1990). How big is the universe of exons? *Science, 250*, 1377–1382.

23. Li, W. H., Gu, Z., Wang, H., & Nekrutenko, A. (2001). Evolutionary analyses of the human genome. *Nature, 409*, 847–849.

24. International Human Genome Sequencing Consortium. (2001). Initial sequencing and analysis of the human genome. *Nature, 409*, 860–921.

25. Ikeo, K., Takahashi, K., & Gojobori, T. (1992). Evolutionary origin of a Kunitz-type trypsin inhibitor domain inserted in the amyloid beta precursor protein of Alzheimer's disease. *Journal of Molecular Evolution, 34*, 536–543.

26. Long, M., Rosenberg, C., & Gilbert, W. (1995). Intron phase correlations and the evolution of the intron/exon structure of genes. *Proceedings of the National Academy of Sciences of the United States of America, 92*, 12495–12499.

27. de Souza, S. J., Long, M., Schoenbach, L., Roy, S. W., & Gilbert, W. (1996). Intron positions correlate with module boundaries in ancient proteins. *Proceedings of the National Academy of Sciences of the United States of America, 93*, 14632–14636.

28. Vibranovski, M. D., Sakabe, N. J., de Oliveira, R. S., & de Souza, S. J. (2005). Signs of ancient and modern exon-shuffling are correlated to the distribution of ancient and modern domains along proteins. *Journal of Molecular Evolution, 61*, 341–350.

29. Patthy, L. (2003). Modular assembly of genes and the evolution of new functions. *Genetica, 118*, 217–231.

30. Kawashima, T., Kawashima, S., Tanaka, C., Murai, M., Yoneda, M., Putnam, N. H., Rokhsar, D. S., Kanehisa, M., Satoh, N., & Wada, H. (2009). Domain shuffling and the evolution of vertebrates. *Genome Research, 19*, 1393–1403.

31. Masuyama, W. (2009). *Evolutionary analysis of protein domains in mammals*. Ph.D. dissertation, Department of Genetics, School of Life Science, Graduate University for Advanced Studies.
32. Mendel, G. (1866). Versuche uber Pflanzenhybriden. Verhandlungen des Naturforschenden Verenines, Abhandlungen, Brunn, *4,* 3–47.
33. Yoshiura, K., et al. (2006). A SNP in the ABCC11 gene is the determinant of human earwax type. *Nature Genetics, 38*, 324–330.
34. Saitou, N. (2007). *Genomu Shinkagaku Nyumon (written in Japanese, meaning 'Introduction to evolutionary genomics')*. Tokyo: Kyoritsu Shuppan.
35. Budd, A. (2012). Protein structure and function. *Common Bioinformatics Teaching Resources* at   http://www.embl.de/~seqanal/courses/commonCourseContent/commonProteinStructure-FunctionExercises.html
36. Osawa, S. (1995). *Evolution of the genetic code*. Oxford: Oxford University Press.
37. Toh, H. (2004). *Bioinformatics for the analysis of protein function (in Japanese)*. Tokyo: Kodansha Scientific.

Mutation

2

Chapter Summary

Mutations, the fundamental sources of evolution, are described in detail. They include nucleotide substitutions, insertions/deletions, recombinations, gene conversions, gene duplications, and genome duplications. Mutation rate estimates and methods to estimate mutation rates are also discussed.

2.1 Classification of Mutations

2.1.1 What Is Mutation?

Any change of nucleotide sequences in one genome can be considered as "mutation" in the broad sense. In the traditional view, mutations occur when genes are transmitted from parent to child. If we focus on diploids, miosis may be the only chance for mutations. However, DNA replications also occur during mitosis and DNA damage can happen at any time without DNA replication. Therefore, indication of time unit, per generation, per year, or per replication, is important when we discuss mutation rate differences.

The classic unit of mutation was gene, because gene was defined as a unit for particular phenotype or particular function. Now we know that DNA is the material basis of inheritance, and any modification of nucleotide sequences should be considered as "mutation." In the early days of molecular genetics, the term "point mutation" was often used. Change of one nucleotide may correspond to point mutation. However, this includes nucleotide substitution and deletion or insertion of one nucleotide. A change involving more than one nucleotide may also be considered as point mutation if they are contiguous. Because of this uncertainty, we should not use this term anymore. Mutations should be classified by their structural characteristics. Table 2.1 shows a list of mutations. Types of DNA polymorphisms caused by various types of mutations are also listed in Table 2.1, and they will be discussed in Chap. 4.

N. Saitou, *Introduction to Evolutionary Genomics*, Computational Biology 17,
DOI 10.1007/978-1-4471-5304-7_2, © Springer-Verlag London 2013

Table 2.1 List of mutation types

Type	Polymorphism
Nucleotide substitution	SNP (single nucleotide polymorphism)
Insertion and deletion	Indel
Gene conversion (allelic)	SNP-like
Gene conversion (paralogous)	SNP-like
Repeat number change	STR (short tandem repeat) or microsatellite
Single crossover	Recombination
Unequal crossover	CNV (copy number variation)
Double crossover	SNP-like
Inversion	Inversion polymorphism
Chromosomal translocation	Translocation polymorphism
Repeat insertion	Insertion polymorphism
Genome duplication	Genome number polymorphism

2.1.2 Temporal Unit of Mutation

The classic unit of time to measure mutation is one generation: between parents and children, for mutations were believed to occur only at meiosis in germ line. However, somatic mutation, or mutation in somatic cells, does occur. Acquired immune system of vertebrates is known to increase antibody amino acid sequence variation by incorporating somatic mutations [1]. Because mitosis may be involved in creating somatic mutations, the number of cell division in germ line is another important factor for mutation. Haldane ([2]; cited in [3]) already suggested in 1947 that the mutation rate might be much higher in males than in females because the number of germ-cell divisions per generation is much higher in the male germ line than in the female germ line. In fact, the long-term mutation rate for Y chromosomal DNA in mammals is clearly higher than that for autosomes and X chromosome [3, 4]. This difference is easy to be understood if we consider a huge number of sperms and a small number of eggs produced in one generation. Y chromosomes always pass through sperms, while autosomes pass through either sperm or egg with equal probability. An X chromosome has 1/3 and 2/3 probabilities for passing sperm and egg, respectively.

Mutations may not be restricted to cell divisions. They may occur at any time, for any damage to DNA molecules is always the starting point for a mutation. In any case, mechanisms of mutation are not yet understood in detail, and this is a future problem.

2.1.3 Mutations Affecting Small Regions of DNA Sequences

When only a small portion of the DNA sequence is modified, say, one to a few nucleotides, this may be called minute mutation. They are nucleotide substitutions,

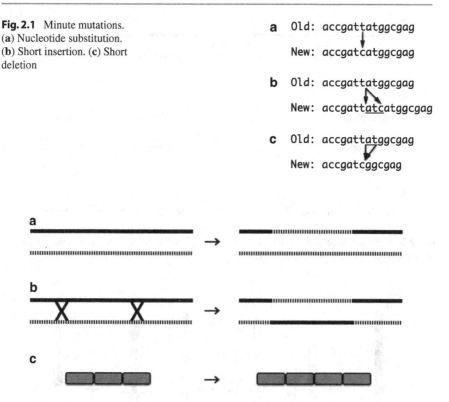

Fig. 2.1 Minute mutations. (a) Nucleotide substitution. (b) Short insertion. (c) Short deletion

a Old: accgattatggcgag

New: accgatcatggcgag

b Old: accgattatggcgag

New: accgattatcatggcgag

c Old: accgattatggcgag

New: accgatcggcgag

Fig. 2.2 Mini-scale mutations. (a) Allelic gene conversion. (b) Double crossover. (c) STR number change

short insertions, and short deletions. Figure 2.1 shows a schematic view of these minute mutations. Nucleotide substitutions will be discussed in detail at Sect. 2.2 and insertions and deletions at Sect. 2.3.

Mutations affecting somewhat larger regions of DNA sequences may be called mini-scale mutations (Fig. 2.2). Allelic gene conversion, double crossover, and short tandem repeat (STR) number change are included in this category.

2.1.4 Mutations Affecting Large Regions of DNA Sequences

The physical order of nucleotide sequence is modified in recombination, paralogous gene conversion, and inversion (see Fig. 2.3). Chromosomal level changes of DNA sequences are classified into inversion, translocation, and fusion as shown in Fig. 2.4. The human chromosomes have been well studied, and more detailed description will be given in Chap. 10. The largest type of mutation is genome duplication or polyploidization.

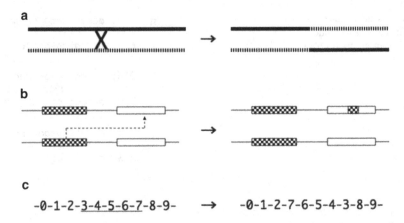

Fig. 2.3 Mutations affecting the physical order of nucleotide sequences. (**a**) Single crossover. (**b**) Paralogous gene conversion. (**c**) Inversion

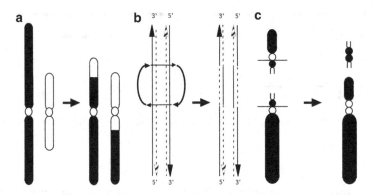

Fig. 2.4 Mutations affecting the large area of one chromosome (Based on [58])

2.2 Nucleotide Substitutions

2.2.1 Basic Characteristics of Nucleotide Substitutions

Nucleotide substitutions are mutual interchanges of four kinds of nucleotides or bases. Figure 2.5 shows all possible 12 nucleotide substitutions in DNA sequences. If a substitution is between chemically similar bases (see Chap. 1), i.e., between purines (adenine and guanine) or pyrimidines (cytosine and thymine), it is called transition. If a substitution is between a purine and a pyrimidine, it is called transversion.

It was predicted that transition should occur in higher frequency than transversion, because transitions have four possible intermediate mispair states, while transversions have only two such states. Figure 2.6 shows six possible intermediate mispairings. If we start from adenine–thymine-type normal base pairing, transition (A–T to G–C)

Fig. 2.5 Pattern of
nucleotide substitution

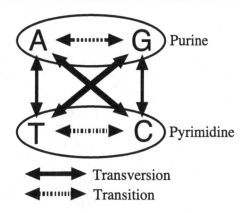

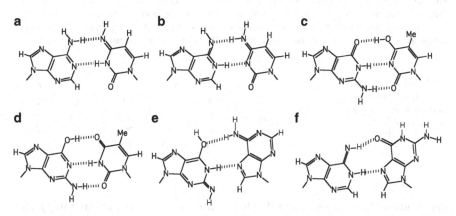

Fig. 2.6 Six possible intermediate mispairings (Based on Ref. [5]). (**a**) Adenine and imino-
cytosine. (**b**) Imino-adenine and cytosine. (**c**) Guanine and enol-thymine. (**d**) Enol-guanine and
thymine. (**e**) Imino-adenine and syn-adenine. (**f**) Imino-adenine and syn-guanine (based on [5])

can occur through (a) to (d) intermediate mispairings, while transversion (A–T to T–A)
can occur only through (e) or (f) intermediate mispairings [5]. This is the basis of
higher transitions than transversions.

Absolute rates of mutations for 12 kinds of directions are not easy to estimate,
because we need to directly compare parental and offspring genomes, and the rate
of fresh or de novo mutations in eukaryotes is usually quite low. Instead, we can
compare evolutionarily closely related sequences. Relative mutation rates of six
pairs of bases (A<==>G, C<==>T, A<==>T, A<==>C, G<==>T, and G<==>C)
can be estimated by comparing many numbers of SNPs (single nucleotide polymorphisms) in one species. However, we need the closely related out-group species
when the directionality of mutation comes in. Figure 2.7 shows how to estimate
direction of mutation in this way.

The pattern of nucleotide substitutions in the human genome was estimated
using the scheme of Fig. 2.7. More than 30,000 human SNP data determined for

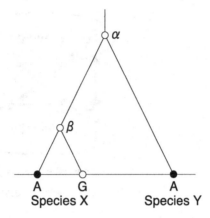

Fig. 2.7 Estimation of mutation direction in species X by using orthologous sequence data for species Y. Species X is polymorphic at some ne nucleotide site, and two nucleotides (A and G) coexist, while the corresponding nucleotide position for species Y is A. Using parsimony principle, ancestral nodes α and β are estimated to be both A. Therefore, the mutation direction is A to G.

Table 2.2 Pattern of nucleotide substitutions in the human genome (From [6])

	A	T	C	G
A	–	2.9	3.6	14.0
T	2.8	–	15.1	3.5
C	4.4	20.3	–	4.5
G	19.6	4.5	4.9	–

Unit: %

chromosome 21 were used, and the direction of mutation was inferred using the chimpanzee chromosome 22 as the out-group sequence [6]. We standardized these frequencies using Gojobori et al.'s (1982; [7]) method, so the sum becomes 100 %. The result is shown in Table 2.2. First of all, transitions shown in top-right to down-left diagonal are much more frequent than transversions. Among transitions, G=>A and C=>T are more frequent than their reverse directions (A=>G and T=>C). This corresponds to the fact that the mammalian genomes, here represented by the human genome, have about 40 % G+C proportion, or are A+T rich.

Transitions are known to be quite high in animal mitochondrial DNA. Kawai and Saitou (unpublished) analyzed complete mitochondrial DNA genome sequences of 7,264 human individuals and observed 4,939 substitutions with inferred direction among 2,179 fourfold degenerate synonymous sites. This result shows that transitions are about 30 times higher than transversions. Among transitions, C=>T changes are less than the other three directions, and C=>A changes are most abundant among transversions.

Evolutionary rates of nucleotide substitution are expected to be equal to the mutation rate of nucleotide substitution types, if the genomic region in question is evolving in purely neutral fashion (see Chap. 4). This characteristic is used for estimate mutation rates of various organisms, as shown in Sect. 2.6.3.

Table 2.3 Examples of various types of substitutions in the protein coding region

Synonymous substitution
All three possible substitutions at third position of a codon
GCT (Ala) → GCA (Ala), GCC (Ala), GCG (Ala)
Transitional type substitution at third position of a codon
AAT (Asn) → AAC (Asn)
Substitution at first position of a codon
CGA (Arg) → AGA (Arg)
Nonsynonymous substitution
All three possible substitutions at second position of a codon
GCT (Ala) → GAT (Asp), GGT (Gly), GTT (Val)
Transvertional type substitutions at third position of a codon
AAT (Asn) → AAA (Lys), AAG (Lys)
Substitution at first position of a codon
CGT (Arg) → AGT (Ser), GGT (Gly), TGT (Cys)
Nonsense substitution
Substitution at first position of a codon
CAA (Gln) → TAA (stop)
Substitution at second position of a codon
TCG (Ser) → TAG (stop)
Substitution at third position of a codon
TGG (Trp) → TGA (stop)
Stop codon to amino acid codon substitution
Substitution at first position of a codon
TAA (stop) → GAA (Glu)
Substitution at second position of a codon
TAG (stop) → TTG (Leu)
Substitution at third position of a codon
TGA (stop) → TGC (Cys)

2.2.2 Nucleotide Substitutions in Protein Coding Regions

When a nucleotide substitution occurs in a protein coding region, it may be either synonymous, nonsynonymous, or nonsense substitution; see the standard genetic code table shown in Table 1.1. Synonymous substitution may also be called silent substitutions in protein coding regions, and nonsynonymous substitution that seems to be used by Gojobori (1983; [8]) for the first time may also be called missense mutation or amino acid replacing mutation.

Amino acid sequence will not be changed when a synonymous substitution occurs, while amino acid will be changed when a nonsynonymous substitution happens. A nonsense substitution will change an amino acid codon to stop codon and will shorten the amino acid sequence. Change from stop codon to amino acid codon will elongate proteins. Table 2.3 shows examples of these four types of nucleotide substitutions.

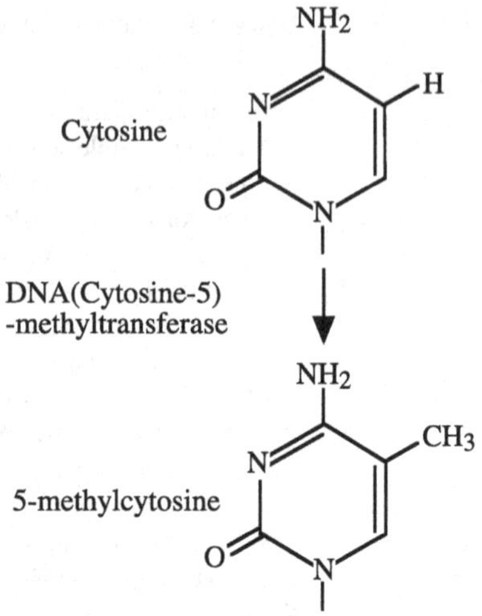

Fig. 2.8 Methylation of
cytosine

2.2.3 Methylation Creates "Fifth" Nucleotide

DNA methyltransferase may recognize 5′-cytocine-guanine-3′ (CpG) and modify
cytosine to 5-methylcytocine in mammalian genomes (Fig. 2.8). Because the CpG
dinucleotide is complementary to itself, this double-strand DNA is methylated in
both strands. This methylation is known to be related to imprinting and is an epigen-
etic phenomenon. Interestingly, CpG islands, regions with high CpG density just
upstream of protein coding genes, are rarely methylated.

2.3 Insertion and Deletion

2.3.1 Basic Characteristics of Insertions and Deletions

The length of DNA does not change with nucleotide substitutions, while it is well
known that genome sizes vary from organism to organism. It is thus clear that there
exist mutations changing length of DNA. They are generically called "insertion" or
"deletion" when the DNA length increases or decreases, respectively. When mutational directions are not known, combinations of insertions and deletions may be
called gaps or indels. When the gap length is only one, this gap or indel polymorphism may be included as a special case of SNP (single nucleotide polymorphism).
In real nucleotide sequence data analysis, insertions and deletions are detected only
after multiple alignment of homologous sequences. The relationship of insertion
and deletion with sequence alignment techniques will be discussed in Chap. 14.

A special class of insertions and deletions is repeat number changes. If repeat
unit length is very short (less than 10 nucleotides), it is called STRs (short tandem
repeats) or microsatellites. In contrast, "minisatellites" or VNTRs (variable number

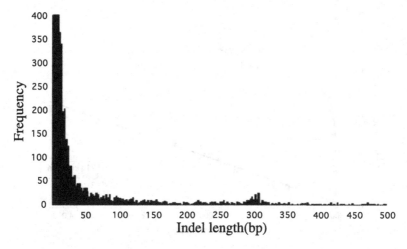

Fig. 2.9 Length distribution of indels in the human/chimpanzee lineages (From The International Chimpanzee Chromosome 22 Consortium 2004; [6])

of tandem repeats) have typically repeat unit lengths of 10–100 nucleotides. Because of their importance on DNA polymorphism studies, let us divide insertions and deletions into two types, unique sequence and repeat sequence, and we discuss them independently.

2.3.2 Insertions and Deletions of Unique Sequences

The length distribution of indels in the human or the chimpanzee lineages is shown in Fig. 2.9. Single nucleotide changes are most frequent, and the frequency quickly drops as the length becomes longer. It should be noted that there is a small peak around 300 bp. This is mostly due to Alu sequence insertions. Figure 2.10 shows data similar to those of Fig. 2.9 with mutational directions. Directions (either insertion or deletion) for each indel position were estimated by checking situations in gorilla and orangutan genomes [6]. Interestingly, the human genome experienced more insertions than the chimpanzee genome, especially for Alu sequences, as shown in Fig. 2.11. In contrast, the length distribution patterns for deletions do not differ significantly between human and chimpanzee genomes.

Minute length gaps or indels can be studied by examining multiple aligned orthologous (see Chap. 3) nucleotide sequences of closely related species. If they are located in the genomic regions under purely neutral evolution, their evolutionary rates can be considered as the mutation rates (see Chap. 4 for this rationale). Saitou and Ueda (1994; [9]) estimated mutation rates of insertions or deletions, for the first time, using primate species noncoding genomic regions, and they found molecular clocks (rough constancy of the evolutionary rates) both in mitochondrial and nuclear DNAs (see Fig. 2.12). The rate (approximately 2.0/kb/Myr) for mitochondrial DNA

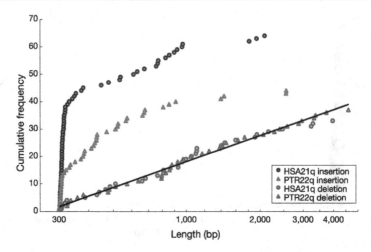

Fig. 2.10 Length distribution (more than 300 bp; cumulative) of insertions and deletions in the human and chimpanzee lineages (From The International Chimpanzee Chromosome 22 Consortium 2004; [6])

was found to be much higher than that (approximately 0.2/kb/Myr) for nuclear DNA. Because the rate of nucleotide substitutions in nuclear genome of primates is approximately 1×10^{-9}/site/year, the rate of insertions and deletions is about 1/5 of substitution. Ophir and Graur (1997; [10]) compared hundreds of functional genes and their processed pseudogenes in human and mouse nuclear genomes and found that deletions are more than two times more frequent than insertions. They also estimated that the rate of insertions is 1/100 of that of nucleotide substitutions. When chimpanzee chromosome 22, corresponding to human chromosome 21, was sequenced in 2004 [6], insertions and deletions were carefully analyzed. A total of 68,000 indels were found from the human–chimpanzee chromosomal alignment. More than 99 % of them are shorter than 300 bp. The human chromosome 21 long arm is 33.1 Mb, and chimpanzee chromosome 22 long arm is 32.8 Mb. If we take their average, 33.0 Mb, as compared length, and if we assume that the human and chimpanzee divergence time is 6 million years, then the overall rate of insertions and deletions in human and chimpanzee lineages becomes 0.38×10^{-9}/site/year (= 68,000/30Mb/6MY). This estimate is about two times higher than that obtained by Saitou and Ueda [9] using much smaller sequence data of primates. See Sect. 2.6.2 for more discussion.

2.3.3 Insertions and Deletions of Repeat Sequences

There are many studies on mutation mechanism of STRs or microsatellites. DNA slippage is commonly accepted to be the major mutation mechanism of microsatellites [60]. Factors affecting STR slippage include repeat number, locus length, motif

Fig. 2.11 A phylogenetic
tree of human-specific and
chimpanzee-specific Alu
sequences (From The
International Chimpanzee
Chromosome 22 Consortium
2004; [6])

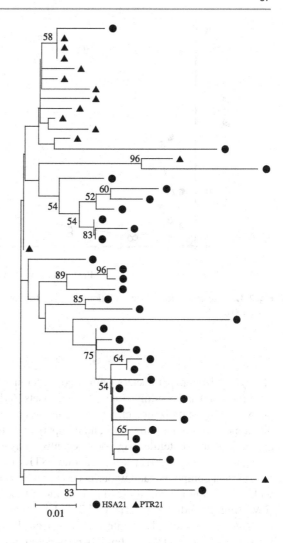

size, motif structure, type, and chromosomal location [61]. Among all those factors, repeat number is the strongest factor positively correlated to mutation rates of STRs [62].

Perfection status of STRs is one of the remaining factors which significantly affects STR mutation rates (e.g., [63]). Four types of status are often recognized: (1) perfect, STR purely composed of only one kind of motif; (2) imperfect, STR having one base pair which does not match the repetitive sequence; (3) interrupted, STR with short sequences within the repetitive sequences; and (4) composite (also called compound), STR with two distinctive, consecutive repetitive sequences linked. However, this classification is somehow idealistic that many of the micro-satellites could not fall into any of the four categories in real cases. Algorithms searching STRs such as Sputnik, RepeatMasker, and Tandem Repeat Finder

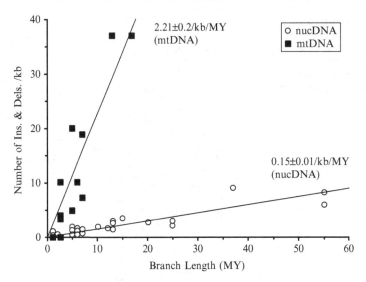

Fig. 2.12 The constancy of the evolutionary rates of insertions and deletions (From Saitou and Ueda 2004; [9])

(TRF) use different parameters, with major difference when dealing with mismatches (interruptions) and resulting nonuniform datasets [64]. Interruptions inside STRs are known to have stabilization effect, where mutation rate is greatly lowered. However, direct measure on mutation rates and comparison between imperfect microsatellites and perfect microsatellites were only recently analyzed [65].

Ngai and Saitou (2012; [66]) redefined STRs into four groups: perfect, imperfect, perfect compound, and imperfect compound. A perfect STR is defined as locus with a perfect repetitive run with its own motif type, abbreviated as the locus motif (LM) (Fig. 2.13a). An imperfect STR is defined as locus with a repetitive run that contains interruptions. Each interruption is from 1 to 10 bp long (Fig. 2.13b). When interruption is more than 10 bp, it is considered as a perfect locus. Besides perfect and imperfect, a locus could also be either compound or noncompound. A compound microsatellite is defined as locus which contains repetitive sequence composed of a non-locus motif, abbreviated as non-LM, where the repeat number passes the threshold value (say 3 repeats) and is within 10-bp flanking region of the locus (Fig. 2.13c, d).

Recently, direct sequence comparison of STR repeat numbers was conducted for parent–offspring pairs [67]. They examined ~2,500 STR loci for ~85,000 Icelanders and found more than 2,000 de novo repeat change-type mutations. Father-to-offspring mutations were three times higher than those for mother-to-offspring, and the mutation rate doubled as the father's age changed from 20 to 58, while no age effect was observed for mother. Mutation rate estimates per locus per generation for dinucleotide and trinucleotide STRs were 2.7×10^{-4} and 10.0×10^{-4}, respectively.

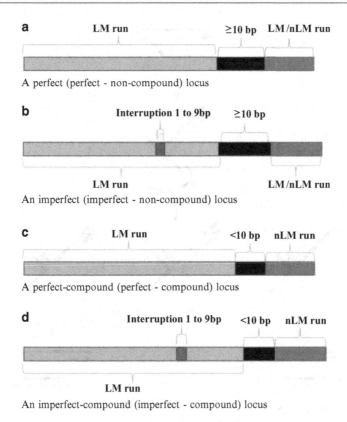

Fig. 2.13 Four types of STR loci (From Ngai and Saitou 2012; [66])

2.4 Recombination and Gene Conversion

Recombination was discovered by Thomas Hunt Morgan and his colleagues in the early twentieth century. The concept of "gene conversion" was first proposed by Winkler in 1930 [11], but it was not fully accepted for a long time, until studies on fungi clearly showed conversion events [12, 13]. Holliday (1964; [14]) proposed the "Holliday structure" model (Fig. 2.14) to connect gene conversion, or nonreciprocal transfer of DNA fragment, and recombination.

The general definition of recombination is reconnection of different nucleotide molecules. There are two types of recombination: homologous and nonhomologous. Homologous recombination usually occurs through crossing-over during meiosis, as already discussed using Figs. 2.2b and 2.3a. We restrict our discussion in this section only to eukaryotes, for "recombination" in prokaryotes are quite different.

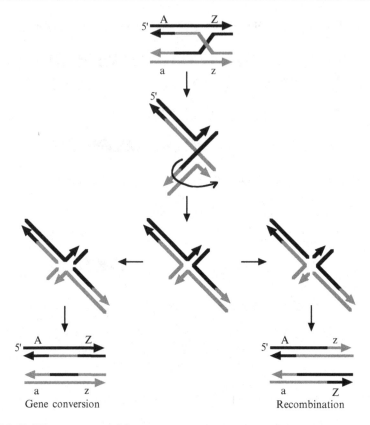

Fig. 2.14 Holliday structure model

2.4.1   Nature of Gene Conversion

Early studies on gene conversion were mostly restricted to fungal genetics. As molecular evolutionary studies of multigene family started, unexpected similarity of tandemly arrayed rRNA genes was found [15]. This phenomenon was termed "concerted evolution," and gene conversion or unequal crossing-over was proposed to explain this characteristic of some multigene families (e.g., [16]). New statistical methods were developed to detect gene conversion between homologous sequences [17, 18]. Program GENECONV developed by Sawyer [19] became the standard tool for analyzing gene conversions. We now know that conversion can occur in any genomic region irrespective of genes (DNA regions having function) or nongenic regions (e.g., [20]). However, "gene conversion" as technical jargon is currently widely accepted, and I follow this nomenclature. Gene conversion can be classified into two types: intragenic or between alleles and intergenic or between duplicated genes.

Fig. 2.15 Branches of
phylogenetic trees where
duplication events were used
in three studies. Branch
length proportions in terms
of evolutionary time are after
Ezawa et al. (2010; Ref. [27])

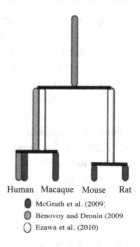

Human Macaque Mouse   Rat

● McGrath et al. (2009)
◐ Benovoy and Drouin (2009
○ Ezawa et al. (2010)

2.4.2 Gene Conversion on Duplog Pairs

After human genome sequencing was "completed" [21], genomes of mouse [22], rat [23], rhesus macaque [24], and other mammalian species were determined. Because genome sequences of these four species (human, macaque, mouse, and rat) are in good condition than those of other mammalian species, genome-wide analyses of gene conversion among duplogs (duplicated genes or DNA regions in one genome) were confined to these species. Boney and Drouin (2009; [25]) studied human duplogs, while Macgrath et al. (2009; [26]) studied young duplogs after speciation of human–macaque and mouse–rat, and Ezawa et al. (2010; [27]) used duplogs before these two speciations (see Fig. 2.15).

A number of duplog pairs used in three studies are 55,050, 3,996, and 1,121 for [25, 26], and [27], respectively. The total number (55,050) of duplog pairs used by [25] was much larger than the number (27,350) of known protein coding gene sequences they used. It is possible that many of these duplog pairs were counted more than once. The human genome has gene families with many copy numbers. For example, the number of functional olfactory receptor genes was estimated to be 388 [28]. Multiple countings of these large-sized multigene families could be a reason to reach such large duplog pairs. Independent duplog sets are preferable for statistical tests, and Ezawa et al. [27] carefully eliminated multiple countings, which were included in their previous study [29]. It is not clear whether McGrath et al. [26] also excluded double counting from their "Methods" section, but they focused on young duplogs (see Fig. 2.1), and two or more duplications in these relatively short evolutionary times for one gene may not be frequent.

It is interesting to compare frequency of gene duplication between primate and rodent lineages. A total of 549 and 363 duplog pairs were extracted from human and rhesus macaque genomes, respectively, in [27], while 1,913 and 1,171 pairs were found from mouse and rat genomes, respectively. Ezawa et al. [27] found 430 and

691 duplog pairs from primate and rodent lineages, respectively. Because primates and rodents started to diverge about 80–95 million years ago [27], we can compare the total number of duplications for these two lineages with the same evolutionary time: 886 (=[549+363]/2+430) for primate lineage and 2,233 (=[1,913+1,171]/2+691) for rodent lineage. It seems that the rodent lineage has more than two times higher rate of gene duplication. However, it is possible that gene duplication is more proportional to the number of nucleotide substitutions rather than evolutionary time. If we use neutral nucleotide divergence data shown in Fig. 2.2b of [27], duplication events can be normalized as 6,329 (=886/0.14) for primates and 8,270 (=2,233/0.27) for rodents per one nucleotide substitution per site. These two values are not much different. This suggests that the rate of gene duplication in mammals is more or less proportional to nucleotide substitutions. Because tandem gene duplication is usually assumed to start from unequal crossing-over that happens in meiosis, it is possible that nucleotide substitutions also happen during meiosis. In any case, it is clear that we need to have ample knowledge on mammalian duplogs if we are interested in intergenic gene conversion in mammalian genomes.

When Winkler [11] proposed gene conversion in 1930, it was a deviation from the Mendelian ratio. Later, detailed observations on baker's yeast and Neurospora [12, 13] established gene conversion, and Holliday's [14] model transformed gene conversion from phenomenon to mechanism. Nowadays several enzymes are known to be involved in DNA strand exchanges [30]. Abundant genome sequence data and their computational analyses again turned gene conversion or more flatly homogenization of homologous sequences from mechanism to phenomenon. We should be careful of any prejudice to a particular phenomenon when we try to interpret them with certain mechanism. One phenomenon, such as homologous sequence homogenization, may occur not only via gene conversion but with some other mechanisms, including one unknown to us at this moment. It is obvious that we should grasp molecular mechanism of gene conversion, including enzymatic machineries.

2.5 Gene Duplication

2.5.1 Classification of Duplication Events

Duplication of DNA fragment can happen in any region of chromosomes, but historically duplicated regions containing protein coding genes were the focus of research on duplication. Therefore, when we mention "gene duplication," nongenic regions may also be included. Under this broad meaning, gene duplication can be classified into four general categories: (1) tandem duplication, (2) RNA-mediated duplication, (3) drift duplication [31], and (4) genome duplication.

Tandem duplication results in two homologous genes in close proximity with each other in the same chromosome via unequal crossing-over (see Fig. 2.16a), while RNA-mediated duplication can create duplicate copies, complementary to original RNA molecules, far from the original gene with the help of reverse transcriptase

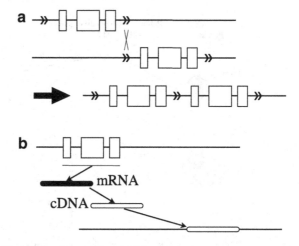

Fig. 2.16 Mechanisms of gene duplications.
(**a**) Unequal crossing over.
(**b**) Retrotransposition

(see Fig. 2.16b). Retrotransposons including SINEs (short interspersed elements) and LINEs (long interspersed elements) are in this category. Alu and L1 sequences are representatives of SINEs and LINEs in the human genome, respectively. This type of duplication is also characterized by intronless sequence, when mature mRNAs, formed after introns are spliced, are reverse transcribed. Intronless copies are also called "processed genes" because they went through processing called "splicing." Protein coding genes require appropriate enhancers and promoters to be transcribed. Therefore, these processed genes are most probably not transcribed and functionless. Most probably, they are "dead on arrival," that is, they become nonfunctional immediately after duplication. Because of this nature, these duplicates are often called "processed pseudogenes." It should be noted that immature mRNA before splicing may have the possibility of reverse transcription.

2.5.2 Drift Duplication

Ezawa et al. (2011; [31]) recently proposed a new category of gene duplication and named it "drift" duplication. Its physical distance distribution appears to peak around a few hundred Kb for vertebrates and a few dozen Kb for invertebrates, which is in between those of tandem duplication (short range) and retrotransposition (long range, i.e., mostly unlinked). Drift duplications are almost randomly oriented, with the frequency ratio of head-to-tail:tail-to-tail:head-to-head ≈ 2:1:1, as opposed to tandem duplications due to unequal crossing-overs, which are mostly head-to-tail. A drift duplication can also create multi-exon duplogs, as opposed to retrotransposition, whose products are mostly intronless. Retrotransposition is also drifting in a sense; however, it always passes through the RNA stage. This is the clear difference from drift duplication. With this name, "drift," Ezawa et al. [31] also implied that even some interchromosomal duplications may be attributed from drift duplication,

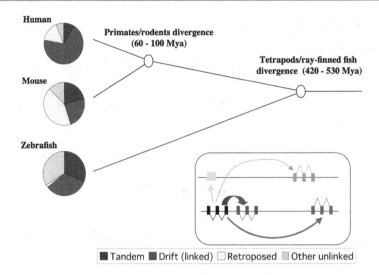

Fig. 2.17 Proportions of duplication types in vertebrate genomes (From Ezawa et al. 2011; [31])

though RNA-mediated duplications may be more frequent among interchromosomal duplications. DNA molecules for drift duplications are usually much larger than those for RNA-mediated duplications and may not be able to move to different chromosomes easily. This conjecture should be examined in future studies. Figure 2.17 shows proportions of duplication types in vertebrate genomes.

2.5.3 Genome Duplication

The last type of duplog generation is genome duplication. It is called "polyploidization" in plants, and this was probably why the word "genome" was coined by Hans Winkler (1920; [32]), a botanist. Genome duplication has never been demonstrated in prokaryotes (Bacteria and Archaea) and is rare in eukaryotes except for plants and some vertebrates (see Chap. 8). Genome duplication is important when we consider the origin of vertebrates, but after two-round genome duplications, the mammalian lineage has not experienced further genome duplication.

2.6 Estimation of Mutation Rates

2.6.1 Direct and Indirect Method

Estimation of mutation rates is quite important, for the mutation is the ultimate source of evolution. The natural way to estimate the mutation rate is to compare nucleotide sequences between parents and offsprings. This is called direct method. However, it was difficult for a long time to directly compare a large number of nucleotide sequences. Therefore, examination of various phenotypes was used to estimate the mutation rate. Even now, this kind of study is conducted, e.g., Watanabe et al. (2009; [33]).

Because of the difficulty of direct estimates of mutation rates from comparison of large nucleotide sequences, various methods were used to indirectly estimate mutation rates, considering the balance among mutation, selection, and random genetic drift (see Chap. 3 of Nei (1987; [34]) for review). However, a more straightforward way is to compare neutrally evolving nucleotide sequences of relatively closely related species, as we will see in Sect. 2.6.3.

2.6.2   Direct Method

Various types of mutations were described in this chapter, and accumulation of these mutations through a long time period is the source of evolution. Therefore, the rate of mutation per generation, or mutation rate, is closely related to the rate of evolution. Classically, the temporal unit of mutation rate is one generation. This is because sudden changes of phenotypes between parent and children were fundamental for mutations in the classic sense. Estimation of mutation rates through comparison of parents and offsprings is called "direct" method. In this case, a temporal unit of the mutation rate is one generation.

The rate of mutation for human was estimated for the first time by using data on achondroplasia, an autosomal dominant Mendelian trait (OMIM #100800). This inherited disease is caused by a mutation that occurred in the gene coding for the fibroblast growth factor receptor-3 (FGFR3) at the short arm of human chromosome 4. The most frequent type is nonsynonymous substitution at the 380th codon, changing glycine to arginine. Haldane (1949; [35]) used data collected in Copenhagen, the capital of Denmark; 10 babies were achondroplastic out of 94,075 newborn babies. Two of them inherited the mutant gene from parents, for one parent was also achondroplastic. Therefore, the number of fresh or de novo mutations was eight. The mutation rate is thus estimated to be 4.3×10^{-5} (= 8/[94,075 × 2]) per generation per gene. An alternative estimate, 1.4×10^{-5}, based on 7 mutants out of 242,257 births [36], is about 1/3 of the estimate originally obtained by Haldane [35].

As already mentioned in this chapter, there are various units of mutation rates for both temporal and spacial situations. Temporal units include one generation, one meiosis, one cell division, and 1 year, while typical spacial units are one gene, one nucleotide, 1 kb, or the whole genome. Because one cell division corresponds to one generation for unicellular asexual organisms such as prokaryotes, one cell division may be a good universal temporal unit for the mutation rate. However, naturally occurring radiations such as cosmic ray or background radiation as well as chemical mutagens may affect an organism at any time. Therefore, a physical time, such as 1 year, may also be a universal temporal unit of the mutation rate. Because some types of mutations may occur only during meiosis, one meiosis may be suitable for this kind of mutations. In the case of achondroplasia, the physical unit of the mutation rate was one gene that consists of up to more than one million nucleotides.

In the 1960s, the use of protein electrophoresis, particularly using starch gel (Smithies, 1995; [37]), became popular for detecting protein variations for many organisms. Because the amino acid sequence information is closely related to the nucleotide sequence (see Chap. 1), this technique should give much better estimate of mutation rates at the nucleotide sequence level. A mass scale study for estimating the mutation rate was conducted using protein electrophoresis for Japanese individuals who were exposed to various degrees of radiation from atomic bomb explosions at Hiroshima and Nagasaki on August 6 and 9, 1945, respectively. The Atomic Bomb Casualty Commission (ABCC) was created by the US government, and many human individuals were examined. One of the final results, published in 1986 led by James V. Neel [38], reported 3 mutations out of 539,170 gene transmissions from parent to offspring. The mutation rate was then estimated to be 1×10^{-8} per nucleotide site per generation. This value is about half of the estimate obtained by comparison of nucleotide sequences shown in the next paragraph.

DNA sequencing (see Chap. 11) became popular in the 1980s, but a huge effort of nucleotide sequencing was necessary to estimate the mutation rates by directly comparing parents and offspring. Therefore, this type of studies was started to be published in this century. Kondrashov (2002; [39]) assembled nucleotide sequence data for 20 Mendelian genetic disease-causing genes and estimated the human nuclear mutation rate to be 1.78×10^{-8} per nucleotide site per generation. The majority (1.7×10^{-8}) of them were nucleotide substitutions, and insertion-type mutations were 1/3 of deletion-type mutations. If we consider one generation of modern-day human as 30 years, the rate of substitution type mutation per nucleotide site per year becomes 0.56×10^{-9}. The mutation rate of insertions and deletions was estimated to be 8×10^{-10} per nucleotide site per generation, and it corresponds to 3×10^{-11}/site/year. This is 13 times lower compared to that $(3.8 \times 10^{-10}$/site/year) obtained from the comparison of human and chimpanzee genomic sequences [6]. Probably the total amount of compared nucleotide sequences was too small to obtain a reliable estimate.

Recently, thanks to the so-called second-generation sequencer (see Chap. 11), genome-wide comparison of parents and offspring was made, and the mutation rate was estimated to be ~1.1×10^{-8} per nucleotide site per generation [40, 41]. Because one generation is about 30 years in human, the rate becomes ~0.4×10^{-9} per nucleotide site per year. During the final process of editing this book, two additional papers [58, 59] were published on this matter, and both showed 1.2×10^{-8} per nucleotide site per generation, which is slightly higher than the estimate obtained by two previous studies [40, 41].

So far, we discussed mutation rates of human using different kinds of data. Let us move to other animal species. More than 40-Mb nucleotide sequences were determined using the PCR-direct sequencing method, after accumulating mutations for hundreds of generations in *Caenorhabditis elegans* [42]. A total of 30 mutations (13 substitution type, 13 insertion type, and 4 deletion type) were discovered. The net mutation rate was estimated to be 2.1×10^{-8}/site/generation. This is similar to the value estimated for human, but insertion-type mutations are more than three times higher than deletion types. This is a good contrast to the estimate based on the indirect method, where deletions are preponderant. This suggests that deletion-type mutations may be somewhat more advantageous than insertion-type mutations so as to keep the genome size smaller. A larger-scale study using second-generation sequencers (both 454 and Solexa) found 391 substitutional-type mutations out of 584-Mb sequences, resulting in the mutation rate to be 2.7×10^{-9}/site/generation [43]. If we assume that the average number of germ-line cell division per generation is 8.5 in this species, the mutation rate becomes 3.2×10^{-10}/site/cell division. This value is more or less similar to those for *S. cerevisiae* [44], *Drosophila melanogaster* [45], and human [43].

As for *Drosophila melanogaster*, a total of 37 fresh mutations were found through examination of 20-Mb sequences by combining DHPLC (denaturing high-performance liquid chromatography) and nucleotide sequencing [45]. The net mutation rate was estimated to be 8.4×10^{-9}/site/generation. Using Illumina sequencing technology, Keightley et al. (2009; [46]) sequenced three *Drosophila melanogaster* lines which accumulated mutations after 262 generations. They obtained the mutation rate to be 3.5×10^{-9}/site/generation, based on 174 de novo mutations out of 72-Mb sequences.

It is not clear how many generations may pass for 1 year in wild conditions for *Drosophila melanogaster*, but probably ten generations may correspond to 1 year (Dr. Takano-Shimizu Toshiyuki, personal communication). If so, the mutation rate becomes 8.4×10^{-8}/site/year. This is more than 100 times higher than that for human.

2.6.3 Indirect Methods

According to the neutral theory, the evolutionary rate (λ) is equal to the mutation rate in the genome region under pure neutral evolution (see Chap. 4). If we apply this idea, we can estimate the mutation rate by estimating the long-term evolutionary rate, under the simple equation:

$$D[i,j] = D[j,i] \tag{2.1}$$

where D is the evolutionary distance and T is the divergence time between the two lineages (see Fig. 2.18). This procedure is called the indirect method. For example, the substitutional difference (D) between human and chimpanzee is 1.23 % [47]. If we assume that the divergence time (T) of these two species is 6 million years, $\lambda = D/[2T] = 1 \times 10^{-9}$/site/year. Because application of the direct method usually takes a large amount of resources, this indirect method has been widely used from the advent of the molecular evolutionary studies. For example, Wu and Li (1985; [48]) showed that rodents seem to have higher mutation rate than primates.

Some cautions should be taken for this method. First of all, some genomic regions may not be under pure neutral evolution, but under purifying selection, as in the case of conserved noncoding sequences (e.g., Takahashi and Saitou, 2012; [71]). In this case, the mutation rate may be underestimated. Another problem is that the estimate is the long-term average of mutation rates for the two lineages. Generation

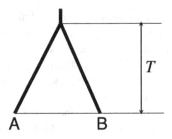

Fig. 2.18 A schematic phylogenetic tree for two lineages, A and B

times greatly vary from species to species. We are seeing only the average pattern of long-term evolution. This is a clear difference from results of the direct method, where a snapshot of the current populations is obtained.

2.6.4 Mutation Rates of Bacteria, Organella, and Viruses

Mutation rate estimates in prokaryotes can differ more than ten times between those using the direct method and those using the indirect method. Assuming the divergence time between *E. coli* and *Salmonella* to be about 100 million years ago, the long-term mean evolutionary rate was estimated to be 4.5×10^{-9}/site/year [49]. This was based on the number (0.9) of synonymous substitution per site, and it can be equated to be the mutation rate. On the other hand, the direct estimate based on lacZ revertants was ~5×10^{-10}/site/generation [49]. Because the number of generations per year for *E. coli* is at least 100, this mutation rate becomes ~5×10^{-8}/site/year.

Organella such as mitochondria and chloroplasts in eukaryotes have their own DNA and replicate them independently from nuclear DNA. Mutation rates of nuclear DNA and organella DNA are thus usually different. The mutation rate of animal mitochondria is more than ten times higher than that of nuclear DNA. However, among-lineage rate heterogeneity is considerable. There are also reports on time dependency. According to Burridge et al. (2008; [68]), pedigree-based mutation rate is the fastest (e.g., 5.1×10^{-7}/site/year) in human, followed by estimates based on 10,000-year-old ancient DNA (3.4–4.4×10^{-7}/site/year), and the slowest rate was obtained from phylogenetic estimates derived from Neogene primate divergences (0.5–2.4×10^{-7}/site/year). In birds, Millar et al. (2008; [69]) used Adélie penguins (*Pygoscelis adeliae*) whose ancient DNA samples were available. Direct mother–offspring mutation rate estimate and indirect estimate using 37,000-year-old ancient DNA samples gave 5.5×10^{-7}/site/year and 8.6×10^{-7}/site/year, respectively. Because these two rates are more or less the same, time dependency does not seem to exist in birds. In fish, however, Burridge et al. (2008; [68]) reported that the mtDNA substitution rates of galaxiid fishes from calibration points younger than 200 kyr (2–11×10^{-8}/site/year) were faster than those (0.8–5×10^{-8}/site/year) based on older calibration points, indicating the existence of time dependency.

In contrast, the mutation rate of plant mitochondria is about 1/10 of that of nuclear DNA. Interestingly, the mutation rate of plant chloroplast DNA is intermediate between those for mitochondrial DNA and nuclear DNA. The reason for such differences is not known.

The so-called RNA viruses have RNA genomes, and some of them have their own RNA replicase. Their replication error rates or mutation rates can be quite high than those of DNA genome organisms. Hanada et al. (2004; [50]) estimated rates of synonymous substitutions for 49 RNA viruses and showed that many of their rates were $\sim 10^{-3}$/site/year, though the whole range was heterogeneous, from 1×10^{-7}/site/year to 1×10^{-2}/site/year. Because the error rate of RNA replication is $\sim 10^{-5}$ per site per replication, variation of replication cycles per unit of time seems to contribute to heterogeneity of the mutation rate per site per year in RNA viruses [50].

2.7 Mutation Affecting Phenotypes

2.7.1 What Is Phenotype?

Various conditions of organisms are collectively called "phenotypes." Phenotypes include macroscopic characters such as seed surface form (round or wrinkled) studied by Mendel (1866; [51]), human height, amino acid sequence of proteins, or even DNA sequences themselves. Genetics has been the science of connecting genotypes and phenotypes. Genotypes are straightforward and objective; one genotype corresponds to a specific nucleotide sequence. In contrast, phenotypes are products of human recognition. It is true that the operational definition of one phenotype is always possible, such as human head length. It is, however, not clear what kind of biological significance exists in head length variation. We should be careful about the subjective nature of phenotypes. It may be ideal to only consider phenotypes which has one-to-one correspondence with one genotype.

2.7.2 Mutations in Protein Coding Region

Mutations are changes in nucleotide sequences, but they may change amino acid sequences if they happen in protein coding regions. We already discussed three types of nucleotide substitutions occurring in protein coding regions at Sect. 2.2.2. Even if amino acid is changed, some changes may not affect protein function. In this case, this mutant may be selectively neutral, as we will see in Chap. 4. When the amino acid change occurred in regions important for protein function, often such changes will reduce protein function. One example is the nonsynonymous mutation at ABCC11 gene in the human genome (see Fig. 2.19). Amino acid change from glycine to arginine essentially nullified transporter function of this protein (Yoshiura et al. 2006; [52]).

Another classic example is the emergence of sickle-cell anemia that is resistant to malaria [53]. Hemoglobin is composed of heme (porphyrin) and globin (protein),

Fig. 2.19 Functional differences of two proteins coded in ABCC11 gene (From Yoshiura et al. 2006; [52])

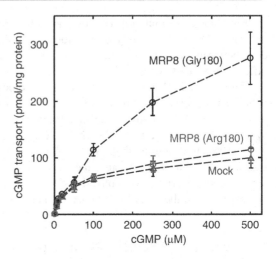

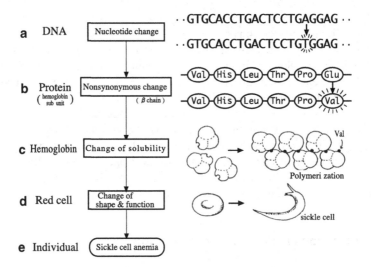

Fig. 2.20 Sickle cell anemia caused by one nonsynonymous substitution (From Saitou 2007; [70])

and its main function in animals is to transport oxygen to the entire body. Figure 2.20a shows normal hemoglobin A gene sequence and mutated hemoglobin S sequence. A to T nonsynonymous change caused Glu to Val change at position 6; see Fig. 2.20b. This position is not a part of the heme pocket where several amino acids are anchoring heme, but is located in globin surface. This change created a slight hump in the protein surface and can be connected to a hollow; see Fig. 2.20c. This produces linear globin polymer, and polymerization continues until all mutant proteins are connected. These polymers will form fibers and eventually cross the red cell body.

```
A allele   aaggatgtcctcgtggtgaccccttggctggctggctcccattgtctgggaggg
O1 allele  aaggatgtcctcgtggt-accccttggctggctggctcccattgtctgggaggg
A protein  LysAspValLeuValValThrProTrpLeuAlaProIleValTrpGluGlyThr
O protein  LysAspValLeuValVal ProLeuGlyTrpLeuProLeuSerGlyArgAla
```

Fig. 2.21 O-1 allele of the ABO blood group gene sequence (Based on Ref. [54])

This is why normally round red cell turns to sickle shape; see Fig. 2.20d. Sickle cells do not easily go through capillary vessels, and anemia will be the final result for the human individual. This is a good example that even a single nucleotide change can affect the whole individual.

If insertions or deletions occur, they will shift protein coding frames unless their numbers are in multiplication of 3. These mutations are thus called "frameshift" mutations, and most of them will no longer produce functional proteins. One such example from the ABO blood group O allele [54] is shown in Fig. 2.21. It should be noted that this example is rather unique, for the O-1 allele that cannot produce the functional enzyme (A or B) has high frequency in human populations. If one protein is indispensable for organism, such frameshift-type mutations will be quickly eliminated from populations, as we will see in Chap. 5.

Another loss-of-function-type mutation is insertion of transposons. Mendel (1866; [51]) studied seven characters of pea (*Pisum sativum*), and one of them is seed shape. Round type was dominant to wrinkled type. After more than 100 years of his study, a British group discovered that a 0.8-kb-long transposon insertion caused functional protein gene to be nonfunctional [55]. This gene encodes a starch-branching enzyme, and nonfunctional gene somehow causes pea skin to be wrinkled.

2.7.3 Mutations in Noncoding Region

Protein expressions are controlled in various ways. One of them is transcription control, and there are two types; trans and cis. DNA sequences responsible for cis control is often called "cis-regulatory element." Some of these functions were discovered by their loss-of-function-type mutations that affected phenotypes (e.g., [56]). One classic example of temporal control of gene expression is lactose tolerance. Mammalian babies, by definition, drink mother's milk as their main food source, and lactose is abundant in milk. They express enzyme called lactase, and this cuts lactose into glucose and galactose. After the lactation period, gene expression of lactase is stopped. In some human individuals, however, lactase expression continues to adulthood, called lactose tolerance. Through association studies, it was found that single nucleotide polymorphism is located at one intron of the adjacent gene to lactose gene, LCT, is the origin of the lifetime expression of lactase gene [57].

References

1. Kato, L., Stanlie, A., Begum, N. A., Kobayashi, M., Aida, M., & Honjo, T. (2012). An evolutionary view of the mechanism for immune and genome diversity. *Journal of Immunology, 188*, 3559–3566.
2. Haldane, J. B. S. (1947). The mutation rate of the gene for haemophilia, and its segregation ratios in males and females. *Annals of Eugenics 13*, 262–271. (cited by Ref. [3])
3. Crow, J. F. (1997). The high spontaneous mutation rate: Is it a health risk? *Proceedings of the National Academy of Sciences of the United States of America, 94*, 8380–8386.
4. Miyata, T., Hayashida, H., Kuma, K., Mitsuyasu, K., & Yasunaga, T. (1987). Male-driven molecular evolution: A model and nucleotide sequence analysis. *Cold Spring Harbor Symposia on Quantitative Biology, 52*, 863–867.
5. Topal, M. D., & Fresco, J. R. (1976). Complementary base pairing and the origin of substitution mutations. *Nature, 263*, 285–289.
6. The International Chimpanzee Chromosome 22 Consortium. (2004). DNA sequence and comparative analysis of chimpanzee chromosome 22. *Nature, 429*, 382–388.
7. Gojobori, T., Li, W.-H., & Graur, D. (1982). Patterns of nucleotide substitution in pseudogenes and functional genes. *Journal of Molecular Evolution, 18*, 360–369.
8. Gojobori, T. (1983). Codon substitution in evolution and the "saturation" of synonymous changes. *Genetics, 105*, 1011–1027.
9. Saitou, N., & Ueda, S. (1994). Evolutionary rate of insertions and deletions in non-coding nucleotide sequences of primates. *Molecular Biology and Evolution, 11*, 504–512.
10. Ophir, R., & Graur, D. (1997). Patterns and rates of indel evolution in processed pseudogenes from humans and murids. *Gene, 205*, 191–202.
11. Winkler, H. (1930). *Die Konversion der Gene*. Jena: Verlag von Gustav Fischer (written in German).
12. Lindegren, C. C. (1953). Gene conversion in *Saccharomyces*. *Journal of Genetics, 51*, 625–637.
13. Michell, L. B. (1955). Aberrant recombination of pyridoxine mutants of Neurospora. *Proceedings of the National Academy of Sciences of the United States of America, 41*, 215–220.
14. Holliday, R. A. (1964). Mechanism for gene conversion in fungi. *Genetic Research Cambridge, 5*, 282–304.
15. Brown, D. D., Wensink, P. C., & Jordan, E. (1972). A comparison of the ribosomal DNA's of *Xenopus laevis* and *Xenopus mulleri*: The evolution of tandem genes. *Journal of Molecular Biology, 63*, 57–73.
16. Eickbush, T. H., & Eickbush, D. G. (2007). Finely orchestrated movements: Evolution of ribosomal RNA genes. *Genetics, 175*, 477–485.
17. Stephens, C. (1985). Statistical methods of DNA sequence analysis: Detection of intragenic recombination or gene conversion. *Molecular Biology and Evolution, 2*, 539–556.
18. Sawyer, S. A. (1989). Statistical tests for detecting gene conversion. *Molecular Biology and Evolution, 6*, 526–538.
19. Sawyer, S. A. (1999). *GENECONV: A computer package for the statistical detection of gene conversion*. Available at http://www.math.wustl.edu/~sawyer
20. Kawamura, S., Saitou, N., & Ueda, S. (1992). Concerted evolution of the primate immunoglobulin a-gene through gene conversion. *Journal of Biological Chemistry, 267*, 7359–7367.
21. International Human Genome Sequencing Consortium. (2001). Initial sequencing and analysis of the human genome. *Nature, 409*, 860–921.
22. Mouse Genome Sequencing Consortium. (2002). Initial sequencing and comparative analysis of the mouse genome. *Nature, 420*, 520–562.
23. Rat Genome Sequencing Project Consortium. (2004). Genome sequence of the Brown Norway rat yields insights into mammalian evolution. *Nature, 428*, 493–521.
24. Rhesus Macaque Genome Sequencing and Analysis Consortium. (2007). Evolutionary and biological insights from the rhesus macaque genome. *Science, 316*, 222–234.
25. Benovoy, D., & Drouin, G. (2009). Ectopic gene conversions in the human genome. *Genomics, 93*, 27–32.

26. McGrath, C. L., Casola, C., & Hahn, M. W. (2009). Minimal effect of ectopic gene conversion among recent duplicates in four mammalian genomes. *Genetics, 182,* 615–622.
27. Ezawa, K., Ikeo, K., Gojobori, T., & Saitou, N. (2010). Evolutionary pattern of gene homogenization between primate-specific paralogs after human and macaque speciation using the 4-2-4 method. *Molecular Biology and Evolution, 27,* 2152–2171.
28. Nei, M., Niimura, Y., & Nozawa, M. (2008). The evolution of animal chemosensory receptor gene repertoires: Roles of chance and necessity. *Nature Reviews Genetics, 9,* 951–963.
29. Ezawa, K., OOta, S., & Saitou, N. (2006). Genome-wide search of gene conversions in duplicated genes of mouse and rat. *Molecular Biology and Evolution, 23,* 927–940.
30. Liu, Y., & West, S. C. (2004). Happy Hollidays: 40th anniversary of the Holliday junction. *Nature Reviews Molecular Cell Biology, 5,* 937–944.
31. Ezawa, K., Ikeo, K., Gojobori, T., & Saitou, N. (2011). Evolutionary patterns of recently emerged animal duplogs. *Genome Biology and Evolution, 3,* 1119–1135.
32. Winkler, H. (1920). *Verbreitung und Ursache der Parthenogenesis im Pflanzen- und Tierreiche.* Jena: Fischer (written in German).
33. Watanabe, Y., et al. (2009). Molecular spectrum of spontaneous de novo mutations in male and female germline cells of *Drosophila melanogaster. Genetics, 181,* 1035–1043.
34. Nei, M. (1987). *Molecular evolutionary genetics.* New York: Columbia University Press.
35. Haldane, J. B. S. (1949). The rate of mutation of human genes. *Hereditas, 35,* 267–273.
36. Gardner, R. J. (1977). A new estimate of the achondroplasia mutation rate. *Clinical Genetics, 11,* 31–38.
37. Smithies, O. (1995). Early days of gel electrophoresis. *Genetics, 139,* 1–4.
38. Neel, J. V., Satoh, C., Goriki, K., Fujita, M., Takahashi, N., Asakawa, J., & Hazama, R. (1986). The rate with which spontaneous mutation alters the electrophoretic mobility of polypeptides. *Proceedings of the National Academy of Sciences of the United States of America, 83,* 389–393.
39. Kondrashov, A. S. (2002). Direct estimates of human per nucleotide mutation rates at 20 loci causing Mendelian diseases. *Human Mutation, 21,* 12–27.
40. Roach, J. C., Glusman, G., Smit, A. F., Huff, C. D., Hubley, R., Shannon, P. T., Rowen, L., Pant, K. P., Goodman, N., Bamshad, M., Shendure, J., Drmanac, R., Jorde, L. B., Hood, L., & Galas, D. J. (2010). Analysis of genetic inheritance in a family quartet by whole-genome sequencing. *Science, 328,* 636–639.
41. Conrad, D. F., et al. (2011). Variation in genome-wide mutation rates within and between human families. *Nature Genetics, 43,* 712–715.
42. Denver, D. R., Morris, K., Lynch, M., & Thomas, W. K. (2004). High mutation rate and predominance of insertions in the *Caenorhabditis elegans* nuclear genome. *Nature, 430,* 679–682.
43. Denver, D. R., Dolan, P. C., Wilhelm, L. J., Sung, W., Lucas-Lledó, J. I., Howe, D. K., Lewis, S. C., Okamoto, K., Thomas, W. K., Lynch, M., & Baer, C. F. (2009). A genome-wide view of *Caenorhabditis elegans* base-substitution mutation processes. *Proceedings of the National Academy of Sciences of the United States of America, 106,* 16310–16314.
44. Lynch, M., et al. (2008). A genome-wide view of the spectrum of spontaneous mutations in yeast. *Proceedings of the National Academy of Sciences of the United States of America, 105,* 9272–9277.
45. Haag-Liautard, C., Dorris, M., Maside, X., Macaskill, S., Halligan, D. L., Charlesworth, B., & Keightley, P. D. (2007). Direct estimation of per nucleotide and genomic deleterious mutation rates in *Drosophila. Nature, 445,* 82–85.
46. Keightley, P. D., Trivedi, U., Thomson, M., Oliver, F., Kumar, S., & Blaxter, M. L. (2009). Analysis of the genome sequences of three *Drosophila melanogaster* spontaneous mutation accumulation lines. *Genome Research, 19,* 1195–1201.
47. Fujiyama, A., Watanabe, H., Toyoda, A., Taylor, T. D., Itoh, T., Tsai, S.-F., Park, H.-S., Yaspo, M.-L., Lehrach, H., Chen, Z., Fu, G., Saitou, N., Osoegawa, K., de Jong, P. J., Suto, Y., Hattori, M., & Sakaki, Y. (2002). Construction and analysis of a human-chimpanzee comparative clone map. *Science, 295,* 131–134.
48. Wu, C.-I., & Li, W.-H. (1985). Evidence for higher rates of nucleotide substitution in rodents than in man. *Proceedings of the National Academy of Sciences of the United States of America, 82,* 1741–1745.

49. Ochman, H. (2003). Neutral mutations and neutral substitutions in bacterial genomes. *Molecular Biology and Evolution, 20,* 2091–2096.
50. Hanada, K., Suzuki, Y., & Gojobori, T. (2004). A large variation in the rates of synonymous substitution for RNA viruses and its relationship to a diversity of viral infection and transmission modes. *Molecular Biology and Evolution, 21,* 1074–1080.
51. Mendel, G. (1866). Versuche uber Pflanzenhybriden (written in German). *Verhandlungen des Naturforschenden Verenines, Abhandlungen, Brunn, 4,* 3–47.
52. Yoshiura, K., et al. (2006). A SNP in the ABCC11 gene is the determinant of human earwax type. *Nature Genetics, 38,* 324–330.
53. Branden, C., & Tooze, J. (1991). *Introduction to protein structure* (p. 40). New York: Garland.
54. Yamamoto, F., Clausen, H., White, T., Marken, J., & Hakomori, S. (1990). Molecular genetic basis of the histo-blood group ABO system. *Nature, 345,* 229–233.
55. Bhattacharyya, M. K., Smith, A. M., Ellis, T. H., Hedley, C., & Martin, C. (1990). The wrinkled-seed character of pea described by Mendel is caused by a transposon-like insertion in a gene encoding starch-branching enzyme. *Cell, 60,* 115–122.
56. Wray, G. A. (2007). The evolutionary significance of *cis*-regulatory mutations. *Nature Reviews Genetics, 8,* 206–216.
57. Enattah, N. S., et al. (2002). Identification of a variant associated with adult-type hypolactasia. *Nature Genetics, 30,* 233–237.

Additional Citations Not Ordered According to Text Locations

58. Kong, A., et al. (2012). Rate of de novo mutations and the importance of father's age to disease risk. *Nature, 488,* 471–475.
59. Campbel, C. D. (2012). Estimating the human mutation rate using autozygosity in a founder population. *Nature Genetics, 44,* 1277–1283.
60. Ellegren, H. (2004). Microsatellites: Simple sequence with complex evolution. *Genetics, 5,* 435–445.
61. Bhargava, A., & Fuentes, F. F. (2010). Mutational dynamics of microsatellites. *Molecular Biotechnology, 44*(3), 250–266.
62. Kelkar, Y. D., Tyekucheva, S., Chiaromonte, F., & Makova, K. D. (2008). The genome-wide determinants of human and chimpanzee microsatellite evolution. *Genome Research, 18,* 30–38.
63. Oliveira, E. J., Padua, J. G., Zucchi, M. I., Vencovsky, R., & Vieira, M. L. (2006). Origin, evolution and genome distribution of microsatellites. *Genetics and Molecular Biology, 29*(2), 294–307.
64. Leclercq, S., Rivals, E., & Jarne, P. (2007). Detecting microsatellites within genomes: Significant variation among algorithms. *BMC Bioinformatics, 8,* 125.
65. Boyer, J. C., Hawk, J. D., Stefanovic, L., & Farber, R. A. (2008). Sequence-dependent effect of interruptions on microsatellite mutation rate in mismatch repair-deficient human cells. *Mutation Research, 640,* 89–96.
66. Ngai, M. Y., & Saitou, N. (2012). The effect of perfection status on mutation rates of microsatellites in primates (Unpublished).
67. Sun, J. X., et al. (2012). A direct characterization of human mutation based on microsatellites. *Nature Genetics, 44,* 1161–1165.
68. Burridge, C. P., et al. (2008). Geological dates and molecular rates: Fish DNA sheds light on time dependency. *Molecular Biology and Evolution, 25,* 624–633.
69. Millar, C. D., et al. (2008). Mutation and evolutionary rates in Adélie Penguins from the Antarctic. *PLoS Genetics, 4,* e1000209.
70. Saitou, N. (2007). *Genomu Shinkagaku Nyumon (written in Japanese, meaning 'Introduction to evolutionary genomics').* Tokyo: Kyoritsu Shuppan.
71. Takahashi, M., & Saitou, N. (2012). Identification and characterization of lineage-specific highly conserved noncoding sequences in mammalian genomes. *Genome Biology and Evolution, 4,* 641–657.

Phylogeny

3

Chapter Summary

DNA replications generate phylogenies. Therefore, phylogenetic relationship of DNAs is fundamental for those of individuals, genes, and species. Their relationships and differences are discussed as well as the biologically important concepts such as gene genealogy, paralogy, orthology, and horizontal gene transfer. Basic concepts of trees and networks are then explained including mathematical definition, number of possible tree topologies, and description of trees and networks. Biological implications of trees and networks such as fission and fusion of species and populations and the relationship with taxonomy are also discussed.

3.1 DNA Replications Generate Phylogenies

Charles Darwin was probably one of the first persons who imagined the phylogenetic relationship of species. It may be noted that Fig. 3.1 is the only figure included in his "Origin of Species" [1]. The time arrow goes from bottom to top in this figure, for older species were found from lower strata if we consider fossils. One important message of this tree is the existence of many extinct lineages. Nowadays, we usually consider ancestral sequences of present-day DNA sequences alone. However, most of ancestral lineages became extinct, and very few DNA lineages happened to continue for long times.

Darwin described evolution as "descent with modification." In modern terms, modification is mutations accumulated in the genomes of diversified organisms. Self-replication is one of the most important characteristics of life, and it relies on DNA replication. Successive DNA replications generate a genealogical relationship of DNAs or genes; thus, we call it "gene genealogy." Semiconservative replication of DNA double helix (see Fig. 1.5) automatically produces bifurcating tree of genes. Figure 3.2 shows an imaginary phylogenetic relationship of seven 10 bp nucleotide sequences. Their common ancestral sequence was "AAGCATCTCG," and 11

N. Saitou, *Introduction to Evolutionary Genomics*, Computational Biology 17,
DOI 10.1007/978-1-4471-5304-7_3, © Springer-Verlag London 2013

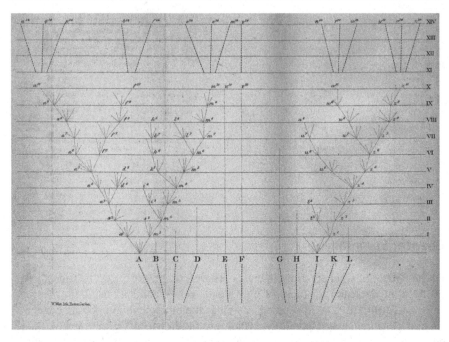

Fig. 3.1 Schematic representation of species phylogeny by Darwin (1859; Ref. [1])

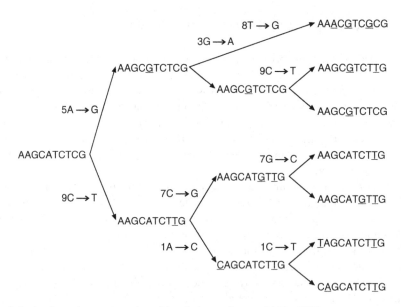

Fig. 3.2 A schematic representation of the phylogenetic tree of 10 bp DNA sequence

nucleotide substitutions (see Chap. 2) accumulated in this phylogenetic tree. For example, "5A→G" means the substitution from A to G at position 5. This change is inherited to all three descendant DNA sequences. Sometimes, identical changes will take place in different times, as in the case of "9C→T" substitutions. Successive changes at the same position may happen, as shown for position 1. Reverse changes can also occur, as in the case of C→G→C at position 7.

In this chapter, we will consider phylogenies in three categories: individuals, genes, and species. Individuals are unit of life for both single-cellular and multicellular organisms. We therefore first consider phylogeny or genealogy of individuals at various situations. We then move to gene genealogy that has been the main focus of molecular evolutionary studies. There are many aspects in gene genealogy or gene phylogeny. The third category is species or populations. Before the advent of molecular evolutionary studies, study of species phylogeny was the main stream of phylogenetics.

It should also be noted that various terms such as genealogy, phylogeny, and tree are interchangeably used in this chapter. Although their meanings are overlapped, there is some nuance. "Genealogy" is usually meant for relatively shallow time scale, while "phylogeny" is often used for long-term evolution. "Tree" is quite general, yet dry. Mathematically, a tree does not have any reticulate relationship that may be included in one genealogical or phylogenetic relationship.

3.2 Genealogy of Individuals

3.2.1 Genealogy of Haploid Individuals

In single-cellular organisms such as bacteria or yeasts, one individual is one cell. This situation is depicted in Fig. 3.3. There are three cells at generation 0, but many cell deaths, designated as "d," happened, and only two cells that are descendants of cell A remain at generation 4. In any case, repeated cell divisions create genealogy of cells. This is also true for cells originated from fertilized egg of multicellular organisms. All cells in one individual are descendants of that single cell, as depicted in Fig. 3.4, where *C. elegans* P1 cell genealogy or lineage is shown. Consideration of cell genealogy in one individual is also important when we deal with somatic mutations.

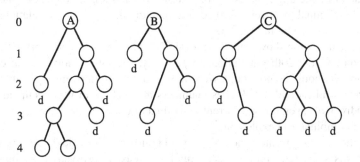

Fig. 3.3 Example of cell genealogy

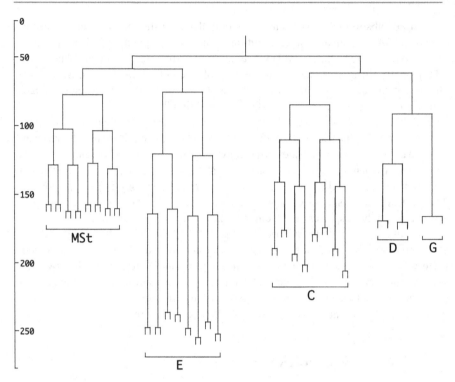

Fig. 3.4 Cell genealogy of *C. elegans* (Based on [2])

3.2.2 Genealogy of Diploid Individuals

As for sexually reproducing diploid organisms, the relationship of gene genealogy and individual genealogy becomes quite complex. By definition, sexually reproducing diploid organism is produced from two individuals: mother and father. Mother and father also had their own parents. The number of one person's ancestor is thus doubled as one goes back one generation, and there are 2^n individuals for one individual at nth-generation ancestor. In return, contribution of one nth-generation ancestral individual to current (0th) generation individual is 2^{-n}. This relationship is graphically represented in Fig. 3.5.

Simple calculation shows that there are 1,024 and 1,048,576 ancestors at the 10th generation and the 20th generation, respectively, for single individual at current generation. It should be noted that we simply counted the maximum possible number of ancestral individuals. If these numbers are real, earth was occupied by so many individuals even for one current individual of one species. This clearly shows power of combinatorial explosion.

In reality, earth is limited and only a small number of individuals can live in this planet. This seemingly paradoxical situation is easily resolved if we introduce matings between relatives, or inbreeding. In fact, if two individuals of Fig. 3.5 are

Fig. 3.5 Genealogy of one diploid individual

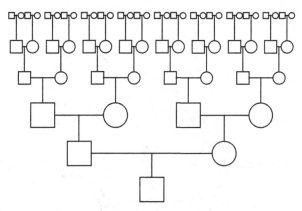

Individual in question

first cousins, then paternal or maternal grandparents of these two individuals should be identical. By this way, the actual number of ancestral individuals is grossly reduced. In other words, all sexually reproducing organisms are extensively conducting inbreeding to some extent.

We then ask the maximum number of individuals who really existed and contributed their genetic materials to offsprings. This number can be obtained by considering the minimum unit of inheritance, that is, one nucleotide. If the genome of one diploid organism consists of N nucleotides, then it has 2N nucleotides for each cell. Therefore, 2N is the maximum number of individuals who really existed. There are ~3×10^9 bp DNA in the human genome. Thus, the maximum number of ancestors really existed for any human individual is about 6×10^9. It looks like a vast number; however, 2^{33} (=8,589,934,592) already exceeds this number. Therefore, before 33 generations ago, or approximately more than 1,000 years ago, we should have our authentic ancestors who failed to inherit any nucleotide to us. Saitou [3] proposed to call such individuals as "null ancestor."

In human population studies, the so-called individual tree is sometimes produced by comparing sharing of genes among individuals. For example, Fig. 3.6 shows a phylogenetic tree of human individuals including five persons whose genomes were sequenced [4]. Other individuals are from the HapMap data [5]. Unfortunately, this tree is misleading, for this gives wrong impression that all the human individuals compared in this figure originated from the common ancestral human individual. In reality, human is diploid, and autosomal genes have many ancestral genes as already discussed.

3.2.3 Paternal or Maternal Genealogy of Diploid Individuals

Eukaryote cells contain organelles (mitochondria and chloroplasts), and they have their own DNA (see Chap. 8). These organelle DNA genomes replicate themselves, and their genealogy is different from that of the nuclear genome. In most

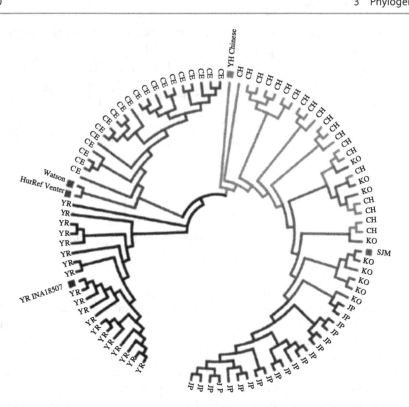

Fig. 3.6 A phylogenetic tree of human individuals (Based on Ref. [4])

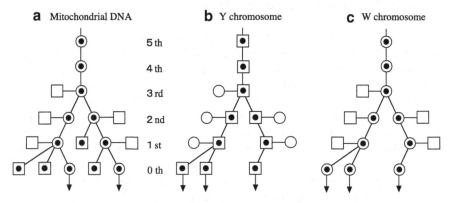

Fig. 3.7 Gene genealogy for animal mitochondrial DNA (**a**) and for mammalian Y chromosomes (**b**) W chromosome (**c**)

mitochondria, their genome DNA is inherited only through the mother. This maternal inheritance is graphically shown in Fig. 3.7a. All the paternal line individuals in this case are null ancestors, though paternal mtDNA is inherited in rare occasion, called paternal leakage.

In any case, if we ignore males, mtDNA gene genealogy becomes identical with the female genealogy. Therefore, all mitochondrial DNAs contained in the current-day individuals of one species should have the single common ancestral mitochondrial DNA in one female. Because of this nature of maternal genealogy, that ancestral female having the common ancestral mitochondrial DNA is sometimes called "mitochondrial Eve" when mass media reports human evolutionary studies. Because Eve was the wife of Adam in Genesis of the Old Testament, this naming can give wrong impression that the modern human in fact started from only one male and one female, as the myth of Garden of Eden was saying. In reality, there were many females, probably at least thousands, when the common ancestral population of the modern human emerged somewhere in Africa (see Chap. 10). To avoid this misinterpretation, this common ancestral woman is sometimes called "lucky mother."

The mode of inheritance for chloroplasts in plants is complicated. Although maternal inheritance was observed in many angiosperm species, there are a considerable number of reports of paternal and biparental inheritance [6]. It seems to be not easy to generalize the mode of inheritance for chloroplasts.

Y chromosome is one of the sex chromosomes in mammals and restricted to males (see Chap. 9), for XX individuals are females and XY are males. This characteristic of Y chromosome results in the paternal lineage inheritance, as shown in Fig. 3.7b. It is clear that a similar situation as in mitochondrial DNA applies to the Y chromosome genealogy and male genealogy. In birds, sex chromosomes are ZW system, and male and females are in ZZ and ZW chromosome modes. In this case, W chromosome lineages are corresponding to the female genealogy (see Fig. 3.7c). If we consider gene genealogy of autosomal genes, however, the situation is completely different, as we will see in Sect. 3.3.2.

It should be noted that the expression as "maternal and paternal lineages" is somewhat misleading for mitochondrial DNA and Y chromosomes, respectively. It is true that one mammalian male individual's mitochondrial DNA is inherited from the mother and its Y chromosome is from the father. However, these two molecules descended from the maternal grandmother and paternal grandfather, respectively. Maternal grandfather's Y chromosomes and paternal grandmother's mitochondrial DNA were not inherited to this male. If we go back further, contributions of "maternal" and "paternal" mitochondrial DNA and Y chromosomes are gradually reduced, as depicted in Fig. 3.8.

3.3 Gene Genealogy

3.3.1 When a Gene Tree Is Identical to a Genome Tree

Saitou [7] defined genome as "the maximum unit of self-replicating body," and there are many genes in one genome. Genomes of haploid organisms create phylogenetic relationships. Nowadays, many complete bacterial genome sequences are available, and the construction of "genome tree" just started. However, we have to be careful to interpret this kind of genome tree. It is well known that horizontal gene transfer often

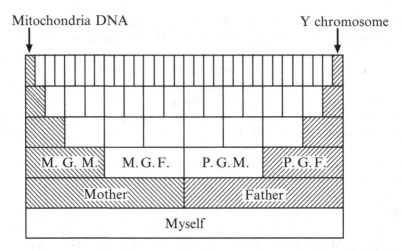

Fig. 3.8 Contributions of maternal and paternal lineages of mitochondrial DNA and Y chromosomes. M.G.M., M.G.F., P.G.M., and P.G.F. designate maternal grand mother, maternal grand father, paternal grand mother, and paternal grand father, respectively

happens among bacterial species (see Sect. 3.3.6 and Chap. 7). Therefore, a genome tree is at best the phylogenetic relationship of majority of genes.

Animal mitochondrial DNA (mtDNA) is considered to be not recombining; thus, the phylogenetic tree of mtDNA genomes is essentially a gene tree. For example, 53 human complete mtDNA genomes were compared, and clear phylogenetic tree was reconstructed [8], as shown in Fig. 3.9.

When diploid organisms are considered, a genome tree no longer exists, as mentioned in Sect. 3.2.2. Therefore, the phylogenetic relationship of a particular gene or particular DNA region has been constructed. However, it is the time to consider the evolutionary history of the whole genome in the diploid organisms. This is the future prospect.

3.3.2 Gene Phylogenies in Diploids

Let us now focus on a particular gene or the DNA region of one genome and consider the past history of this gene for many individuals of one population. It should be noted that a single common ancestral molecule always exists irrespective of the number of present-day individuals.

Figure 3.10 shows gene genealogies for two different genes in the same genome of the same population. Rectangle indicates individual, and two circles inside rectangle are genes transmitted from the mother and from the father. Individuals 1 and 2 share one grand parent, and they are first cousins. The last common ancestral gene for four genes of individuals P and Q appears by going back five generations for locus A, while it takes ten generations for locus B. It should be noted that the same two individuals were considered in Fig. 3.10. This difference in gene genealogy comes

Fig. 3.9 Human mitochondrial DNA genome trees for 53 individuals (Based on [8])

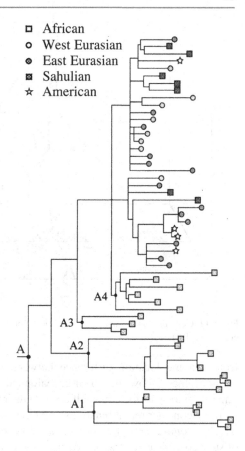

☐ African
○ West Eurasian
⊘ East Eurasian
⊠ Sahulian
★ American

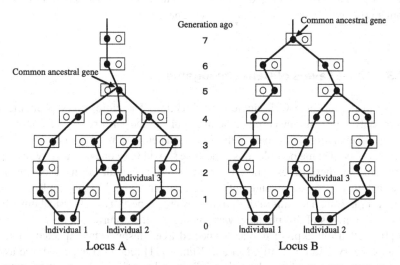

Fig. 3.10 Alternative gene genealogies for four autosomal genes of two diploid individuals

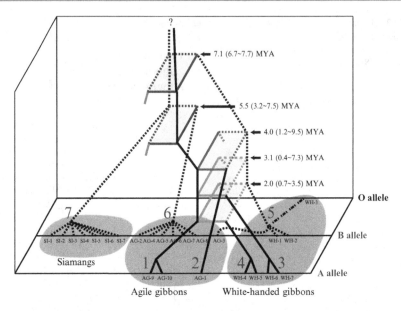

Fig. 3.11 Gene phylogeny with recombinations for gibbon ABO blood group genes (From Kitano et al. 2009; [9])

from genome location difference between genes A and B. Genes in different chromosomes follow their own genealogical history. Even if two genes are in the same chromosome, they may differentiate their genealogies through recombinations.

If we consider recombinations, we have to draw ancestral recombination graph which is nontree structure, for there are now reticulations. Figure 3.11 is an example of such complicated gene phylogeny, where five ancient recombinations were estimated in gibbon ABO blood group genes [9]. These recombinations were detected through phylogenetic network analysis (see Chap. 18).

3.3.3   Three Layers of Gene Phylogenies

We live in the universe with single time direction. Therefore, every phenomenon occurs only once somewhere, sometime during the history of this universe. Organismal evolution is just part of this history. Therefore, if we consider the evolutionary history of some homologous genes, we should have one unique phylogenetic relationship for these genes, where many mutations accumulated in various branch. Let us call this as "temporal and mutational gene phylogeny." An example is shown in Fig. 3.12a. Branch lengths of this phylogeny are proportional to physical time. Full circles on branches of this phylogeny are mutations that occurred on these branches. Nei [10] called this type of tree as "expected tree," inspired by expected value in statistics theory. More recently, Nei and Kumar [11] redefined expected tree as "a tree that can be constructed by using infinitely long sequences or the expected

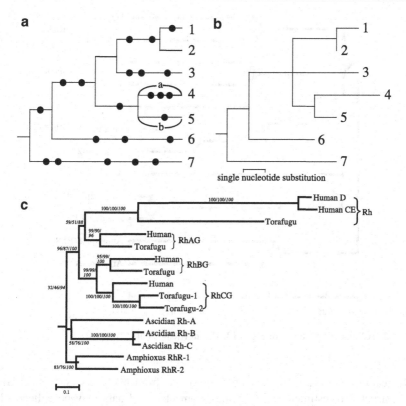

Fig. 3.12 Three kinds of gene phylogenies. (**a**) Temporal and mutational gene phylogeny. (**b**) Mutational gene phylogeny. (**c**) Estimated gene phylogeny (From Ref. [13])

number of substitutions for each branch." Under this definition, however, its branch lengths will become proportional to evolutionary times only when molecular clock, or the constancy of evolutionary rate, holds. "Expected tree" concept is therefore valid only in this restricted sense. Hedges [12] proposed to call a phylogenetic tree whose branch lengths are proportional to evolutionary time as "time tree."

Unfortunately, absolute divergence times of branching points are often quite difficult to infer. Even relative divergence times based on number of mutations are not easy, for we have to assume local or global constancy of the evolutionary rates. It is still true that mutation events are essential in reconstructing phylogenetic relationships of genes. Therefore, consideration of only mutation accumulation is still important. We thus consider "mutational gene phylogeny" (Fig. 3.12b). Branch lengths are now proportional to the numbers of mutations accumulated in each branch. A gene phylogeny is created through accumulation of DNA replications; thus, the true gene phylogeny should be strictly bifurcating. Because some branches may not accumulate mutations by chance, mutational gene phylogenies may be multifurcating, as depicted in Fig. 3.12b. It may be noted that Nei [10] and Nei and Kumar [11] called this type of gene phylogeny, a tree based on the actual number of substitution, as "realized" gene tree.

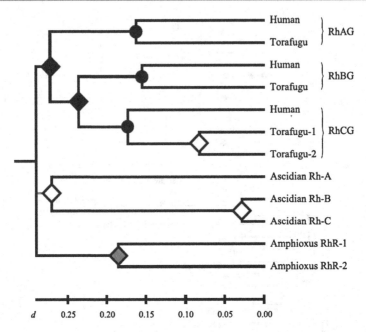

Fig. 3.13 A linialized gene tree of Fig. 3.12c (From Kitano et al. 2010; [13])

In reality, we have to infer gene phylogenies from nucleotide sequence data. Reconstructed or estimated phylogenies from observed data may be collectively called "estimated" gene phylogenies. Figure 3.12c shows an example of the estimated gene phylogeny by Kitano et al. [13]. Branch lengths of estimated gene phylogenies may be proportional to either physical time or mutational events, depending on phylogeny constructing methods used. Branch lengths are proportional to the number of amino acid substitutions in Fig. 3.12c that was made by using the neighbor-joining method [14]; see Chap. 16 for this method. Because of limitation of the information, estimated gene trees are often unrooted trees; see Sect. 3.5.1 for definition of "unrooted tree."

3.3.4 Orthology and Paralogy

Branching nodes of a gene phylogeny may correspond to speciation or gene duplication. The estimated gene tree shown in Fig. 3.12c has five proteins for human. Therefore, at least four gene duplications are necessary to explain this pattern. Torafugu also has five homologous genes, and there are four human–torafugu clusters: Rh, RhAG, AhBG, and RhCG. Because the Rh gene lineage has much longer branches than other lineages, a linearized tree [15] was constructed without the Rh gene lineage, as shown in Fig. 3.13. Human (a tetrapod) and torafugu (a teleost fish) diverged about 500 million years ago, and this speciation event is depicted with

Fig. 3.14 Two possible trees regarding gene duplications (From Saitou 2007; [46]). (**a**) Gene duplication was misinterpreted as speciation. (**b**) Two gene duplications occurred after speciation

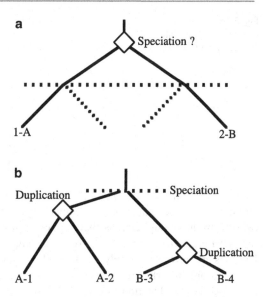

three full circles in this tree. Two gene duplications that produced three Rh-related genes (RhAG, RhBG, and RhCG) are represented as full diamonds. There are four other gene duplications in this tree, shown in open diamonds.

Detection of gene duplication is easy when two duplicate genes are in the same genome, such as human gene for RhAG and RhBG in Fig. 3.13. Ezawa et al. [16] proposed to call such pair as "duplog." When we consider homologous genes from different species, however, it is possible to misinterpret the gene duplication event as speciation. This is schematically shown in Fig. 3.14a. Two homologous genes 1 and 2 were produced by a gene duplication, then speciation produced species A and B. We expect to observe four genes: 1-A, 1-B, 2-A, and 2-B. Let us consider a situation that two of them happen to be missing, either by deletion of these gene regions from their genomes or simply by lack of sequence data. If genes 1-B and 2-A are missing, as shown in Fig. 3.14a, two remaining genes, 1-A and 2-B, may be considered as the product of speciation, though in reality they were generated by a gene duplication.

It is clear that we should distinguish "orthology," that is, homology of genes reflecting the phylogenetic relationship of species, from "paralogy," that is, homology of genes caused by gene duplication. These two terms were proposed by Fitch [17], who was inspired by ortho, meta, and para terms in organic chemistry. Both orthologous and paralogous genes are homologous with each other. Thus, homologous gene pairs [1-A and 1-B] and [2-A and 2-B] are "orthologous," while four other pairwise relationships such as [1-A and 2-B] and [1-A and 2-A] are "paralogous." In the case of Fig. 3.14b, [A-1 and A-2] and [B-3 and B-4] are paralogous, while four other homologous pairwise relationships such as [A-1 and B-3] and [A-2 and B-4] are orthologous.

Paralogous gene pairs existing in one species genome is "duplog" [16], while paralogous pairs from different species may be called allo-duplog [16]. Wolfe [18] proposed to call duplicated genes created through genome duplications as

Fig. 3.15 Duplogs
transformed to be "alleles"
(Based on Saitou 2007; [46])

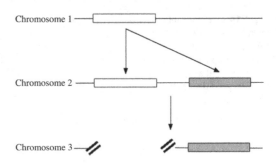

"ohnologs" after Susumu Ohno. All ohnologs in one species genome are duplogs. Duplogs are somewhat related to "inparalogs," paralogs in a given lineage that all evolved by gene duplications that happened after the radiation (speciation) event that separated the given lineage from the other lineage under consideration [19]. Duplogs are, however, simply any kind of paralogs found in one species genome.

Genes are linearly ordered on chromosomes as explained in Chap. 1. Each of them has its own location in certain chromosome. This location is called "locus" (Fig. 3.15). One individual has its own DNA sequence on a particular locus. If this DNA sequence is different from that residing in other individual or homologous chromosome of the same individual, these different DNA sequences are called "alleles." As discussed in Sect. 3.3.2, these alleles will eventually coalesce at some ancestral individual DNA. Although this process usually takes place within one species, it may be still useful to distinguish orthology and paralogy among alleles.

If nothing special happens, all alleles of one locus in one species should be considered as orthologous with each other. When duplication happens, a duplog pair is formed in the same chromosome, and these genes are often in propinquity (chromosome 2 of Fig. 3.15). These genes are different loci and are paralogous. It is then possible for the original locus, shown in white box, to be deleted from the chromosome. This deletion-type chromosome 3 of Fig. 3.15 now has only duplicated locus, shown in shaded box. Although genes in the original locus and those in the duplicated locus are paralogous, they look like alleles of the same locus, if chromosome 2 disappears from the population. This situation is somewhat similar to that of Fig. 3.14a. We have to be careful when considering "alleles," for orthology and paralogy may be intermingled.

3.3.5   Gene Conversions May Twist the Gene Tree

When gene conversions and/or unequal crossovers (see Chap. 2) occurred between duplogs, these homogenization effects may twist the phylogenetic relationship of the gene phylogeny. Kawamura et al. [20] determined the nucleotide sequences of primate immunoglobulin alpha genes 1 and 2 and constructed their phylogenetic tree using UPGMA (see Chap. 18 for this method), as shown in Fig. 3.16a. This tree suggests that independent gene duplications occurred in

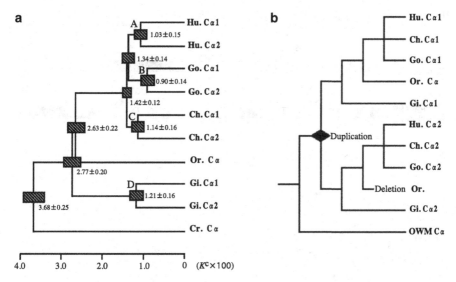

Fig. 3.16 Phylogenetic relationship of hominoid immunoglobulin A gene region (From Ref. [20]). (**a**) Spurious tree when nucleotide sequence data were simply used for tree construction. (**b**) Proposed tree where one gene duplication occurred before speciation of human and great apes

human, chimpanzee, gorilla, and gibbon lineages. However, detailed examinations of nucleotide sequences suggested that gene duplication occurred only once in the common ancestor of human and apes more than 15 million years ago and one gene was deleted in the orangutan lineage, as shown in Fig. 3.16b. High similarities observed in human, chimpanzee, gorilla, and gibbon duplogs were created by homogenizations, probably via gene conversions. Therefore, the tree shown in Fig. 3.16a is spurious.

It is now becoming clear that copy number variation (CNV) is ubiquitous. This means that gene duplications frequently occur in any gene region. We have to face gene duplications before and after speciation. One example is hominoid Rh blood group genes. There are two Rh blood group genes in the human genome, as shown in Fig. 3.12c. The gene duplication to produce CE and D duplog pair occurred before the divergence of human, chimpanzee, gorilla, and orangutan [21], and these duplogs were known to experience homogenization in each species lineage [22].

Kitano et al. (unpublished) recently determined five chimpanzee BAC clone sequences that contain Rh blood group genes. They found three genes in chimpanzees (Fig. 3.17a). Because of synteny, human Rh-D (shown as Hosa-D in Fig. 3.17) seems to be orthologous to Patr-1.1 and human Rh-CE (shown as Hosa-CE in Fig. 3.17) with Patr-2.1 and/or Patr-3.1 (Fig. 3.17a). However, the phylogenetic network for these five genes and two other chimpanzee genes (Fig. 3.17b) shows that human Rh-D and Rh-CE seem to cluster with Patr-3.1 and Patr-2.1, respectively. Either a high degree of homogenization took place in two human Rh blood group genes or some rearrangements occurred in multiple chimpanzee genes. It is thus clear that ortholog relationships should be inferred after close examinations.

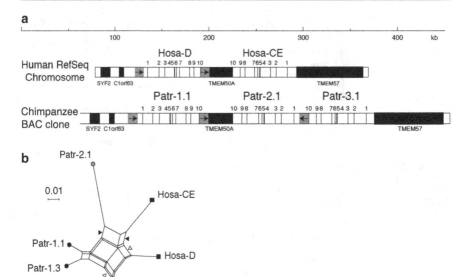

Fig. 3.17 Comparison of human and chimpanzee Rh blood group genes (from Kitano et al. unpublished). (**a**) Genomic structure. (**b**) Phylogenetic network of two human and five chimpanzee sequences

3.3.6 Horizontal Gene Transfer and Xenology

Prokaryote genomes are known to have many genes transferred from other bacterial species. This phenomenon is called "horizontal gene transfer," for it does not go through ordinary cell division. If we find homologous genes from one genome, they are duplog pair and are assumed to be generated via gene duplication. In bacterial genomes, however, it is possible that one gene copy is authentic host gene while the other copy is horizontally transferred. In this case, these pair may be called xenolog [23]. We should be careful that a totally different meaning also exists for "xeonology"; see http://www.xenology.info/.

Sawada et al. [24] showed that one gene responsible for pathogenicity horizontally transferred within bacteria *Pseudomonas syringae* lineages. Nakamura et al. [25] analyzed many bacterial genomes applying Bayesian inferences and found that horizontal gene transfers probably occurred quite frequently between species with large divergence.

Horizontal gene transfer is very rare in multicellular organisms, but plant mitochondrial genomes are known to experience horizontal gene transfers between species with varied divergence [26]. Draft genome sequences of ascidian *Ciona intestinalis* also revealed that its genes involved in cellulose metabolism system were horizontally transferred from bacteria and fungi [27]. If many more plant and animal genome sequences are determined in the near future, there will be more examples of horizontal gene transfers in multicellular organisms.

3.4 Species Phylogeny

3.4.1 Two Layers of Species Trees

One species may split into two. This is called speciation, and their topological relationship is similar to cell division. Repeated speciation events should produce a bifurcating phylogenetic tree of species. Under this assumption, we should expect one phylogenetic tree for a set of species. Nei [10] called this as "expected species tree." This naming, as in the case of "expected gene tree," again came from statistical thinking. However, the phylogenetic relationship of species is a historically unique phenomenon, and the tree is more like a "sample" in modern statistics theory. This view should hold not only to species tree but any evolutionary event. Therefore, we should be careful when we apply modern statistics theory to evolutionary data. We will go back to this point in Chap. 16.

We discussed three layers of gene phylogenies in Sect. 3.3.3. In the case of species phylogeny, the situation is less complex, for a gene tree is directly reflecting history of DNA replications, while a species tree is gross simplification of course of differentiation of populations. Therefore, speciation time is always not clear, in contrast to clear DNA replication event. With this caution in mind, a tree representation of species is useful. We thus introduce two layers of species tree: true tree and estimated tree. Branch lengths in both cases are proportional to evolutionary time. Figure 3.18a shows an example of the true tree for five species, while a tree shown in Fig. 3.18b is reconstructed from observed data and is an estimated tree.

There are two topological differences between true and estimated species trees. There is a short branch α as the common ancestral species for species 1 and 2 in the true tree, but it disappeared in the estimated tree, and three species [1, 2, and 3] started to diverge simultaneously. This situation is called trifurcation. Because the branch α was short, the available data could not find clear evidence to reconstruct this branch. Branch β is also short, but in this case, an erroneous estimation was made in Fig. 3.18b; now species 4 is shown to be more closely related to species group [1, 2, and 3]. Now new but erroneous branch γ was created instead of true branch β.

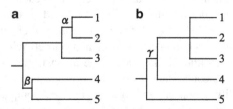

Fig. 3.18 Two layers of species tree (Based on Saitou 2007). (**a**) Expected or true species tree. (**b**) Estimated species tree

Fig. 3.19 A mitochondrial
DNA genome tree as
approximation of a species
tree. Human, chimpanzee,
and bonobo, five individuals
each were used

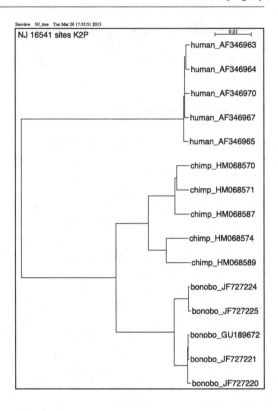

3.4.2   Gene Tree as the Approximation of Species Tree

If we go back to the gene genealogy of a certain gene (locus) of one organism, we will
reach single gene or the most recent common ancestor (MRCA) of this gene for this
species. This gene lineage can be further traced back as the single line, and eventually
this lineage will coalesce with the one from closely related species. Figure 3.19 depicts
this situation. There are three species (bonobo, chimpanzee, and human), and their
phylogenetic relationship, or species tree, and the combined gene genealogy of five
gene copies each sampled from the three species are shown. The gene genealogy is for
complete mitochondrial DNA sequences. Kimura's (1980; [28]) two-parameter method
(see Chap. 15) was used for estimating number of substitutions and MISHIMA (Kryukov
and Saitou, 2010; [29]) was used for multiple alignment, and the neighbor-joining
method [14] was used for tree construction. SeaView 4 [15] was used for computation
and drawing this tree. This combined gene genealogy roughly corresponds to the species phylogeny. Gene genealogy is a sort of gene tree; however, it usually indicates gene
tree within species or that among closely related species.

   Although there are some inconsistencies between gene trees and species trees,
molecular data, such as amino acid sequences or nucleotide sequences, can be easily produced with a huge amount compared to morphological data. Nowadays
nucleotide sequence data became standard to estimate the phylogenetic relationship
of species, and this study field is called "molecular phylogenetics."

Fig. 3.20 Two possible gene genealogies for three species

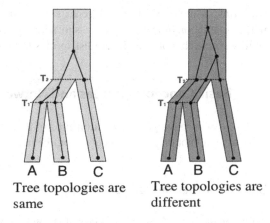

Tree topologies are same

Tree topologies are different

3.4.3   Gene Tree May Be Different from Species Tree

If two speciation events between X and Y are close enough, the topological relationship of a gene tree may become different from that of species tree, as shown in Fig. 3.20. Although species A and B are more closely related than to C, genes from species B and C are more closely related than to that from species A in the case of the gene genealogy at the right side. This can happen when speciation times T_1 and T_2 are similar and the ancestral population size N was large. Nei [1986; in "Evolutionary perspectives and the new genetics," 1987] showed that the probability P of obtaining a gene genealogy different from the species tree is

$$P = (2/3)\exp[-T/2N], \tag{3.1}$$

where T is $T_2 - T_1$. It should be noted that there are three possible tree topologies for gene trees for three species as we will see in the next section. The probability of no coalescence during T_1 and T_2 is $\exp[-T/2N]$, using the coalescent theory (see Chap. 4). In this situation, three genes from species A, B, and C coexist in the common ancestor of these three species (right tree of Fig. 3.20). Because three possible gene trees are all equally likely, we multiply 2/3 to obtain the probability that a gene genealogy is different from the species tree.

How can we reduce the probability to obtain erroneous tree? The only way is to use many independently evolving genes and choose the tree topology with the highest support. Saitou and Nei (1986; [31]) computed probabilities of obtaining the correct species tree with various parameters and found that we need 3 or 7 genes if $T/2N = 2$ or 1, respectively, to obtain the correct species tree with probability higher than 0.95. If $N = 10,000$, $T = 20,000$ generations when $T/2N = 1$. This corresponds to roughly 400,000 years. If we apply this scheme to the human–chimpanzee–gorilla trichotomy problem, we have high confidence that the trichotomy was now solved. Kitano et al. (2004; [32]) used 103 protein coding gene sequences for human, chimpanzee, gorilla, and orangutan, and 49 genes showed the human–chimpanzee clustering, while only 24 and 21 genes supported two alternative topologies. Even if T, or speciation time difference, is short, now we are sure that chimpanzee (and its sibling species bonobo) is the closest species to human.

3.5 Basic Concepts of Trees and Networks

In this section, phylogenetic relationships of genes and species will be mathematically treated. Not only trees but also nontree networks will be discussed.

3.5.1 Mathematical Definition

A phylogenetic tree is a special class of graph in graph theory. A graph is composed of nodes and edges. Nodes are abstract entities, such as nucleotide sequences, genes, proteins, genomes, populations, or species in evolutionary studies. Edges represent certain relationships between nodes. In evolutionary studies, edges are almost always the ancestor–descendant relationship, and the word "branch" is usually used instead of edge. While edges show only topological relationship of nodes, branches are expected to include lengths (usually nonnegative) that represent mutational changes or evolutionary times, as we discussed in Sect. 3.3.3.

A tree can be either rooted (see Fig. 3.21a) or unrooted (see Fig. 3.21b). A rooted tree has a root, or the position of common ancestor (R in Fig. 3.21a), and there will be unique path from the root to any other node, and this direction is of course that of time. We already introduced many rooted trees in this chapter, such as Fig. 3.13. A phylogenetic tree in ordinary sense is a rooted tree. Unfortunately, however, many methods for building phylogenetic trees produce unrooted trees. This is why we discuss unrooted trees in this chapter. If we eliminate the root node R from Fig. 3.21a, we have an unrooted tree shown in Fig. 3.21b. Angles between branches have no information in both rooted and unrooted trees.

There should be only one path between any two nodes for a tree. Nodes are divided into external and internal ones, and they are shown in full circles with

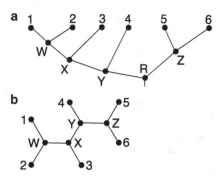

Fig. 3.21 Trees for six OTUs. (**a**) A rooted tree. (**b**) An unrooted tree

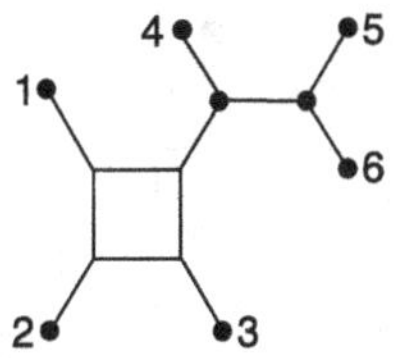

Fig. 3.22 Example
of nontree networks
for six OTUs

numeral labels 1–6 and open circles with alphabetical labels W–Z, respectively, in both trees of Fig. 3.21. External nodes and internal nodes correspond to "operational taxonomic units" (OTUs) and "hypothetical taxonomic units" (HTUs) in numerical taxonomy. Internal nodes are sometimes called "ancestral nodes." Branches in a tree are also divided into external and internal ones. An external branch connects an external node and one internal node, such as branch 1W and 6Z in Fig. 3.21a, b. It may be easy to see that one external node and one external branch have one-to-one correspondence. Therefore, external branches do not play any role in defining the topological relationship of a tree. An internal branch connects two internal nodes such as branches WX and XY in Fig. 3.21a, b. In contrast to external branches, internal nodes are essential for defining the tree topology. For example, branch WX indicates the grouping of nodes 1–2 and 3–6, and branch XY corresponds to grouping of nodes 1–3 and 4–6. This indicates that a branch is like a bridge connecting two separate parts. Therefore, a branch in one tree corresponds to "split" in graph theory, for a split is partition of one set into two groups.

A set of splits forms a tree if these splits are mutually compatible or if they are in nested structure. In Fig. 3.21b, split 12–3456 is nested in split 123–456, and these two splits are further nested in split 1234–56. These three splits correspond to branches WX, XY, and YZ, respectively. When some splits in one set are incompatible, the resulting graph is no longer a tree, and reticulation is produced. Because splits 12–3456 and 13–2456 are not compatible, a rectangle appears, as shown in Fig. 3.22. This graph is no longer a tree, for there are more than one possible path for some pair of nodes. This kind of graph is extension of tree. When all nodes are connected via edges (or branches), this graph is called "network" in graph theory. Trees are special class of networks, for all nodes are connected in any tree. When a reticulated structure exists in a network, we call it nontree network. We will see examples of nontree networks produced from certain nucleotide sequence data in later sections of this chapter. Dress et al. [33] gave detailed discussions on mathematical treatments of trees and networks.

An unrooted tree does not have a root, but it can be converted to rooted tree if root position is specified. Figure 3.23a shows three possible unrooted tree topologies for four OTUs, and these three unrooted trees and rooted trees in Fig. 3.23b have one-to-one correspondence. If we designate root of each tree in Fig. 3.23b as node R, this is topologically identical with node 4 of Fig. 3.23a. This relation between rooted and unrooted trees is used for "out-group" method of rooting as follows. When we are interested in determining phylogenetic relationship among three

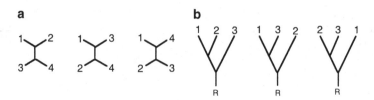

Fig. 3.23 Three possible unrooted trees for 4 OTUs and their corresponding rooted trees

Fig. 3.24 Sequential addition of new OTUs

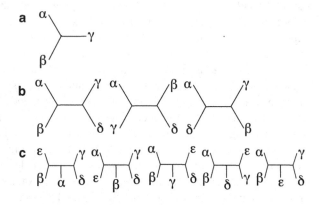

sequences (or species) 1–3, we will add another one [4], that has been known to be out-group to 1–3. After building an unrooted tree, we can easily convert that unrooted tree to rooted tree.

When two external nodes are connected only through one internal node, they are called "neighbors." OTU pairs 1 and 2 as well as OTUs 5 and 6 are neighbors in trees in Fig. 3.21a, b, while OTUs 3 and 4 are not neighbors. The total number of neighbors can be different depending on topology of unlabeled trees.

3.5.2 The Number of Possible Tree Topologies

How many branching patterns, or possible tree topologies, exist for n OTUs? To answer this question, we start from a tree with two OTUs, α and β. There is only one branch connecting these two OTUs. If we add the third OTU, γ, to this 2-OTU tree, we still have a single tree topology for three OTUs, as shown in Fig. 3.24a. There are three branches, and the fourth OTU, δ, can be attached to one of the three branches. We thus have three possible trees for four OTUs (Fig. 3.24b). This was already mentioned in Fig. 3.23a. We then proceed to attach the fifth OTU, ε, to one of the three 4-OTU trees. Because each tree has four branches, there are four different possibilities for each tree topology for 5-OTU trees. Figure 3.24c shows five topologies generated from the left-most tree of Fig. 3.24b. In total, there are 15 ($= 3 \times 5$) possible tree topologies for 5-OTU trees.

Table 3.1 Number of possible unrooted tree topologies for up to 20 OTUs

Number of OTUs	Possible number of unrooted trees
3	1
4	3
5	15
6	105
7	945
8	10, 395
9	135, 135
10	2, 027, 025
11	34, 459, 425
12	654, 729, 705
13	13, 749, 310, 575
14	316, 234, 143, 225
15	7, 905, 853, 580, 625
16	213, 458, 046, 676, 875
17	6, 190, 283, 353, 629, 375
18	191, 898, 783, 962, 510, 625
19	6, 332, 659, 870, 762, 850, 625
20	221, 643, 095, 476, 699, 771, 875

It is now easy to see that a general equation for the possible number of topologies for bifurcating unrooted trees [N_unrooted(n)] for n OTUs (n > 2) is given by

$$N\_unrooted(n) = 1 \times 3 \times 5 \times ... \times (2n - 5). \tag{3.2}$$

An analytical formula for Eq. 3.2 is given as

$$N\_unrooted(n) = (2n - 5)! / \left[2^{n-3} (n-3)! \right], \tag{3.3}$$

where exclamation mark [!] means factorial. Because there is a one-to-one correspondence between an unrooted tree for n OTUs and a rooted tree for n−1 OTUs, the possible number of topologies for bifurcating rooted trees [N_rooted(n)] for n OTUs (n > 1) is

$$N\_rooted(n) = 1 \times 3 \times 5 \times ... \times (2n - 3). \tag{3.4}$$

An analytical formula for Eq. 3.4 is given as

$$N\_rooted(n) = (2n - 3)! / \left[2^{n-2} (n-2)! \right]. \tag{3.5}$$

Equations 3.3 and 3.5 were first shown in biological literature by Cavalli-Sforza and Edwards [34]. Table 3.1 gives the possible numbers of unrooted bifurcating tree topology up to 20 OTUs. The number of possible tree topologies rapidly increases

Fig. 3.25 An example of
unrooted tree for nine OTUs

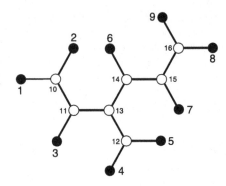

with increasing number of OTUs. Possible numbers of multifurcating trees are given by Felsenstein [35]. It is clear that search of the true phylogenetic tree for many OTUs is a very difficult problem. This is why so many methods have been proposed for building phylogenetic trees.

3.5.3   How to Describe Trees and Networks

When we pick up one tree topology for a set of OTUs, we have to describe it. Let us consider one unrooted tree for 9 OTUs shown in Fig. 3.25 as a worked out example. One way of describing a tree is to list all branches. For this purpose, we need to label not only external nodes (OTUs) but also internal nodes. In Fig. 3.25, internal nodes were labeled as 10, 11, ... 16. Under this labeling, Eq. 3.6 gives Tree_description as the list of all branches:

$$
\begin{aligned}
\text{Tree}\_\text{description} = \big[& (1,10), (2,10), (3,11), (4,12), (5,12), \\
& (6,14), (7,15), (8,16), (9,16), (10,11), \\
& (11,13), (12,13), (13,14), (14,15), (15,16) \big]
\end{aligned}
\tag{3.6}
$$

There are 2N−3 branches for completely bifurcating trees of N OTUs, and 15 branches are listed in Eq. 3.6. Branch lengths can be given after the node pair information.

There is arbitrariness in this procedure, for any labels can be used for internal nodes. It may be better not to use the IDs of internal nodes. With this in mind, I wrote a computer program using a recursive nature of neighbors for the neighbor-joining method [14]. We start from one neighbor, say OTUs 1 and 2. We join them and name the combined OTU using smaller ID OTU. In this case, the combined OTU will be called OTU 1. Then this new OTU 1 and OTU 3 are found to be neighbors. By this way, description of tree shown in Fig. 3.25 can be given by

$$
\text{Tree}\_\text{description} = \big[(1,2), (1,3), (4,5), (3,4), (8,9), (3,6) \big]
\tag{3.7}
$$

Here, six neighbors are listed, corresponding to the six internal branches. There are N−3 internal branches for a bifurcating tree of N OTUs. Because the same number such as 1 appears more than once depending on the order of joining neighbors, it is sometimes confusing. To avoid this confusion, new ID may be supplied automatically to the new OTU. Because there are 9 external nodes, we start from 10 as

$$\text{Tree\_description} = \left[(1,2), (10,3), (4,5), (11,12), (8,9), (13,6) \right] \quad (3.8)$$

We can consider these new OTUs as internal nodes. The correspondence between the internal node designations 10–16 in Fig. 3.25 is 10, 11, 12, 13, 15, none, and 14 for Eq. 3.8. Internal node 15 in Fig. 3.25 is not necessary for description of the tree topology in this case. However, we also need branch length informations for describing a tree. Therefore, let us attach a trio of nodes to designate trifurcation:

$$\text{Tree\_description} = \left[(1,2), (10,3), (4,5), (11,12), \right.$$
$$\left. (6,14), (7,15), (8,9,15) \right] \quad (3.9)$$

The final internal node 16 is still missing in this description, but we can interpret numbers 1–15 as external branches corresponding to OTUs 1–9 and internal branches with the direction from designated internal node (10–15) to the root node 16. This procedure is used in the computer program for the neighbor-joining method written by myself.

We eventually returned to the situation using internal node description, though the number of list is smaller than that for the list of branches given in Eq. 3.6. We would like to now introduce widely used tree description procedure called Newick format. The Newick Standard for representing trees in computer-readable form makes use of the correspondence between trees and nested parentheses, noticed in 1857 by Arthur Cayley, an English mathematician [37]. The idea is the same as utilizing the recursive nature of neighbors used for Eq. 3.7. Let us assume one neighboring pair A of OTUs i and j. A can be written as (i, j). We then consider this pair A as combined OTU to form neighbor with OTU k. We then have the new neighbor B = (A, k). If we insert A, B = ((i, j), k). Using this recursive nature, Eq. 3.9 can be modified to

$$\text{Tree\_description} = \left[\left(\left(\left((1,2),3 \right), (4,5) \right), 6 \right), 7 \right), 8, 9 \right] \quad (3.10)$$

The basic structure is trifurcation $[[\alpha, \beta, \gamma]]$, and each element $((\alpha, \beta, \text{ and } \gamma))$ is described as a neighbor or one node. In the case of Eq. 3.10, $\alpha = ((((1,2),3),(4,5)),6), 7)$, $\beta = 8$, and $\gamma = 9$. The common way of Newick format corresponding to Eq. 3.10 is $(((((1,2),3),(4,5)),6), 7), 8, 9)$;. There is still arbitrariness in Newick format representation, for one internal node is chosen to start the radiation into three directions. Internal node 16 of Fig. 3.25 was chosen in the case of Eq. 3.10. If we choose

Table 3.2 A splits matrix that describes the tree shown in Fig. 3.25

	1	2	3	4	5	6	7	8	9
A	+	+	−	−	−	−	−	−	−
B	+	+	+	−	−	−	−	−	−
C	+	+	+	−	−	+	+	+	+
D	+	+	+	+	−	−	−	−	−
E	+	+	+	+	+	+	−	−	−
F	+	+	+	+	+	+	+	−	−

internal node 13, the corresponding Newick format is $(((1,2),3)),(4,5),(6,(7),(8,9))));$.
The Newick format is widely used in phylogenetic tree drawing softwares.

Let us now consider the tree description method without arbitrariness. We should
have one-to-one correspondence between tree topology and description. This is list
of splits. As we discussed in Sect. 3.5.1, one split corresponds to one internal branch
of a tree. If we describe a split corresponding to internal branches 10–11 of Fig. 3.25
as 12-3456789, the tree shown in Fig. 3.25 is described as

$$\text{Tree\_description} = [12 - 3456789, 123 - 456789, 45 - 1236789,$$
$$12345 - 6789, 123456 - 789, 1234567 - 89] \quad (3.11)$$

This list can be shown in the form of matrix. Table 3.2 shows the matrix of splits
corresponding to the tree shown in Fig. 3.25. Rows 1–9 are OTUs 1–9, while splits
A–F are those listed in Eq. 3.11. Plus and minus signs were used to describe attribu-
tion of two groups formed by the split. For simplicity, OTU 1 always belongs to the
group with plus sign.

3.5.4   Static Networks and Dynamic Trees

How many splits are possible for N OTUs? Because each OTU can be assigned
either plus or minus sign, there are 2^N possibilities. However, each split has two possible configurations. For example, split A of Table 3.2 is identical with $--++++++$
pattern. Therefore, the total number should be divided by 2, resulting the number to
be 2^{N-1}. There are some trivial splits among them. One is grouping N versus 0
OTUs. This corresponds to an array of all plus or all minus signs. There are N other
trivial splits where only one OTU has a different sign. This split corresponds to
exterior branch of a tree. For example, branch 1–10 of Fig. 3.25 can be represented
with split $+---------$. In conclusion, the total number of nontrivial splits becomes
$2^{N-1} - N - 1$.

If we relax the condition of set of splits to form trees, we will have nontree network, as already discussed in Sect. 3.5.2. In Table 3.3, split G was added to splits
A–F of Table 3.2. Because split G is incompatible with split A, we need a reticulation, as shown in Fig. 3.26. Compared to a tree shown in Fig. 3.25, we now have one
parallelogram formed by splits A and G in Fig. 3.26.

Table 3.3 A splits matrix for nine OTUs that are not mutually compatible or nested

	1	2	3	4	5	6	7	8	9
A	+	+	−	−	−	−	−	−	−
B	+	+	+	−	−	−	−	−	−
C	+	+	+	−	−	+	+	+	+
D	+	+	+	+	−	−	−	−	−
E	+	+	+	+	+	+	−	−	−
F	+	+	+	+	+	+	+	−	−
G	+	−	+	−	−	−	−	−	−

Fig. 3.26 A nontree network corresponding to the split matrix shown in Table 3.3

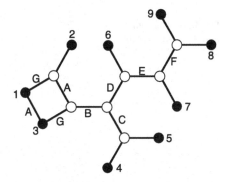

These nontree networks are static, for they are visual representations of list of splits including incompatible ones. Trees can also be static. For example, a minimum spanning tree [36] is the optimum connection of N nodes distributed in one plane. In evolution, however, rooted trees are no longer static, but dynamic because of the time arrow.

3.5.5   Relationship Between Trees

Let us now discuss the relationships between different topologies in unrooted trees. Figure 3.27 shows a tree with a trifurcation. The trifurcating node connects external nodes 7, 8, and 9. This tree was produced after eliminating the internal branch 15–16 of Fig. 3.25. Alternatively, we can set the branch length 15–16 as zero. We can then attach alternative internal nodes different from Fig. 3.25. The resulting two new tree topologies can be described using the Newick format:

$$\text{Tree A} = \left(\left(\left(\left((1,2),3 \right),(4,5) \right),6 \right),8,(7,9) \right); \tag{3.12}$$

$$\text{Tree B} = \left(\left(\left(\left((1,2),3 \right),(4,5) \right),6 \right),9,(7,8) \right); \tag{3.13}$$

Fig. 3.27 A tree with one trifurcation

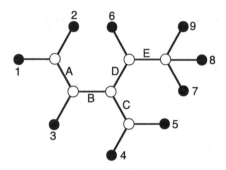

Fig. 3.28 The completely multifurcating tree for n OTUs

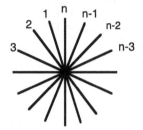

where 1–9 are external nodes of Fig. 3.27. Accordingly the tree of Fig. 3.25, or tree C, is represented as $(((((1,2),3),(4,5)),6),7,(8,9))$;, using internal node 15 as the starting node. The topologies among external nodes 1–6 are identical among these three trees. Although there was no explicit description, it is now clear that exclusively bifurcating trees A, B, and C are interconnected through the trifurcating tree shown in Fig. 3.27. Robinson and Folds [38] called the two kinds of operations to eliminate one internal node and to add a new internal node with nonzero distance as "contraction" and "expansion," respectively. The total number of contractions and expansions to transform one tree to another tree can be used to define "topological distance" between any two trees with the same number of external nodes. Because we need the same number of contractions and expansions to transform one exclusively bifurcating tree to another one, the topological distance between any two exclusively bifurcating trees is an even number.

Because of the one-to-one correspondence between an unrooted tree and a rooted tree, topological distance can also be defined for rooted trees.

What is the maximum topological distance for trees with N external nodes or OTUs? First of all, the two trees to be compared should be both exclusively bifurcating, for these have the maximum number (N−3) of internal branches. If we eliminate all these internal branches, both trees will become the completely multifurcating tree, shown in Fig. 3.28. Only N external branches are connected between corresponding external nodes and the sole internal node. This tree is sometimes called starlike tree or explosion tree. If the two exclusively bifurcating trees in question share no internal branches, the maximum topological distance, 2(N−3), is attained.

Fig. 3.29 The hypersphere of tree topologies: from the star tree to completely bifurcating tree when N = 7. Trees 1–6 are exclusively bifurcating, while trees 7–10 have one trifurcating node. Trees 1–3 can be reduced to tree 10 if one external branch is omitted

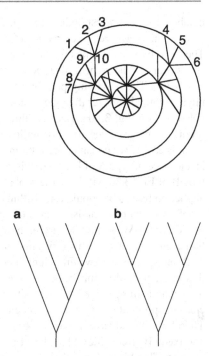

Fig. 3.30 Two possible unlabeled rooted trees for four species

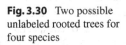

We can imagine a hypersphere in which the completely multifurcating tree is located in its center, while all the exclusively bifurcating trees are distributed on its surface. Various degrees of multifurcating trees are located in the inner surfaces hierarchically. Figure 3.29 shows one example of such hypersphere when N = 5. Because there are only 15 possible exclusively bifurcating trees and 10 trifurcating trees, the hypersphere is shown in circles.

We are usually interested in the phylogenetic relationship of a given set of OTUs in evolutionary studies, and trees or networks are labeled. In graph theory, unlabeled trees or networks are also considered. For example, two trees in Fig. 3.30a, b have different topologies even if labels 1–4 are dropped. This unlabeled pattern is called "topology" in graph theory. In evolutionary studies, however, unlabeled trees are not interesting, and different branching patterns in labeled trees are called "topology."

3.6 Biological Implications of Trees and Networks

3.6.1 Fission and Fusion of Species and Populations

Speciation as a random process was considered already in the early twentieth century by Yule [39], cited in 34). This is now called "Yule process." When n species will become n+1 species, all n species have equal probability for fission. There are 15 possible topologies for rooted tree of four species, but these are divided into 12 and 3 unlabeled type, corresponding to Fig. 3.30a, b topologies, respectively. These two types of unlabeled trees are produced from single three-OTU tree, shown in Fig. 3.23b. There are two possibilities for this tree to become a four-species tree. One is that fission occurs at the right-most species. This event will produce the unlabeled tree of Fig. 3.30b. The remaining possibility is that fission occurs in either species that are phylogenetically closer than the right-most species. This event will produce the unlabeled tree of Fig. 3.30a. Because three species in Fig. 3.23b are equal in terms of producing new species, the probability of the first event is 1/3 and that for the second event is 2/3. If we consider the total number of tree topologies for Fig. 3.30a, b, one labeled tree of type A of Fig. 3.30 will have the probability [2/3]/12 = 1/18, and that of type B of Fig. 3.30 will have the probability [1/3]/3 = 1/9. This argument can also be applied to fission of populations within one species. These probabilities for each tree topology under random fission are closely related with those used in coalescent theory (see Chap. 4). We should be careful, however, to apply this Yule process to real speciation events, because each species may have quite different probability for fission.

After speciation, two incipient species may not form reproductive barriers. In this case, hybridization between these new species can occur, as estimated for human and chimpanzee lineages [40]. There are often substructures within one species, and populations within one species have, by definition, no reproductive barrier. Therefore, it is always possible for two genetically differentiated populations to be admixed. If a considerable admixture occurred, the new hybrid population will emerge. This kind of population fusion probably often happened in a species with wide geographical distributions, such as human.

3.6.2 Importance of Branch Length

We so far discussed mainly on topology or branching pattern of phylogenetic trees. Here the importance of branch lengths is stressed. In the middle of the twentieth century, great apes, which consist of chimpanzee, bonobo, gorilla, and orangutan, were considered to be monophyletic, and the divergence time between human and great apes was estimated to be about 15 million years ago. This was because Ramapithecus, an extinct hominoid species, was believed to be phylogenetically close to the modern human at that time.

Using the microcomplement fixation techniques in immunology, Sarich and Wilson [41] estimated that human and chimpanzee are only 5 million years diverged in 1967. After that epoch-making paper, many molecular and some morphological data were produced, and now the latest estimate for the divergence time between human and chimpanzee is about 6 million years [42]. Interestingly, the divergence time between the Old World monkeys and hominoids (human and apes) has not changed drastically, about 30 million years. The tree topology is obviously identical, yet many researchers, including myself, were involved in this problem of divergence time or branch length.

A branch length can become zero. Therefore, a trifurcating tree of Fig. 3.27 is considered as a special case of the bifurcating tree shown in Fig. 3.25 where the internal branch length for nodes 15 and 16 is zero. This relationship seems obvious, yet some methods for phylogenetic tree reconstruction have to specify the number of variables first. In this situation, two trees shown in Figs. 3.25 and 3.27 should be clearly distinguished in terms of parameter numbers (see Chap. 16). However, we should remember that all trees are interconnected with each other in the hypersphere shown in Fig. 3.29.

3.6.3 Trees and Taxonomy

Taxonomy, or classification of organism, has been developed much earlier than the evolution concept emerged. Linne unified the organismal taxonomy by using scientific binomen in the eighteenth century. Any taxonomical structure is rigidly nested or hierarchical structure, so as phylogenetic trees. In classic taxonomy, there are seven obligatory taxa: kingdom, phylum (in animal taxonomy) or division (in plant taxonomy), class, order, family, genus, and species. When one would like to describe new species, at least these seven levels should be clarified. For example, human's place in these obligatory taxa is as follows: kingdom Animalia, phylum Chordata, class Mammalia, order Primates, family Hominidae, genus *Homo*, and species *Homo sapiens*. It should be noted that species description is not noun but adjective for the genus name, following the grammar of Latin language. This combination of genus name with species adjective, i.e., Latin or scientific binomen, expanded the possibility of organism names. When Linne was active in the eighteenth century, that expansion was enough to describe all known species at that time. However, now we know there are a large number of species, and binomen is no longer suitable. Therefore, in the DDBJ/EMBL/GenBank International Nucleotide Sequence Database, all lists of taxa are shown for each organism. For example, human is described as

Eukaryota, Metazoa, Chordata, Craniata, Vertebrata, Euteleostomi, Mammalia, Eutheria, Euarchontoglires, Primates, Haplorrhini, Catarrhini, Hominidae, and Homo.

In 1958, Julian Huxley [43] (cited in [44]) proposed two taxonomic concepts: grade and clade. "Grade" corresponds to a shaky evolutionary level, and it is no longer used in taxonomy. The word "clade" was coined after an ancient Greek word κλαδος (klados) meaning branch. This word was later used for new taxonomic area called "cladistics." A clade is groups of genes or organisms which share a common ancestor, and its synonyms are cluster, monophyletic group, and sister group. It should be noted that Willi Hennig [45] is considered to initiate cladistic thinking.

Although there are many problems in cladistic thinkings, there are some useful terms proposed in this field. Let us consider a rooted tree shown in Fig.3.21a. Nodes or OTUs 1 and 2 are one clade, so as 1–3, 1–4, and 5–6. Because all the six OTUs share the common ancestor R, OTUs 1–6 also form one clade. In contrast, OTUs 3 and 4 are not clade, but they are paraphyletic, and this situation is called paraphyly. A paraphyletic grouping is prohibited in usual cladistics: however, it can be considered as a complementary set. For example, the paraphyletic groups 3 and 4 in Fig. 3.21a are complementary to monophyletic groups 1 and 2 within the larger taxonomic groups 1–4. In fact, we are familiar with this kind of paraphyletic groupings, such as invertebrates. "Polyphyly" is also prohibited in cladistics to consider one taxonomic group. For example, OTUs 2 and 4 and 1 and 6 are polyphyletic in Fig. 3.21a. In any occasions, they should not be considered as the biologically meaningful group.

The "out-group" concept was already discussed in Sect. 3.5.1, and it is closely related to "basal." For example, OTU 4 may be called basal within the monophyletic groups 1–4.

References

1. Darwin, C. (1859). *On the origin of species*. London: John Murray.
2. Deppe, U., et al. (1978). Cell lineages of the embryo of the nematode *Caenorhabditis elegans*. *Proceedings of the National Academy of Sciences of the United States of America, 75*, 376–380.
3. Saitou, N. (1995). A genetic affinity analysis of human populations. *Human Evolution, 10*, 17–33.
4. Ahn, S. M., et al. (2011). *Genome Research, 16*, 1622–1629.
5. International HapMap Project Home Page: http://hapmap.ncbi.nlm.nih.gov/
6. Hansen, A. K., et al. (2007). *American Journal of Botany, 94*, 42–46.
7. Saitou, N. (2004). *"Genomu to Shinka" (in Japanese, meaning 'Genome and evolution' in English)*. Tokyo: Shin-yosha.
8. Ingman, M., Kaessman, H., Paabo, S., & Gyllensten, U. (2000). Mitochondrial genome variation and the origin of modern humans. *Nature, 408*, 708–713.
9. Kitano, T., Noda, R., Takenaka, O., & Saitou, N. (2009). Relic of ancient recombinations in gibbon ABO blood group genes deciphered through phylogenetic network analysis. *Molecular Phylogenetics and Evolution, 51*, 465–471.
10. Nei, M. (1987). *Molecular evolutionary genetics*. New York: Columbia University Press.
11. Nei, M., & Kumar, S. (2000). *Molecular evolution and phylogenetics*. Oxford/New York: Oxford University Press.
12. Hedges, S. B., Dodley, J., & Kumar, S. (2006). TimeTree: A public knowledge-base of divergence times among organisms. *Bioinformatics, 22*, 2971–2972.
13. Kitano, T., Satou, M., & Saitou, N. (2010). Evolution of two Rh blood group-related genes of the amphioxus species *Branchiostoma floridae*. *Genes & Genetic Systems, 85*, 121–127.
14. Saitou, N., & Nei, M. (1987). The neighbor-joining method: A new method for reconstructing phylogenetic trees. *Molecular and Biological Evolution, 4*, 406–425.
15. Takezaki, N., Rzhetsky, A., & Nei, M. (1995). Phylogenetic test of the molecular clock and linearized trees. *Molecular and Biological Evolution, 12*, 823–833.
16. Ezawa, K., Ikeo, K., Gojobori, T., & Saitou, N. (2011). Evolutionary patterns of recently emerged animal duplogs. *Genome Biology and Evolution, 3*, 1119–1135.
17. Fitch, W. M. (1970). Distinguishing homologous from analogous proteins. *Systematic Zoology, 19*, 99–113.
18. Wolfe, K. (2000). Robustness – It's not where you think it is. *Nature Genetics, 25*, 3–4.
19. Sonnhammer, E. L. L., & Koonin, E. V. (2002). Orthology, paralogy, and proposed classification for paralog subtypes. *Trends in Genetics, 18*, 619–620.
20. Kawamura, S., Saitou, N., & Ueda, S. (1992). Concerted evolution of the primate immunoglobulin alpha gene through gene conversion. *Journal of Biological Chemistry, 267*(11), 7359–7367.
21. Kitano, T., Sumiyama, K., Shiroishi, T., & Saitou, N. (1998). Conserved evolution of the Rh50 gene compared to its homologous Rh blood group gene. *Biochemical and Biophysical Research Communications, 249*, 78–85.

22. Kitano, T., & Saitou, N. (1999). Evolution of Rh blood group genes have experienced gene conversions and positive selection. *Journal of Molecular Evolution, 49*, 615–626.
23. Koonin, E. V., Makarova, K. S., & Aravind, L. (2001). Horizontal gene transfer in prokaryotes: Quantification and classification. *Annual Review of Microbiology, 55*, 709–742.
24. Sawada, H., Suzuki, F., Matsuda, I., & Saitou, N. (1999). Phylogenetic analysis of *Pseudomonas syringe* pathovar suggests the horizontal gene transfer of *argK* and the evolutionary stability of hrp gene cluster. *Journal of Molecular Evolution, 49*, 627–644.
25. Nakamura, Y., Itoh, T., Matsuda, H., & Gojobori, T. (2004). Biased biological functions of horizontally transferred genes in prokaryotic genomes. *Nature Genetics, 36*, 760–766.
26. Archibald, J. M., & Richards, T. A. (2011). Gene transfer: Anything goes in plant mitochondria. *BMC Biology, 8*, 147.
27. Dehal P. et al. (2002). The Draft Genome of *Ciona intestinalis*: Insights into Chordate and Vertebrate Origins. *Science, 298*, 2157–2167.
28. Kimura, M. (1980). A simple method for estimating evolutionary rates of base substitutions through comparative studies of nucleotide sequences. *Journal of Molecular Evolution, 16*, 111–120.
29. Kryukov, K., & Saitou, N. (2010). MISHIMA – A new method for high speed multiple alignment of nucleotide sequences of bacterial genome scale data. *BMC Bioinformatics, 11*, 142.
30. Gouy, M., Guindon, S., & Gascuel, O. (2010). SeaView version 4: a multiplatform graphical user interface for sequence alignment and phylogenetic tree building. *Molecular Biology and Evolution, 27*, 221–224.
31. Saitou, N., & Nei, M. (1986). The number of nucleotides required to determine the branching order of three species, with special reference to the human-chimpanzee-gorilla divergence. *Journal of Molecular Evolution, 24*, 189–204.
32. Kitano, T., Liu, Y.-H., Ueda, S., & Saitou, N. (2004). Human specific amino acid changes found in 103 protein coding genes. *Molecular Biology and Evolution, 21*, 936–944.
33. Dress, A., Huber, K. T., Koolen, J., Moulton, V., & Spillner, A. (2011). *Basic phylogenetic combinatorics*. Cambridge: Cambridge University Press.
34. Cavalli-Sforza, L. L., & Edwards, A. (1967). Phylogenetic analysis. Models and estimation procedures. *American Journal of Human Genetics, 19*, 233–257.
35. Felsenstein, J. (1978). The number of evolutionary trees. *Systematic Zoology, 27*, 27–33.
36. Courant, R., Robbins, H., & Stewart, I. (1996). *What is mathematics?* Oxford: Oxford University Press.
37. http://evolution.genetics.washington.edu/phylip/newicktree.html
38. Robinson, D. F., & Foulds, L. R. (1981). Comparison of phylogenetic trees. *Mathematical Biosciences, 53*, 131–147.
39. Yule, G. U. (1924). A mathematical theory of evolution, based on the conclusions of Dr. J. C. Willis, F.R.S. *Philosophical Transaction of Royal Society of London Series B, 213*, 21–87.
40. Patterson, N., Richter, D. J., Gnerre, S., Lander, E. S., & Reich, D. (2006). Genetic evidence for complex speciation of humans and chimpanzees. *Nature, 441*, 1103–1108.
41. Sarich, V. M., & Wilson, A. C. (1967). Immunological time scale for hominoid evolution. *Science, 158*, 1200–1204.
42. Saitou, N. (2005). Evolution of hominoids and the search for a genetic basis for creating humanness. *Cytogenetic and Genome Research, 108*, 16–21.
43. Huxley, J. (1958). *Evolutionary process and taxonomy with special reference to grades* (pp. 21–38). Uppsala: Uppsala University Arsskr.
44. Simpson, G. G. (1961). *Principles of animal taxonomy*. New York: Columbia University Press.
45. Wikipedia on Willi Henning: http://en.wikipedia.org/wiki/Willi_Hennig
46. Saitou, N. (2007b). *Genomu Shinkagaku Nyumon (written in Japanese, meaning 'Introduction to evolutionary genomics')*. Tokyo: Kyoritsu Shuppan.

Neutral Evolution

4

Chapter Summary

Neutral evolution is the default process of the genome changes. This is because our world is finite and the randomness is important when we consider history of a finite world. The random nature of DNA propagation is discussed using branching process, coalescent process, Markov process, and diffusion process. Expected evolutionary patterns under neutrality are then discussed on fixation probability, rate of evolution, and amount of DNA variation kept in population. We then discuss various features of neutral evolution starting from evolutionary rates, synonymous and nonsynonymous substitutions, junk DNA, and pseudogenes.

4.1 Neutral Evolution as Default Process of the Genome Changes

It is now established that the majority of mutations fixed during evolution are selectively neutral, as amply demonstrated by Kimura (1983; [1]) and by Nei (1987; [2]). Reports of many genome sequencing projects routinely mention neutral evolution in the twenty-first century, e.g., mouse genome paper in 2002 ([3]) and chicken genome in 2004 ([4]). We thus discuss neutral evolution as one of the basic processes of genome evolution in this chapter.

Neutral evolution is characterized by the egalitarian nature of the propagation of selectively neutral mutants. For example, let us consider a bacterial plaque that is clonally formed. All cells in one plaque are homogeneous, or have the identical genome sequences, if there are no mutations during the formation of that plaque. Because of identicalness in genome sequences, there will be no change of genetic structure for this plaque. Let us assume that three cells at time 0 in Fig. 3.2 are in this clonal plaque. Their descendant cells at time 4 also have the same genome sequences, though the number of offspring cells at that time varies from 0 to 4. This variation is attributed to nongenetic factors, such as heterogeneous distribution of nutrients. However, the most significant and fundamental factor is randomness, as we will see in Sect. 4.2.2 on branching process.

Mutation is the ultimate source of diversity of organisms. If a mutation occurring in some gene modifies gene function, there is a possibility of heterogeneity in terms of number of offsprings. This is the start of natural selection that will be discussed in Chap. 5. However, some mutations may not change gene function, and although they are somewhat different from parental type DNA sequences, mutants and parental or wild types are equal in terms of offspring propagation. We meet the egalitarian characteristic of the selectively neutral mutants. If all members of evolutionary units, such as DNA molecules, cells, individuals, or populations, are all equal, the frequency change of these types is dominated by random events. It is therefore logical that randomness is the most important factor in neutral evolution.

4.1.1 Our World Is Finite

Randomness also comes in when abiotic phenomena are involved in organismal evolution. Earthquakes, volcanic eruptions, continental drifts, meteorite hits, and many other geological and astronomical events are not the outcome of biotic evolution, and they can be considered to be stochastic from organismal point of view.

Before proposal of the neutral theory of evolution in 1968 by Kimura ([5]), randomness was not considered as the basic process of evolution. Systematic pressure, particularly natural selection, was believed to play the major role in evolution. This view is applicable if the population size, or the number of individuals in one population, is effectively infinite. However, the earth is finite, and the number of individuals is always finite. Even this whole universe is finite. This finiteness is the basis of the random nature of neutral evolution as we will see in later sections of this chapter.

4.1.2 Unit of Evolution

Nucleotide sequences reside genetic information, and one gene is often treated as a unit of evolution in many molecular evolutionary studies. A cell is the basic building block for all organisms except for viruses. It is thus natural to consider cell as unit of evolution. One cell is equivalent to one individual in single-cell organisms. In multicellular organisms, by definition, one individual is composed of many cells, and a single cell is no longer a unit of evolution. However, if we consider only germ-line cells and ignore somatic cells, we can still discuss cell lineages as the mainstream of multicellular organisms as in the case for single-cell organisms. Alternatively, clonal cells of one single-cell organism can be considered to be one individual. Cellular slime mold cells form a single body with many cells, or each cell may stay independently, depending on environmental conditions [6]. We therefore should be careful to define cell or individual.

Organisms are usually living together, and multiple individuals form one "population." We humans are sexually reproducing, and it seems obvious for us to consider one mating group. In classic population genetics theory, this reproduction unit is called Mendelian population, after Gregor Johann Mendel, father of genetics. From individual point of view, the largest Mendelian population is its species. Asexually reproducing organisms are not necessary to form a population, and multiple individuals observed in proximity, which are often recognized as one population, may be just an outcome of past life history of the organism, and each individual may reproduce independently. Gene exchanges also occur in asexually reproducing organisms, including bacteria. Therefore, by extending species concept, bacterial cells with similar phylogenetic relationship are called species. Population or species is also defined for viruses, where each virus particle is assumed as one "individual."

However, we have to be careful to define individuals and populations. One tree, such as cherry tree, is usually considered to be one individual, for it starts from one seed. Unlike most animal organisms, trees or many plant species can use part of their body to start new "individual." This asexual reproduction prompted plant population biologist John L. Harper to create terms genet (genetic individual) and ramet (physiological individual) [7]. We should thus be careful about the number of "individuals" especially for asexually reproducing organisms.

4.2 How to Describe the Random Nature of DNA Propagation

We discuss the four major processes to mathematically describe the random characteristics of DNA transmission. The first two, branching process and coalescent process, are considering the genealogical relationship of gene copies, while the latter two, Markov process and diffusion process, treat temporal changes of allele frequencies.

4.2.1 Gene Genealogy Versus Allele Frequency Change

For organisms to evolve and diverge, we need changes, or mutation. Supply of mutations to the continuous flow of self-replication of genetic materials (DNA or RNA) is fundamental for organismal evolution. This process is most faithfully described in phylogenetic relationship of genes. Because every organism is product of eons of evolution, we are unable to grasp full characteristics of living beings without understanding the evolutionary history of genes and organisms. It is thus clear that reconstruction of phylogeny of genes is essential not only for study of evolution but also for biology in general. In another words, gene genealogy is the basic descriptor of evolution.

Fig. 4.1 Schematic gene genealogy for some locus of a population. Open circles and full circles designate two different alleles, and star is mutation. TIme scale is in terms of generation, where N is the number of individuals. Autosomal locus of a diploid organism is assumed

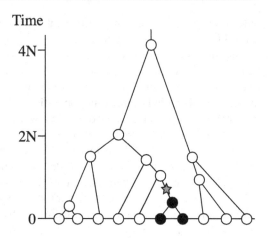

It should be emphasized that the genealogical relationship of genes is independent from the mutation process when mutations are selectively neutral. A gene genealogy is the direct product of DNA replication and always exists, while mutations may or may not happen within a certain time period in some specific DNA region. Therefore, even if many nucleotide sequences·happened to be identical, there must be genea-logical relationship for those sequences. However, it is impossible to reconstruct the genealogical relationship without mutational events. In this respect, search of muta-tional events from genes and their products is also important for reconstructing phylogenetic trees. Advancement of molecular biotechnology made it possible to routinely produce gene genealogies from many nucleotide sequences.

Figure 4.1 shows a schematic gene genealogy for 10 genes. There are two types of genes that have small difference in their nucleotide sequences, depicted by open and full circles. Both types are located in the same location in one particular chro-mosome of this organism. This location is called "locus" (plural form is "loci"), after a Latin word meaning place, and one type of nucleotide sequence is called "allele," using a Greek word αλλο meaning different. Open circle allele, called allele A, is ancestral type, and full circle allele, called allele M, emerged by a mutation shown as a star mark. The numbers of gene copies are 8 and 2 for alleles A and M. We thus define allele frequencies of these two alleles as 0.8 (=8/10) and 0.2 (=2/10). Allele frequency is sometimes called gene frequency. It should be noted that these frequencies are exact values if there are only 10 genes in the population in question. If these 10 genes were sampled from that population with many more genes, two values are sample allele frequency.

Because all these 10 genes are homologous at the same locus, they have the common ancestral gene. Alternatively, only descendants of that common ancestral gene are considered in the gene genealogy of Fig. 4.1. There are, however, many genes which did not contribute to the 10 genes at the present time. If we consider these genes once existed, the population history may look like Fig. 4.2. In this

Fig. 4.2 Relationship between gene genealogy and allele frequency change (From Saitou 2007; [55])

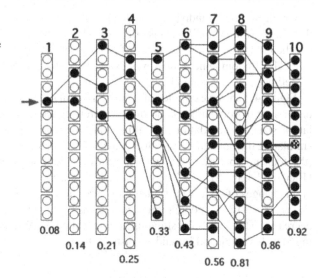

figure, gene genealogy starting from full circle gene at generation 1 is embedded with other genes coexisted at each generation but became extinct. If we consider the whole population, it is clear that allele frequency changes temporarily, and many genes shown in open circle did not contribute to the current generation.

How can this allele frequency change occur? Natural selection does influence this change (see Chap. 5), but the more fundamental process is the random genetic drift. This occurs because a finite number of genes are more or less randomly sampled from the parental generation to produce the offspring generation. This simple stochastic process is the source of random fluctuation of allele frequencies through generations.

The random genetic drift can be described as follows. Let us focus on one particular diploid population with $N[t]$ individuals at generation t. We consider certain autosomal locus A, and the total number of genes on that locus at generation t is $2N[t]$. There are many alleles in locus A, but let us consider one particular allele A_i with n_i gene copies. By definition, allele frequency p_i for allele A_i is $n_i/2N[t]$. When one sperm or egg is formed via miosis, one gene copy is included in that gamete from locus A. If male and female are assumed to be more or less the same allele frequency, the probability to have allele A_i in that gamete is p_i. This procedure is a Bernoulli trial, and the offspring generation at time $t+1$ will be formed with $2N[t+1]$ Bernoulli trials. Because all these trials are expected to be independent, we have the following binomial distribution to give the probability Prob[k] of having k gene copies among $2N[t+1]$ genes in the offspring generation:

$$\text{Prob}[k] = {}_{2N[t+1]}C_k p_i^k \left(1 - p_i\right)^{2N[t+1]-k} \tag{4.1}$$

where xCy is the possible combinations to choose y out of x. If we continue this binomial distribution for many generations, the random genetic drift will occur.

When the number of individuals in that population, or population size, is quite large, this fluctuation is small because of "law of large numbers" in probability theory, yet the effect of random genetic drift will never disappear under finite population size. The random genetic drift was extensively studied by Sewall Wright and was sometimes called Wright effect.

Figure 4.3 shows examples of computer simulations for the random genetic drift under a set of very simple conditions: discrete generations, haploid, constant population size, no population structure, and no recombination. The perl script for simulating the random genetic drift is available at this book website. Population size (the total number of individuals or genes in one population) varies in Fig. 4.3a (1000) and 4.3b (10,000). The initial allele frequency was set to be 0.2, and the temporal changes of up to 1,000 generations are shown. In each case, 5 replications are shown. Clearly, as population size increases, fluctuation of allele frequencies decreases. This simplified situation is often called the Wright–Fisher model, honoring Sewall Wright and Ronald A. Fisher ([8]).

4.2.2 Branching Process

Francis Galton, a half cousin of Charles Darwin, was interested in extinction probability of surnames. He was thus trying to compute probability of surname extinction. He himself could not reach appropriate answer, so he asked some mathematicians. Eventually he was satisfied with a solution given by H. W. Watson, who used generating function, and they published a joint paper in 1874 [9]. Because of this history, the mathematical model considered by them is sometimes called "Galton–Watson process," but usually it is called "branching process" (see [10] for detailed description of this process). It may be noted that surnames have been studied in human genetics (e.g., [11]) and in anthropology (e.g., [12]), for their transmissions often coincide with Y chromosome transmissions.

Fisher (1930; [13]) applied this process to obtain the probability of mutants to be ultimately fixed or become extinct. Later in 1940s, when physicists in the USA developed the atomic bomb, the branching process was used to analyze behavior of neutron number changes (see [14]).

The distribution of transmission probability of gene copies from parents to offsprings is the basis of the branching process. The number of individuals in the population is usually not considered, for this process is mainly applied for the shallow genealogy of mutant gene copies within the large population. In a sense, the branching process is a finite small world in an infinite world.

A Poisson process is the default probability distribution for the gene copy transmission under random mating. Let us explain why the Poisson process comes in. We assume a simple reproduction process where one haploid individual can reproduce one offspring n times during its life span, and the probability, p, of reproduction is uniform at each time unit (see Fig. 4.4). The probability Prob[k] of having k offspring during the n times is given by the following binomial distribution:

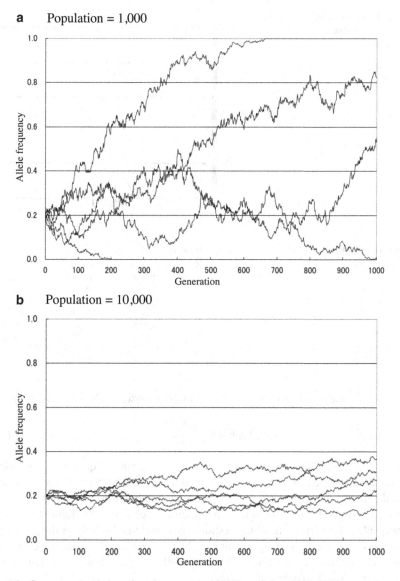

a   Population = 1,000

b   Population = 10,000

Fig. 4.3 Computer simulation of random genetic drift (From Saitou 2007; [55])

$$\text{Prob}[k] = {}_nC_k p^k (1-p)^{n-k} \qquad (4.2)$$

Equation 4.2 is equivalent to Eq. 4.1, though the meanings of parameters are somewhat different. The mean, m, of this binomial distribution is

$$m = np \qquad (4.3)$$

Fig. 4.4 From binomial
distribution to Poisson
distribution

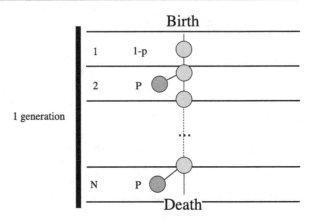

Table 4.1 Prob[k] values for
various m values

Prob[k]			
K	m = 1.0	m = 1.5	m = 2.0
0	0.368	0.223	0.135
1	0.368	0.335	0.271
2	0.184	0.251	0.271
3	0.061	0.126	0.180
4	0.015	0.047	0.090
5	0.003	0.014	0.036

Let us increase n and decrease p while keeping m constant. The limit, n = ∞, gives

$$\text{Prob}[k] = \frac{m^k e^{-m}}{k!} \tag{4.4}$$

where e (= 2.718281828459…) is basis of natural logarithm. Equation 4.4 is called Poisson distribution, after French mathematician Siméon Denis Poisson. When m = 1, the mutant gene is expected to keep its copy number, while m>1 or m<1 corresponds to positive or negative selection situations (see Chap. 5). Table 4.1 shows Prob[k] values for various m values. It should be noted that Prob[0], or the probability of transmitting no offspring, is quite high. Even for m = 2, where the expected number of offspring is two times, Prob[0] is ~0.135 even though the gene copy number explosion is expected to occur.

Fisher [13] showed that the mutant is destined to become extinct for m≤1. When m = 1, one may expect this is a stable situation and the mutant will continue to survive in the population. The population size is assumed to be infinite in the usual branching process, and this causes the mutant gene copy with m = 1 to become extinct. However, we live in finite environment, and the branching process under infinite population size is not appropriate when we consider the long-term evolution. When m>1, the mutant is advantageous, and the probability of survival becomes positive, as we will see in Chap. 5. Readers interested in application of the branching process to fates of mutant genes should refer to Crow and Kimura (1970; [15]).

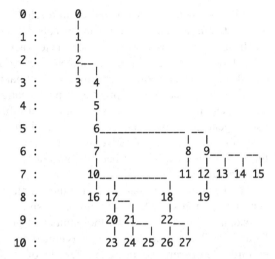

Fig. 4.5 Examples of branching process when m = 1

Although the Poisson process is usually assumed in a random mating population, the real probability distribution of gene copy number may be different. In human study, pedigree data are used to estimate the gene transmission probability. A Kalahari San population (!Kung bushman) was reported to have a bimodal distribution of gene transmission, where the variance is larger than mean (Howell, 1979; [16]). Interestingly, a Philippine Negrito population was shown to have an approximate Poisson distribution with mean 1.05 (Saitou et al. 1988; [17]).

Figure 4.5 shows an example of the branching process with m = 1. A Monte Carlo method was used to generate this genealogy. The perl script for simulating the branching process is available at this book website.

4.2.3 Coalescent Process

Mutant gene transmission follows with the time arrow in the branching process. In another way, it is a forward process. However, as we saw, most of gene lineages become extinct, and it is not easy to track the lineage which will eventually propagate in the population. Now let us consider a genealogy only for sampled genes. It is natural to look for their ancestral genes, finally going back to the single common ancestral gene. This is viewing a gene genealogy as the backward process. When two gene lineages are joined at their common ancestor, this event is called "coalescence" after Kingman (1982; [18]). It should be noted that Hudson [19] and Tajima [20] independently invented essentially the same theory in 1983.

Let us consider Fig. 4.1 again. Left most two gene copies coalesce first, followed by coalescence of two mutant genes shown in full circles. At this moment, there are eight lineages left, and one of them experienced mutation, shown in a star. After six more coalescent events, at around 2N generations ago, there are only two lineages.

Then it took another ~2N generations to reach the final 9th coalescence. If there is no population structure in this organism, called "panmictic" situation, and if there is no change in population size (N), the time to reach the last common ancestral gene, or coalescent time, is expected to be 4N generations ago, according to the coalescent theory of an autosomal locus for diploid organisms.

The simplest coalescent process is pure neutral evolution. Even if mutations accumulate, they do not affect survival of their offspring lineages. Because of this nature, gene genealogy and mutation accumulation can be considered separately. If natural selection, either negative (purifying) or positive, comes in for some mutant lineages, this independence between generation of gene genealogy and mutation accumulation no longer holds.

Another important assumption for the simplest coalescent process is the constant population size, N. In diploid organism, the number of gene copies for an autosomal locus is 2N, while the number of gene copies for haploid organism locus is N. The former situation is assumed explicitly or implicitly in many literatures. However, the original lifestyle of organisms is haploid, and many organisms today are haploids. Therefore, we consider the situation in haploid organisms first. It should be noted that the constant population size is more or less expected if we consider a long-term evolution. Otherwise, the species will become extinct or will have exponential growth. Though we, *Homo sapiens*, in fact experience population explosion, this is a rather rare situation among many species. In short-term evolution, population size is expected to fluctuate for any organism. Therefore, assumption of the constant population size is not realistic and is only for mathematical simplicity. We have to be careful about this sort of too simplistic assumptions inherent in many evolutionary theories. There are some more simplifications in the original coalescent theory: discrete generation and random mating. Random mating means that any gene copy is equal in terms of gene transmission to the next generation, and there is no subpopulation structure within the population of N individuals in question. These assumptions were also used for the Wright–Fisher model.

Let us first consider the coalescent of only two gene copies. What is the probability, Prob[2→1, 1], for 2 genes to coalesce in one generation? If we pick up one of these two gene copies arbitrarily, this gene, say, G1, should have its parental gene, PG1, in the previous generation. Another gene, G2, also has its parental gene PG2. Because all genes are equal in terms of gene transmission probability under our assumption, all N genes, including G1, can be PG2. We should remember Fig. 4.5, where multiple offsprings may be produced from one individual during one generation. Therefore, to have one offspring G1 does not affect the probability of having another offspring, for these reproductions are independent. It is then obvious that

$$\text{Prob}\left[2 \rightarrow 1,1\right] = \frac{1}{N} \tag{4.5}$$

The probability of the complementary event, i.e., no coalescence, can be written as Prob[2→2, 1] and

$$\text{Prob}[2 \rightarrow 2,1] = 1 - \left(\frac{1}{N}\right) \tag{4.6}$$

We now move to slightly more complicated situation. What is the probability, Prob[2→1, t], for 2 genes to coalesce exactly after t generations? The coalescent event must occur only after no coalescence of (t−1) generations. Thus,

$$\text{Prob}[2 \rightarrow 1,t] = \left[1 - \left(\frac{1}{N}\right)\right]^{t-1} \cdot \left[\frac{1}{N}\right] \tag{4.7}$$

When N is large, $[1-(1/N)]^{t-1}$ can be approximated as $e^{-t/N}$. Then

$$\text{Prob}[2 \rightarrow 1,t] \sim \left[\frac{1}{N}\right] e^{-\frac{t}{N}} \tag{4.8}$$

We can obtain the mean, Mean[2→1, t], and the variance, Var[2→1, t], of the time, t, for coalescence, using this geometric distribution:

$$\text{Mean}[2 \rightarrow 1,t] = \Sigma_{t=1,\infty} \, t \cdot \left[\frac{1}{N}\right] \cdot \left[1 - \left(\frac{1}{N}\right)\right]^{t-1} \tag{4.9}$$

After some transformations,

$$\text{Mean}[2 \rightarrow 1,t] = N \tag{4.10}$$

The variance of this exponential distribution is

$$\text{Var}[2 \rightarrow 1,t] = \Sigma_{t=1,\infty} (t-N)^2 \cdot \left[\frac{1}{N}\right] \cdot \left[1 - \left(\frac{1}{N}\right)\right]^{t-1} \tag{4.11}$$

It can be shown that

$$\text{Var}[2 \rightarrow 1,t] = N(N-1) \tag{4.12}$$

When N>>1, v [2→1, t] ~ N^2. Therefore, the standard deviation of t is ~N generations, same as its mean. When a diploid autosomal locus is assumed, mean and variance are 2N and $(2N)^2$, respectively.

Let us now consider the coalescent process for n genes sampled from the population of N individuals. We assume n << N. The first step is the probability for two of n gene copies to coalesce during t generations. The probability of three gene copies to coalesce in one generation, is $(1/N)^2$. If N is large, $(1/N)^2 \sim 0$, and we can ignore coalescence of more than 2 genes in one generation, and focus on coalescence of the only pair of genes. Because there are $_nC_2$ [= n(n−1)/2] possible combinations to choose two out of n genes,

$$\text{Prob}[n \rightarrow n-1,1] = {_nC_2} \cdot \left[\frac{1}{N}\right]. \tag{4.13}$$

We can thus generalize Eq. 4.7 to consider the probability that 2 genes among n genes sampled are coalesced in one generation as

$$\text{Prob}[n \rightarrow n-1,t] = \left[1 - \left(\frac{{}_nC_2}{N}\right)\right]^{t-1} \cdot \left[\frac{{}_nC_2}{N}\right] \tag{4.14}$$

The mean of t under this distribution is

$$\text{Mean}[n \rightarrow n-1,t] = \frac{N}{{}_nC_2} = \frac{2N}{n(n-1)}. \tag{4.15}$$

We can then obtain the mean or expected time of coalescence from the current generation of n genes to single common ancestral gene by summing the means above:

$$\text{Mean}[n \rightarrow 1,t] = \Sigma_{i=2,n} \cdot \frac{2N}{i(i-1)} \tag{4.16}$$

$$= 2N\left[1 - \left(\frac{1}{n}\right)\right] \tag{4.17}$$

If n is large,

$$\text{Mean}[n \rightarrow 1,t] \sim 2N \tag{4.18}$$

When diploid autosomal genes are considered, this approximate mean becomes 4N, and the variance of the coalescent time, when n is large, is given by Tajima [20]:

$$\text{Var}[n \rightarrow 1,t] \sim 16N^2\left(\frac{\pi^2}{3-3}\right) \tag{4.19}$$

If n is not much different from N, or almost exhaustive sampling was conducted, the possibility of coalescence of three or more gene copies together at one gene copy within one generation is no longer negligible, and Eq. 4.13 and later do not hold any more. We need to consider "exact" coalescence. The following explanation is after Fu (2006; [21]). If we consider a randomly mating population with constant size N, each gene copy at the present population was sampled from N gene copies of the previous generation with replacement. Therefore, if we choose one particular gene copy, say, copy ID 1, from the present population, the probability of its transmission from a specific gene copy of the previous generation is 1/N. Then the probability of gene copy ID 2 from the present population not sharing the same parental copy with copy ID 1 is 1−[1/N]. We then go to the next situation in which gene copy ID 3 from the present population shares the parental gene copy with neither ID 1 nor ID 2. Its probability becomes 1−[2/N]. Applying a similar argument for IDs 4 to n (n≤N), the probability, Prob[n→n, 1], that none of gene copy at the present generation shares the parental gene copy at the previous generation becomes

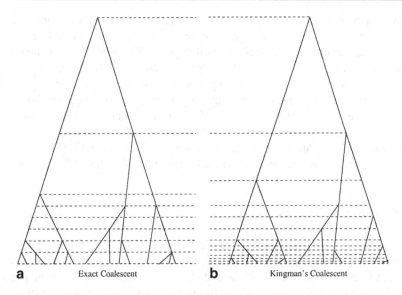

a   Exact Coalescent   b   Kingman's Coalescent

Fig. 4.6 Comparison of exact and Kingman coalescence (From [21])

$$\text{Prob}[n \rightarrow n,1] = \Pi_{k=1,n-1}\left(1-\left[\frac{k}{N}\right]\right)^{N} \tag{4.20}$$

$$= \frac{N_{[n]}}{N^{n}} \tag{4.21}$$

$$N_{[n]} = N(N-1)(N-2)...(N-n+1) \tag{4.22}$$

Therefore, the probability corresponding to Eq. 4.14 under the exact coalescent in which n gene copies at the present generation will coalesce to m (<n) ancestral gene copies at t generations ago becomes

$$\text{Prob}[n \rightarrow m,t] = \left[1 - \frac{N_{[n]}}{N^{n}}\right] \cdot \left[\frac{N_{[n]}}{N^{n}}\right]^{t-1} \tag{4.23}$$

Generally speaking, the coalescent time for exact process is shorter than the approximation, or Kingman coalescence, first given by Kingman (1982). Figure 4.6 shows examples of gene genealogies of the same sample size under the exact coalescence and Kingman coalescence (reproduced from Fu 2006).

Unlike the treatment of allele frequency changes to be discussed in later sections, the coalescent generation time is given in terms of the total number of population in the coalescent theory. Because of this, we can check the implicit assumption of the constant population size. For example, the total number of human population as of 2011 A.D. is over 7 billion. If we apply the coalescent theory under the constant population model, the expected number of generations for coalescence of an autosomal gene, 4N, is 27 billion generations. If one generation is 20 years, the expected coalescent time becomes 540 billion years! This value is far greater than the start of this universe, i.e., Big Bang, approximately 14 billion years ago. This seemingly paradoxical situation simply comes from the population explosion, which violates the assumption of constant population size. To overcome this problem, the "effective population size" is often used. Modern human is estimated to have ca. 10,000 as the effective population size (e.g., Takahata 1993; [59]).

Recent developments of the coalescent theory are discussed in [22] and [23]. We will discuss various applications of the coalescent theory in Chap. 17.

4.2.4 Markov Process

We now move to the treatment of allele frequency changes. For simplicity, a constant population size (N) is assumed. We also consider haploid organism as before. Let us consider one particular allele A_i, and the number of gene copies at generation t is denoted as i. Allele frequency for this allele at generation t is i/N. Then the probability of having j gene copies among N genes in the next generation (t+1) becomes

$$\text{Prob}[i \rightarrow j] = {}_N C_j \left[\frac{i}{N}\right]^j \left(1 - \left[\frac{i}{N}\right]\right)^{N-j} \tag{4.24}$$

This is the transition probability of i to j gene copies from generation t to t+1. For simplicity, let us denote Prob[i→j] as $P_{i,j}$ ($0 \le i, j \le N$). Then we can have the transition probability matrix **P** as

$$
\mathbf{P} = \begin{vmatrix}
P_{0,0} & P_{1,0} & P_{2,0} & P_{3,0} & P_{4,0} & \cdots & P_{N-2,0} & P_{N-1,0} & P_{N,0} \\
P_{0,1} & P_{1,1} & P_{2,0} & P_{3,0} & P_{4,0} & \cdots & P_{N-2,0} & P_{N-1,0} & P_{N,1} \\
P_{0,2} & P_{1,2} & P_{2,2} & P_{3,2} & P_{4,2} & \cdots & P_{N-2,2} & P_{N-1,2} & P_{N,2} \\
\cdot & \cdot & \cdot & \cdot & \cdot & \cdots & \cdot & \cdot & \cdot \\
P_{0,N-2} & P_{1,N-2} & P_{2,N-2} & P_{3,N-2} & P_{4,N-2} & \cdots & P_{N-2,N-2} & P_{N-1,0} & P_{N,N-2} \\
P_{0,N-1} & P_{1,N-1} & P_{2,N-1} & P_{3,N-1} & P_{4,N-1} & \cdots & P_{N-2,N-1} & P_{N-1,N-1} & P_{N,N-1} \\
P_{0,N} & P_{1,N} & P_{2,N} & P_{3,N} & P_{4,N} & \cdots & P_{N-2,N} & P_{N-1,N} & P_{N,N}
\end{vmatrix} \tag{4.25}
$$

We can derive the probability, Prob[i/N, t+1], of having allele frequency i/N at generation t+1, using this transition probability matrix and the probability at generation t as follows. At the initial generation (t=0), let us assume that there are k ($1 \le k \le N-1$) gene copies in the population. Then Prob[k/N,0] = 1 and Prob[i/N,0]=0 ($0 \le i \le N$, $i \ne k$):

Fig. 4.7 Example of Markov process. (**a**) N = 1,000, initial frequency = 0.5. (**b**) N = 1,000, initial frequency = 0.1. G is generation

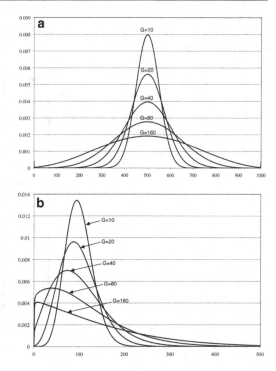

$$\text{Prob}\left[\frac{1}{N}, t+1\right] = \Sigma_{j=0,N}\text{Prob}\left[\frac{1}{N}, t\right] \cdot P_{j,i} \qquad (4.26)$$

Figure 4.7 shows some examples of Markov process when N = 1,000 and initial frequencies to be 0.5 (case a) and 0.1 (case b). The perl script for computing the Markov process is available at this book website. In the past, the Markov process was not extensively used, for it requires a large number of computations. Thanks to the great advancement of computational powers, we can now obtain allele frequency spectrum for a relatively large number of genes. It may be interesting to apply this exact Markov process for various realistic situations in the future.

4.2.5   Diffusion Process

There are various mathematical models which can describe the evolutionary changes of allele frequency. The diffusion equation is the most widely used method. It can easily combine the stochastic effect such as random genetic drift and deterministic effect such as selection. The random genetic effect alone is discussed in this section, and natural selection will be discussed in Chap. 5.

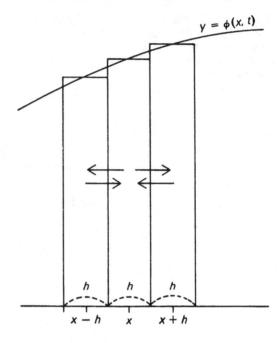

Fig. 4.8 Explanation of diffusion model (From Kimura 1964; [25])

The starting point is the binomial distribution, the basic process for the random genetic drift (see Sect. 4.2.1). We assume that the population size is constant, and haploid organism is considered. The binomial distribution in Eq. 4.1 can be written as

$$\text{Prob}[k] = {}_N C_k p_i^{\ k} (1-p_i)^{N-k} \tag{4.27}$$

Let us note that the mean (m) and variance (v) of this distribution are

$$m = Np \tag{4.28}$$

$$v = Np(1-p) \tag{4.29}$$

This process can be approximated by a differential equation, a Fokker–Planck equation for the random genetic drift:

$$\frac{\delta\varphi(p \to x;t)}{\delta t} = \left[\frac{1}{4N}\right]\left[\frac{\delta^2}{\delta p^2}\left\{p(1-p)\varphi(p \to x;t)\right\}\right] \tag{4.30}$$

Figure 4.8 explains the basic concept of Eq. 4.41 on the change of allele frequency class, based on Kimura (1955; [24]). Let us consider a very small range of length h, and histograms of many rectangles approximate the probability density function $\varphi(p \to x;t)$. Each rectangle has the width h and the height given by the value of $\varphi(p \to x;t)$ at allele frequency x, at the middle of the rectangle unit. We also consider a very short time Δt, so the change of allele frequency during that time period

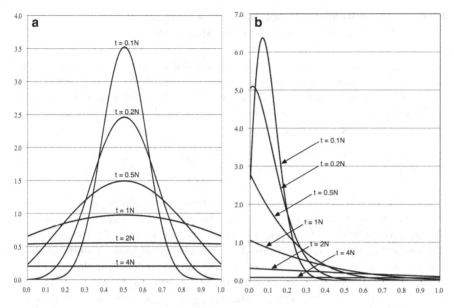

Fig. 4.9 Diffusion process. (**a**) Initial frequency = 0.5. (**b**) Initial frequency = 0.1

is restricted to at most to adjacent range, either left or right. If we take limits (h→zero and Δt→zero), differential equation (4.30) is obtained.

The exact solution for this equation, for probability density distribution of allele frequency x at generation time t, starting from initial frequency p, is

$$\delta(p \rightarrow x; t) = \Sigma_{i=1,\infty} p(1-p)i(i+1)(2i+1)F(1-i,i+2,2;p)$$
$$\cdot F(1-i,i+2;x)\exp\left[\frac{-1(i+1)t}{4N}\right] \qquad (4.31)$$

F(a,b,c;z) in Eq. 4.31 is a hypergeometric function:

$$F(a,b,c;z) = \Sigma_{n=0,\infty} \quad \frac{\left\{a_{[n]} \cdot z^n\right\}}{\left\{b_{[n]} \cdot c_{[n]} \cdot n!\right\}} \qquad (4.32)$$

This solution was given by Kimura in 1955 [24]. Interested readers should refer to [15] and [25] for detailed explanation of the diffusion process.

Figure 4.9 shows the probability density changes for various generations when the initial allele frequency is 0.5 and 0.1. The perl script for computing the diffusion process is available at this book website. Initially, at time zero, all probability density is concentrated at the initial allele frequency. This is Dirac's delta function. As the random genetic drift starts to operate, allele frequency will start to diffuse. After a long time, probability density becomes flat and low, and majority of probabilities will be residing at either allele frequency being 0 or 1.

4.2.6   A More Realistic Process of Allele Frequency Change of Selectively Neutral Situation

In reality, the population size is not only finite but also not constant. Therefore, a more realistic process of frequency change of selectively neutral mutant alleles is as follows. Let us denote the total gene copy number of the population at generation t as N[t] and the gene copy number of a selectively neutral allele A at generation t as $N_A[t]$. Then the allele frequency at generation t, Freq_A[t], becomes

$$\text{Freq}\_A[t] = \frac{N_A[t]}{N[t]} \tag{4.33}$$

We need to consider a finite population in finite maximum population size, or carrying capacity. Then the population size fluctuation can be approximated by a Markov process with constant global population size or carrying capacity. The problem is that the carrying capacity itself will change depending on the change of environment. In the case of human, environment includes technological innovation. We need to redefine the probability transition matrix. Because an extinct population cannot produce new population, $P_{0,j}$ (0<j≤N_max) remains zero or maintains its characteristic as the absorbing barrier. In contrast, $P_{N\_max,j}$ (0≤ j≤N_max) is not zero. This is because we are not considering frequency change, but the population size fluctuation is considered. Unfortunately, population genetics theory so far does not consider this more realistic dynamics of populations. It is left for future developments.

4.3   Expected Evolutionary Patterns Under Neutrality

We discuss three categories when the pure neutral evolution is occurring: fixation probability, the evolutionary rate, and the amount of DNA variation. Because the majority of eukaryotic genome is evolving in this fashion, the understanding of the pure neutral evolutionary process is quite important for evolutionary genomics.

4.3.1   Fixation Probability

As stated at the beginning of this chapter, neutral evolution is characterized by the egalitarian nature of the propagation of mutants. Therefore, all genes at one generation have the same potential to leave offsprings. If one population is destined to continue for a long time, eventually fixation of one gene will occur. Because any of N genes in the initial generation can become the common ancestor of later generations, the fixation probability of one gene in a population of N genes is 1/N. In an autosomal locus of diploid organisms, the fixation probability becomes 1/2N.

In reality, we do not know if one population in question at this time will continue to survive in later time. Therefore, the absolute fixation probability, Prob_fixation, of one gene should be

$$\text{Prob}\_\text{fixation} = \text{Prob}\_\text{existence} \cdot \left[\frac{1}{N}\right] \tag{4.34}$$

Prob_existence is the probability of existence of that population for a certain long time. Unfortunately, we do not know this probability, and almost always it is implicitly assumed to be unity. Thus,

$$\text{Prob}\_\text{fixation} = \frac{1}{N} \tag{4.35}$$

4.3.2   Rate of Evolution

If a gene fixation occurs in one population, there will be no change of allele frequency, though the gene genealogy will grow as time goes on. We definitely need mutations for evolution to proceed. If a mutation happens, the population of N genes with only one allele will again become polymorphic with a single copy mutant allele and N-1 copies of the original allele. If all genes in later generations will become descendants of this mutant gene, now gene substitution is attained. Evolution of one gene or one locus can be seen as the accumulation of mutations. Therefore, the rate of gene substitution is equated as the rate of evolution.

Let us define the mutation rate per gene locus per generation as μ. Considering all N genes in this population, $N\mu$ mutants appear in every generation. During T generations, the total amount of mutant genes becomes $N\mu T$, under the assumption of the constant population size. Because the fixation probability for each mutant gene is $1/N$, the total number of mutant genes fixed during T generation is $N\mu T \cdot [1/N] = \mu T$. The rate, λ, or speed of evolution in terms of continuous mutant fixation is thus

$$\lambda = \frac{\mu T}{T} = \mu \tag{4.36}$$

Equation 4.36 was first shown by Kimura and Ohta (1971; [26]). This explanation assumes the constant population size for a long time. We can relax this assumption to obtain Eq. 4.36. Figure 4.10 shows a schematic gene genealogy for a single lineage during time T. The vertical axis represents the whole population, and the population size can vary. Full circles are mutations accumulated in this single lineage, and thin lines represent increase of allele frequency for each mutant. The total number of mutations accumulated during time T is μT. Therefore, the evolutionary rate, λ, of gene substitutions per generation should

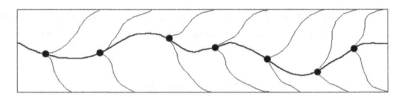

Fig. 4.10 Single lineage gene genealogy (From Saitou 2007 [55])

be μT/T = μ. This argument applies to any time irrespective of population size change. Even if speciation occurs, it does not affect this argument based on the single gene lineage. This is why we can consider the long-term evolution. Of course, any gene at the present population can be the starting point for the single lineage genealogy. This generality comes from the egalitarian nature of the selectively neutral mutant gene copies.

If the mutation rate, μ, does not change for a long time and for diverse group of organisms, we can estimate the mutation rate by estimating the evolutionary rate in the neutrally evolving genomic region. This is the basis of the indirect method for estimating mutation rates discussed in Chap. 2.

4.3.3 Amount of DNA Variation Kept in Population

If we consider a relatively long DNA fragment, say, composed of n nucleotides, as "locus," there are 4^n possible alleles in this locus. If we consider a 1-kb-long DNA fragment, $n = 1,000$, and $4^{1,000}$ is more than 10^{600}. Considering this enormous possibility of alleles for even a short DNA fragment, Kimura and Crow (1964; [27]) proposed the infinite allele model. All new mutations are different with each other in this model. The phylogenetic relationship of alleles is not considered in the infinite allele model. Kimura (1969; [28]) thus proposed the infinite site model where an infinitely long DNA sequence is considered. Now new mutations appear by substituting one nucleotide site, which was not changed before. In this sense, this model is similar to the infinite allele model, but now accumulation of nucleotide substitutions can be considered with the infinite site model. This means that a genealogical relationship of alleles is behind this model. In either case, the expected heterozygosity, H, under these two models is

$$H = \frac{4N_e\mu}{\left\{1 + 4N_e\mu\right\}} \tag{4.37}$$

N_e is effective population size and μ is mutation rate per locus per generation. The numerator of Eq. 4.37, $4N_e\mu$, which is often denoted as M or θ, should be identical with the nucleotide diversity, π, per nucleotide site ([29]; Kimura 1968).

4.4 DNA Polymorphism

When we compare gene copies of one locus in one organism, nucleotide sequences may be slightly different because various types of mutation may accumulate. In this case, this locus has genetic or DNA polymorphism. We will classify DNA polymorphisms according to type of mutation (see Table 2.1 of Chap. 2). In classic evolutionary studies, "polymorphism" applies only to one species; however, definition of species is often ambiguous, and there is no clear difference between within-species genetic polymorphism and between-species genetic differences. Therefore, when multiple closely related species are compared, nucleotide sites which have variations are sometimes called polymorphic.

Traditionally, one locus may be called polymorphic if the major allele frequency is equal to or less than 0.99, while it is called monomorphic if the major allele frequency is more than 0.99. However, nowadays we often have sample size of more than 1,000, and if some nucleotide sequences were found to be different from the major allele, this locus may be called polymorphic, even if the frequency of the major allele is more than 0.99.

Although there are no essential differences between haploid and diploid genomes in terms of the random genetic drift, patterns of genetic composition of alleles per locus, called "genotypes," are different with each other. If there are two alleles, A_1 and A_2, at a locus, the possible genotypes are the same as alleles for haploids. In diploids, there are three genotypes, or possible combination of alleles: A_1A_1, A_1A_2, and A_2A_2. Genotypes with single type of allele are called "homozygotes" and those with two types of alleles are called "heterozygotes," after Greek words oμo and έτερος meaning same and different, respectively. In general, if there are N types of alleles in one population, the possible number of homozygotes and heterozygotes are N and $N(N-1)/2$, respectively.

4.4.1 Single Nucleotide Polymorphism (SNP)

DNA polymorphism observed at one nucleotide, the smallest unit of DNA molecule, is called "single nucleotide polymorphism," or SNP. The majority of SNP is created via nucleotide substitution-type mutation, but sometimes 1-nucleotide length insertion or deletion is also included as SNP. An SNP locus is usually biallelic. In nucleotide substitution-type SNPs, there are usually only two nucleotides in the population, for the mutation rate of nucleotide substitution is quite low. However, if we sample many individuals, such as for medical studies of humans, we may encounter SNP loci with three or four nucleotide alleles. There is gap or no-gap allele for single nucleotide indel SNPs.

SNPs observed in protein coding regions may be called cSNPs, and SNPs found in noncoding genomic region may be called gSNPs. There are synonymous and nonsynonymous cSNPs.

If we can estimate the ancestral SNP alleles, we can distinguish typical two alleles into ancestral and derived (mutated) alleles. If one allele has allele frequency lower than 0.5, it is called "minor" allele. Many databases for SNP are available, such as dbSNP (http://www.ncbi.nlm.nih.gov/projects/SNP/).

4.4.2 Insertions and Deletions (Indel)

Insertion-type and deletion-type (often abbreviated as indel) mutations create indel DNA polymorphisms. Broadly speaking, repeat number polymorphism and copy number polymorphism to be discussed later are also in this type; however, non-repeat type indels are usually called as indel polymorphism. When the gap length is one, this indel polymorphism may be included in SNP, as mentioned above.

Insertions and deletions are detected as gaps in multiple alignments (see Chap. 16). Therefore, if nucleotide sequences are misaligned, we have incorrect indel information. Nucleotide sequences within the same species are expected to have quite high homology; however, if we are not aware of microinversions, misalignment will occur and often gaps are observed.

4.4.3 Repeat Number Polymorphism

When insertions or deletions occur within the repeat sequences, they are called repeat polymorphisms. Short repeat sequences of 1–5 nucleotides as unit are called "short tandem repeat" (STR) polymorphism or microsatellite DNA polymorphism. When the repeat length is longer, it is called "variable number of tandem repeat" (VNTR) polymorphism or minisatellite DNA polymorphism.

4.4.4 Copy Number Variation

If the repeat unit is much bigger, say, at least a few kilobases, it is called "copy number variation" (CNV). A classic example is the Rh blood group D+/D- poly-morphism. Recently, many genes in the human genome were found to have CNV-type polymorphism [30, 31]. If CNV haplotype of more copy number is fixed in the population, the original gene is duplicated. Therefore, frequent occurrence of CNVs suggests high frequency of gene duplications.

4.5 Mutation Is the Major Player of Evolution

Mutations can be classified into deleterious, neutral, and advantageous ones according to their effects to organisms (see Chap. 5). Figure 4.11 shows the semiquantitative proportions of these three categories for a typical mammalian genome. Because the majority of the mammalian genome is noncoding, mutations occurring in this region are selectively neutral. If a mutation occurs in DNA regions where important

Fig. 4.11 Comparison
between total mutations and
surviving mutations under the
strict neutral theory

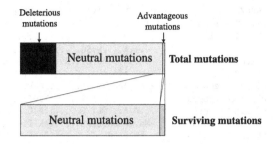

genetic informations such as protein amino acids and some RNA sequences are coded, this mutation may be deleterious, and the mutant individual may have less possibility of transmitting that gene to the offsprings. In contrast, although in rare occasions, some DNA changes will cause that mutant individual to have more off-springs than those without mutant genes. This type of mutants is called "advantageous." In any case, when mutations occur, selectively neutral mutants are dominating. If we consider a long-term evolution, only a small fraction of these mutations will survive. Because deleterious mutations will soon disappear from the population (see Chap. 5), only neutral and advantageous mutants will survive in the population for a long time.

Because the fixation probability for advantageous mutants is higher than that for selectively neutral mutants, the proportion of advantageous mutations among the surviving mutations may be slightly higher than their proportion when they were produced. However, the majority of mutations surviving for long evolutionary time are selectively neutral. This is a clear difference from the prediction made by researchers who advocated the dominant power of natural selection in 1960s and 1970s. As we will see, the fixation of selectively neutral mutations via stochastic effects is the main power of evolution, and the natural selection to choose advantageous mutations has only a limited contribution, although natural selection to eliminate deleterious mutations is quite effective to keep the current genetic entity. In short, natural selection is mostly conservative, and the chance effects, including the fixation of selectively neutral mutations, are really responsible for creative nature of evolution.

4.6   Evolutionary Rate Under the Neutral Evolution

We considered the fate of selectively neutral mutants in Sect. 4.3. In reality, there are deleterious and advantageous mutations. Because the fraction of advantageous mutations are expected to be much smaller than that of deleterious ones, we consider only neutral and deleterious mutations. Let us denote the fraction of neutral mutations as f. This fraction has the rate of evolution identical with the mutation rate μ. The remaining fraction, 1−f, is deleterious mutations, and all of them are

```
P68871 Homo       MVHLTPEEKSAVTALWGKVNVDEVGGEALGRLLVVYPWTQRFFESFGDLSTPDAVMGNPK
P68222 Macaca     .........N...T...............................S.........
P02034 Ateles     .....G.......T...............................S.........
P60524 Canis      -....A...L.SG...............I........D...........S.A.
P02062 Equus      -.Q.SG..A..L..D...EE....................D......N.G....
P02070 Bos        --M..A...A...F....K.......................A....N...
P02084 Elephas    -.N..AA..TQ..N.......K.L.....S.........R....H.....A..LH.A.
P02106 Macropus   -....A..N.I.S.....AIEQT.........I.....S...DH.....NAK...A...
P02112 Gallus     ...W.A...QLI.G.......A.C.A...A...I.........A...N..S.T.IL...M
P07432 Xenopus    -MG..AHDRQLINST....CAKTI.K.......WT......Y.S...N.NSA...FH.EA
P02139 Cyprinus   -.EW.DA.R..II.....L.P..L.P...A.C.I.........A.Y.N..S.A.I.....
                  :  ::  :   *.*:     * ***.* * .**** *:*  :*:*... *:: *

P68871 Homo       VKAHGKKVLGAFSDGLAHLDNLKGTFATLSELHCDKLHVDPENFRLLGNVLVCVLAHHFG
P68222 Macaca     .................N..........Q................K..........
P02034 Ateles     ...........................Q...........................
P60524 Canis      ..........NS.....KN........K.................K..........
P02062 Equus      ..........HS.GE.VH.........A.....................V...R...
P02070 Bos        ..........DS..N.MK...D.....A.............K......V...RN..
P02084 Elephas    .L...E...TS.GE..K.........D.......................I...R...
P02106 Macropus   .L...A...V..G.AIKN.........K............K...II.IC..E...
P02112 Gallus     .R.....TS.G.AVKN...I.N..SQ...........DI.II...A..S
P07432 Xenopus    .A...E..VTSIGEAIK.M.DI..YY.Q..KY.SET.....C..KRF.GC.SIS..RQ.H
P02139 Cyprinus   .A...RT.E.GLMRAIKDM..I.A.Y.P..VM.SE......D.....ADCITVCA.MK..
                  * *** .*  .:  .: ..*::* :: ** *.:.***** **: ...:    * :*

P68871 Homo       -KEFTPPVQAAYQKVVAGVANALAHKYH
P68222 Macaca     -.....Q.....................
P02034 Ateles     -.....QL....................
P60524 Canis      -.....Q.....................
P02062 Equus      -.D...EL..S.................
P02070 Bos        -.....VL..DF.............R..
P02084 Elephas    -.....D.....E...............
P02106 Macropus   -....IDT.V.W..L.............
P02112 Gallus     -.D...EC...W..L.RV..H...R...
P07432 Xenopus    -E.Y..ELH...EHLFDAI.D..GKG..
P02139 Cyprinus   PSG.S.N..E.W..FLSV.V...KRQ..
                  . ::   :  :::..  :..** : **
```

Fig. 4.12 Multiple alignment of 11 vertebrate β-globin sequences

assumed to be not fixed and do not contribute to gene substitution. Thus, the evolutionary rate λ becomes

$$\lambda = f \cdot \mu + (1 - f) \cdot 0 = f\mu \qquad (4.38)$$

The value of f varies from the genomic region to region, as we will see in this section.

4.6.1 Molecular Clock

Vertebrate hemoglobin consists of globin (protein) and heme (porphyrin), and Fe ion in heme will attach to oxygen. There are two major globin gene families, α and β. Figure 4.12 shows the multiple alignment of 11 vertebrate β globin amino acid

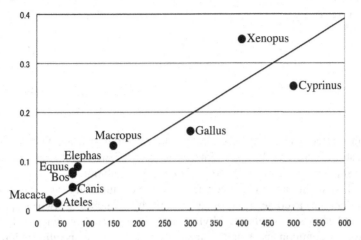

Fig. 4.13 Approximate linearity or molecular clock for vertebrate β-globin (Based on data of Fig. 4.12)

sequences (see Chap. 14 for the procedure). Amino acid sequence names are composed of UniProt accession number and genus. Only human amino acid sequence is fully written at the top, and the amino acids of remaining sequences are shown only when they are different from the corresponding human amino acid. If amino acid of nonhuman species globin is identical, dot (.) is given. For example, horse β globin amino acid sequence is different from that of human at 24 sites out of 146 total amino acids. This proportion, p (0.164 = 24/146), can be used to estimate the number, d, of amino acid substitutions per amino acid site:

$$d = -\log_e(1-p) \tag{4.39}$$

This number is often called evolutionary distance, and d stands for "distance." Please see Chap. 15 for derivation of this equation. In any case, using this equation, d becomes 0.18 from p. Evolutionary distances between human and the other 10 vertebrate species are plotted in vertical axis of Fig. 4.13. The horizontal axis of this figure represents divergence time between human and corresponding species. Interestingly, evolutionary distances and divergence times are more or less proportional. This rough constancy of the evolutionary rate is often called "molecular clock" after Zuckerkandl and Poring (1965; [32]). It should be noted that evolutionary distances were obtained from molecular data determined in wet experiments, while divergence times were obtained from paleontological studies.

Existence of the molecular clock is easily explained under the neutral theory. If the mutation rate (μ) and the fraction (f) of deleterious mutations are constant for a long evolutionary time, the evolutionary rate λ (=fμ) should be constant according to Eq. 4.38. In contrast, if the evolutionary rate is mainly determined by positive selection, not only mutation rate but also population size and selection coefficients

Fig. 4.14 Divergence
of two lineages

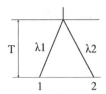

of mutants affect the evolutionary rate, and the latter two are known to vary considerably. Therefore, the approximate constancy of the evolutionary rate is one evidence supporting the neutral theory of molecular evolution.

Even if we do not assume the constancy of the evolutionary rate, it is possible to consider the average rate of evolution by comparing two sequences. Figure 4.14 shows a schematic phylogenetic tree of two sequences, 1 and 2. They have the common ancestor T years ago, and the lineage-specific evolutionary rates, $\lambda 1$ and $\lambda 2$, are given. Thus, the average rate, λ, of evolution between sequences 1 and 2 becomes

$$\lambda = \frac{(\lambda 1 + \lambda 2)}{2} \tag{4.40}$$

Let us denote the evolutionary distance between sequences 1 and 2 as d. Then,

$$d = \lambda \cdot 2T \tag{4.41}$$

We can thus estimate the evolutionary rate:

$$\lambda = \frac{d}{2T} \tag{4.42}$$

If the constancy of the evolutionary rate approximately holds, we can estimate the divergence time:

$$T = \frac{d}{2\lambda} \tag{4.43}$$

This equation is often used because the divergence time of two sequences are frequently unknown, while the molecular data such as amino acid sequences or DNA sequences can be easily determined.

4.6.2   Heterogeneous Evolutionary Rates Among Proteins

The fraction, f, of neutral mutations in Eq. 4.38 may vary in various situations. Let us first consider the heterogeneity among different proteins. Table 4.2 lists the rates of amino acid substitutions per amino acid site per year for 12 proteins. The evolutionary rates considerably vary from 0.01 to 9.0, and the rate for fibrinopeptide is almost 100 times higher than that for histone H4. Histone is the major basic protein of nucleosome

Table 4.2 Rates of amino acid substitutions (From Nei 1987; [2])

Protein	Rate ($\times 10^{-9}$)
Fibrinopeptide	9.0
Growth hormone	3.7
Igγ chain constant region	3.1
Serum albumin	1.9
Globin α chain	1.2
Trypsin	0.59
Lactose dehydrogenase	0.34
Cytochrome C	0.22
Glucagon	0.12
Histone H3	0.014
Ubiquitin	0.010
Histone H4	0.010

Unit: per amino acid per year

that binds DNA, an acid. The very low evolutionary rate for this protein indicates that f is quite small, and majority of amino acid changing mutations are deleterious.

Fibrinopeptide is the leftover of fibrinogen which was cut to fibrin and fibrinopeptide. The main function of blood coagulation is residing in fibrin, and the function of fibrinopeptide is just to keep fibrin not to become fibrous until it is detached from fibrin part. It is thus understandable that many amino acid substitutions on fibrinopeptide gene may be permissible; hence, its f became high.

Because of this relationship between f values and protein functions, it is routine to discuss the importance of one function in terms of its rate of amino acid substitutions. If the rate is slow, the protein may be called "quite important," and it may be "less important" if the rate is relatively high.

4.6.3 Heterogeneous Evolutionary Rates Among Protein Parts

One protein has its specific 3D structure (see Chap. 1), and the functional part is often localized. Figure 4.15 is a 3D structure of hemoglobin, or globin and heme. Globin protein is mostly composed of α helix, and there is heme pocket that grabs heme. Kimura and Ohta [33] estimated the rate of amino acid substitutions for four parts of α and β globins. Table 4.3 shows the results. As expected, the rate at heme pocket, where the oxygen-transporting heme is anchored, is lowest compared to the other three parts.

Domains are often defined for many proteins because of their wide conservations (see Chap. 1). Therefore, it is natural for a domain part with lower evolutionary rate than the remaining part of the protein. For example, hox genes have highly conserved homeobox domain. If we compare amino acid sequences of human and mouse orthologous HoxA1–HoxA5 amino acid sequences, amino acid identities are certainly higher for homeodomain region. Table 4.4 shows the estimated rate of amino acid substitution for this gene using Eq. 4.39. As expected, the evolutionary rate of homeobox domains is quite low compared to the remaining parts.

Fig. 4.15 The 3D structure of globin and heme (pointed by *arrow*)

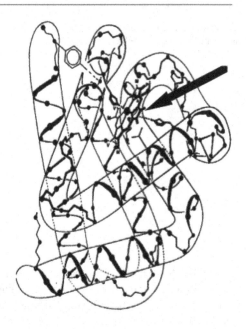

Table 4.3 The rate of amino acid substitutions for various protein components of α and β globins (Data from Kimura and Ohta 1973; [33])

Region	α globin	β globin
Surface	1.37	2.21
Outside	1.00	1.00
Inside	0.69	0.99
Heme pocket	0.49	0.49
Total	1.02	1.33

(Unit: per amino acid per year $\times 10^{-9}$)

Table 4.4 Comparison of amino acid identity between homeodomain and the other regions of HoxA. From Saitou 2007 [55]

	Amino acid identity between human and mouse (%)	
	Homeodomain	
Gene	region	Other regions
HoxA1	99.1	96.2
HoxA2	100	97.6
HoxA3	100	96.3
HoxA4	100	83.0
HoxA5	100	98.6

4.6.4 Heterogeneous Evolutionary Rates Among Organisms

The evolutionary rate is proportional to f and μ. Therefore, if μ, the mutation rate, differs among various lineages, molecular clock no longer holds. This is the case for the rodent lineage and other mammalian lineage, as first clearly shown by Wu and Li [34,

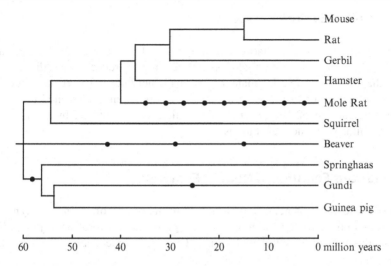

Fig. 4.16 Evolutionary history of rodent α-crystalin. Based on [38]

35]. Hominoid and Old World monkeys diverged at ~30 million years ago. Because human and rhesus macaque genomic distance is ~0.06 [36], the average evolutionary rate in terms of nucleotide substitutions is, from Eq. 4.42, λ[primates] $= 0.06/[2 \times 30$ million] $= 1 \times 10^{-9}$/site/year. The genomic distance between mouse and rat in terms of fourfold degenerate synonymous sites (see Sect. 4.7.1) is ~0.15 [37]. The divergence time between mouse and rat is not well known, so we use a range of 10–20 million years. Then λ[rodents] $= 0.15/[2 \times \{10–20\}$ million] $= 4–8 \times 10^{-9}$/site/year. Because mammalian genomes are mostly consisting of junk DNAs (see Sect. 4.7.2), genome-wide evolutionary rates are approximately mutation rate. It is clear that the mutation rate of rodents is 4–8 times higher than that for primates.

Compared to ordinary DNA genome organisms, genomes of RNA viruses such as influenza virus, SARS, and HIVs (see Chap. 8) are RNA molecules, and their evolutionary rates are million times higher than those of DNA genome organisms.

If the value of f, fraction of neutral mutations, varies among lineages for a particular protein gene, the evolutionary rate obviously changes. In this case, molecular clock no longer holds, yet this variation naturally follows the pattern of neutral evolution. Although the molecular clock is often considered as the important characteristics of the neutral evolution, this comes from the simple relationship shown in Eq. 4.38. Therefore, if f and/or μ changes, the evolutionary rate should change, according to the principle of neutral evolution. Figure 4.16 is the evolutionary history of rodent α crystallin [38]. The amino acid sequence of this protein is identical among mouse, rat, and hamster, and their sequence is identical with that of common ancestor or all rodents. In marked contrast to that situation, nine amino acid substitutions accumulated in the mole rat lineage during 40 million years. Mole rat eye is diminished, and apparently, importance of α crystallin, the major lens protein, is reduced. It is natural to expect higher fraction (f) of selectively neutral mutations for mole rat than other rodents whose eyes are necessary for their existence.

4.6.5   Unit of Evolutionary Rate

We discussed the unit of mutation rate in Chap. 2. Because mutation is the main player of evolution, unit of the evolutionary rate is closely related to that discussion. While the generation time for many organisms is not known, divergence times of some organism groups such as vertebrates have been well documented thanks to paleontological studies. Thus, the rate of evolution is often obtained by Eq. 4.42 and the time unit is years, not generations.

4.7   Various Features of Neutral Evolution

We discuss features of neutral evolution in terms of preponderance of synonymous substitutions to nonsynonymous ones, pure neutral evolution of junk DNA and pseudogenes, and neutral evolution at the macroscopic levels and at genomic levels.

4.7.1   Synonymous and Nonsynonymous Substitutions

If synonymous or nonsynonymous mutations (see Chap. 2) are fixed in populations, these are called synonymous and nonsynonymous substitutions, respectively. Historically, synonymous substitutions are called silent substitutions, and nonsynonymous ones are amino acid replacing substitutions.

If we consider the consequences of synonymous mutations, it is easy to expect that they are selectively neutral with original alleles because produced proteins are identical with each other. Nonsynonymous mutations may become deleterious because they may disrupt or reduce the protein function. As we saw in evolution of fibrinopeptides, it is also possible that the effect of a nonsynonymous substitution may be very minor and essentially selectively neutral. It is therefore a good approximation that f for synonymous mutations is 1, and the evolutionary rate is identical with mutation rate. As for nonsynonymous mutations, f is smaller than 1, and the evolutionary rate of nonsynonymous substitutions is expected to be smaller than that for synonymous substitutions. As we will see in Chap. 5, the evolutionary rate of nonsynonymous substitutions may become larger than the mutation rate when a special type of natural selection is operating, in which any amino acid change is advantageous. In this case, the rate of nonsynonymous substitutions will be higher than that of synonymous substitutions. Figure 4.17 shows a schematic comparison of the rates of synonymous and nonsynonymous substitutions.

Because the number of synonymous substitutions (Ds) and nonsynonymous substitutions (Dn) is simultaneously estimated for the same proteins of different species (or different paralogous genes), comparison of Ds and Dn values is routinely conducted for many studies of genome comparison. Figure 4.18 is such example. In both (A) for mouse and rat and (B) for human and rhesus macaque, Ds > Dn for the majority of protein coding genes.

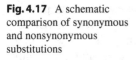

Fig. 4.17 A schematic comparison of synonymous and nonsynonymous substitutions

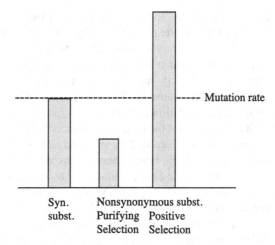

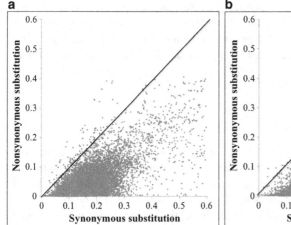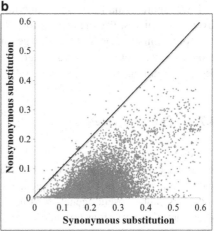

Fig. 4.18 Comparison of synonymous substitutions (*horizontal axis*) and nonsynonymous substitutions (*vertical axis*). (**a**) Comparison between mouse and rat. (**b**) Comparison between human and rhesus macaque

It should be noted that the rate of synonymous substitutions may not be identical with the mutation rate, for biases of codon usages exist ([39]; Ikemura 1985). We will discuss the consequences of these sorts of purifying selection on synonymous substitutions in Chap. 5.

4.7.2   Junk DNA

Susumu Ohno proclaimed the characteristics of mammalian genomes as "So much "junk" DNA in our genome" as early as 1972 [40]. Junk DNA means functionless DNA. In fact, only 1.5 % of the human genome is used for protein coding [41], and the rest are mostly junk. They are interspersed repeats (LINEs and SINEs), microsatellites, other intergenic regions, and introns (see Chap. 10). It is true that a small fraction of noncoding genomic regions are highly conserved [42, 43], and they are expected to have some functions such as enhancers. Even some SINE is known to obtain an important function during the mammalian evolution [44, 45].

It is still true that the majority of noncoding genomic regions are functionless and just junk DNAs. Recently there are some reports of transcriptions on many noncoding regions [46, 47]. However, these results were obtained by problematic ChIP-chip techniques and found to be artifact [48] by checking ChIP-seq techniques.

Because the f value of Eq. 4.38 is 1 for junk DNA and for synonymous sites, their evolutionary rates are expected to be similar, if we ignore heterogeneity of mutation rates in one genome. In fact, the number (~0.15) of nucleotide substitutions per site in intergenic regions for mouse and rat genomes was shown to be quite similar to that of synonymous substitutions ([37]).

If we ignore a small portion of functional DNAs that are highly conserved among diverse organisms, the majority (more than 90%; see Babarinde and Saitou 2013 [56]) of mammalian or all vertebrate genomes are junk DNAs. Therefore, a genome-wide divergence of two species is a good approximation of the consequence of pure neutral evolution.

4.7.3 Pseudogenes

Pseudogenes are DNA sequences which are homologous to functional genes, but themselves are no longer functional. For example, if there are frameshift mutations and/or stop codons in a DNA sequence highly homologous to a known functional gene, it is called "pseudogene," for functional protein is expected to be not formed. Therefore, they are often products of gene duplications. Because of nonfunctional nature of pseudogenes, the pseudogenes should be genuine members of junk DNAs. Figure 4.19 shows one of initial analysis of pseudogene evolution by Li, Gojobori, and Nei (1980; [49]).

There are four types of gene duplication (see Chap. 2). Among them, RNA-mediated duplication produces intronless sequences via reverse transcription of mRNAs. These cDNAs will be integrated to a DNA region unrelated to its place of origin, where a series of gene regulatory sequences exist. Therefore, such cDNA inserts are almost always 'dead on arrival'. We can see a clear enhancement of evolutionary rate for intronless, or processed, pseudogenes for the mouse p53 gene. The estimated numbers of nucleotide substitutions between *M. musculus* and *M. leggada* are 0.0157 and 0.0651 for functional genes and pseudogenes, respectively (data from Table 16.1C; originally from Ohtsuka et al. (1996; [57])).

Nonfunctionalization can happen without gene duplication. Vitamins are molecules that exist in small quantity but essential for organisms, especially human, to survive. By definition, vitamins are not produced by the organism itself, and they

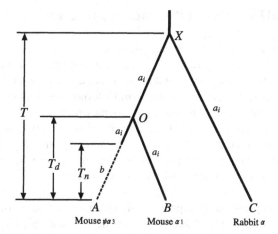

Fig. 4.19 Formation of pseudogene and nonfunctionalization. Gene A is pseudogene which diverged from its functional counterpart (gene A) for T_d years. Gene A became nonfunctional T_n years ago. Gene B in mouse diverged from rabbit counterpart (gene C) T years ago. The evolutionary rate of functional genes is assumed to be a_i (i = 1, 2, or 3) for i-th position of codons, while that of pseudogene is b for all three codon positions (From Li et al. (1980); [49])

should be taken in as a part of food. Their very existences are enigmatic, for these molecules are coming from other organisms which produce them. If vitamins are so important, why are they not produced by a certain species such as human? The neutral theory of evolution easily resolves this paradox. If vitamins are abundant in everyday foods, even the mutants with no ability of producing a certain vitamin are selectively neutral compared to wild types with ability to produce that vitamin through the existing enzymatic pathway.

Vitamin C, or ascorbic acid, is a good example. If appropriate intake of vitamin C is stopped for a long time, human will develop scurvy. King and Jukes (1969; [50]) already predicted that the lack of ascorbic acid production could be explained by assuming the neutral evolution. Not only human but all primates except for prosimians, elephants, guinea pigs, and fruit bats lack the ability of producing ascorbic acid [51]. Medaka, a teleost fish, also does not produce ascorbic acid [52]. In fact, nonfunctionalization of L-gulono-γ-lactone oxidase (enzyme number E.C.1.1.3.8) gene was confirmed by Nishikimi and his collaborators [53].

A more drastic situation of pseudogene formation without gene duplication is found in parasitic bacterial genomes. *Mycobacterium leprae*, the causative bacteria of leprosy, was found to have many pseudogenes in its genome ([54]). This is because this bacterium is hiding deep in host body and receives many nutrients from host.

A gene function is often quite complex, and it is not easy to determine if a "pseudogene" is really nonfunctional. Even if protein is not produced, mRNA or even DNA sequences themselves may still have some function. Therefore, when we discuss about the evolution of pseudogenes, it may be too simplistic to assume that f, fraction of neutral mutations, is 1 for a pseudogene. A "pseudogene" with some function is not surprising, for they were named so only because of sequence comparison.

4.7.4 Neutral Evolution at the Macroscopic Level

So far, we discussed evolution of nucleotide or amino acid sequences and saw that the fixations of selectively neutral mutations are the major process of evolution. It is thus natural to expect that the evolution at the macroscopic or so-called phenotypic level is also following mostly neutral fashion. Unfortunately, this logically derived conjecture seems to be not kept by many evolutionary biologists. Ever since Charles Darwin, many biologists have been enchanted by seemingly powerful positive selection. They are biologists who study macroscopic morphology of organisms, animal behaviors, developmental process, and so on. As we will see in Chap. 5, we should be careful to discuss adaptation without clear demonstration at the molecular level.

It may be still optimistic to expect a rapid expansion of our knowledge on the genetic basis of developmental and behavioral traits in the near future. However, modern biology is proceeding to this direction, and I personally hope that the superficial dichotomy between molecules (genotypes) and phenotypes will disappear sooner or later. Evolutionary genomics is at the foundation of this edifice of modern biology. It should be added that Nei's (2013) recent book "Mutation-driven evolution" ([58]) covers many interesting topics related to this chapter.

References

1. Kimura, M. (1983). *The neutral theory of molecular evolution*. Cambridge: Cambridge University Press.
2. Nei, M. (1987). *Molecular evolutionary genetics*. New York: Columbia University Press.
3. Mouse Genome Sequencing Consortium. (2002). Initial sequencing and comparative analysis of the mouse genome. *Nature, 420*, 520–562.
4. International Chicken Genome Sequencing Consortium. (2004). Sequence and comparative analysis of the chicken genome provide unique perspectives on vertebrate evolution. *Nature, 432*, 695–716.
5. Kimura, M. (1968). Evolutionary rate at the molecular level. *Nature, 217*, 624–626.
6. Bonner, J. T. (2008). *The social amoebae: The biology of cellular slime molds*. Princeton: Princeton University Press.
7. Cook, R. E. (1979). Asexual reproduction: A further consideration. *American Naturalist, 113*, 769–772.
8. Ewens, W. J. (1979). *Mathematical population genetics*. Berlin/New York: Springer.
9. Watson, H. W., & Galton, F. (1874). On the probability of the extinction of families. *Journal of Anthropological Institute, 4*, 138–144.
10. Haccou, P., Jagers, P., & Vatutin, V. A. (2005). *Branching processes: Variation, growth, and extinction of populations*. Cambridge: Cambridge University Press.
11. Crow, J. F. (1989). The estimation of inbreeding from isonymy. *Human Biology, 61*, 935–948.
12. Saitou, N. (1983). An attempt to estimate the migration pattern in Japan by surname data (in Japanese). *Jinruigaku Zasshi, 91*, 309–322.
13. Fisher, R. A. (1930). The distribution of gene ratios for rare mutations. *Proceedings of Royal Society of Edinburgh, 50*, 205–220.
14. Feller, W. (1968). *Introduction to probability theory and its applications* (3rd ed., Vol. 1). New York: Wiley.

15. Crow, J. F., & Kimura, M. (1970). *An introduction to population genetics theory*. New York: Prentice-Hall.
16. Howell, N. (1979). *Demography of the Dobe !Kung*. New York: Academic.
17. Saitou, N., Shimizu, H., & Omoto, K. (1988). On the effect of the fluctuating population size on the age of a mutant gene. *Journal of the Anthropological Society of Nippon, 96*, 449–458.
18. Kingman, J. F. C. (1982). On the genealogy of large populations. *Journal of Applied Probability, 19A*, 27–43.
19. Hudson, R. R. (1983). Testing the constant rate neutral allele model with protein sequence data. *Evolution, 37*, 203–217.
20. Tajima, F. (1983). Evolutionary relationship of DNA sequences in finite populations. *Genetics, 105*, 437–460.
21. Fu, Y.-X. (2006). Exact coalescent for the Wright-Fisher model. *Theoretical Population Biology, 69*, 385–394.
22. Hein, J., Schierup, M. H., & Wiuf, C. (2005). *Gene genealogies, variation, and evolution – a primer in coalescent theory*. Oxford: Oxford University Press.
23. Wakeley, J. (2008). *Coalescent theory: An introduction*. Greenwood Village: Roberts & Co.
24. Kimura, M. (1955). Solution of a process of random genetic drift with a continuous model. *Proceedings of National Academy of Sciences USA, 41*, 144–150.
25. Kimura, M. (1964). Diffusion models in population genetics. *Journal of Applied Probability, 1*, 177–232.
26. Kimura, M., & Ohta, T. (1971). Protein polymorphism as a phase of molecular evolution. *Nature, 229*, 467–469.
27. Kimura, M., & Crow, J. F. (1964). The number of alleles that can be maintained in a finite population. *Genetics, 49*, 725–738.
28. Kimura, M. (1969). The number of heterozygous nucleotide sites maintained in a finite population due to steady flux of mutations. *Genetics, 61*, 893–903.
29. Kimura, M. (1968). Genetic variability maintained in a finite population due to mutational production of neutral and nearly neutral isoalleles. *Genetical Research, 1*, 247–269.
30. Iafrate, A. J., Feuk, L., Rivera, M. N., Listewnik, M. L., Donahoe, P. K., Qi, Y., Scherer, S. W., & Lee, C. (2004). Detection of large-scale variation in the human genome. *Nature Genetics, 36*, 949–951.
31. Sebat, J., Lakshmi, B., Troge, J., Alexander, J., Young, J., Lundin, P., Maner, S., Massa, H., Walker, M., Chi, M., et al. (2004). Large-scale copy number polymorphism in the human genome. *Science, 305*, 525–528.
32. Zuckerkandl, E., & Pauling, L. (1965). Evolutionary divergence and convergence in proteins. In V. Bryson & H. J. Vogel (Eds.), *Evolving genes and proteins* (pp. 97–166). New York: Academic.
33. Kimura, M., & Ohta, T. (1973). Mutation and evolution at the molecular level. *Genetics (Supplement), 73*, 19–35.
34. Wu, C.-I., & Li, W.-H. (1985). Evidence for higher rates of nucleotide substitution in rodents than in man. *Proceedings of the National Academy of Sciences of the United States of America, 82*, 1741–1745.
35. Li, W.-H., & Wu, C.-I. (1987). Rates of nucleotide substitution are evidently higher in rodents than in man. *Molecular Biology and Evolution, 4*, 74–82.
36. Rhesus Macaque Sequencing and Analysis Consortium. (2007). Evolutionary and biomedical insights from the rhesus macaque genome. *Science, 316*, 222–234.
37. Abe, K., Noguchi, H., Tagawa, K., Yuzuriha, M., Toyoda, A., Kojima, T., Ezawa, K., Saitou, N., Hattori, M., Sakaki, Y., Moriwaki, K., & Shiroishi, T. (2004). Contribution of Asian mouse subspecies Mus musculus molossinus to genomic constitution of strain C57BL/6J, as defined by BAC end sequence-SNP analysis. *Genome Research, 14*, 2239–2247.
38. Hendriks, W., Leunissen, J., Nevo, E., Bloemendal, H., & de Jong, W. W. (1987). The lens protein alpha A-crystallin of the blind mole rat, *Spalax ehrenbergi*: Evolutionary change and functional constraints. *Proceedings of the National Academy of Sciences of the United States of America, 84*, 5320–5324.

39. Ikemura, T. (1985). Codon usage and tRNA content in unicellular and multicellular organisms. *Molecular Biology and Evolution, 2*, 13–34.
40. Ohno, S. (1972). So much "junk" DNA in our genome. *Brookhaven Symposium in Biology, 23*, 366–370.
41. International Human Genome Sequencing Consortium. (2004). Finishing the euchromatic sequence of the human genome. *Nature, 431*, 931–945.
42. Bejerano, G., Pheasant, M., Makunin, I., Stephen, S., Kent, W. J., Mattick, J. S., & Haussler, D. (2004). Ultraconserved elements in the human genome. *Science, 304*, 1321–1325.
43. Takahashi, M., & Saitou, N. (2012). Identification and characterization of lineage-specific highly conserved noncoding sequences in mammalian genomes. *Genome Biology and Evolution, 4*, 641–657.
44. Bejerano, G., Lowe, C. B., Ahituv, N., King, B., Siepel, A., Salama, S. R., Rubin, E. M., Kent, W. J., & Haussler, D. (2006). A distal enhancer and an ultraconserved exon are derived from a novel retroposon. *Nature, 441*, 87–90.
45. Sasaki, T., Nishihara, H., Hirakawa, M., Fujimura, K., Tanaka, M., Kokubo, N., Kimura-Yoshida, C., Matsuo, I., Sumiyama, K., Saitou, N., Shimogori, T., & Okada, N. (2008). Possible involvement of SINEs in mammalian-specific brain formation. *Proceedings of the National Academy of Sciences of the United States of America, 105*, 4220–4225.
46. Johnson, J. M., Edwards, S., Shoemaker, D., & Schadt, E. E. (2005). Dark matter in the genome: Evidence of widespread transcription detected by microarray tiling experiments. *Trends in Genetics, 21*, 93–102.
47. Birney, E., Stamatoyannopoulos, J. A., Dutta, A., Guigo, R., Gingeras, T. R., et al. (2007). Identification and analysis of functional elements in 1 % of the human genome by the ENCODE pilot project. *Nature, 447*, 799–816.
48. van Bakel, H., Nislow, C., Blencowe, B. J., & Hughes, T. R. (2010). Most "dark matter" transcripts are associated with known genes. *PLoS Biology, 8*, e1000371.
49. Li, W.-H., Gojobori, T., & Nei, M. (1981). Pseudogenes as paradigm of the neutral evolution. *Nature, 292*, 237–239.
50. King, J. L., & Jukes, T. H. (1969). Non-Darwinian evolution. *Science, 164*, 788–798.
51. Lehninger, A. L. (1975). *Biochemistry*. New York: Worth Publishers.
52. Toyohara, H., Nakata, T., Touhata, K., Hashimoto, H., Kinoshita, M., Sakaguchi, M., Nishikimi, M., Yagi, K., Wakamatsu, Y., & Ozato, K. (1996). Transgenic expression of L-gulono-gamma-lactone oxidase in medaka (*Oryzias latipes*), a teleost fish that lacks this enzyme necessary for L-ascorbic acid biosynthesis. *Biochemical and Biophysical Research Communications, 223*, 650–653.
53. Nishikimi, M., Fukuyama, R., Minoshima, S., Shimizu, N., & Yagi, K. (1994). Cloning and chromosomal mapping of the human nonfunctional gene for L-gulono-gamma-lactone oxidase, the enzyme for L-ascorbic acid biosynthesis missing in man. *Journal of Biological Chemistry, 269*, 13685–13688.
54. Cole, S. T., & others. (2001). Massive gene decay in the leprosy bacillus. *Nature, 409*, 1007–1011.
55. Saitou, N. (2007). *Genomu Shinkagaku Nyumon (written in Japanese, meaning 'Introduction to evolutionary genomics')*. Tokyo: Kyoritsu Shuppan.
56. Babarinde, I. A., & Saitou, N. (2013). Heterogeneous tempo and mode of conserved noncoding sequence evolution among four mammalian orders. *Genome Biology and Evolution* (advance access).
57. Ohtsuka, H., Oyanagi, M., Mafune, Y., Miyashita, N., Shiroishi, T., Moriwaki, K., Kominami, R., & Saitou, N. (1996). The presence/absence polymorphism and evolution of p53 pseudogene within the genus Mus. *Molecular Phylogenetics and Evolution, 5*, 548–556.
58. Nei, M. (2013a). *Mutation-driven evolution*. Oxford: Oxford University Press.
59. Takahatan, N. (1993). Allelic genealogy and human evolution. *Molecular Biology and Evolution, 10*, 2–22.

Natural Selection

5

Chapter Summary

Basic concept of natural selection is first discussed, and purifying (negative) selection is shown to be much more prevalent than positive selection. Natural selection on haploids and diploids is discussed under both large populations and small populations. Natural selection at the genomic level is then described, covering various topics such as gain and loss of genes and purifying selection at synonymous sites and at noncoding regions. Positive selection for ape and human genes and detection of positive selection through genome-wide searches are also discussed.

5.1 Basic Concept of Natural Selection

The fundamental source of evolution is mutation, or any change of genome sequences. Therefore, natural selection is tightly connected with the effect of mutations. If mutations are highly deleterious, they will soon disappear through natural selection. Before Darwin (1859; [1]) proposed the possibility of natural selection as a creative power of evolution, natural selection has been considered as the mechanism to keep status quo, as initially created by the divine power. Elimination of deleterious mutations is now called "negative" selection after Kimura (1983; [2]). When the conservative nature of this process is stressed, it is called "purifying" selection (Lewontin 1974; [3]). To keep the current genetic entity thorough elimination of deleterious mutations is the core of the "struggle for existence."

As we discussed in Chap. 4, most of mutations become extinct simply by chance, and only a small fraction exists for a long evolutionary time. If we focus on these mutations that contribute to evolution for a certain period, say 1,000 generations, the majority is selectively neutral, and only a small fraction is kept through natural selection that favors advantageous mutations. Charles Darwin, together with Alfred Russel Wallace, proposed this type of natural selection as the mechanism of evolution in 1858. This process is therefore sometimes called "Darwinian" selection,

N. Saitou, *Introduction to Evolutionary Genomics*, Computational Biology 17, DOI 10.1007/978-1-4471-5304-7_5, © Springer-Verlag London 2013

but nowadays "positive" selection, again after Kimura ([2]), is commonly used. Negative and positive selection sandwich the selective neutrality, the dominant situation in evolution.

Natural selection is tightly connected with two concepts: adaptation and fitness. Although "adaptation" is widely used in many books on evolution, it is often not quantitatively defined, and we should be careful for using this word in the context of natural selection unless clearly defined. This word is also used to describe various characters of many organisms without testing whether those characters were products of positive selection. For example, "adaptive radiation" has often been used on the evolution of many mammalian lineages (e.g., [4]) as if adaptation caused by positive selection is the main factor for radiation. Recent studies, however, suggest that the geographical isolation caused by the continental drift seems to be the main cause of mammalian radiation (e.g., [5]). Any evolutionary biologist should have a discreet attitude on the usage of "adaptation."

In contrast, "fitness" is quantitatively defined in population genetics theory. There are two kinds of fitness: absolute and relative. Because the basis of evolution is the change of genetic constituents of organisms, natural selection should be discussed in terms of differential rates of reproduction among genotypes. The absolute fitness, $W_{absolute}$, is the mean number of offsprings for one particular genotype. If this is larger than 1, the genotype is expected to increase its offspring if the effect of random genetic drift is negligible. Consideration of the absolute fitness is important, for we can directly discuss the temporal change of population size. It should be noted that the absolute fitness may change without mutations, if the environmental condition changes.

We are often interested in the relative success of one genotype to other ones and consider the relative fitness, $W_{relative}$. The relative fitness of one particular genotype is usually set to be 1, and that of other genotypes are expressed using selection coefficients. Genotypes are identical with alleles in haploid organisms, and $W_{relative}$ for allele i may be written as $1 + s_i$, where s_i is selection coefficient for allele i. Let us assume that the relative fitness of allele 0 is 1 ($s_0 = 0$). If s_i ($i = 1, 2, 3, ...$) is positive, allele i is advantageous compared to allele 0 and is positively selected, while allele i is deleterious compared to allele 0 and is negatively selected when s_i is negative. Allele i is selectively neutral with allele 0 when $s_i = 0$.

Wolch et al. (2001; [6]) estimated selection coefficients of mutations occurred in haploid strains of *Saccharomyces cerevisiae*. Figure 5.1 shows the distribution of selection coefficients. Only 2 (0.6 %) out of 336 mutants had higher relative fitness (selection coefficients were 0.015 and 0.026) than that for wild types, and 130 (39 %) were lethal mutants. Their method could not detect mutations with selection coefficients between −0.01 and +0.01, and these neutral and nearly neutral mutations should be much more abundant than clearly deleterious ones. Joseph and Hall (2004; [7]) conducted a somewhat different experiment for diploid strains of *Saccharomyces cerevisiae*. They used 151 lines and accumulated mutations for ~1,000 generations. Because of the difference in the experimental design, they did not find lethal mutations, and most of the mutations are distributed around the $s = 0$ point, and the shape is similar to the normal distribution. This is clearly different from that shown in Fig. 5.1. Although Joseph and Hall ([7]) estimated that ~6 %

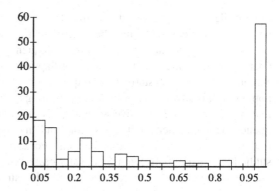

Fig. 5.1 Distribution of mutations in terms of their selection coefficients (Data from [6]). Horizontal axis: selection coefficient. Vertical axis: number of mutations

of the fitness-altering mutations were beneficial (positively selected), there is a possibility that their results were biased because of their experimental design. Perfeito et al. (2007; [8]) measured the genomic mutation rate that generates beneficial mutations and their effects on fitness in *Escherichia coli* and suggested that 0.7 % (1 in 150) of newly arising mutations is beneficial. It is interesting that the two studies on yeast and bacteria both gave very low values for the proportion of advantageous mutations. A review on the distribution of fitness effects of new mutations by Eyre-Walker and Keightley (2007; [9]) also concluded that advantageous mutations are rare.

5.2 Various Types of Positive Selection

Disadvantageous mutations are usually eliminated from the population through negative or purifying selection unless its selective coefficient is small enough so as for the random genetic drift to be more dominant than the natural selection. We thus consider mainly positive selection in this chapter. Positive selection can be divided into directional selection and balancing selection. When a mutant is selectively advantageous than any genes in the current population, the mutant allele frequency is expected to increase, and eventually become fixed, in both haploid and diploid organisms. This type of selection is called "directional" selection. In certain selection schemes, more than one type of alleles may coexist in one population, if a particular balance between the selective force and other factors is attained. This is called "balancing" selection.

5.2.1 Balancing Selection

There are a variety of selection schemes to attain some balance or equilibrium on allele frequencies. Let us assume that the environment is heterogeneous, and a newly arisen mutant allele is more advantageous than the preexisting alleles only in a certain environment. In this case, the mutant may increase its allele frequency to some value, but will never be fixed. However, if the environment changes, the mutant allele which coexisted with the other alleles may no longer be advantageous in any environment, and it may become extinct.

Another type of balancing selection is frequency-dependent selection. Selection coefficient of an allele will change according to its allele frequency in this scheme. If mutant allele, which starts from frequency of 1/N (N is the number of genes in the population), is advantageous to already existing alleles, the mutant allele frequency may increase. If the selective advantage is diminishing as the mutant allele frequency increases, then the increase stops and an equilibrium occurs until new mutation or the environmental change happens. For example, Carius et al. (2001; [128]) studied interaction between a freshwater crustacean and bacteria. They found that there was no strain that was superior to all other strains in both host crustacean and parasite bacteria, and this situation may create the frequency-dependent selection in both hosts and parasites.

Selection schemes discussed above can apply to any organisms; however, overdominant selection, another type of balancing selection, can occur only in diploids, for we need a heterozygote advantage. We will discuss overdominant selection in Sect. 5.3.4.

Historically, balancing selection was once considered to be one of the major mechanisms to keep high polymorphism in diploid populations. For example, the PTC taste sensitivity polymorphism in human discovered by Fox (1932; [10]) was once considered to be under overdominant selection. Fisher, Ford, and Huxley (1939; [11]) reported that chimpanzees and humans both harbor apparently dominant and recessive "taster" and "nontaster" alleles at roughly equal frequencies. They thus argued: "Wherein the selective advantages lie, it would at present be useless to conjecture, but of the existence of a stably balanced and enduring dimorphism determined by this gene there can be no room for doubt" [11]. It is clear that Fisher et al. proposed "transspecific polymorphism" between human and chimpanzee. Because taster allele is dominant to nontaster allele, heterozygote individuals have no selective advantage to taster homozygotes. Other types of selection schemes such as frequency-dependent selection are also not easy to envisage. A more straightforward explanation for coexistence of taster–nontaster genetic polymorphism is parallel loss of taster activity in human and chimpanzee lineages. In fact, soon after the gene (TAS2R38) responsible for the PTC taste polymorphism was discovered (Kim et al. 2003; [12]) and Wooding et al. (2006; [13]) demonstrated that the molecular basis of this variation has arisen twice, independently, in the two species. Interested readers are suggested to read a review on this research by Wooding (2006; [14]).

The scheme of balancing selection implies the existence of a stable equilibrium allele frequency, and multiple alleles will coexist infinitely if the population size is approximated to be infinite. However, the real world is always finite, and there will never be any static equilibrium. This is true even for a fictitious population with an infinite individual size, for the equilibrium will be shifted by change of selection pattern. We have to be careful for any kind of discussion on equilibrium under balancing selection schemes.

5.2.2   Arms Race Between Hosts and Parasites

Under some special circumstances, any change becomes advantageous, and many more mutations accumulate compared to the pure neutral evolution, for mutations are, by definition, genetic changes. One such special case is a sort of "arms race" between the host immune system and parasitic organisms. The idea of arms race or runaway goes back to Fisher (1930; [15]). Some viral protein part may be the target of the host immune system. When a nonsynonymous mutation occurs on a virus protein coding gene, that change may become advantageous, for the change of amino acid could weaken the effect of host antibody or T-cell receptor that fits to the existing viral protein motif. In fact, nonsynonymous substitutions higher than synonymous ones were observed for the spike region of influenza A hemagglutinin protein genes by Ina and Gojobori (1994; [16]). As for the host immune system, changes of amino acids responsible for peptide presentation may become advantageous. It was demonstrated for human and mouse MHC class II molecules by Hughes and Nei (1988; [17]).

Immunoglobulin α protein has a hinge region which connects constant and variable regions, and this hinge is the target of bacterial proteinase. The peptide change of this hinge will reduce the possibility of breaking immunoglobulin α protein that will attack bacteria. In fact, not only nonsynonymous substitutions but also insertions and deletions were shown to be quite frequent on the hinge region (Sumiyama et al. 2003; [18]). This is a good example of arms race between vertebrate host organisms having acquired immune system and parasites (virus or bacteria).

5.3   Natural Selection on Populations with a Large Number of Individuals

We discuss the effect of natural selection when the population size is large and they can be approximated as infinite. This is a rather classic treatment of natural selection. As we will see in Sect. 5.4, it is more realistic to consider the case where the population size is finite. However, this classic theory still holds when the allele frequency of advantageous mutant reached some threshold to overcome the stochastic effect of random fluctuation of allele frequency or the random genetic drift.

5.3.1   A Series of Simplifications in Many Theoretical Models

We need to have a mathematical model to consider the effect of natural selection at one genetic locus. The case for haploid organisms is first discussed as in Chap. 4, followed by situations for diploids. For simplicity, all alleles already existing are assumed to be selectively neutral, and they are treated as one allele called A_0. One mutant, A_1, appears in the population, and this may be deleterious or advantageous relative to allele A_0.

Table 5.1 A simple model of natural selection for haploid organism

Genotype	Absolute fitness	Relative fitness[a]	Frequency at	
			t	t+1
A_0	W_0	1	p_0	p'_0
A_1	W_1	$1+s$	p_1	p'_1

[a]$s \geq -1$

Natural selection operates to each individual, and one individual has many genes in its genome. Because we assume that there are many, effectively infinite, individuals in this population, the influence of all other loci in the genome is assumed to be averaged, and we can consider the natural selection on one particular locus. This is an implicit assumption in many population genetics theories, but this is not a good approximation if we consider a small population size. This is because each individual may have a different set of genotypes for the whole genome.

Another simplification is the model of generation. In human populations, generations are overlapped, and the age difference between mates exists. For simplicity, discrete generation model is considered here. This simplified situation does apply to annual plants in which fertilization takes place only among the same generation.

5.3.2   The Case for Haploids

We first consider the case for haploids. Table 5.1 shows the absolute fitness, relative fitness, and frequency for the two types of genotypes, A_0 and A_1, same as allele designation. Because we are interested in the temporal change of genotype frequency, let p_0 and p_1 be frequencies of genotypes (alleles) A_0 and A_1 at generation t, respectively. Then, given the constant values of absolute fitness, W_0 and W_1, frequency of genotype A_1 at time $t+1$ will become

$$p'_1 = p_1 W_1 / (p_0 W_0 + p_1 W_1). \qquad (5.1)$$

We set the relative fitness of genotype A_0 as 1, and the difference of the absolute fitness $(W_1 - W_0)$ is designated as s, called "selection coefficient." The selection coefficient can be positive, zero, or negative, depending on the advantageous, selectively neutral, or deleterious mutations. The smallest possible value for s is -1, where the relative fitness of individuals with genotype A_1 are zero, or lethal. Then, the relationship of genotype frequencies at generation t and $t+1$ become

$$p'_0 = p_0 / \{p_0 + p_1(1+s)\}, \qquad (5.2a)$$

$$p'_1 = p_1(1+s) / \{p_0 + p_1(1+s)\}. \qquad (5.2b)$$

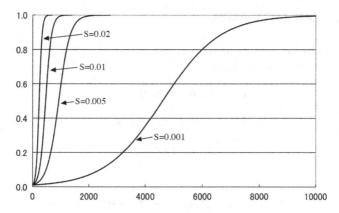

Fig. 5.2 Frequency change of genotype A_1 for haploid organisms. Horizontal axis: generation. Vertical axis: allele frequency

Note that by definition, $p_0 + p_1 = p'_0 + p'_1 = 1$. If we denote the difference of frequency for genotype A_1 between two generations as $\Delta p_1 (= p_1[t+1] - p_1[t])$,

$$\Delta p_1 = p_1(1-p_1)s / (1+sp_1). \tag{5.3}$$

When $s=0$, $\Delta p_1 = 0$, and no change occurs. This is because the population size is assumed to be infinite and the effect of random genetic drift is ignored. This is why this model is unrealistic especially when the mutant allele or genotype frequency is small and it is affected by stochastic effect even for a large population.

When $s>0$, $\Delta p_1 > 0$, the genotype frequency of the mutant A_1 will continuously increase toward 1 (fixation). If $s<0$, $\Delta p_1 < 0$, the mutant will soon disappear from the population irrespective of the value of s, for the initial allele or genotype frequency, p_1, must be very small for any fresh mutant. Figure 5.2 shows the temporal change of p_1 starting from the initial frequency of 0.01 to 0.9999 (essentially fixed) for various values of selection coefficient s. It takes 698, 1,388, 2,768, and 13,811 generations to reach $p=0.9999$ when s is 0.02, 0.01, 0.005, and 0.001, respectively. The selection coefficient of 0.02 (2 %) is a very strong positive selection, and in fact, it takes only ~700 generations for a minor allele with 1 % frequency to be fixed. When the selection coefficient is 0.001, however, it takes 20 times more (~14,000) generations to reach the almost fixed situation. If we consider an organism with large population size, such as bacteria, it may be reasonable to start the allele frequency at 0.0001, or 100 mutant cells in the one million cell population. In this case, it takes 931, 1,852, 3,694, and 18,430 generations to reach $p=0.9999$ when s is 0.02, 0.01, 0.005, and 0.001, respectively. If one generation is 1 hour as in *E. coli* in laboratory, it takes 2.1 years for mutants with 0.0001 allele frequency to be fixed when they are 0.1 % more advantageous than the wild types. This calculation shows that the positive selection may be rather slow even in large populations like bacteria when selective advantage is small.

5.3.3   The Case for Diploids

We now consider the case for diploids. There are three types of genotypes for the two alleles, A_0 and A_1. As in the case of 5.1,

$$P'_{00} = P_{00} W_{00} / \text{Mean\_PW},   \tag{5.4a}$$

$$P'_{01} = P_{01} W_{01} / \text{Mean\_PW},   \tag{5.4b}$$

$$P'_{11} = P_{11} W_{11} / \text{Mean\_PW},   \tag{5.4c}$$

where

$$\text{Mean\_PW} = P_{00} W_{00} + P_{01} W_{01} + P_{11} W_{11}.   \tag{5.5}$$

Table 5.2 shows the absolute fitness, relative fitness, and frequencies for these three genotypes. Because we focus on the change in one generation (t to t + 1), frequency designation was simplified. We have an additional parameter h $(0 \leq h \leq 1)$ for the relative fitness for the heterozygote $A_0 A_1$. This is the heterozygote selection factor to show recessive or dominance in terms of selection pattern. If allele A_0 is completely dominant to allele A_1, the phenotype of heterozygote $A_0 A_1$ is expected to have the same relative fitness with that of $A_0 A_0$. In this case, $h = 0$. In the complete recessive case, $h = 1$, and the relative fitness of heterozygote $A_0 A_1$ is identical with that of $A_1 A_1$. When $h = \frac{1}{2}$, the relative fitness of $A_0 A_1$ is the average of those of two homozygotes, and this situation is called "genic" selection, because the effect of selection is proportional to the number of mutant allele A_1. It should be noted that h is not always identical with the dominance effect of genes for phenotypes, for it is possible for two different phenotypes to have the same fitness.

Although we can consider the temporal changes of genotype frequencies, change of mutant allele frequency is more important in the evolutionary process. If we assume that the diploid population in question is randomly mating, frequencies of three genotypes, $A_0 A_0$, $A_0 A_1$, and $A_1 A_1$, can be approximated by using the binomial distribution as p_0^2, $2 p_0 p_1$, and p_1^2, respectively. This simple relation is often called "Hardy–Weinberg ratio," after two persons who independently showed this relationship in 1908 ([19, 20]). Putting these relationships and relative fitness designations to Eqs. 5.4a, b and c,

Table 5.2   A simple model of natural selection for diploid organisms

Genotype	Absolute fitness	Relative fitness[a]	Frequency at	
			t	t + 1
$A_0 A_0$	W_{00}	1	P_{00}	P'_{00}
$A_0 A_1$	W_{01}	$1 + hs$	P_{01}	P'_{01}
$A_1 A_1$	W_{11}	$1 + s$	P_{11}	P'_{11}

[a]$s \geq -1$, $0 \leq h \leq 1$

$$P'_{00} = p_0^2 / \text{Mean\_PW}, \tag{5.6a}$$

$$P'_{01} = 2p_0 p_1 (1 + hs) / \text{Mean\_PW}, \tag{5.6b}$$

$$P'_{11} = p_1^2 (1 + s) / \text{Mean\_PW}, \tag{5.6c}$$

where

$$\text{Mean\_PW} = p_0^2 + 2p_0 p_1 (1 + hs) + p_1^2 (1 + s). \tag{5.7}$$

If we denote p'_0 and p'_1 as frequencies at generation $t+1$ of alleles A_0 and A_1, respectively,

$$p'_0 = P'_{00} + P'_{01} / 2, \tag{5.8a}$$

$$p'_1 = P'_{11} + P'_{01} / 2. \tag{5.8b}$$

Then, noting the relationship, $p_0 + p_1 = 1$,

$$p'_1 = \{p_1^2 (1 + s) + p_0 p_1 (1 + hs)\} / \{p_0^2 + 2p_0 p_1 (1 + hs) + p_1^2 (1 + s)\} \tag{5.9a}$$

$$= p_1 s (1 + (1 - p_1) hs) / \{1 + 2p_1 (1 - p_1) hs + p_1^2 s\}. \tag{5.9b}$$

Finally, $\Delta p_1 (= p'_1 - p_1)$ becomes

$$\Delta p_1 = p_1 (1 - p_1) \{h \quad (2h - 1) p_1\} s / \{1 + 2hs p_1 + s(1 - 2h) p_1^2\}. \tag{5.10}$$

Figure 5.3 shows temporal allele frequency changes from 0.01 to 0.9999 under various sets of h and s values.

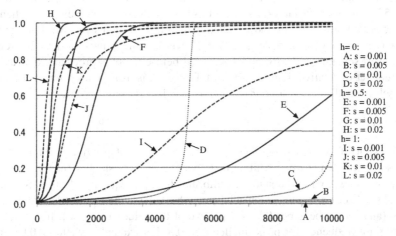

Fig. 5.3 Frequency change of allele A_1 under various h and s values. Horizontal axis: generation. Vertical axis: allele frequency

Table 5.3 Alternative models of natural selection for diploid organism

	Relative fitness		
Genotype	Case 1[a]	Case 2[b]	Case 3[c]
A_0A_0	1	$1-s_0$	1
A_0A_1	$1+s$	1	$1-s$
A_1A_1	$1+2s$	$1-s_1$	1

[a]When genic selection is assumed, $-\frac{1}{2} \leq s$
[b]When heterozygote advantage is assumed, $0 < s_0$, $s_0 \leq 1$
[c]When heterozygote disadvantage is assumed, $0 < s \leq 1$

Under the genic selection, $h = \frac{1}{2}$, and Δp_1 becomes

$$\Delta p_1 = \frac{1}{2}s \cdot p_1(1-p_1)/(1+sp_1). \tag{5.11}$$

This equation is similar with Eq. 5.4. In fact, historically, the relative fitness under the genic selection is set to be 1, $1+s$, and $1+2s$ for three genotypes, A_0A_0, A_0A_1, and A_1A_1, respectively (case 1 of Table 5.3), and Δp_1 becomes identical with Eq. 5.4 in this model. This is expected, for gene effects are additive with no dominance.

5.3.4 Overdominant Selection

It is possible for heterozygotes to be more advantageous than homozygotes. This situation is called "overdominance." A classic example is selection on β globin gene regarding resistance against malaria. Homozygotes of globin S mutant allele causes serious sickle cell anemia (see Chap. 2), while heterozygotes of common allele A and S are more resistant to malaria than AA homozygotes. Because this selection model cannot be represented by the range of h ($0 \leq h \leq 1$) assumed in Sect. 5.3.3, h should be larger than 1 for this case. This is why this selection scheme that heterozygotes are more advantageous than homozygotes is "overdominant." When two homozygotes (A_0A_0, and A_1A_1) have the same relative fitness, the relative fitness scheme (1, $1+hs$, and $1+s$) used before is not appropriate. Instead, a different set of relative fitness, case 2 of Table 5.3, is usually used for overdominant selection. Then in a similar manner as in 5.10, Δp_1 will become

$$\Delta p_1 = p_1(1-p_1)\{s_0 - (s_0 + s_1)p_1\} / \{(1-p_1)^2(1-s_0) + 2p_1(1-p_1) + p_1^2(1-s_1)\}. \tag{5.12}$$

The sign of Δp_1 changes depending on the value of p_1. If it is larger than $s_0/(s_0+s_1)$, Δp_1 is negative, while Δp_1 is positive if p_1 is smaller than $s_0/(s_0+s_1)$. Therefore, $s_0/(s_0+s_1)$ is the stable equilibrium for p_1 (frequency of allele S in the case of sickle cell anemia). In a special case where $s_0 = s_1$, the equilibrium frequency for p_1 (and for p_0) becomes 0.5, irrespective of the value of s_0 (= s_1). If $s_0 < s_1$, the equilibrium frequency for p_1 is smaller than 0.5. For example, when $s_0 = 0.01$ and $s_1 = 0.09$, the equilibrium frequency for p_1 is 0.9. Because frequencies of two alleles (A_0 and A_1) are balanced under the overdominant selection, this is a type of balancing selection (see Sect. 5.4.2).

Let us consider the overdominant selection on β globin with relation to malaria. Allison (1955; [21]) assumed the relative fitness values for AA, AS, and SS genotypes to be 0.80, 1.00, and 0.25, respectively. Then $s_0 = 0.20$ and $s_1 = 0.75$, and the equilibrium S allele frequency becomes 0.21. This equilibrium frequency is slightly higher than that (~0.15) observed in malaria-endemic Africa, suggesting that the real situation was close to the equilibrium.

Genome-wide data now available show that loci under overdominance are quite rare (Bubb et al. 2006; [22]). Notable exceptions are MHC in vertebrates [17], self-incompatibility loci of plants [23], and sex determining locus of honey bee [24]. A complex lineage-specific system is required for all the three cases, and this indicates that the overdominant selection is not a common mechanism for natural selection.

Heterozygotes are advantageous in overdominant selection. When heterozygotes are disadvantageous than homozygotes, what will happen? This is called "underdominance." Biologically, hybrids or introgression of two recently diverged species fits to this situation. A common allele A_1 of species 1 that introgressed to another species can be considered a mutant in species 0 which is fixed with allele A_0. If this locus is responsible for the reproductive barrier, A_0A_1 heterozygotes are expected to produce no or few individuals compared to homozygotes. This situation can be represented as case 3 of Table 5.3. As expected, the equilibrium at frequency of 0.5 is not stable.

5.4 Natural Selection on Populations with Finite Number of Individuals

Any organism has a finite number of individuals. Therefore, to approximate large number of individuals as infinite is clearly oversimplification, and this causes serious miss-prediction in many situations. We therefore need a model of natural selection with a finite population size. Gene genealogy-based models (forward branching process and backward coalescent process) were used for describing the history of selectively neutral alleles in Chap. 4. These can also be used when natural selection exists, but the applicability is limited. As for the allele frequency-based model, diffusion approximation is discussed.

5.4.1 Gene Genealogy Under Natural Selection

When there are heterogeneity in reproductive success among alleles, their relative selection coefficients differ. In this case, the gene genealogical process is no longer independent from mutations. This is a clear contrast from the purely neutral process discussed in Chap. 4. Depending on the effect of mutations, the mutant lineage will have a higher or lower reproduction rate than the wild-type lineage. Fisher [15]

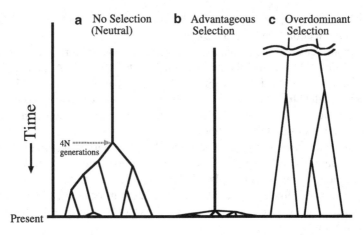

Fig.5.4 Gene genealogies for three types of natural selection. (**a**) Pure neutrality. (**b**) Advantageous mutation. (**c**) Overdominant selection (diploid only)

used the branching process for obtaining the fixation probability of advantageous mutants, while that for disadvantageous and neutral mutants are zero. This is because the fate of a mutant in an infinite population was considered.

While the branching process is producing time-forward gene genealogies, the coalescent theory gives the time-backward gene genealogy for the pure neutral process (see Chap. 4), and the mean time of coalescence is ~2N generations for haploids (Fig. 5.4a). If an advantageous mutation occurs, its offsprings are expected to propagate more quickly than the neutral situation, depending on the degree of reproductive success. The gene genealogy of this advantageous mutation is thus expected to have a shorter coalescent time for the same number of gene copies (Fig. 5.4b). There are many papers on application of the coalescent process to directional selection (e.g., [25, 26]); however, this field seems to need more in-depth studies.

In diploids, heterozygotes may be more advantageous than homozygotes, as we discussed in Sect. 5.3.4. We need multiple allelic lineages for producing heterozygotes, and the gene genealogy under the overdominant selection is expected to have much larger coalescent time than that for the pure neutral evolution. Figure 5.4c schematically shows this pattern. Takahata (1990; [27]) proved that alleles under a strong symmetric balancing selection show an allelic genealogy that is similar to the neutral coalescent process. The only difference between them is the different time scales. The coalescent time estimates based on this theory showed a good agreement with those from simulations given by Takahata and Nei (1985; [28]).

5.4.2 Diffusion Approximation with Natural Selection

We discussed the diffusion process in Chap. 4 by considering the random genetic drift. When natural selection exists as a systematic pressure, we need an additional term to incorporate it to the diffusion equation. The population size, N, and the selection coefficient, s, is assumed to be constant. When the frequency of allele A_1 is p at the initial generation ($t=0$), the conditional probability, $u(p, t)$, that allele A_1 is fixed at generation t can be approximated as

$$\delta u(p, t) / \delta t = \tfrac{1}{2}V[\delta^2 u(p, t) / \delta p^2] + M[\delta u(p, t) / \delta p], \qquad (5.13)$$

where V and M are variance and mean of the change of p per generation, respectively (Kimura, 1962; [29]). This is a Kolmogorov backward equation. The first term of equation (5.13) is for temporal changes of allele frequency due to the random genetic drift, and the second term is for that due to natural selection. Therefore,

$$V = p(1-p)/2N. \qquad (5.14)$$

M depends on the selection scheme, and if we assume a genic selection with selective coefficient s,

$$M = sp(1-p). \qquad (5.15)$$

The boundary conditions are as follows:

$$u(0, t) = 0, \qquad (5.16a)$$
$$u(1, t) = 1. \qquad (5.16b)$$

Then the solution of Eq. 5.13 under these boundary conditions is as follows:

$$u(p, t) = \int_{x=0, p} G(x)dx \, / \int_{x=0, 1} G(x)dx, \qquad (5.17)$$

where

$$G(x) = \exp[-\int (2M / V)dx]. \qquad (5.18)$$

5.4.3   Fixation Probability of a Mutant

We can derive the fixation probability of mutants using Eq. 5.17. Because any new mutant starts from a single copy, the initial allele frequency is 1/N. For simplicity, we consider genic selection. If we consider probability $u(1/N, t)$ when t approaches infinity, the ultimate probability of this mutant to be fixed in the population becomes (Kimura 1957; [30]):

$$\mathrm{Prob}_{\mathrm{fix}}(s, N) = [1 - e^{-s}] / [1 - e^{-Ns}]. \qquad (5.19)$$

When there is no selection, we set selection coefficient s to approach to zero. In this limit, numerator and denominator of 5.14 will approach s and Ns, respectively, and the fixation probability under pure neutrality can be obtained as ratio of s and Ns, namely, 1/N (see Chap. 4).

Fig. 5.5 Comparison of fixation probabilities under various values of selection coefficient s A: $s = -0.02$, B: $s = -0.01$, C: $s = 0$ (pure neutrality), D: $s = +0.01$, E: $s = +0.02$ Horizontal axis: population size. Vertical axis: ratio of fixation probabilities compared to the pure neutral case (C)

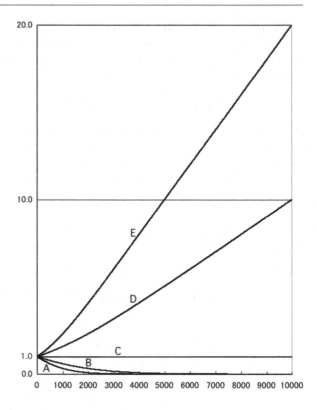

When the extent of selection is weak and when the population size is small, the fate of advantageous and deleterious mutants resemble that of selectively neutral mutants. If $s = -0.001$ and $N = 10,000$, the fixation probability becomes $\sim 4.54 \times 10^{-8}$ or essentially zero. When $s = -0.001$ and $N = 100$, the fixation probability is 0.009 that is only slightly lower than that ($1/N = 0.01$) for selectively neutral mutants. This characteristic is the basis of slightly deleterious hypothesis, as we will see in the next section.

When s is positive, now Darwinian selection can happen. If $s = 0.001$ and $N = 10,000$, the fixation probability becomes ~ 0.001, very similar to the amount of selection coefficient. This is because the denominator of Eq. 5.19 is almost unity when N is large, and the numerator is approximately s when $s \ll 1$. It should be noted that the fixation probability of selectively neutral mutants for $N = 10,000$ is $1/N$ or 0.0001, only 1/10 of that when $s = 0.0001$. However, when $N = 100$, the fixation probability under $s = 0.001$ becomes 0.0105 that is only slightly larger than that (0.01) for selectively neutral mutants. In conclusion, when the value of Ns is much smaller than 1, selectively advantageous or deleterious mutants behave almost like selectively neutral mutants.

Figure 5.5 shows the relationship of the fixation probability and population size under various selection coefficients. The fixation probability ($1/N$) of selectively neutral allele is set to be unity (the straight line C) for comparison with selected alleles. Curves A and B are for selection coefficients −0.02 and −0.01, while those for curves D and E are for selection coefficients +0.01 and +0.02, respectively.

We should be careful for the concept of "fixation." If N is not so large and the nucleotide sequence that corresponds to one locus is short, fixation or monopoly of one allele for all N gene copies of the population may be achieved. However, if N is large and the relatively long nucleotide sequence is considered, mutation (either nucleotide substitution, recombination, or gene conversion) always happens in some nucleotide site of some gene copy, and fixation may never be achieved.

5.4.4 Slightly Deleterious and Nearly Neutral Mutations

We discussed the neutral evolution by dividing effects of mutations into deleterious and selectively neutral ones in Chap. 4. However, the effect of mutations is continuous in terms of selective coefficients. It should also be remembered that the allele frequency change depends on the population size as we saw before. Therefore, depending on the extent of selective coefficient and the population size, the fate of a mutant may be more like selectively neutral or deleterious ones. Ohta (1973; [31]) thus stressed the importance of slightly deleterious mutations in the neutral evolution; see also Ohta (1975; [32]) and Ohta (1987; [33]). Later, Ohta [34, 35] modified her hypothesis as "near-neutral evolution" by incorporating advantageous mutations.

Eyre-Walker and his colleagues studied fates of slightly deleterious mutations [36–38]. They predicted that some slightly deleterious mutations were fixed because the species with the smaller effective population size showed lower constraint and a higher fraction of radical to conservative amino acid substitutions in species with smaller effective population sizes [36]. Balbi et al. (2009; [39]) examined nucleotide sequences of 2,098 orthologous genes of *Escherichia coli* and *Shigella spp.* and concluded that their results are consistent with relaxed selection constraint in *Shigella* due to a reduced effective population size.

Sawyer et al. (2007; [40]) compared nucleotide sequences of 91 protein coding genes for *Drosophila melanogaster* and *D. simulans*. They estimated that ~95 % of nonsynonymous mutations were deleterious, while ~95 % of fixed differences between species are positively selected, though most of them could be regarded as nearly neutral.

5.4.5 Compensatory Mutations

A pair of mutations at different loci (or sites) which are singly deleterious but restore normal fitness in combination may be called compensatory neutral mutations, according to Kimura (1985; [41]). He showed that double mutants occurring in the tightly linked loci can be fixed by random drift, even when the single mutants are definitely deleterious. A molecular example was a pair of nucleotides forming a hydrogen bond in the stem part of ribosomal RNA. Since then, compensatory mutations caught interest to molecular evolutionists. Ohta [42, 43] studied complex effects of compensatory mutations on duplicated genes.

Kulathinal et al. (2007; [44]) examined pathogenic genes in Dipteran genomes. Regardless of the amount of evolutionary divergence from *Drosophila melanogaster,* approximately 10 % of those fixed mutations at these sites carried the same amino acid found in *D. melanogaster.* This finding suggests that compensatory mutations were rapidly fixed in various lineages of Diptera.

Osada and Akashi (2012; [45]) showed that some mutations accumulated in primate mitochondrial protein genes encoded in nuclear DNA could be compensatory adaptive substitutions which prevented fitness decline in mitochondrial protein complexes. They proposed that high mutation rate and small effective population size in primate mitochondrial genomes accelerated these compensatory evolutions.

5.5   Natural Selection at the Genomic Level

It is now possible for us to study the genome-wide pattern of natural selection for analyzing various genome sequences. We first discuss two basic concepts of natural selection at the genomic level, followed by a series of comparative genomics on various aspects of natural selection.

5.5.1   Selective Sweep and Background Selection

When one advantageous mutation occurred on a certain location of the genome, it may spread through the population and may eventually be fixed. If it happens, the surrounding genomic region (the haplotype harboring the advantageous mutation) as a whole will increase its frequency in the population, as long as there is no recombination within that region. This situation is called "hitchhiking" effect, for the surrounding genomic region is assumed to be selectively neutral, and this genomic region is passively driven to increase their frequency, as if hitchhikers can get free ride. Kaplan et al. (1989; [46]) wrote a review on this effect. The hitchhiking length may vary depending on the strength of positive selection and recombination rates. One outcome of the hitchhiking effect is reduction of genetic variation within that region, and this is called "selective sweep" (see Fig. 5.6). However, elimination of a deleterious

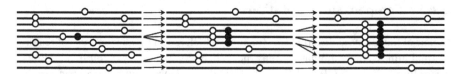

Fig. 5.6 Selective sweep. Horizontal bar, one chromosome; open circle, selectively neutral mutation; full circle, advantageous mutation

mutation also reduces the genetic variation surrounding that mutation, and this is called "background selection" (Charlesworth et al. 1993; [47]). It is not easy to distinguish these two effects (e.g., see Stephan 2010; [48]).

5.5.2 Gain and Loss of Genes

Organisms with small genome size and consequently small gene numbers such as viruses (see Chap. 7) may not be possible to lose even single gene, for almost all genes in such small genomes may be indispensable. For example, the genome of influenza A virus is RNA type, and the nucleotide substitution-type mutation rate is more than ten million times higher than that of mammals (e.g., [49]). Therefore, 10 years for influenza A viruses may be equivalent to 100 million years for mammalian evolution in terms of nucleotide substitution. Even after so many nucleotide substitutions, its genome composition has not changed. In contrast, it is well documented that different orders of mammals have different sets of genes in their genomes due to frequent gene duplications and gene losses including nonfunctionalization (e.g., [50]). Many of them are probably selectively neutral changes, as discussed in Chap. 4; however, positive selection was suggested to be involved in some cases.

The host organism may become advantageous if a gene coding for cell membrane protein become nonfunctional, if that protein is target of the parasite organism to get into the host cell. This logic applies to the deletion mutant of chemokine receptor 5 (CCR5), which is target of HIV, the causative virus for AIDS [51]. Interestingly, this mutant allele frequency was found to be quite high (as high as 14 %) in Northern Europe. Because AIDS is estimated to become infected to human from nonhuman primates in Africa less than 300 years ago, this geographical distribution of the CCR5 deletion mutant is probably caused by some other parasites, such as plague or small pox during the Middle Ages in Europe [51]. Recently, however, this interpretation was questioned [52, 53], for the emergence of this mutation was estimated to be more than 5,000 years ago.

There are three major alleles in the human ABO blood group locus: A1, B, and O. The A and B alleles code for glycosyltransferases which transfer N-acetylgalactosamine and galactose, respectively, to a common precursor called H (for review, see Yamamoto et al. 2012; [54]). The most frequent null O alleles have a point deletion in exon 6, which causes a frameshift, resulting in a truncated protein deprived of any glycosyltransferase activity [54]. Therefore, the human ABO blood group locus may be a rather rare case in which nonfunctional O allele coexists with functional A and B alleles. Although many studies were conducted on the evolutionary mechanism for this locus, such as balancing selection (e.g., [55, 56]), it is still not clear. Recently, Kitano et al. (2012; [57]) discovered that the A allele, possibly once extinct in the human lineage a long time ago, was resurrected by a recombination between B and O alleles less than 300,000 years ago. We may have to reconsider previously proposed evolutionary schemes for the ABO blood group locus based on this finding.

5.5.3 Purifying Selection at Synonymous Sites

The primary role of a protein coding gene is to encode amino acids. Therefore, synonymous sites of codons, which do not change the encoded amino acid, are regarded as evolving neutrally. However, if a certain region of a protein coding gene contains a functional nucleotide element, synonymous sites in the region may have some sort of selective pressure.

One of the well-known factors is the codon bias toward optimum codons. Ikemura (1981; [58]) discovered that optimum codons reflect the composition of genomic tRNA pool. Because optimum codons are advantageous for fast and accurate translation, highly expressed or biologically important genes would have more optimum codons than others [59–61]. Changes from an optimum codon to a nonoptimum codon will be suppressed in such genes. Because optimum codons are similar in closely related species, highly expressed genes tend to show similar codon usage; therefore, synonymous substitution is lowered. In fact, the requirement for translational efficiency or accuracy enhances the optimum codon usage and suppresses nucleotide changes through purifying selection [62–64]. Codon bias toward optimum codons is strong in fast-growing organisms like *Escherichia coli* or *Saccharomyces cerevisiae*, but generally weak in species with slow growth rate and small population size [65, 66].

Another factor is exonic splicing enhancer or silencer, which are splicing signals embedded in exons [67, 68]. Existence of such elements lowers the synonymous substitution [69, 70]. In addition, ultraconserved elements (UCEs), which majorly reside in nonprotein coding regions, sometimes extend to coding regions [71]. In mammals, UCEs are reported to exist near to or overlap with genes associated with nucleotide binding, transcriptional regulation, RNA recognition motif, zinc finger domain, and homeobox domain [71–73]. Hox genes also contain long conserved nucleotide regions other than UCEs outside of the homeobox domain [74].

The existence of such elements would be detected by searching regions of low nucleotide substitution. Suzuki and Saitou (2011; [75]) explored invariant nucleotide sequences in 10,790 orthologous genes of six mammalian species (*Homo sapiens, Macaca mulatta, Mus musculus, Rattus norvegicus, Bos taurus, and Canis familiaris*) and extracted 4,150 sequences whose conservation is significantly stronger than other regions of the gene, and named them significantly conserved coding sequences (SCCSs). SCCSs are observed in 2,273 genes. The genes are mainly involved with development, transcriptional regulation, and the neurons and are expressed in the nervous system and the head and neck organs. No strong influence of conventional factors that affect synonymous substitution was observed in SCCSs. These results imply that SCCSs may have double function as nucleotide element and protein coding sequence and retained in the course of mammalian evolution. Suzuki and Saitou (2013; [76]) expanded their study from mammals to a wide range of eukaryotes (four teleost species, six *Drosophila* species, four *Caenorhabditis* species, three dicot plants, four monocot plants, and four *Saccharomyces* species) and found that genes related with DNA/RNA/nucleotide binding function and protein kinase activity showed significantly high ratio of SCCSs among the genes that contain low synonymous substitution regions in all groups and genes related with ion channels have significantly high number of SCCSs in the mammalian and the teleost groups.

5.5.4 Purifying Selection at Noncoding Regions

Noncoding regions were suspected to be involved in gene regulation from the early days of molecular evolutionary studies [77–79]. Now it is widely accepted that some noncoding regions play important roles in gene regulation (e.g., [80]). Any functionally important nucleotide sequence is expected to be evolutionarily conserved under purifying selection [2, 81]. In fact, 5 % of the human genome is conserved [82], a considerably higher proportion than that (2 %) for the protein coding regions [83]. Now many highly conserved noncoding sequences (HCNSs) among vertebrates have been identified [71, 84, 85], and some of them were reported to function as distal enhancer for neighboring genes [86, 87].

One interesting problem is to identify HCNSs found only in one lineage comprised of very closely related species. The primates are one of the lineages of closely related species compared to the mammals and vertebrates. Sequence comparisons only among primates are likely to capture functional components of the lineage due to shared biological processes [88]. However, this strategy of comparing genomes among closely related species has often been applied only to the very limited regions. Furthermore, the goal of this method was to identify all sequences conserved among species at various levels of divergence, such as vertebrates, mammals, and primates, but not primate-specific HCNSs. In contrast to the lineage-specific phenotypic changes, the HCNSs which are conserved only in one particular lineage have not been well studied.

Takahashi and Saitou (2012; [89]) identified HCNSs that were conserved in a particular lineage (either in primates or in rodents) and compared characteristics of the lineage-specific HCNSs with those conserved among mammals and vertebrates. They used human and marmoset genomes for detecting primate-specific HCNSs, while mouse and rat genomes were used for detecting rodent-specific HCNSs. Derived allele frequency analysis of primate-specific HCNSs showed that these HCNSs were under purifying selection, indicating that they may harbor important functions. They selected the top 1,000 largest HCNSs and compared the lineage-specific HCNS-flanking genes (LHF genes) with UCE (ultraconserved element)-flanking genes. Interestingly, the majority of LHF genes were different from UCE-flanking genes. This lineage-specific set of LHF genes was more enriched in protein binding function. Conversely, the number of LHF genes which were also shared by UCEs was small but significantly larger than random expectation, and many of these genes were involved in anatomical development as transcriptional regulators, suggesting that certain groups of genes preferentially recruit new HCNSs in addition to old HCNSs which are conserved among vertebrates. This group of LHF genes might be involved in the various levels of lineage-specific evolution among vertebrates, mammals, primates, and rodents. If so, the emergence of

Table 5.4 Genes found to be under positive selection in primate evolution

Gene	Reference
ABO	Saitou and Yamamoto [55]
BRCA1	Huttley et al. [129]
FOXP2	Enard et al. [91], Zhang et al. [92]
IgA hinge region	Sumiyama et al. [130]
LCR16a	Johnson et al. [131]
Lysozyme	Messier and Stewart [132]
Protamine 1 & 2	Wyckoff et al. [133]
Rh50	Kitano et al. [134]
Rh blood group	Kitano and Saitou [135]
RNASE2 & 3	Zhang et al. [136]
TNP2	Wyckoff et al. [133]

HCNSs in and around these two groups of LHF genes developed lineage-specific characteristics. These results of Takahashi and Saitou [89] provide new insight into lineage-specific evolution through interactions between HCNSs and their LHF genes. Recently, Babarinde and Saitou (2013; [138]) extracted conserved noncoding sequences (CNSs) specific to primates, rodents, carnivores, and cetartiodactyls. They used more relaxed thresholds than Takahashi and Saitou (2012; [89]), and found that abundance of CNSs varied among lineages with primates and rodents having highest and lowest number of CNSs respectively, while carnivores and cetartiodactyls had intermediate values. These CNSs cover 1.3~5.5 % of the mammalian genomes and have signatures of selective constraints which are stronger in more ancestral than the recent ones. With CNSs shown to cluster around genes involved in nervous systems and the higher number of primate CNSs, their result suggests that CNSs may be involved in the higher complexity of the primate nervous system.

5.5.5   Positive Selection for Ape and Human Genes

We now move to discuss positive selection. Because the author is familiar with studies on primates, let us first consider positive selection in primates. Hughes and Nei [17, 90] showed that both MHC class I and class II genes experienced positive selection, probably of the overdominant type, through the comparison of synonymous and nonsynonymous nucleotide substitutions. This d_N/d_S test is quite powerful to detect positive selection based on the neutral theory [2], and many studies were conducted to detect such selection. However, one drawback of this test is that it can detect only a limited type of genes under positive selection, and to obtain a statistically significant result, we need to have many substitutions. Therefore, a protein gene must have many variable amino acids, a precondition that may be satisfied in protein coding genes that are involved in the immune system, such as MHC. If a single amino acid change were responsible in enhancing fitness, however, it would be difficult to detect it through the Dn/Ds test.

Table 5.4 lists the primate genes found to be under the influence of positive selection in relatively recent studies. The majority of protein coding genes have lower nonsynonymous substitutions than synonymous ones (see Chap. 4), and only a small proportion is under positive selection. Although most of the genes listed in Table 5.4 were found to have experienced positive selection thorough the Dn/Ds test, the FOXP2 gene studied by Enard et al. (2002; [91]) and Zhang et al. (2002; [92]) did not show significantly higher Dn than Ds.

The FOXP2 gene was initially found in mouse [93], and its highly homologous human ortholog was shown to be responsible for hereditary orofacial dyspraxia associated with dysphasia [94]. Interestingly, two nonsynonymous nucleotide substitutions at codon 304 (ACC:Thr → AAC:Asn) and at codon 326 (AAT:Asn → AGT:Ser) occurred after the human and chimpanzee divergence, and the DNA polymorphism study of modern human populations suggested that the FOXP2 gene region experienced a selective sweep after mutations occurred within modern humans [91, 92]. However, one of the two amino acid substitutions which occurred in the human lineage after the divergence of the human–chimpanzee common ancestor also occurred in the carnivores [92]. It is therefore not yet clear if these amino acid changes were truly responsible for the emergence of language. There was a report that FOXP2 of Neanderthals also had these two amino acid substitutions (Krause et al. 2007; [95]). If this finding is true, two nonsynonymous substitutions occurred in the common ancestor of modern humans and Neanderthals. Coop et al. (2008; [96]) suggested two alternative scenarios for this: low rates of gene flow between modern humans and Neanderthals and contamination of modern human DNA in the putative Neanderthal genome. Mouse FOXP2 gene knockout experiment by Fujita et al. (2008; [97]) and human sequence knock-in mice experiments by Enard et al. (2009; [98]) both showed some phenotypic differences in vocalization. Konopka et al. (2009; [99]) reported that the FoxP2 gene is responsible for the human-specific transcriptional regulation of the central nervous system.

Pollard et al. (2006; [100], 2006; [101]) found a series of short DNA sequences which are highly conserved in vertebrates but show accelerated evolution only in humans and named them HARs (human accelerated regions). Prabhakar et al. (2006; [102]) and Bird et al. (2007; [103]) also conducted similar genome-wide studies. Prabhakar et al. (2008; [104]) found one such sequence from the human genome HACNS1 which includes 119-bp HAR2 and showed it to act as a limb bud enhancer with enhanced limb enhancer activity specifically in human. This change was caused by 13 human-specific substitutions within that region, and they interpreted that accumulation of these positively selected substitutions created multiple novel transcription factor binding sites (gain of function) and that the deposition of those facilitated the human-specific enhanced activity [104].

However, a GC-biased gene conversion may be an alternative explanation for fixation of such mutations causing loss of function in a repressor element within HACNS1 [105–107]. GC-biased gene conversion is a consequence of DNA double-strand break repair between homologous chromosomal regions, and the alleles from one chromosome are converted to the other with a bias of A or T to G or C [108]. Neutral or even deleterious alleles could be fixed by GC-biased gene conversions [109]. It is possible that the 13 human-specific substitutions were caused by a GC-biased gene conversion and resulted in a disruption of repressor function of the 81 bp region (loss of function), which may eventually enhance the activity of human HACNS1.

To evaluate the function of HACNS1, Sumiyama and Saitou (2011; [110]) performed transgenic mouse assay by using the HACNS1 construct lacking the 13 human-specific substitutions. The deleted construct showed similar enhancer activity to the intact human HACNS1. This result suggests that the function of HACNS1 is not an activating enhancer but rather a disrupted repressor. If so, loss of function in the HACNS1, possibly via a GC-biased gene conversion, not via positive selection, played an important role in human-specific evolution.

Kryukov et al. (2005; [111]) reported that the selective pressure affecting the evolution of regulatory elements in the hominid lineage was significantly relaxed compared with that of the rodent lineage. Keightley et al. (2005; [112]) suggested that regulatory elements in hominids may be diverging at a neutral evolutionary rate. All these studies discussed in this section revealed the difficulty in detecting evidence for positive selection in one lineage. We therefore should be careful for any study which jumped to a conclusion of accelerated evolution without carefully examining an alternative neutral evolution scenario.

5.5.6   Detection of Positive Selection Through Genome-Wide Searches

Endo et al. (1996; [113]) conducted a pioneering study of a large-scale analysis of synonymous and nonsynonymous substitutions for mammals before the genome sequence era. Out of 3,595 homologous nucleotide sequences, they found that only 17 (0.5 %) genes showed higher values for nonsynonymous substitutions than synonymous ones. Those genes are candidates on which positive selection may operate, and nine genes were the surface antigens of parasites or viruses.

After rice genome sequences were determined (see Chap. 8), a series of studies were conducted to discover key genes responsible for rice domestication. One example is loss of seed shattering. Konishi et al. (2006; [114]) discovered that the qSH1 gene, a quantitative trait locus of seed shattering in rice, encodes a BEL1-type homeobox gene, and they demonstrated that an SNP in the regulatory region of the qSH1 gene caused loss of seed shattering owing to the absence of abscission layer formation. Haplotype analysis and association analysis in various rice collections suggested that it was a target of artificial selection during rice domestication. Another example on rice domestication gene is qsW5 (QTL for seed width on chromosome 5). This gene is involved in the determination of grain width in rice. Shomura et al. (2008; [115]) found that a deletion in qSW5 resulted in a significant increase in grain size. This trait might have been selected by ancient humans to increase rice production.

Xia et al. (2009; [116]) constructed a silkworm genetic variation map from 40 domesticated and wild silkworm genome sequencing and identified ~16 million SNPs, indels, and structural variations. Through a series of statistical analyses on these variations, they found that the domesticated silkworms were genetically differentiated from the wild ones while maintaining a large genetic variability. Interestingly, the third eigenvector separated the two high silk-producing Japanese domesticated strains from the other domesticated strains in their PCA analysis.

It may be interesting to analyze SNPs responsible for this separation, since these could be artificially selected by ancient people on Japanese Archipelago after introducing silkworms from China.

It should be noted that artificial selection is very effective on particular set of genes responsible for one phenotype which was picked up by people. In this case, a small number of genes may have very high selection coefficients. This was probably why Darwin (1859; [1]) put "Variation Under Domestication" as Chap. 1 of his book, "Origin of Species." In natural conditions, however, concerted effects of many genes are related to reproduction and death of one organism, which are directly connected to natural selection. Therefore, it is rare for one mutant to be highly positively selected. See also Nei (2013) [138] on clear difference between artificial selection and natural selection.

Bazykin et al. (2004; [117]) compared 9,390 orthologous mouse and rat genes and found that 28,196 codons showed two-nucleotide difference out of ~3 million codons. By comparing with out-group orthologous sequences, they estimated locations of two substitutions in a parsimonious way, either on the mouse lineage or on the rat lineage. If the two substitutions were independent and equally common in rat and mouse lineages, expected proportions of their locations would be 25 %, 50 %, and 25 % for mouse–mouse, mouse–rat, and rat–rat locations, respectively. Although these proportions were approximately achieved when one or both substitutions at a codon were synonymous, mouse–mouse and rat–rat patterns were significantly much higher than expected when both substitutions were nonsynonymous. They argued that this pattern cannot be explained by correlated mutation or episodes of relaxed negative selection, but instead indicates that positive selection acts at many sites of rapid, successive amino acid replacement.

Let us now discuss about positive selection on modern humans. Adaptation to a high-altitude environment has long been an interesting subject in human genetics. Recently, three independent genome-wide studies [118–120] found that the EPAS1 gene showed the highest differentiation between Han Chinese and Tibetans who are believed to have adapted to high altitude. This gene codes a transcription factor HIF-2α involved in the induction of genes regulated by oxygen. The EGLN1 gene, which encodes HIF-prolyl hydroxylase 2, was the second-best differentiated gene. EGLN1 catalyzes the posttranslational formation of 4-hydroxyproline in HIF-α proteins.

Moreno-Estrada et al. (2009; [121]) compared ~11,000 human genes with their orthologs in chimpanzee, mouse, rat, and dog and found 11 genes as showing the signatures of positive selection on the human lineage through a branch-site likelihood method [122]. These genes were then analyzed for signatures of recent positive selection using SNP data in modern humans. One SNP every 5–10 kb inside each candidate gene and up to around 30 kb in both upstream and downstream flanking regions plus additional SNPs around 200 kb in both flanking regions were selected, and a total of 223 SNPs were typed for 39 worldwide populations from the HGDP–CEPH diversity panel. They also analyzed 4,814 SNP data distributed along 2 Mb centered on each gene from the HGDP–CEPH panel (Li et al. 2008; [123]). Through examination of allele frequency spectrum, population differentiation, and the maintenance of long unbroken haplotypes, they found signals of recent adaptive phenomena in only one gene region. The signal of recent positive selection may come from a neighboring gene CD5, which codes a transmembrane receptor expressed in the T-cell surface. This careful study suggests that most of positively selected genes in modern humans are involved in the immune system. It is not surprising, for our ancestors have suffered many sorts of infectious diseases, and the human immune system-related genes confronted the battle against bacteria, virus, or parasitic eukaryotes. When we discuss positive selection on modern humans, we should consider natural selection on the interaction with other organisms, before attempting to apply other types of natural selection such as sexual selection.

If we look at studies on *Drosophila*, various reports estimated that a high proportion of protein coding gene fixed mutations experienced positive selection. For example, Bierne and Eyre-Walker (2004; [124]) estimated that $5 \sim 45$ % of amino acid substitutions were driven by positive selection in the divergence between *D. simulans* and *D. yakuba*, and Shapiro et al. (2007; [125]) estimated that ~30 % of the amino acid substitutions between *D. melanogaster* and its close relatives were adaptive. Andolfatto (2007; [126]) also estimated that ~50 % of divergent amino acids between *D. melanogaster* and *D. simulans* were driven to fixation by positive selection. However, we should be careful for these estimates, for they are often based on simplified statistical model. Problems of multiple alignments of *Drosophila* genome sequence comparison was also pointed out by Markova-Raina and Petrov (2011; [127]). They reported that the rate of false positives is unacceptably high on inferences of positive selection using site-specific models of molecular evolution. In any case, any statistical inference is not a conclusive evidence but a mere possibility. We should eventually demonstrate biological effect of a mutation for demonstrating positive selection.

References

1. Darwin, C. (1859). *On the origin of species*. London: John Murray.
2. Kimura, M. (1983). *The neutral theory of molecular evolution*. Cambridge: Cambridge University Press.
3. Lewontin, R. (1974). *The genetic basis of evolutionary change*. New York: Columbia University Press.
4. Sato, J. J., Wolsan, M., Minami, S., Hosoda, T., Shinaga, S. H., Hiyama, K., Yamaguchi, Y., & Suzuki, H. (2009). Deciphering and dating the red panda's ancestry and early adaptive radiation of Musteloidea. *Molecular Phylogenetics and Evolution, 53*, 907–922.
5. Nishihara, H., Maruyama, S., & Okada, N. (2009). Retroposon analysis and recent geological data suggest near-simultaneous divergence of the three superorders of mammals. *Proceedings of the National Academy of Sciences of the United States of America, 106*, 5235–5240.
6. Wloch, D. M., Szafraniec, K., Borts, R. H., & Korona, R. (2001). Direct estimate of the mutation rate and the distribution of fitness effects in the yeast *Saccharomyces cerevisiae*. *Genetics, 159*, 441–452.
7. Joseph, S. B., & Wall, D. W. (2004). Spontaneous mutations in diploid *Saccharomyces cerevisiae*: More beneficial than expected. *Genetics, 168*, 1817–1825.
8. Perfeito, L., Fernandes, L., Mota, C., & Gordo, I. (2007). Adaptive mutations in bacteria: High rate and small effects. *Science, 317*, 813–815.

9. Eyre-Walker, A., & Keightley, P. D. (2007). The distribution of fitness effects of new mutations. *Nature Reviews Genetics, 8*, 610–681.
10. Fox, A. L. (1932). The relationship between chemical constitution and taste. *Proceedings of the National Academy of Sciences of the United States of America, 18*, 115–120.
11. Fisher, R. A., Ford, E. B., & Huxley, J. (1939). Taste-testing the anthropoid apes. *Nature, 144*, 750.
12. Kim, U., et al. (2003). Positional cloning of the human quantitative trait locus underlying taste sensitivity to phenylthiocarbamide. *Science, 299*, 1221–1225.
13. Wooding, S., et al. (2006). Independent evolution of bitter-taste sensitivity in humans and chimpanzees. *Nature, 440*, 930–934.
14. Wooding, S. (2006). Phenylthiocarbamide: A 75-year adventure in genetics and natural selection. *Genetics, 172*, 2015–2023.
15. Fisher, R. A. (1930). *The genetical theory of natural selection.* Oxford: Oxford University Press.
16. Ina, Y., & Gojobori, T. (1994). Statistical analysis of nucleotide sequences of the hemagglutinin gene of human influenza A viruses. *Proceedings of the National Academy of Sciences of the United States of America, 91*, 8388–8392.
17. Hughes, A., & Nei, M. (1988). Pattern of nucleotide substitution at major histocompatibility complex class I loci reveals overdominant selection. *Nature, 335*, 167–170.
18. Sumiyama, K., Ueda, S., & Saitou, N. (2002). Adaptive evolution of the IgA hinge region in primates. *Molecular Biology and Evolution, 19*, 1093–1099.
19. Hardy, G. H. (1908). Mendelian proportions in a mixed population. *Science, 28*, 49–50.
20. Weinberg, W. (1908). Uber den Nachweis der Verbung beim Menschen. *Jahreschefte des Vereins fur Vaterlandische Naturkunde in Wurttemburg, 64*, 368–382.
21. Allison, A. C. (1955). Aspects of polymorphism in Man. *Cold Spring Harbor Symposia on Quantitative Biology, 20*, 239–251.
22. Bubb, K. L., Bovee, D., Buckley, D., Haugen, E., Kibukawa, M., Paddock, M., Palmieri, A., Subramanian, S., Zhou, Y., Kaul, R., Green, P., & Olson, M. (2006). Scan of human genome reveals no new loci under ancient balancing selection. *Genetics, 173*, 2165–2177.
23. Takebayashi, N., Brewer, P. B., Newbigin, E., & Uyenoyama, M. K. (2003). Patterns of variation within self-incompatibility loci. *Molecular Biology and Evolution, 20*, 1778–1794.
24. Cho, S., Huang, Z. Y., Green, D. R., Smith, D. R., & Zhang, J. (2006). Evolution of the complementary sex-determination gene of honey bees: Balancing selection and trans-species polymorphisms. *Genome Research, 16*, 1366–1375.
25. Hudson, R. R., & Kaplan, N. L. (1988). The coalescent process in models with selection and recombination. *Genetics, 120*, 831–840.
26. Teshima, K. M., & Innan, H. (2012). The coalescent with selection on copy number variants. *Genetics, 190*, 1077–1086.
27. Takahata, N. (1990). A simple genealogical structure of strongly balanced allelic lines and trans-species evolution of polymorphism. *Proceedings of the National Academy of Sciences of the United States of America, 87*, 2419–2423.
28. Takahata, N., & Nei, M. (1985). Gene genealogy and variance of interpopulational nucleotide differences. *Genetics, 110*, 325–344.
29. Kimura, M. (1962). On the probability of fixation of mutant genes in a population. *Genetics, 47*, 713–719.
30. Kimura, M. (1957). Some problems of stochastic processes in genetics. *Annals of Mathematical Statistics, 28*, 882–901.
31. Ohta, T. (1973). Slightly deleterious mutant substitutions in evolution. *Nature, 246*, 96–98.
32. Ohta, T. (1976). Role of very slightly deleterious mutations in molecular evolution and polymorphism. *Theoretical Population Biology, 10*, 254–275.
33. Ohta, T. (1987). Very slightly deleterious mutations and the molecular clock. *Journal of Molecular Evolution, 26*, 1–6.
34. Ohta, T. (1992). The nearly neutral theory of molecular evolution. *Annual Review of Ecology and Systematics, 23*, 263–286.

35. Ohta, T. (2002). Near-neutrality in evolution of genes and gene regulation. *Proceedings of the National Academy of Sciences of the United States of America, 99*, 16134–16137.
36. Eyre-Walker, A., Keightley, P. D., Smith, N. G., & Gaffney, D. (2002). Quantifying the slightly deleterious mutation model of molecular evolution. *Molecular Biology and Evolution, 19*, 2142–2149.
37. Charlesworth, J., & Eyre-Walker, A. (2008). The McDonald-Kreitman test and slightly deleterious mutations. *Molecular Biology and Evolution, 25*, 1007–1015.
38. Eyre-Walker, A., & Keightley, P. D. (2009). Estimating the rate of adaptive molecular evolution in the presence of slightly deleterious mutations and population size change. *Molecular Biology and Evolution, 26*, 2097–2108.
39. Balbi, K. J., RochaE, P. C., & Feil, E. J. (2009). The temporal dynamics of slightly deleterious mutations in *Escherichia coli* and *Shigella spp. Molecular Biology and Evolution, 26*, 345–355.
40. Sawyer, S. A., Parsch, J., Zhang, Z., & Hartl, D. L. (2007). Prevalence of positive selection among nearly neutral amino acid replacements in Drosophila. *Proceedings of the National Academy of Sciences of the United States of America, 104*, 6504–6510.
41. Kimura, M. (1985). The role of compensatory neutral mutations in molecular evolution. *Journal of Genetics, 64*, 7–19.
42. Ohta, T. (1988). Evolution by gene duplication and compensatory advantageous mutations. *Genetics, 120*, 841–847.
43. Ohta, T. (1989). Time for spreading of compensatory mutations under gene duplication. *Genetics, 123*, 579–584.
44. Kulathinal, R. J., Bettencourt, B. R., & Hartl, D. L. (2004). Compensated deleterious mutations in insect genomes. *Science, 306*, 1553–1554.
45. Osada, N., & Akashi, H. (2012). Mitochondrial-nuclear interactions and accelerated compensatory evolution: Evidence from the primate cytochrome C oxidase complex. *Molecular Biology and Evolution, 29*, 337–346.
46. Kaplan, N. L., Hudson, R. R., & Langley, C. H. (1989). The "hitchhiking effect" revisited. *Genetics, 123*, 887–899.
47. Charlesworth, B., Morgan, M. T., & Charlesworth, D. (1993). The effect of deleterious mutations on neutral molecular variation. *Genetics, 134*, 1289–1303.
48. Stephan, W. (2010). Genetic hitchhiking versus background selection: The controversy and its implications. *Philosophical Transactions of the Royal Society B, 365*, 1245–1253.
49. Hanada, K., Suzuki, Y., & Gojobori, T. (2004). A large variation in the rates of synonymous substitution for RNA viruses and its relationship to a diversity of viral infection and transmission modes. *Molecular Biology and Evolution, 21*, 1074–1080.
50. Gregory, T. R. (Ed.). (2005). *The evolution of the genome.* Burlington: Elsevier Academic.
51. Stephens, J. C., et al. (1998). Dating the origin of the *CCR5-D32* AIDS-resistance allele by the coalescence of haplotypes. *American Journal of Human Genetics, 62*, 1507–1515.
52. Sabeti, P. S., et al. (2005). The case for selection at CCR5-Δ32. *PLoS Biology, 3*, e378.
53. Hedrick, P. W., & Verrelli, B. C. (2006). 'Ground truth' for selection on *CCR5-Δ32. Trends in Genetics, 22*, 293–296.
54. Yamamoto, F., Cid, E., Yamamoto, M., & Blancher, A. (2012). ABO research in the modern era of genomics. *Transfusion Medicine Reviews, 26*, 103–118.
55. Saitou, N., & Yamamoto, F. (1997). Evolution of primate ABO blood group genes and their homologous genes. *Molecular Biology and Evolution, 14*, 399–411.
56. Calafell, F., Roubinet, F., Ramírez-Soriano, A., Saitou, N., Bertranpetit, J., & Blancher, A. (2008). Evolutionary dynamics of the human ABO gene. *Human Genetics, 124*, 123–135.
57. Kitano, T., Blancher, A., & Saitou, N. (2012). The functional A allele was resurrected via recombination in the human ABO blood group gene. *Molecular Biology and Evolution, 29*, 1791–1796.
58. Ikemura, T. (1981). Correlation between the abundance of Escherichia coli transfer RNAs and the occurrence of the respective codons in its protein genes: A proposal for a synonymous codon choice that is optimal for the *E. coli* translational system. *Journal of Molecular Biology, 151*, 389–409.

59. Akashi, H. (1994). Synonymous codon usage in *Drosophila melanogaster*: Natural selection and translational accuracy. *Genetics, 136*, 927–935.
60. Kurland, C. G. (1991). Codon bias and gene expression. *FEBS Letters, 285*, 165–169.
61. Hershberg, R., & Petrov, D. A. (2008). Selection on codon bias. *Annual Review of Genetics, 42*, 287–299.
62. Ikemura, T. (1985). Codon usage and tRNA content in unicellular and multicellular organisms. *Molecular Biology and Evolution, 2*, 13–34.
63. Sharp, P. M., & Li, W. H. (1987). The rate of synonymous substitution in enterobacterial genes is inversely related to codon usage bias. *Molecular Biology and Evolution, 4*, 222–230.
64. Akashi, H. (2003). Translational selection and yeast proteome evolution. *Genetics, 164*, 1291–1303.
65. Eyre-Walker, A. C. (1991). An analysis of codon usage in mammals: Selection or mutation bias? *Journal of Molecular Evolution, 33*, 442–449.
66. Sharp, P. M., Averof, M., Lloyd, A. T., Matassi, G., & Peden, J. F. (1995). DNA sequence evolution: The sounds of silence. *Philosophical Transactions of the Royal Society of London. Series B, Biological Sciences, 349*, 241–247.
67. Reed, R. (1996). Initial splice-site recognition and pairing during pre-mRNA splicing. *Current Opinion in Genetics and Development, 6*, 215–220.
68. Blencowe, B. J. (2000). Exonic splicing enhancers: Mechanism of action, diversity and role in human genetic diseases. *Trends in Biochemical Sciences, 25*, 106–110.
69. Takahashi, A. (2009). Effect of exonic splicing regulation on synonymous codon usage in alternatively spliced exons of Dscam. *BMC Evolutionary Biology, 9*, 214.
70. Parmley, J. L., & Hurst, L. D. (2007). Exonic splicing regulatory elements skew synonymous codon usage near intron-exon boundaries in mammals. *Molecular Biology and Evolution, 24*, 1600–1603.
71. Bejerano, G., et al. (2004). Ultraconserved elements in the human genome. *Science, 304*, 1321–1325.
72. Schattner, P., & Diekhans, M. (2006). Regions of extreme synonymous codon selection in mammalian genes. *Nucleic Acids Research, 34*, 1700–1710.
73. Lareau, L. F., Inada, M., Green, R. E., Wengrod, J. C., & Brenner, S. E. (2007). Unproductive splicing of SR genes associated with highly conserved and ultraconserved DNA elements. *Nature, 446*, 926–929.
74. Lin, Z., Ma, H., & Nei, M. (2008). Ultraconserved coding regions outside the homeobox of mammalian Hox genes. *BMC Evolutionary Biology, 8*, 260.
75. Suzuki, R., & Saitou, N. (2011). Exploration for functional nucleotide sequence candidates within coding regions of mammalian genes. *DNA Research, 18*, 177–187.
76. Suzuki, R., & Saitou, N. (2013). Highly conserved nucleotide sequences within protein coding genes of eukaryotes. unpublished.
77. Zuckerkandl, E., & Pauling, L. (1965). Evolutionary divergence and convergence in proteins. In V. Bryson & H. J. Vogel (Eds.), *Evolving genes and proteins* (pp. 97–166). New York: Academic.
78. Britten, R. J., & Davidson, E. H. (1971). Repetitive and non-repetitive DNA sequences and a speculation on the origins of evolutionary novelty. *Quarterly Review of Biology, 46*, 111–138.
79. King, M. C., & Wilson, A. C. (1975). Evolution at two levels in humans and chimpanzees. *Science, 188*, 107–116.
80. Carroll, S. B. (2005). Evolution at two levels: On genes and form. *PLoS Biology, 3*, e245.
81. Nei, M. (1987). *Molecular evolutionary genetics*. New York: Columbia University Press.
82. Mouse Genome Sequencing Consortium. (2002). Initial sequencing and comparative analysis of the mouse genome. *Nature, 420*, 520–562.
83. International Human Genome Sequencing Consortium. (2004). Finishing the euchromatic sequence of the human genome. *Nature, 431*, 931–945.
84. Ahituv, N., Rubin, E. M., & Nobrega, M. A. (2004). Exploiting human–fish genome comparisons for deciphering gene regulation. *Human Molecular Genetics, 13*, R261–R266.

85. Siepel, A., et al. (2005). Evolutionarily conserved elements in vertebrate, insect, worm, and yeast genomes. *Genome Research, 15*, 1034–1050.
86. Pennacchio, L. A., et al. (2006). In vivo enhancer analysis of human conserved non-coding sequences. *Nature, 444*, 499–502.
87. Woolfe, A., et al. (2005). Highly conserved non-coding sequences are associated with vertebrate development. *PLoS Biology, 3*, e7.
88. Boffelli, D., et al. (2003). Phylogenetic shadowing of primate sequences to find functional regions of the human genome. *Science, 299*, 1391–1394.
89. Takahashi, M., & Saitou, N. (2012). Identification and characterization of lineage-specific highly conserved noncoding sequences in mammalian genomes. *Genome Biology and Evolution, 4*, 641–657.
90. Hughes, A. L., & Nei, M. (1989). Nucleotide substitution at major histocompatibility complex class II loci: Evidence for overdominant selection. *Proceedings of the National Academy of Sciences of the United States of America, 86*, 958–962.
91. Enard, W., Przeworski, M., Fisher, S. E., Lai, C. S., Wiebe, V., Kitano, T., Monaco, A. P., & Paabo, S. (2002). Molecular evolution of FOXP2, a gene involved in speech and language. *Nature, 418*, 869–872.
92. Zhang, J., Webb, D. M., & Podlaha, O. (2002). Accelerated protein evolution and origins of human-specific features. Foxp2 as an example. *Genetics, 162*, 1825–1835.
93. Shu, W., Yang, H., Zhang, L., Lu, M. M., & Morrisey, E. E. (2001). Characterization of a new subfamily of winged-helix/forkhead (Fox) genes that are expressed in the lung and act as transcriptional repressors. *Journal of Biological Chemistry, 276*, 27488–27497.
94. Lai, C. S. L., Fisher, S. E., Hurst, J. A., Vargha-Khadem, F., & Monaco, A. P. (2001). A forkhead-domain gene is mutated in a severe speech and language disorder. *Nature, 413*, 519–523.
95. Krause, J., et al. (2007). The derived FOXP2 variant of modern humans was shared with Neanderthals. *Current Biology, 17*, 1908–1912.
96. Coop, G., Bullaughey, K., Luca, F., & Przeworski, M. (2008). The timing of selection at the human FOXP2 gene. *Molecular Biology and Evolution, 25*, 1257–1259.
97. Fujita, E., Tanabe, Y., Shiota, A., Ueda, M., Suwa, K., Momoi, M. Y., & Momoi, T. (2008). Ultrasonic vocalization impairment of Foxp2 (R552H) knockin mice related to speech-language disorder and abnormality of Purkinje cells. *Proceedings of the National Academy of Sciences of the United States of America, 105*, 3117–3112.
98. Enard, W., et al. (2009). A humanized version of Foxp2 affects cortico-basal ganglia circuits in mice. *Cell, 137*, 961–971.
99. Konopka, G., et al. (2009). Human-specific transcriptional regulation of CNS development genes by FOXP2. *Nature, 462*, 213–217.
100. Pollard, K. S., et al. (2006). An RNA gene expressed during cortical development evolved rapidly in humans. *Nature, 433*, 167–172.
101. Pollard, K. S., et al. (2006). Forces shaping the fastest evolving regions in the human genome. *PLoS Genetics, 2*, e168.
102. Prabhakar, A., Noonan, J. P., Pääbo, S., & Rubin, E. M. (2006). Accelerated evolution of conserved noncoding sequences in humans. *Science, 314*, 786.
103. Bird, C. P., et al. (2007). Fast-evolving noncoding sequences in the human genome. *Genome Biology, 8*, R118.
104. Prabhakar, S., et al. (2008). Human-specific gain of function in a developmental enhancer. *Science, 321*, 1346–1350.
105. Galtier, N., & Duret, L. (2007). Adaptation or biased gene conversion? Extending the null hypothesis of molecular evolution. *Trends in Genetics, 23*, 273–277.
106. Galtier, N., Duret, L., Glemin, S., & Ranwez, V. (2009). GC-biased gene conversion promotes the fixation of deleterious amino acid changes in primates. *Trends in Genetics, 25*, 1–5.
107. Katzman, S., Kern, A. D., Pollard, K. S., Salama, S. R., & Haussler, D. (2010). GC-biased evolution near human accelerated regions. *PLoS Genetics, 6*, e1000960.

108. Strathern, J. N., Shafer, B. K., & McGill, C. B. (1995). DNA synthesis errors associated with double-strand-break repair. *Genetics, 140*, 965–972.

109. Galtier, N., Piganeau, G., Mouchiroud, D., & Duret, L. (2001). GC-content evolution in mammalian genomes: The biased gene conversion hypothesis. *Genetics, 159*, 907–911.

110. Sumiyama, K., & Saitou, N. (2011). Loss-of–function mutation in a repressor module of human-specifically activated enhancer HACNS1. *Molecular Biology and Evolution, 28*, 3005–3007.

111. Kryukov, G. V., Schmidt, S., & Sunyaev, S. (2005). Small fitness effect of mutations in highly conserved non-coding regions. *Human Molecular Genetics, 14*, 2221–2229.

112. Keightley, P. D., Lercher, M. J., & Eyre-Walker, A. (2005). Evidence for widespread degradation of gene control regions in hominid genomes. *PLoS Biology, 3*, e42.

113. Endo, T., Ikeo, K., & Gojobori, T. (1996). Large-scale search for genes on which positive selection may operate. *Molecular Biology and Evolution, 13*, 685–690.

114. Konishi, S., Izawa, T., Lin, S. Y., Ebana, K., Fukuta, Y., Sasaki, T., & Yano, M. (2006). An SNP caused loss of seed shattering during rice domestication. *Science, 312*, 1392–1396.

115. Shomura, A., Izawa, T., Ebana, K., Ebitani, T., Kanegae, H., Konishi, S., & Yano, M. (2008). Deletion in a gene associated with grain size increased yields during rice domestication. *Nature Genetics, 40*, 1023–1028.

116. Xia, Q., et al. (2009). Complete resequencing of 40 genomes reveals domestication events and genes in silkworm (Bombyx). *Science, 326*, 433–436.

117. Bazykin, G. A., Kondrashov, F. A., Ogurtsov, A. Y., Sunyaev, S., & Kondrashov, A. S. (2004). Positive selection at sites of multiple amino acid replacements since rat-mouse divergence. *Nature, 429*, 558–562.

118. Beall, C. M., et al. (2010). Natural selection 11 on EPAS1 (HIF2alpha) associated with low hemoglobin concentration in Tibetan 12 highlanders. *Proceedings of the National Academy of Sciences of the United States of America, 107*, 11459–11464.

119. Yi, X., et al. (2010). Sequencing of 50 human exomes reveals adaptation to high altitude. *Science, 329*, 75–78.

120. Xu, S., Li, S., Yang, Y., Tan, J., Lou, H., Jin, W., Yang, L., Pan, X., Wang, J., Shen, Y., Wu, B., Wang, H., & Jin, L. (2011). A Genome-wide search for signals of high altitude adaptation in Tibetans. *Molecular Biology and Evolution, 28*, 1003–1011.

121. Moreno-Estrada, A., Tang, K., Sikora, M., Marquès-Bonet, T., Casals, F., Navarro, A., Calafell, F., Bertranpetit, J., Stoneking, M., & Bosch, E. (2009). Interrogating 11 fast-evolving genes for signatures of recent positive selection in worldwide human populations. *Molecular Biology and Evolution, 26*, 2285–2297.

122. Zhang, J., Nielsen, R., & Yang, Z. (2005). Evaluation of an improved branch-site likelihood method for detecting positive selection at the molecular level. *Molecular Biology and Evolution, 22*, 2472–2479.

123. Li, J. Z., et al. (2008). Worldwide human relationships inferred from genome-wide patterns of variation. *Science, 319*, 1100–1104.

124. Bierne, N., & Eyre-Walker, A. (2004). The genomic rate of adaptive amino acid substitution in Drosophila. *Molecular Biology and Evolution, 21*, 1350–1360.

125. Shapiro, J. A., Huang, W., Zhang, C., Hubisz, M. J., Lu, J., Turissini, D. A., Fang, S., Wang, H. Y., Hudson, R. R., Nielsen, R., Chen, Z., & Wu, C.-I. (2007). Adaptive genic evolution in the Drosophila genomes. *Proceedings of the National Academy of Sciences of the United States of America, 104*(7), 2271–2276.

126. Andolfatto, P. (2007). Hitchhiking effects of recurrent beneficial amino acid substitutions in the Drosophila melanogaster genome. *Genome Research, 17*, 1755–1762.

127. Markova-Raina, P., & Petrov, D. (2011). High sensitivity to aligner and high rate of false positives in the estimates of positive selection in the 12 *Drosophila* genomes. *Genome Research, 21*, 863–874.

128. Carius, H. J., Little, T. J., & Ebert, D. (2001). Genetic variation in a host-parasite association: Potential for coevolution and frequency-dependent selection. *Evolution, 55*, 1136–1145.

129. Huttley, G. A., Easteal, S., Southey, M. C., Tesoriero, A., Giles, G. G., McCredie, M. R., Hopper, J. L., & Venter, D. J. (2000). Adaptive evolution of the tumour suppressor BRCA1 in humans and chimpanzees. Australian Breast Cancer Family Study. *Nature Genetics, 25*, 410–413.

130. Sumiyama, K., Saitou, N., & Ueda, S. (2002). Adaptive evolution of the IgA hinge region in primates. *Molecular Biology and Evolution, 19*, 1093–1099.

131. Johnson, M. E., Viggiano, L., Bailey, J. A., Abudul-Rauf, M., Goodwin, G., Rocchi, M., & Eichler, E. E. (2001). Positive selection of a gene family during the emergence of humans and African apes. *Nature, 413*, 514–519.

132. Messier, W., & Stewart, C. B. (1997). Episodic adaptive evolution of primate lysozymes. *Nature, 385*, 151–154.

133. Wyckoff, G. J., Wang, W., & Wu, C. I. (2000). Rapid evolution of male reproductive genes in the descent of man. *Nature, 403*, 304–309.

134. Kitano, T., Sumiyama, K., Shiroishi, T., & Saitou, N. (1998). Conserved evolution of the Rh50 gene compared to its homologous Rh blood group gene. *Biochemical and Biophysical Research Communications, 249*, 78–85.

135. Kitano, T., & Saitou, N. (1999). Evolution of the Rh blood group genes has experienced gene conversions and positive selection. *Journal of Molecular Evolution, 49*, 615–626.

136. Zhang, J., Rosenberg, H. F., & Nei, M. (1998). Positive Darwinian selection after gene duplication in primate ribonuclease genes. *Proceedings of the National Academy of Sciences of the United States of America, 95*, 3708–3713.

137. Babarinde, I. A., & Saitou, N. (2013). Heterogeneous tempo and mode of conserved noncoding sequence evolution among four mammalian orders. *Genome Biology and Evolution* (advance access).

138. Nei, M. (2013b). *Mutation-driven evolution.* Oxford: Oxford University Press.

Part II

Evolving Genomes

A Brief History of Life

6

Chapter Summary

The evolutionary history of organisms is summarized in this chapter, starting from the origin of life. Evolution of the genetic code is discussed with reference to tRNAs, and estimation of the phylogenetic relationship of metabolic pathways before the diversification of prokaryotes and eukaryotes is given. The history of eukaryotes follows with special reference to multicellular lineages, finally reaching us human beings.

6.1 Origin of Life

Big Bang, or the start of this universe, is currently estimated to be about 13.7 billion (or giga) years ago [1]. The most abundant atom in the universe is hydrogen (H), followed by He, O, C, Ne, N, Mg, Si, Fe, S, Ar, Al, Ca, Na, Ni, Cr, P, Mn, Cl, K, and Cl [2]. Organisms are mostly composed of six atoms (C, H, N, O, P, and S), and all of them are abundant in the universe. Our solar system with Sun as the star is the second or later generation in the history of this universe, and its origin is about 4.6 billion years ago [3]. The earth, the third planet of our solar system, was established also at the same time. The sea was formed when the earth was cooled down. Figure 6.1 shows the history of the universe and the solar system at a glance.

We do not know the exact age when the premordial life form emerged on earth, but probably between 4.0 and 3.8 billion years ago [4]. Life originated most probably at sea. It is not clear if it happened in the middle of ocean, or sea surface, or sea bed, or on the sea rock. There is an alternative idea that life on earth originated outside of earth. This is called "panspermia" hypothesis [5]. Because meteorites fall on earth every year even now, some key catalyst-like molecules could come from space via meteorites. Recently there are some reports that organic molecules are formed outside of the earth (e.g., [6]). However, it does not automatically prove that the life originated outside the earth. Life is much more complex than mere

N. Saitou, *Introduction to Evolutionary Genomics*, Computational Biology 17, DOI 10.1007/978-1-4471-5304-7_6, © Springer-Verlag London 2013

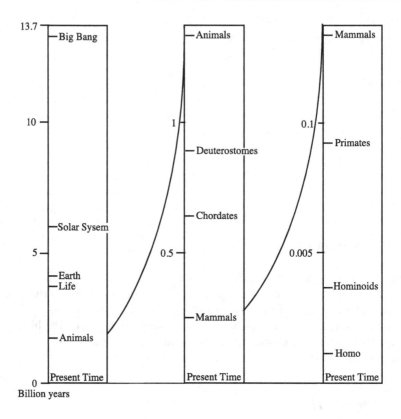

Fig. 6.1 History of the universe at a glance

combination of organic molecules. The author is personally inclined to the idea that the life on earth originated on this planet earth.

The cell is the basic unit of life, and it contains many sorts of molecules, such as proteins, nucleotides, lipids, and carbohydrates. This heterogeneity is one of important characteristics of life. Formation of these molecules is the necessary step for the start of life, and this process is called "chemical evolution." "Vital Dust" written by Christian de Duve (1995; [7]) gave many interesting ideas about this period.

The two major characteristics of life are metabolism and inheritance. Metabolism is divided into catabolism and anabolism, and many biochemical pathways are involved in both phases. Inheritance is transmission of genetic information from parents to offsprings, and most organisms use DNA as the material basis of transmission. Because the present-day organisms have both characteristics, it is not easy for us to imagine a prototype life with only one characteristic. Therefore, if one molecule can do both metabolism and inheritance, we may incline to hope this molecule as the origin of real life. This idea is called "RNA world" by Walter Gilbert in 1986 [8], because RNA has such dual aspects. Some

Fig. 6.2 Representative hypotheses on the origin of life

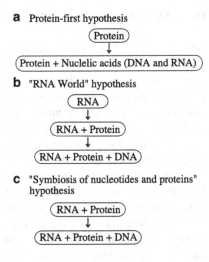

a Protein-first hypothesis

> (Protein)
> ↓
> (Protein + Nuclelic acids (DNA and RNA))

b "RNA World" hypothesis

> (RNA)
> ↓
> (RNA + Protein)
> ↓
> (RNA + Protein + DNA)

c "Symbiosis of nucleotides and proteins" hypothesis

> (RNA + Protein)
> ↓
> (RNA + Protein + DNA)

viruses use RNA as the material basis of genetic information, and some RNAs, called ribozyme, are known to have enzymatic activity. Various RNA molecules, such as rRNAs and tRNAs, are also important in many cellular functions. The RNA world hypothesis became quite popular, and it is sometimes considered as the inevitable initial condition of life.

We still do not know the situation of the ancestral earth when the life originated. Some other molecules such as proteins could have been formed before nucleotides emerged. The central dogma of molecular biology tells us that nucleotides and proteins are tightly linked in any cell. This can be described as "symbiosis of nucleotides and proteins." The RNA world hypothesis cannot explain how this symbiosis started. As mutations are the basis of diversification of life forms, a series of rare events similar to accumulation of mutations might have been necessary to create this unique coexistence of nucleotides and proteins, later with carbohydrates and lipids. If this is the case, all primordial proteins have nucleotide-binding functions that are seen nowadays only in DNA- and/or RNA-binding proteins. Figure 6.2 shows three hypotheses on the origin of life. Panel (A) is the classic protein-first hypothesis, and panel (B) is the "RNA world" hypothesis, and panel (C) is "symbiosis of nucleotides and proteins" hypothesis, in which the period of RNA molecules alone considered in the RNA world hypothesis is missing. In any case, the origin of life is still a conundrum for modern biology.

Viruses do not have cells, and they are quite diverse (see Chaps. 7 and 8 for bacterial and eukaryotic viruses, respectively). Some viruses use RNA as the molecular basis of the genome unlike cellular organisms which exclusively use DNA for their genomes. There are two main hypotheses about the origin of viruses, and they can be called virus early and virus late. The virus-early hypothesis is related to the RNA world where RNA viruses are considered to be molecular fossils [9]. In the traditional concept, viruses are viewed as degenerated cellular organisms in the virus-late hypothesis.

6.2   Evolution of the Genetic Code

So far, there is no exception in any cellular organisms and viruses that 64 kinds of triplet codons are used for translating RNA nucleotide sequence information to protein amino acid sequence information. This means that the last common ancestor of all organisms probably used this 64 codon system. However, the standard codon table often has four-codon families in which the third nucleotide of one codon has no information to determine amino acid. This poses a possibility that the third position of a triplet codon had no information, essentially period. If so, there existed a 16-codon system prior to the 64-codon system. One possible 16-codon system is shown in Table 6.1. This was deduced based on a series of parsimonious arguments. First, there are eight codon quartets which correspond to the same amino acid and only the third nucleotide of the codon is different, such as UCn (n = A, C, G, or U) for Ser. These eight boxes of codons are assigned to the amino acid they correspond in the 64-codon table. Two amino acids, Met and Trp, have only one codons, ATG and TGG, respectively. We eliminate these two amino acids from the codon table. Because Met is also used for initiation of protein synthesis, elimination of Met implies that stop codons are also relatively new. We thus free three codons, TAA, TAG, and TGA, from correspondence to termination of protein synthesis. If we apply these arguments, 11 out of 16 codon quartets are assigned to one amino acid. Remaining five quartets correspond to two amino acids, depending on chemical nature (purine or pyrimidine). TTn codon quartets may be assigned to Phe, for CTn quartets are already assigned to the other amino acid, Leu. Table 6.1 was made by considering these arguments above.

The 16-codon table is already complicated, and the initial codons could have been only four kinds, corresponding to only four nucleotides at the second position of one triplet codon: nAn, nCn, nGn, and nUn (n = A, C, G, or U). If we recall that 20 amino acids used for proteins can be classified into four chemically similar group (see Table 1.3 in Chap. 1), a plausible primordial genetic code assigning only four amino acids is shown in Table 6.2. They are Leu/Val/Ile, Ser/Thr/Pro, Lys/Asp, and Arg corresponding to nTn, nCn, nAn, and nGn, respectively, and these amino acids are representatives with residue groups of nonpolar, polar, negatively charged, and positively charged, respectively.

An amino acid is covalently bonded to a particular tRNA with the aid of an aminoacyl synthetase. Figure 6.3 shows phylogenetic trees of class I and II aminoacyl-tRNA synthetases corresponding to the 20 amino acids (Nagel and Doolittle, 1995;

Table 6.1 A possible 16-codon table

Codon	A.A.	Codon	A.A.	Codon	A.A.	Codon	A.A.
UUn	Phe	UCn	Ser	UAn	Tyr	UGn	Cys
CUn	Leu	CCn	Pro	CAn	His/Gln	CGn	Arg
AUn	Ile	ACn	Thr	AAn	Asn/Lys	AGn	Ser/Arg
GUn	Val	GCn	Ala	GAn	Asp/Glu	GGn	Gly

Table 6.2 A possible
4-codon table

nUn = Leu/Val/Ile	(Nonpolar R group amino acid)
nCn = Ser/Thr/Pro	(Polar R group amino acid)
nAn = Lys/Asp	(Negatively charged R group amino acid)
nGn = Arg	(Positively charged R group amino acid)

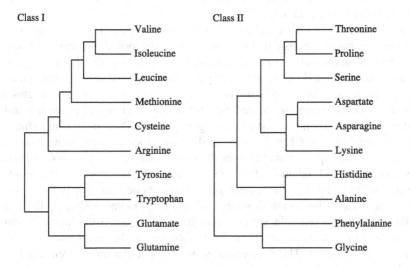

Consensus trees based on various alignments for the two groups of aminoacyl synthetases.

Fig. 6.3 Phylogenetic trees of classes I and II aminoacyl synthetase (Based on [10])

[10]). A later study (Woese et al. 2000; [11]) showed that the Gln and Asn synthetase emerged from Glu and Asp synthetase, respectively. Interestingly, ancestral amino-acyl-tRNA synthetases corresponding to two initial codons (nGn, and nTn) are located in class I enzymes, and ancestral aminoacyl-tRNA synthetases corresponding to the other two initial codons (nAn, nCn) are located in class II enzymes. However, if we map the nucleotides at the second position of the 16 codons shown in Table 6.1, the most parsimonious interpretation of the roles of class I and II aminoacyl-tRNA synthetases emerges. That is, class I and class II enzymes were responsible for purines (A and G) and pyrimidines (C and U), respectively. If we accept this argument, the first nucleotides, probably RNA, were composed of only two types, one purine and one pyrimidine. This is reasonable, for purine and pyrimidine bonding is the essence of the double-helix formation. Because ATP is most abundant in the current organisms, it may be more plausible to consider adenine (A) and its corresponding uracil (U) as the initial bases. And the two types of bases formed the primordial RNA molecules. Of course, this is only a speculation but may be an interesting first step to think about the origin of genetic codes.

6.3   Establishment of Cell

The basis of life is coexistence of a variety of molecules that are always changing. This phenomenon is called metabolism. Enzymes are the core of metabolism. One organism has many enzymes, and together with their substrates, they form complex metabolism networks. We will discuss various databases on this network in Chap. 13. These networks should have the products of the long-term evolution, and it is interesting how they evolved.

Tomiki and Saitou (2004; [12]) tried to infer the phylogenetic relationship of the five electron transfer energy metabolism systems: photosynthesis, aerobic respiration, denitrification, sulfur respiration, and sulfur metabolism. Because all these four systems exist in three domains of life, they must have evolved in the common ancestor of eubacteria, archaea, and eukaryotes. The five energy metabolism systems are thought to be evolutionarily related because of the similarity of electron transfer patterns and the existence of some homologous proteins. Protein domains defined in the Pfam database (see Chap. 13) were used for constructing phylogenetic trees, and these trees were superimposed to infer the evolutionary tree of metabolic systems, based on OOta and Saitou's (1999; [13]) method. Figure 6.4 shows the result of superimposition of phylogenetic trees for seven metabolic pathways: photosynthesis (P), aerobic respiration, denitrification (N_{DN}), nitrate assimilation (N_{AS}), sulfur respiration (S_R), dissimilation in sulfur metabolism (S_{DS}), and assimilation in sulfur metabolism (S_{AS}). Five possible branching locations are shown for aerobic respiration as 1–5 in Fig. 6.4. The three pathways involved in dissimilation emerged after the divergence of eubacteria and the archaea/eukaryote common ancestor.

The phylogenetic tree shown in Fig. 6.4 suggests that assimilation first emerged, and in particular, photosynthesis is basal (diverged first) among the energy metabolism pathways. Aerobic respiration might be utilized after the increase in oxygen by photosynthesis, while an ancestral sulfur and/or nitrogen metabolism system coexisted with that. Therefore, branching position 1 may be most plausible among the five possibilities for the emergence of the aerobic respiration system [12]. In fact, the atmospheric oxygen was present at significant levels around 2.3 billion years ago [14].

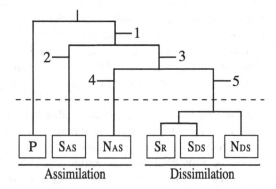

Fig. 6.4 Phylogenetic tree of seven metabolic pathways (Based on Tomiki and Saitou 2004; [12])

Single-cell organisms are majority of the present-day life forms, including Eubacteria, Archaea, and eukaryotes, and their common ancestor was most probably single cell. As present-day cells, the ancestral cells were expected to be composed of lipid bilayer cell membrane, DNAs, RNAs, proteins, and many metabolites. The total number of genes may be much smaller than thousands that are typical for present-day bacterial genomes. Prokaryotes is divided into Eubacteria and Archaea (Fig. 6.5; from [15]). Genomes of prokaryotes will be discussed in Chap. 7.

6.4 Emergence of Eukaryotes

Eukaryotes are phylogenetically more close to Archaea but their genomes are quite different from those of Archaea and Eubacteria, as we will see in Chap. 8. Eukaryotes were originally single cell, and soon after the ancestral eukaryotic organism diverged from the Archaea lineage, a proteobacteria sneaked into the premordial eukaryotic cell. This kind of parasitism may also frequently occur among prokaryotes, but their cellular system does not allow alien cells to survive within their cell. Probably because the nucleus membrane was developed and cell size became much bigger in eukaryotic ancestors, it was now possible for some Eubacterial organisms, and possibly Archaea too, to live within the much bigger eukaryotic cell. Only one type of eubacterial parasites succeeded to proliferate within the most of eukaryotes, and they became mitochondria. Mitochondria are now in intracellular symbiosis with host eukaryotes. A similar event happened in the common ancestor of green plants. We will discuss the characteristics of eukaryote genomes in Chap. 8.

Most of eukaryotes are still in the form of single cell, and they are generically called protists. Eukaryotes diverged considerably and produced many existing lineages. Figure 6.6a is a consensus phylogeny of the major groups of eukaryotes based on amino acid sequence and ultrastructural data (Baldauf 2008; [16]). Figure 6.6b shows the phylogenetic relationship of major lineages of eukaryotes proposed by Nozaki (2005; [17]). Basal eukaryotes were deduced based on the paralogous comparison of the α- and β-tubulin amino acid sequences. After determining the root position, actin and EF1α sequences were also included, and a total of 1,525 amino acid sequences were used [17]. Amoebozoa (Physarum and Dictyostelium) are basal, followed by opisthokonta (animals and fungi). It should be noted that the phylogenetic trees shown in Fig. 6.6a, b are somewhat inconsistent, and the overall phylogeny of all major eukaryotic lineages is still on the way.

6.5 Multicellularization

There are three major lineages of multicellular eukaryotes: plants, fungi, and animals. Figure 6.7 shows a phylogenetic tree of major land plant lineages based on Palmer et al. (2004; [18]). Angiosperms and gymnosperms are both monophyletic, and they constitute seed plants. Ferns are sister group of seed plants, and with

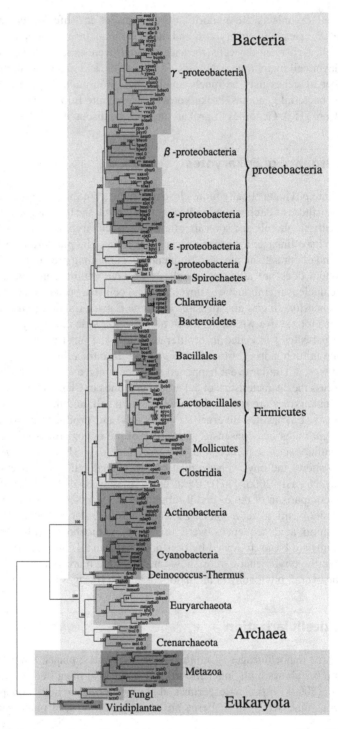

Fig. 6.5 Phylogenetic relationship of eubacteria, archaea, and eukaryotes (From Fukami et al. 2007; [15])

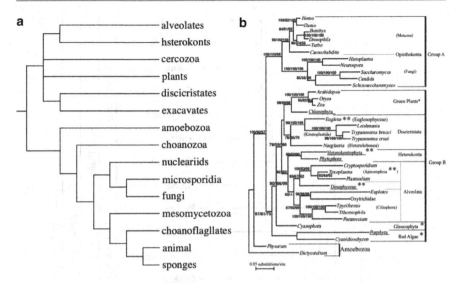

Fig. 6.6 Two possible phylogenetic relationships of major eukaryotic lineages. (**a**) Based on [16]. (**b**) From Nozki (2005); [17]

fernlike lycophytes, three groups constitute vascular plants. Finally, bryophytes (mosses, liverworts, and hornworts) is basal, although they are not monophyletic.

Fungi may be multicellular organisms with the most loose body plan. In fact, some fungal lineages returned to single-cell stages, and they are called "yeasts." Two fungi organisms, *Saccharomyces cerevisiae* and *Schizosaccharomyces pombe*, often considered as model eukaryote organisms, are both yeasts, and their ancestors were once multicellular. We therefore should be careful to treat them as if they represent the common ancestral situation of all eukaryotes, for they are both phylogenetically fungi. Figure 6.8 shows a phylogenetic tree of 42 fungal species constructed using a concatenated alignment of 38,000 amino acid sites from 153 universally distributed fungal genes [19]. Zygomycota is basal, followed by Basidiomycota. *Schizosaccharomyces pombe* stands at a unique phylogenetic position; basal to Ascomycota which is divided to Pezizomycotina and Saccharomycotina. *Saccharomyces cerevisiae* belongs to Saccharomycotina.

It is still not clear how animals emerged from a single-cellular ancestral protist. Choanoflagellates are phylogenetically the closest single-cell eukaryote species to animals [20]. Because fungi and animals are phylogenetically closer among eukaryote lineages, it is possible that the common ancestor of fungi and animals already had a set of genes which were later used for true multicellularization. Because yeasts are single cellular, we should compare genomes of modest multicellular fungi and extremely multicellular animals. This problem is left to future studies.

Traditionally, each animal phylum represents unique body plan. Scores of animal phyla were classified into only a few groups based mostly on developmental stages: Porifera (sponges), Radiata (Cnidaria such as jellyfish, corals, starfish, and hydra

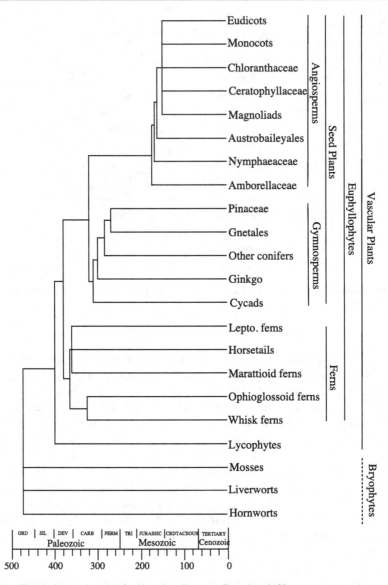

Fig. 6.7 The phylogenetic tree of major plant lineages (Based on [18])

and Ctenophora), Protostomia, and Deuterostomia. After gastrulation, blastopore becomes mouth and anus in Protostomia and Deuterostomia, respectively. Some molecular data suggested the monophyly of Arthropoda and Priapulida (nematodes and tardigrades) as Ecdysozoa as shown in Fig. 6.9 (based on [21]). Both Protostomia and Deuterostomia are monophyletic in this figure; thus, it is not clear which body plan was ancestral in the common ancestor of protostomes and deuterostomes.

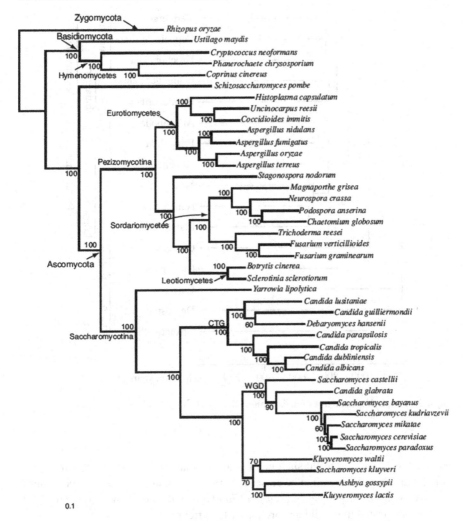

Fig. 6.8 A phylogenetic tree of 42 fungal species (From Fitzpatrick et al. 2006; [19])

6.6 From Origin of Chordates to Emergence of Modern Human

The author of this book is mainly interested in human evolution. Therefore, this book is inevitably human centric. This last section of this chapter thus focuses on the lineage going to modern humans.

One branch of deuterostomes formed notochord at an early stage of their life history, and their descendants are now classified as phylum Chordata. Ancestral chordates probably looked like modern-day amphioxus (Cephalochordata), for their

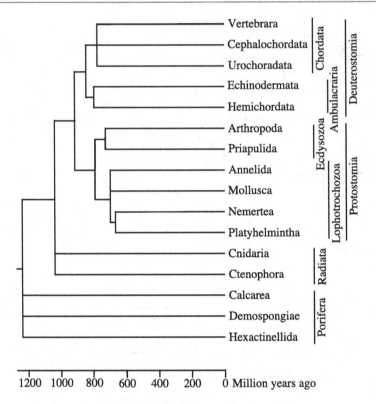

Fig. 6.9 The phylogenetic tree of major animal lineages (Based on [21])

morphology is similar to Pikaia, one of Burgess Shale fossil species. Traditionally, amphioxus was considered to be closer to vertebrates than ascidians (Urochordata); however, genome sequences of ascidian *Ciona intestinalis* [22] and amphioxus [23] now demonstrated that ascidians are closer to vertebrates.

Vertebrates emerged about 500 million years ago, and the common ancestor did not have jaw, like present-day jawless vertebrates (Agnatha), lamprey and hagfish. Jawed vertebrates (Gnatha) later evolved, and cartilaginous fish (Chondrichthyes) and teleost fish (Actinopterygii) further diverged. Tetrapods, or land vertebrates, emerged about 400 million years ago. The phylogenetic relationship of jawless vertebrates, cartilaginous fish, bony fish, and tetrapods shown in Fig. 6.10 (based on [24]) is well supported by the molecular data of 31 nuclear protein coding genes [25].

The first tetrapod was amphibian, and then amniotes followed. Amniotes have amnion in their eggs and these were tolerant to dry conditions. Reptiles, birds, and mammals are included in amniotes, and dinosaurs flourished during the Mesozoic. The common ancestor of mammals and birds diverged about 300 million years ago, and primordial mammals emerged about 150 million years ago.

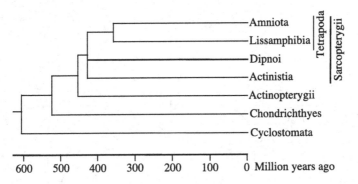

Fig. 6.10 The phylogenetic tree of major vertebrate lineages (Based on [24])

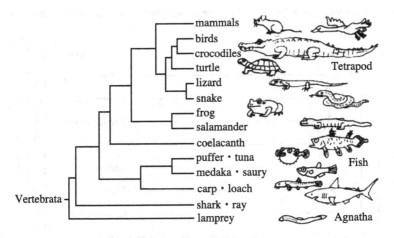

Fig. 6.11 The phylogenetic tree of tetrapods (Based on [26])

Monotremes are basal in mammals, and then marsupials and eutherians (placental mammals) diverged. Figure 6.11 (based on [26]) shows the phylogenetic tree of the major tetrapod lineages.

Radiation of placental mammals started to occur before the Cenozoic started, following the continental drift (see Fig. 6.12; after [27]). Edentates evolved in South America, while elephants, elephant shrew, aardvark, golden mole, hyrax, and tenrec evolved in Africa, and now they are called Afrotheria (see Fig. 6.13; based on [28]). It is now established that an asteroid hit at the tip of the Yucatan Peninsula 65 million years ago caused mass extinction, including extinction of dinosaurs [29]. Disappearance of dinosaurs led creation of open niches, and this prompted the further diversification of mammalian families in each order.

Primates probably evolved in tropical forests of Laurasia, and prosimians (Strepsirrhini) and simians (Haplorhini) diverged about 75 million years ago [30]. Many present-day prosimians live in Madagascar Island, which separated from

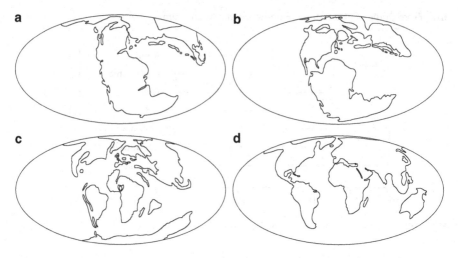

Fig. 6.12 Continental drift (Based on Paleomap Project; http://www.scotese.com/earth.htm)

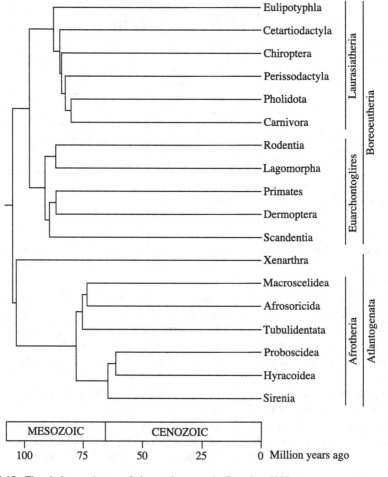

Fig. 6.13 The phylogenetic tree of placental mammals (Based on [28])

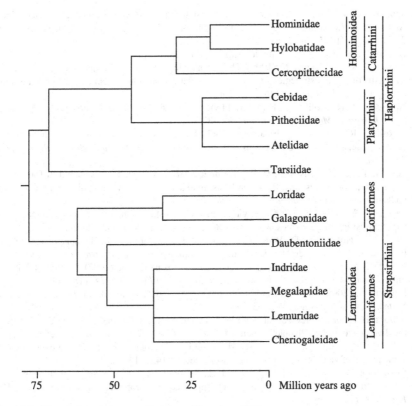

Fig. 6.14 The phylogenetic tree of primates (Based on [30])

African continent about 50 million years ago. The human lineage goes from Haplorhini, Catarrhini, superfamily Hominoidea, family Hominidae, and genus Homo. The second word of "Homo sapiens" means clever in Latin language, while genus name Homo means human in Latin. Figure 6.14 presents the phylogenetic tree of the major primate lineages (based on [30]).

References

1. Hinshaw, G., et al. (2009). Maps, & basic results. *Astrophysical Journal Supplement Series, 180*, 225.
2. Anders, E., & Ebihara, M. (1982). Solar-system abundances of the elements. *Geochimica et Cosmochimica Acta, 46*, 2363–2380.
3. Tilton, G. G. (1988). Age of the solar system. In *Meteorites and the early solar system* (pp. 259–275). Tucson: University of Arizona Press.
4. Sleep, N. H., Zahnle, K. J., Kasting, J. F., & Morowitz, H. J. (1989). Annihilation of ecosystems by large asteroid impacts on the early Earth. *Nature, 342*, 139–142.
5. Naganuma, T. (2011). *In search of origin of life to space; ark of panspermia (in Japanese)*. Tokyo: Kagaku Dojin.

6. Super-complex organic molecules found in interstellar space. Royal Astronomical Society News Archive (http://www.ras.org.uk/news-and-press/157-news2010/1853-complex-space-molecules)
7. de Duv, C. R. (1995). *Vital dust*. New York: Basic Books.
8. Gilbert, W. (1986). The RNA world. *Nature, 316*, 618.
9. Weiner, A. M., & Maizels, N. (1999). The genomic tag hypothesis: Modern viruses as molecular fossils of ancient strategies for genomic replication, and clues regarding the origin of protein synthesis. *The Biological Bulletin, 196*, 327–330.
10. Nagel, G. M., & Doolittle, R. F. (1995). Phylogenetic analysis of the aminoacyl-tRNA synthetases. *Journal of Molecular Evolution, 40*, 487–498.
11. Woose, C. R., Olsen, G. J., Ibba, M., & Soll, D. (2000). Aminoacyl-tRNA synthetases, the genetic code, and the evolutionary process. *Microbiology and Molecular Biology Reviews, 64*, 202–236.
12. Tomiki, T., & Saitou, N. (2004). Phylogenetic analysis of proteins associated in four major energy metabolism systems: Photosynthesis, oxidative phosphorylation, nitrogen metabolism and sulfur metabolism. *Journal of Molecular Evolution, 59*, 158–176.
13. OOta, S., & Saitou, N. (1999). Phylogenetic relationship of muscle tissues deduced from superimposition of gene trees. *Molecular Biology and Evolution, 16*, 856–867.
14. Bekker, A., Holland, H. D., Wang, P. L., Rumble, D., III, Stein, H. J., Hannah, J. L., Coetzee, L. L., & Beukes, N. J. (2004). Dating the rise of atmospheric oxygen. *Nature, 427*, 117–120.
15. Fukami-Kobayashi, K., Minezaki, Y., Tateno, Y., & Nishikawa, K. (2007). A tree of life based on protein domain organizations. *Molecular Biology and Evolution, 24*, 1181–1189.
16. Baldauf, S. (2008). An overview of the phylogeny and diversity of eukaryotes. *Journal of Systematics and Evolution, 46*, 263–273.
17. Nozaki, H. (2005). A new scenario of plastid evolution: Plastid primary endosymbiosis before the divergence of the "Plantae", emended. *Journal of Plant Research, 118*, 247–255.
18. Palmer, J. D., Soltis, D. E., & Chase, M. W. (2004). The plant tree of life: An overview and some points of view. *American Journal of Botany, 91*, 1437–1445.
19. Fitzpatrick, D. A., Logue, M. E., Stajich, J. E., & Butler, G. (2006). A fungal phylogeny based on 42 complete genomes derived from supertree and combined gene analysis. *BMC Evolutionary Biology, 6*, 99.
20. Ruiz-Trillo, I., Roger, A. I., Burger, G., Gray, M. W., & Lang, B. F. (2008). A phylogenomic investigation into the origin of Metazoa. *Molecular Biology and Evolution, 25*, 664–672.
21. Blair, J. E. (2009). Chapter 24: Animals (Metazoa). In S. B. Hedges & S. Kumar (Eds.), *The timetree of life* (pp. 223–230). Oxford: Oxford University Press.
22. Dehal, P., et al. (2002). The draft genome of *Ciona intestinalis*: insights into chordate and vertebrate origins. *Science, 298*, 2157–2167.
23. Putnam, N. H., Butts, T., Ferrier, D. E. K., Furlong, R. F., Hellsten, U., Kawashima, T., Robinson-Rechavi, M., Shoguchi, E., Terry, A., Yu, J. K., Benito-Gutiérrez, E., Dubchak, I., Garcia-Fernàndez, J., Grigoriev, I. V., Horton, A. C., de Jong, P. J., Jurka, J., Kapitonov, V., Kohara, Y., Kuroki, Y., Lindquist, E., Lucas, S., Osoegawa, K., Pennacchio, L. A., Salamov, A. A., Satou, Y., Sauka-Spengler, T., Schmutz, J., Shin-I, T., Toyoda, A., Gibson-Brown, J. J., Bronner-Fraser, M., Fujiyama, A., Holland, L. Z., Holland, P. W. H., Satoh, N., & Rokhsar, D. S. (2008). The amphioxus genome and the evolution of the chordate karyotype. *Nature, 453*, 1064–1071.
24. Hedges, S. B. (2009). Chapter 39: Vertebrates (Vertebrata). In S. B. Hedges & S. Kumar (Eds.), *The timetree of life* (pp. 309–314). Oxford: Oxford University Press.
25. Takezaki, N., Figueroa, F., Zaleska-Rutczynska, Z., & Klein, J. (2003). Molecular phylogeny of early vertebrates: Monophyly of the Agnathans as revealed by sequences of 35 genes. *Molecular Biology and Evolution, 20*, 287–292.
26. Saitou, N. (2009). Chapter 5: From origin of life to emergence of. In N. Saitou (Ed.), *Graphical guide of human evolution*. Tokyo: Kodansha {in Japanese}.
27. PALEOMAP project (http://www.scotese.com/earth.htm)

28. Murphy, W. J., & Eizirik, E. (2009). Chapter 71: Placental mammals (Eutheria). In S. B. Hedges & S. Kumar (Eds.), *The timetree of life* (pp. 471–474). Oxford: Oxford University Press.

29. Shulte, P., et al. (2010). The Chicxulub asteroid impact and mass extinction at the Cretaceous-Paleogene boundary. *Science, 327,* 1214–1218.

30. Steiper, M. E., & Young, N. M. (2009). Chapter 74: Primates (Primates). In S. B. Hedges & S. Kumar (Eds.), *The timetree of life* (pp. 482–486). Oxford: Oxford University Press.

Prokaryote Genomes

7

Chapter Summary

The world of prokaryotes (Bacteria and Archaea) is much more diverse than eukaryotes. After glancing the diversity of prokaryotes and their genome sequencing efforts, the basic structure of prokaryote genomes is discussed using *Escherichia coli* as an example, followed by discussions on GC content heterogeneity, horizontal gene transfer, codon usage, and plasmids. Finally, we discuss prokaryotic metagenomes.

7.1 Diversity of Prokaryotes and Their Genome Sequencing Efforts

Contrary to the popular notion that organisms with complex body plans are adapted to the earth's environment, the most successful organisms are definitely prokaryotes. Prokaryotes are divided into Bacteria and Archaea. Fox et al. (1977; [1]) found that 16S ribosomal RNA sequences of methanogenic bacteria were quite different from those of bacterial species known at that time. Based on that finding, Woese and Fox (1977; [2]) proposed to divide the living beings into eubacteria, archaebacteria, and eurkaryotes. This trifurcation is now established as three "domains" of life, with slightly different names: Bacteria, Archaea, and Eukarya (Woese et al. 1990; [3]). Bacteria and Archaea are both prokaryotes and share many features, yet they differ in various aspects. For example, genes related to DNA replications and gene expressions of Archaea are more similar with those of eukaryotes, while genes related to metabolism are more similar between Bacteria and Archaea than those of eukaryotes. The molecular structure of lipids of cell membranes is also different between Bacteria and Archaea (e.g., see Koga 2012; [4]).

Figure 6.5 (from Fukami-Kobayashi et al. 2007; [5]) shows a rough sketch of the major groups of Bacteria and Archaea. There are two large groups in Bacteria: Proteobacteria and Firmicutes. Proteobacteria are subdivided into five groups

(α, β, γ, δ, and ϵ), and there are four subgroups in Firmicutes: Bacillales, Lactobacillales, Mollicutes, and Clostridia. Other groups belonging to Bacteria are Spirochaetes, Chlamydiae, Bacteroidetes, Actinobacteria, Cyanobacteria, and Deinococcus–Thermus. Archaea are divided into the two major groups: Euryarchaeota and Crenarchaeota. It should be noted that Fig. 6.5 is an unrooted tree constructed by using the neighbor-joining method [6], and the position of the root (the common ancestor of the all present-day organisms) is not known. There were many studies to determine this root position (e.g., Iwabe et al. 1989; [7]), but we have not yet reached the decisive answer. However, one thing is clear: eukaryotes were derived from some sort of hybridization of Bacteria and Archaea. Fukami-Kobayashi et al. (2007; [5]) also produced a phylogenetic network from the same distance matrix used for producing the tree shown in Fig. 6.5. There were many reticulations, which suggested horizontal gene transfers (see Sect. 7.4). For example, Viridiplantae of Eukaryota and Cyanobacteria of Bacteria share a relatively long split. This was caused by transfer of genes of ancient cyanobacteria which formed chloroplasts to the eukaryotic plant ancestor.

The bacterial genome sequencing project was initiated by a group of Japanese bacteriologists in 1989 [8]. They targeted *Escherichia coli*, and small pieces of its genome were determined step by step [9–14]. However, the first bacterial species whose genome was determined was *Haemophilus influenzae* in 1995 by Craig Venter's group [15]. This bacteria was considered to be an agent of influenza when it was found in 1892 by Richard Friedrich Johannes Pfeiffer [16], and its genome size (1.8–1.9 Mb) is much smaller than that of *Escherichia coli*. Venter's group used the whole-genome shotgun method originally proposed by Sanger et al. (1980, [17]); see Chap. 11. Interestingly, this whole-genome shotgun method was initially not favored by many researchers, and a grant proposal based on this method was rejected from National Institutes of Health [18]. Within 5 years after the completion of the first bacterial genome sequence by Fleischmann et al. (1995; [15]), 30 complete genomes of bacteria were determined [18]. For example, Blattner et al. (1997; [19]) reported a complete genome sequence of *Escherichia coli*, while the Japanese team was struggling to complete their *Escherichia coli* genome sequence at that time. This clearly shows the importance of leadership, challenging spirit, and affection for a large-scale genome sequence production. Gregory and DeSalle (2005; [18]) presented highlights of 10 genome sequences determined until 2004. As of October 2012, ~7,500 bacterial genome sequences (including fragmented draft versions) are already available (Dr. Kirill Kryukov, personal information), and it is a matter of time for reaching 10,000 complete bacterial genomes.

Figure 7.1 shows the genome size distribution of 84 Archaea and 1,043 Bacteria (data from GCDB [20]). Both Archaea and Bacteria show bimodal distributions. This may be caused by sequencing efforts concentrated to some specific species, such as *Escherichia coli*. Koonin (2011; [21]) also noted two peaks of genome sizes for Archaea. The Archaea species with the smallest and the largest genomes are *Nanoarchaeum equitans* (0.49 Mb) and *Methanosarcina acetivorans* (5.75 Mb), respectively, and the Bacteria species with the smallest and the largest genomes are Candidatus *Hodgkinia cicadicola* Dsem (0.14 Mb) and *Sorangium cellulosum* So ce 56 (13.03 Mb), respectively (data from GCDB [20]).

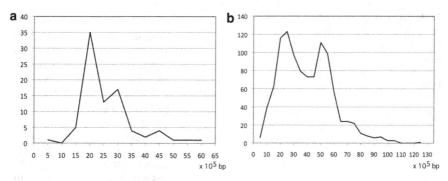

Fig. 7.1 Genome size distribution of 1,043 Bacteria and 84 Archaea (Data from GCDB [20]). (**a**) Archaea. (**b**) Bacteria

7.2   The Basic Genome Structure of Prokaryotes

Escherichia coli has been studied for a long time in molecular genetics and molecular biology as a model prokaryote organism. We therefore use this bacteria as an example. The genome size of the *Escherichia coli* substrain MB1655 of K12 strain is 4,639,675 bp (International Nucleotide Sequence Database accession number U00096.2). There are ~4,300 protein coding genes which account for 87.8 % of the genome. The average length of one protein coding gene is thus ~950 bp, which is able to code a protein with ~300 amino acids. Classification of these protein genes is shown in Table 7.1. Genes for enzymes occupy more than 1/3 of the total protein coding genes, followed by those for transporters, regulators, membrane proteins, etc. It is interesting that there exist ~300 genes of external origins via horizontal gene transfers (see Sect. 7.4) and ~100 pseudogenes in the *Escherichia coli* genome.

There are 22 rNRA genes assorted in seven ribosomal RNA operons, 88 transfer RNA genes, and 61 miscellaneous RNA genes in this genome (from PEC database at http://www.shigen.nig.ac.jp/ecoli/pec/genes.jsp), and they consist of 0.8 % of the genome. There are various types of repeat sequences in this genome. The largest are five Rhs dispersed elements (each length is 5.7–9.6 kb), which alone occupy 0.8 % of the total genome size. Other repeat sequences are REP, IRU, Box C, RSA, Ter, LDR, and iap, and they comprise ~0.5 % of the genome [19]. *Escherichia coli* genomes often contain "insertion sequences (ISs)," which are less than 2.5-kb DNA segment, capable of inserting at multiple sites in a target molecule [22]. K-12, a representative laboratory strain, is known to have a small number of ISs, yet there are 10 groups of ISs in this strain [19]. If we combine various kinds of repeat sequences, they occupy about 10 % of the *Escherichia coli* K12 strain, and they can be considered as "junk DNAs." The situation of *Escherichia coli* is a representative view of all prokaryote genomes. Unlike genomes of many eukaryotes, prokaryotic genomes are mostly composed of protein coding sequences, and noncoding regions are limited.

Table 7.1 Classification of protein genes of *Escherichia coli* K12 strain (Based on GenProtEC database at http:// genprotec.mbl.edu/overview. html)

Protein type	No. (%)
Enzyme	1,503 (34.8)
Transporter	594 (13.8)
Regulator	408 (9.5)
Membrane protein	249 (5.8)
Protein factor	212 (4.9)
Structural protein	125 (2.9)
Carrier protein	116 (2.7)
Cell process	55 (1.3)
Lipoprotein	46 (1.1)
Leader peptide	11 (0.3)
Unknown function	598 (13.9)
External origin	305 (7.1)
Pseudogene	95 (2.2)
Total	4,317

Fig. 7.2 Comparison of the genome structures of *Escherichia coli* K12 and O157 strains (From Hayashi et al. 2001; [23])

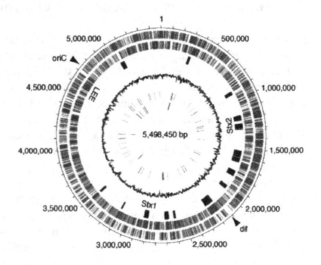

Hayashi et al. (2001; [23]) determined the genome sequence of the *Escherichia coli* O157 strain, a major food-borne infectious pathogen. Its genome size was ~860 kb larger than that of K-12, and a total of 4.1-Mb regions were highly conserved between the two strains. The remaining 1.4-Mb regions were O157 specific, and most of them are horizontally transferred DNAs. Figure 7.2 (from [23]) shows the comparison of *Escherichia coli* O157 and K12 strains. It is clear that there is a huge sequence difference even within the same bacterial "species." This situation is quite different from diploid and sexually reproducing eukaryotic species such as humans. Therefore, we should not equate a bacterial species with an animal species.

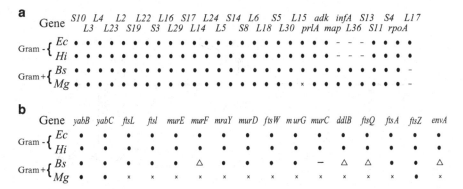

Fig. 7.3   The gene-order comparison (slightly simplified from Fig. 5 of Watanabe et al. 1997; [24])

Roughly speaking, a prokaryote genome is composed of protein coding genes, structural RNA coding genes, short repeat sequences, insertion sequences, long repeat sequences, and short gene expression controlling regions. Sequentially arranged genes are often simultaneously transcribed and express proteins which coordinate in prokaryote genomes. Such genes are called "operons." Watanabe et al. (1997; [24]) compared locations of orthologous genes among the genome sequences of *Haemophilus influenzae, Escherichia coli* K12, *Mycoplasma genitalium, and Bacillus subtilis* (abbreviated as Hi, Ec, Mg, and Bs, respectively, in Fig. 7.3) They found that dynamic rearrangements occurred very frequently, yet several highly conservative regions were kept, probably through strong structural constraints. Figure 7.3a, b (from [24]) show two regions with extensively conserved gene orders. Figure 7.3a is composed of three consecutive operons, and most of the genes are conserved for all four bacteria. Figure 7.3b shows genes responsible for cell-wall synthesis and cell division. Many genes are missing only in the *Mycoplasma genitalium* genome. *Mycoplasma genitalium* is parasitic, living within primate cells, and its genome size is only 580 kb (Fraser et al. 1995; [25]). Many genes are expected to be deleted or become nonfunctional for parasitic bacteria, because they can rely on host organisms in various aspects.

Buchnera, an endocellular bacterial symbiont of aphids, has also a small genome with 641 kb (Shigenobu et al. 2000; [26]). This genome contains genes for the biosyntheses of amino acids essential for the hosts, but those for nonessential amino acids are missing, suggesting complementarity and syntrophy between the host and the symbiont [26]. *Mycobacterium leprae*, an obligate intracellular bacterium responsible for leprosy, is phylogenetically closely related to *Mycobacterium tuberculosis*, the causative agent of tuberculosis. Cole et al. (2001; [27]) determined the 3.27-Mb complete genome sequence of *Mycobacterium leprae*, and by comparing the 4.41-Mb genome sequence of *Mycobacterium tuberculosis*, they found that more than half of protein coding genes were pseudogenes. These three examples indicate that parasitic bacteria tend to have smaller genome sizes [18] and many pseudogenes. The neutral evolution (Chap. 4) clearly operates behind these changes.

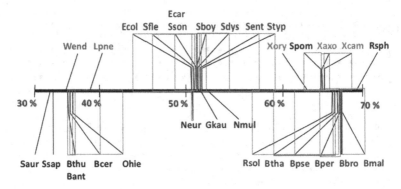

Fig. 7.4 Distribution of GC content for some representative bacterial species (Data from [29])

Most of the prokaryote genomes are circular as in *Escherichia coli*; however, some species have linear genomes, such as *Borrelia burgdorferi,* the Lyme disease agent [28], and *Streptomyces avermitilis* [29]. Both circular and linear genome structures were observed in *Agrobacterium tumefaciens* [18]. It is an interesting problem how circular genome structure can change to a linear one or vice versa.

Many bacterial genomes are composed of more than one circular structure. Smaller ones are often called "plasmids," but the difference between the longest structure and shorter ones is not clear.

The diversity of prokaryotes is much higher than that of eukaryotes in many respects, and the genomic structures which are thought to be specific to eukaryotes, such as polyploidy, may be found at least in some prokaryote species [18]. In any case, eukaryotes can be considered to be some sort of strange prokaryotic group, and convergent evolution between the eukaryotic lineage and some prokaryotic lineage can happen.

7.3   GC content and Oligonucleotide Heterogeneity

There are extensive variations in the GC contents among bacterial genomes, as first demonstrated by Sueoka in 1962 [30]. Figure 7.4 shows the distribution of GC content for some representative bacterial species (data from [31]). Such a large diversity in the GC content was probably caused by varied patterns of nucleotide substitutions. These differences may be caused by changes in the DNA replication mechanisms and proofreading mechanisms. Moran (2002; [32]) showed a positive correlation between the genome size and GC content (see also [18]). Figure 7.5 shows their relationship using the genome data for 1,043 Eubacteria and 84 Archaea listed in GCDB (Genome Composition DataBase) developed by Kryukov et al. (2012; [20]). Rocha and Danchin (2002; [33]) analyzed not only bacterial genome sequences but also bacterial phages, plasmids, and insertion sequences, which might be regarded as "intracellular pathogens," and showed that those pathogens have

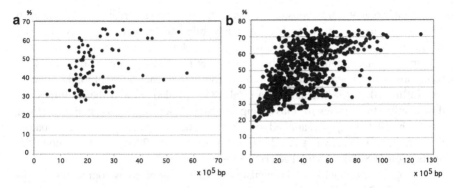

Fig. 7.5 Correlation between the genome size and GC content (Data from GCDB; [31]). (**a**) Archaea. (**b**) Bacteria

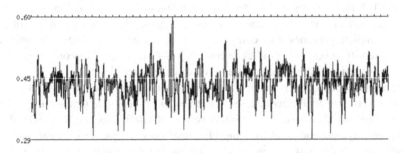

Fig. 7.6 The genome-wide distribution of the GC content for *Bacteroides fragilis* (From Saitou 2007; [34])

significantly low GC content. Based on this finding, they proposed that the GC content differences resulted from the higher energy cost and limited availability of G and C over A and T/U. If we consider Bacteria genomes with less than 2-Mb size, the positive correlation between the GC content and the genome size seems to be real; however, the correlation almost disappears for species with more than 2-Mb genome sizes. Interestingly, the GC content of 75 % is the upper limit for Bacteria. A number of Archaea genomes are limited, and there is no clear correlation between the GC content and the genome size, but they also seem to have the GC content upper limit at ~65 %.

Although the GC content is highly heterogeneous among prokaryote species, within-genome variation is small. For example, Fig. 7.6 (from Saitou 2007; [34]) shows the genome-wide distribution of the GC content for *Bacteroides fragilis*, whose genome size is 5.2 Mb. The GC content fluctuation seems to be essentially random around the genomic average of 45 %, except for some very low or very high GC content regions. This is a good contrast with some vertebrate genomes in which within-genome GC content variation, so-called isochore, was observed (see Chap. 9).

The GC content can be considered as mononucleotide pair frequency. We can also consider frequencies of more than one nucleotide pair frequencies in any genome. Karlin and Ladunga (1994; [35]) initiated the analysis of short oligonucleotides in genome sequences. Samuel Karlin's group continued this area of studies, but they were mostly restricted to analysis of dinucleotide frequencies (e.g., [36, 37]) except for Karlin (2005; [38]). Nakashima et al. [39, 40] also showed the usefulness of dinucleotide frequencies for classifying organisms. Self-organizing map (SOM) of Kohonen (1990; [41]) was applied to oligonucleotide frequency data of some bacterial genomes by Kanaya et al. (2001; [42]). Abe et al. (2003; [43]) expanded the application of SOM to many bacterial organisms.

Phylogenetic relationship of organisms is usually estimated by comparing homologous genes. The 16S rRNA gene has been often used for classification of bacteria (e.g., [1–3]). However, the phylogenetic tree constructed using a single gene may not truly reflect the whole evolutionary history of organisms in question. Moreover, the 16S rRNA classification has been useful only for taxa above the rank of species so that the 16S rRNA analysis does not have high ability for the group lower than species. An alternative way of estimating the phylogenetic relationship of prokaryotes is to use complete genomes, e.g., comparison of the presence and absence of orthologous or family genes, or overall gene content [44–47], presence of conserved insertions and deletions [48, 49], or conservation of gene order [50–53].

Pride et al. (2003, [54]) demonstrated a level of congruence of phylogenetic trees based on tetranucleotide frequencies with trees based on single genes, such as 16S rRNA. These studies revealed the effectiveness of tetranucleotide frequency data. However, phylogenetic analysis using oligonucleotide frequencies longer than tetranucleotides is not studied well. In addition, most of the analyses using oligonucleotide frequency focused on the classification of bacteria, not on the estimation of phylogenetic relationships. Tahakashi et al. (2009; [31]) thus investigated at which level of taxonomic rank the oligonucleotide frequency is the most effective in estimating the phylogenetic relationship of bacteria. They performed phylogenetic analysis by using the Euclidean distances calculated from the di- to deca-nucleotide frequencies in 36 complete bacterial genomes and computed topological distances (measured by the partition metric of Rzhetsky and Nei [55]) of oligonucleotide frequency-based trees with those obtained by using multiply aligned 16S rRNA sequences and concatenated seven gene sequences. When bacterial species with similar GC content were compared, topological distances were small at genus and family level (see also Chap. 16). These results suggest that oligonucleotide frequency is useful not only for classification of bacteria but also for estimation of their phylogenetic relationships for closely related species.

7.4 Horizontal Gene Transfer

DNAs are usually transmitted from parents to offsprings. Prokaryotes are single-cell organisms, and they simply divide their cells into two. The initial cells before cell division are parental cells and two cells after cell division are daughter cells. This sort of

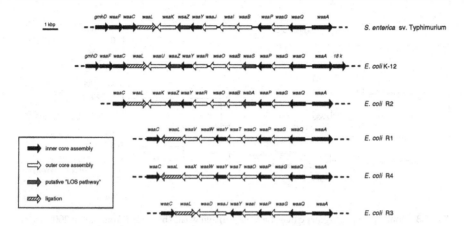

Fig. 7.7 The genetic organization of the core oligosaccharide biosynthesis regions (From Heinrichs et al. 1998; [57])

DNA transmission is called "vertical." When a DNA molecule of a different lineage or different species integrates to one cell, this is called "horizontal" (or lateral) transmission. If horizontally transferred DNA is not expressed in host cell, it is essentially a pseudogene. This DNA may soon accumulate mutations and will disappear from the acceptor cell lineage probably through deletion-type mutations. If a functional unit of DNA, or "gene," is horizontally transmitted, it may be expressed and will have higher chance of survival in the acceptor cell lineage. Probably because of purifying selection, most of horizontally transmitted DNAs detected from prokaryote genomes are genes, and this phenomenon is called "horizontal gene transfer." However, this does not necessarily mean that a "gene" is the unit of horizontal transfer. In any case, horizontal gene transfer is occurring quite frequently among prokaryote genomes (e.g., [21]).

A horizontally transferred gene has, by definition, a different evolutionary history compared to that of the host cell. Demonstration of such difference automatically detects the horizontal gene transfer event. Sawada et al. (1999; [56]) constructed phylogenetic trees of four genes (gyrB, hrpL, hrpS, and rpoD) of 56 *Pseudomonas syringae* strains. Trees produced from these four genes all showed clear three clusters, while the strain distribution of a pathogenicity-related gene *argK* is limited to only some strains of two clusters, and the nucleotide sequences of the *argK* gene found in strains belonging to different clusters were identical. It should be noted that these clusters differ by ~0.6 synonymous substitutions per nucleotide site, and the *argK* genes found in two different clusters are also expected to accumulate a similar amount of synonymous changes if they are orthologous. This clearly suggests the horizontal gene transfer of the *argK* gene.

Gram-negative bacterial lipopolysaccharides provide characteristic components of their outer membranes. In Enterobacteriaceae, the core oligosaccharide links a highly conserved lipid A to the antigenic O-polysaccharide (Heinrichs et al. 1998; [57]). The genetic organization of the core oligosaccharide biosynthesis regions of *Escherichia coli* and *Salmonella* is shown in Fig. 7.7 (from [57]).

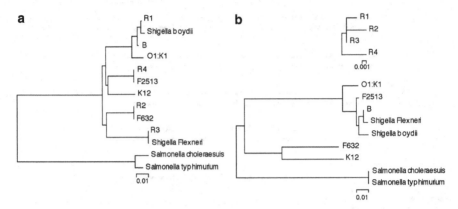

Fig. 7.8 Contrasting phylogenetic trees for waaC (**a**) and waaL (**b**) genes (From Saitou 2007; [34])

R1–R4 were pathovars found from different human patients. These regions are composed of a series of glycosyltransferase genes responsible for *E. coli* cell-wall outer structure. Figure 7.8a, b show phylogenetic trees for waaC and waaL genes, respectively (from Saitou 2007; [34]). While sequences of all *E. coli* strains are homologous for the waaC gene (Fig. 7.8a), there are two distinct groups for waaL gene, and these two groups have no homology (Fig. 7.8b), even though they are located at the same genomic position. Interestingly, one group of waaL gene is contained in all four genomes (R1–R4) found from patients. It therefore seems that an alien glycosyltransferase gene with no apparent sequence homology with the original waaL gene was horizontally transferred and substituted the waaL gene. This transferred gene may trigger the disease.

Nakamura et al. (2004; [58]) developed a method for detecting horizontally transferred genes and their possible donors by applying a Bayesian inference with training models for nucleotide composition. Their method gives the average posterior probability for each gene sequence, and a low probability suggests recent horizontal transfer. They estimated that 14 % of open reading frames in 116 prokaryotic complete genomes experienced recent horizontal transfers. They also showed that genes involved in cell surface, DNA binding, and pathogenicity-related functions are more prone to horizontal transfer.

Since horizontal gene transfers are so ubiquitous, Abby et al. (2012; [59]) utilized horizontal gene transfer events to choose the best species tree. They used a dataset of 12,000 gene families distributed in 336 genomes of 16 bacterial and archaeal phyla. Necessary numbers of horizontal gene transfer events for each gene tree were counted under eight different species tree candidates, and the best tree was chosen parsimoniously. Species trees estimated by using ribosomal RNA gene sequences, which are traditionally used genes for bacterial taxonomy, were shown to be often significantly different from the best tree. They also estimated the rate of horizontal gene transfer as 2–4 % per branch of one phylum per gene family. Because tree topology changes were the basis of these estimates, the real rate of horizontal gene transfer should be much higher, if we include those which do not affect the gene tree topology.

It is now clear that archaeal and bacterial genomes contain genes from multiple sources (e.g., Doolittle, 1999; [60]). We have to face the network structure of bacterial genomes instead of the traditional view of phylogenetic trees. A tree structure is simple and easy to understand, and the phylogenetic relationship of sexually reproductive diploid organisms is mostly represented as a tree. Most of us humans therefore prefer trees than networks. However, if prokaryotes are typical organisms on the earth, the reticulated evolution is the default. Different cells may interchange their DNA fragments sporadically, and some of them may persist for a long time as horizontally transferred genes. This rather chaotic nature can be viewed as "republic of genes," as Saitou (2004; [61]) originally proposed to describe a genome.

7.5 Codon Usage

The neutral theory predicts that synonymous sites of codons are evolving faster than nonsynonymous sites because of the smaller selective pressure, and this is true in general (see Chap. 4). However, several factors are known to influence on a certain region of a coding sequence and suppress synonymous substitution. One of such factors is the codon bias toward optimum codons. Optimum codons reflect the composition of genomic tRNA pool, as first shown by Ikemura (1981; [62]; see also [63, 64]). Because optimum codons are advantageous for fast and accurate translation, highly expressed or biologically important genes would have more optimum codons than others. Changes from an optimum codon to a non-optimum codon will be suppressed in such genes. Because optimum codons are similar in closely related species, highly expressed genes tend to show similar codon usage; therefore, synonymous substitution is lowered. In fact, the requirement for translational efficiency or accuracy enhances the optimum codon usage and suppresses nucleotide changes through purifying selection. Codon bias toward optimum codons is strong in fast-growing organisms such as *Escherichia coli* (e.g., Sharp and Li, 1987; [65]). It should be noted that codon usage bias also exists in eukaryotes [63, 64, 66].

Table 7.2 shows the codon usage pattern of three bacteria: *Bacillus subtilis subsp. natto, Escherichia coli O157:H7 str. Sakai*, and *Rhodobacter capsulatus*. Data were retrieved from Codon Usage Database (http://www.kazusa.or.jp/codon/) maintained by Yasukazu Nakamura. For example, frequencies of four codons for the four-codon box family coding for Val (GUn; n = A, C, G, or U) differ greatly among the three species. Frequencies for GUC and GUG are quite high (40 % and 53 % of this box, respectively) for a high GC content genome of *Rhodobacter capsulatus*, while those are only 16 % and 28 % for a low GC content genome of *Bacillus subtilis*. These differences can be interpreted by AT=>GC mutation pressure and natural selection toward optimum codons [63].

Table 7.2 Codon usage pattern of three bacterial species

	A	B	C		A	B	C		A	B	C		A	B	C
UUU	22.1	10.3	26.4	UCU	8.7	1.4	17.1	UAU	16.5	10.4	23.1	UGU	5.3	1.4	3.6
UUC	15.8	25.0	15.3	UCC	9.0	7.6	9.3	UAC	12.3	7.9	12.6	UGC	6.4	8.6	2.4
UUA	13.8	0.3	21.9	UCA	8.2	0.8	16.8	UAA	1.9	0.3	2.7	UGA	1.1	2.5	0.0
UUG	12.9	7.4	16.2	UCG	8.8	21.6	5.7	UAG	0.3	0.2	0.3	UGG	15.3	14.1	16.5
CUU	11.4	16.8	18.9	CCU	7.3	1.9	11.1	CAU	12.8	13.0	15.6	CGU	20.3	5.4	9.3
CUC	10.5	14.0	7.5	CCC	5.6	20.1	5.4	CAC	9.3	8.3	7.2	CGC	20.9	36.7	7.5
CUA	3.9	0.3	8.4	CCA	8.4	0.9	7.2	CAA	14.6	5.6	21.0	CGA	3.9	1.5	5.4
CUG	50.9	69.0	17.4	CCG	22.5	29.2	11.1	CAG	29.5	23.5	19.5	CGG	6.4	25.4	2.7
AUU	29.4	5.2	32.4	ACU	9.1	1.3	7.8	AAU	19.1	5.8	24.3	AGU	9.4	1.2	8.1
AUC	23.8	42.5	24.3	ACC	22.7	30.3	5.1	AAC	21.6	13.6	21.6	AGC	16.0	12.2	15.6
AUA	5.6	0.4	8.4	ACA	8.2	1.3	22.8	AAA	34.0	6.5	52.2	AGA	3.0	0.6	11.1
AUG	27.1	25.6	23.1	ACG	15.1	20.6	16.2	AAG	11.1	21.0	24.3	AGG	1.9	1.6	2.1
GUU	18.0	4.8	21.0	GCU	15.4	4.5	20.4	GAU	32.8	23.6	32.4	GGU	24.1	6.0	13.8
GUC	14.7	28.4	11.7	GCC	25.2	59.6	11.7	GAC	19.2	29.8	20.4	GGC	27.9	57.2	21.6
GUA	10.9	0.5	19.5	GCA	20.7	4.8	24.3	GAA	39.2	21.9	46.5	GGA	9.0	2.9	26.4
GUG	26.1	37.9	19.8	GCG	32.2	77.6	20.1	GAG	18.9	32.9	17.7	GGG	11.9	26.7	8.7

(A) *Escherichia coli* O157, (B) *Rhodobacter capsulatus*, (C) *Bacillus subtilis* subsp. Natto

7.6 Metagenomes

If we consider the ecosystem of the earth as a whole, all organisms depend on each other. In another words, every species is symbiont of the other species. Symbiosis in the classic sense is simply an extreme case. It is therefore necessary to keeping in mind the symbiosis view when we study evolution of genomes in one ecosystem, or ecome (see Chap. 12). Metagenomes are just one incomplete example of ecomes.

It has been well known that the majority of bacteria are unculturable. When nucleotide sequencing of PCR amplicons became easier than culturing bacteria, so-called environmental samples such as soil or gut flora were started to be studied. They are now generically called "metagenomes." Craig Venter's group, who pioneered various nucleotide sequencing projects including the Human Genome Project, determined more than one gigabase nucleotide sequences from filtered sea water collected in the Sargasso Sea and found more than 1.2 million new genes from at least 1,800 species (Venter et al. 2004; [67]). Abe et al. (2005; [68]) applied SOM analysis [41] for those massive data and showed that oligonucleotide frequency data can be used to distinguish many prokaryote species because oligonucleotide frequencies vary significantly among their genomes. Venter's group extended their metagenome sequencing, called "Global Ocean Sampling," by collecting 41 locations of sea surface water from Northwest Atlantic through Eastern Tropical Pacific (Rusch et al. 2007; [69]). They determined a total of 6.3 billion bp sequences. Yooseph et al. (2007; [70]) analyzed 17.4 million amino acid sequence data translated from these massive nucleotide sequences with four existing datasets: NCBI nonredundant database, ORFs identified from 222 prokaryotic genome projects deposited to NCBI, ORFs identified from 72 EST assemblies datasets constructed by TGI (The Genome Institute), and peptide sequences from 12 animal species retrieved from Ensembl. They estimated that the majority (90.8 %) of peptide sequences estimated from ocean surface water samples are belonging to Bacteria, and the remaining part is more or less equally shared by Archaea, eukaryotes, and viruses. In contrast, eukaryote sequences are a majority (64 %) in the existing four databases, showing a clear bias in biologists who determine genome sequences of organisms. Although the protein sequences newly estimated from the Global Ocean Sampling [69] were slightly higher than the total sequences of the existing datasets, newly found ~4,000 protein family clusters occupy only 23.4 % of the total clusters. Yet, this shows the high and hidden diversity of metagenomes.

Metagenome studies are now expanding greatly, especially on human gut microbiomes (e.g., Kurokawa et al. 2007; [71]). Recently, Yatsunenko et al. (2012; [72]) studied genomic sequences of bacteria in feces from 531 human individuals sampled from three geographical locations (Venezuela, Malawi, and the USA) as well as a wide age range (from a 1-year-old baby to 86-year-old person). Ironically, however, their major results were not so surprising; native Venezuelans and rural Malawis were similar and they are clearly different from metropolitan people of the USA, and the bacterial diversity quickly goes up from 1-year to 3-year babies, then reaches the plato. New approaches to analyze these massive sequence data are definitely necessary so as to find out the diversity of the universe of metagenomes.

References

1. Fox, G. E., Magrum, L. J., Balch, W. E., Wolfe, R. S., & Woese, C. R. (1977). Classification of methanogenic bacteria by 16S ribosomal RNA characterization. *Proceedings of the National Academy of Sciences of the United States of America, 74*, 4537–4541.
2. Woese, C. R., & Fox, G. E. (1977). Phylogenetic structure of the prokaryotic domain: The primary kingdoms. *Proceedings of the National Academy of Sciences of the United States of America, 74*, 5088–5090.
3. Woese, C. R., et al. (1990). Towards a natural system of organisms: Proposal for the domains Archaea, Bacteria, and Eucarya. *Proceedings of the National Academy of Sciences of the United States of America, 87*, 4576–4579.
4. Koga, Y. (2012). Archaea. In *Encyclopedia of evolution*. Tokyo: Kyoritsu Shuppan (in Japanese).
5. Fukami-Kobayashi, K., Minezaki, Y., Tateno, Y., & Nishikawa, K. (2007). A tree of life based on protein domain organizations. *Molecular Biology and Evolution, 24*, 1181–1189.
6. Saitou, N., & Nei, M. (1987). The neighbor-joining method: A new method for reconstructing phylogenetic trees. *Molecular Biology and Evolution, 4*, 406–425.
7. Iwabe, N., Kuma, K., Hasegawa, M., Osawa, S., & Miyata, T. (1989). Evolutionary relationship of archaebacteria, eubacteria, and eukaryotes inferred from phylogenetic trees of duplicated genes. *Proceedings of the National Academy of Sciences of the United States of America, 86*, 9355–9359.
8. Mori, H., et al. (1997). Post-sequencing genome analysis of *Escherichia coli* (in Japanese). *Tanpakushitsu-Kakusan-Koso, 46*, 1977–1985.
9. Yura, T., Mori, H., Nagai, H., Nagata, T., Ishihama, A., Fujita, N., Isono, K., Mizobuchi, K., & Nakata, A. (1992). Systematic sequencing of the *Escherichia coli* genome: Analysis of the 0–2.4 min region. *Nucleic Acids Research, 20*, 3305–3308.
10. Fujita, N., Mori, H., Yura, T., & Ishihama, A. (1994). Systematic sequencing of the *Escherichia coli* genome: Analysis of the 2.4–4.1 min (110,917–193,643 bp) region. *Nucleic Acids Research, 22*, 1637–1639.
11. Oshima, T., et al. (1996). A 718-kb DNA sequence of the *Escherichia coli* K-12 genome corresponding to the 12.7–28.0 min region on the linkage map. *DNA Research, 3*, 137–155.
12. Aiba, H., et al. (1996). A 570-kb DNA sequence of the *Escherichia coli* K-12 genome corresponding to the 28.0–40.1 min region on the linkage map. *DNA Research, 3*, 363–377.
13. Itoh, T., et al. (1996). A 460-kb DNA sequence of the *Escherichia coli* K-12 genome corresponding to the 40.1–50.0 min region on the linkage map. *DNA Research, 3*, 379–392.
14. Yamamoto, Y., et al. (1997). Construction of a contiguous 874-kb sequence of the *Escherichia coli* – K12 genome corresponding to 50.0–68.8 min on the linkage map and analysis of its sequence features. *DNA Research, 4*, 91–113.
15. Fleischmann, R. D., Adams, M. D., White, O., Clayton, R. A., Kirkness, E. F., Kerlavage, A. R., Bult, C. J., Tomb, J. F., Dougherty, B. A., Merrick, J. M., McKenney, K., Sutton, G., FitzHugh, W., Fields, C., Gocayne, J. D., Scott, J., Shirley, R., Liu, L.-I., Glodek, A., Kelley, J. M., Weidman, J. F., Phillips, C. A., Spriggs, T., Hedblom, E., Cotton, M. D., Utterback, T. R., Hanna, M. C., Nguyen, D. T., Saudek, D. M., Brandon, R. C., Fine, L. D., Fritchman, J. L., Fuhrmann, J. L., Geoghagen, N. S. M., Gnehm, C. L., McDonald, L. A., Small, K. V., Fraser, C. M., Smith, H. O., & Venter, J. C. (1995). Whole-genome random sequencing and assembly of *Haemophilus influenzae* Rd. *Science, 269*, 496–512.
16. http://en.wikipedia.org/wiki/Richard_Friedrich_Johannes_Pfeiffer
17. Sanger, F., Coulson, A. R., Barrell, B. G., Smith, A. J. H., & Roe, B. A. (1980). Cloning in single-stranded bacteriophage as an aid to rapid DNA sequencing. *Journal of Molecular Biology, 143*, 161–178.
18. Gregory, T. R., & DeSakke, R. (2005). Chapter 10: Comparative genomics in Prokaryotes. In T. R. Gregory (Ed.), *The evolution of the genome*. Burlington: Elsevier.
19. Blattner, F. R., Plunkett, G., 3rd, Bloch, C. A., Perna, N. T., Burland, V., Riley, M., Collado-Vides, J., Glasner, J. D., Rode, C. K., Mayhew, G. F., Gregor, J., Davis, N. W., Kirkpatrick, H. A.,

Goeden, M. A., Rose, D. J., Mau, B., & Shao, Y. (1997). The complete genome sequence of *Escherichia coli* K-12. *Science, 277*, 1453–1462.

20. Kryukov, K., Sumiyama, K., Ikeo, K., Gojobori, T., & Saitou, N. (2012). A new database (GCD) on genome composition for eukaryote and prokaryote genome sequences and their initial analyses. *Genome Biology and Evolution, 4*, 501–512.

21. Koonin, E. V. (2011). *The Logic of chance*. Upper Saddle River: Pearson Education.

22. Mahillon, J., & Chandler, M. (1998). Insertion sequences. *Microbiology and Molecular Biology Reviews, 62*, 725–774.

23. Hayashi, T., Makino, K., Ohnishi, M., Kurokawa, K., Ishii, K., Yokoyama, K., Han, C. G., Ohtsubo, E., Nakayama, K., Murata, T., Tanaka, M., Tobe, T., Iida, T., Takami, H., Honda, T., Sasakawa, C., Ogasawara, N., Yasunaga, T., Kuhara, S., Shiba, T., Hattori, M., & Shinagawa, H. (2001). Complete genome sequence of enterohemorrhagic *Escherichia coli* O157:H7 and genomic comparison with a laboratory strain K-12. *DNA Research, 8*, 11–22.

24. Watanabe, H., Mori, H., Itoh, T., & Gojobori, T. (1997). Genome plasticity as a paradigm of eubacteria evolution. *Journal of Molecular Evolution, 44*(Suppl 1), S57–S64.

25. Fraser, C. M., Gocayne, J. D., White, O., Adams, M. D., Clayton, R. A., Fleischmann, R. D., Bult, C. J., Kerlavage, A. R., Sutton, G., Kelley, J. M., Fritchman, R. D., Weidman, J. F., Small, K. V., Sandusky, M., Fuhrmann, J., Nguyen, D., Utterback, T. R., Saudek, D. M., Phillips, C. A., Merrick, J. M., Tomb, J. F., Dougherty, B. A., Bott, K. F., Hu, P. C., Lucier, T. S., Peterson, S. N., Smith, H. O., Hutchison, C. A., 3rd, & Venter, J. C. (1995). The minimal gene complement of *Mycoplasma genitalium*. *Science, 270*, 397–403.

26. Shigenobu, S., Watanabe, H., Hattori, M., Sakaki, Y., & Ishikawa, H. (2000). Genome sequence of the endocellular bacterial symbiont of aphids *Buchnera sp.* APS. *Nature, 407*, 81–86.

27. Cole, S. T., et al. (2001). Massive gene decay in the leprosy bacillus. *Nature, 409*, 1007–1011.

28. Ferdows, M. S., & Barbour, A. G. (1989). Megabase-sized linear DNA in the bacterium *Borrelia burgdorferi*, the Lyme disease agent. *Proceedings of the National Academy of Sciences of the United States of America, 86*, 5969–5973.

29. Omura, S., Ikeda, H., Ishikawa, J., Hanamoto, A., Takahashi, C., Shinose, M., Takahashi, Y., Horikawa, H., Nakazawa, H., Osonoe, T., Kikuchi, H., Shiba, T., Sakaki, Y., & Hattori, M. (2001). Genome sequence of an industrial microorganism *Streptomyces avermitilis*: Deducing the ability of producing secondary metabolites. *Proceedings of the National Academy of Sciences of the United States of America, 98*, 12215–12220.

30. Sueoka, N. (1962). On the genetic basis of variation and heterogeneity of DNA base composition. *Proceedings of the National Academy of Sciences of the United States of America, 48*, 582–592.

31. Takahashi, M., Kryukov, K., & Saitou, N. (2009). Estimation of bacterial species phylogeny through oligonucleotide frequency distances. *Genomics, 93*, 525–533.

32. Moran, N. A. (2002). Microbial minimalism: Genome reduction in bacterial pathogens. *Cell, 108*, 583–586.

33. Rocha, E. P. C., & Danchin, A. (2002). Base composition bias might result from competition for metabolic resources. *Trends in Genetics, 18*, 291–294.

34. Saitou, N. (2007). *Introduction to evolutionary genomics (in Japanese)*. Tokyo: Kyoritsu Shuppan.

35. Karlin, S., & Ladunga, I. (1994). Comparisons of eukaryotic genome sequences. *Proceedings of the National Academy of Sciences of the United States of America, 91*, 12832–12836.

36. Karlin, S., & Mrazek, J. (1997). Compositional differences within and between eukaryotic genomes. *Proceedings of the National Academy of Sciences of the United States of America, 94*, 10227–10232.

37. Karlin, S., Mrazek, J., & Campbell, A. (1997). Compositional biases of bacterial genomes and evolutionary implications. *Journal of Bacteriology, 179*, 3899–3913.

38. Karlin, S. (2005). Statistical signals in bioinformatics. *Proceedings of the National Academy of Sciences of the United States of America, 102*, 13355–13362.

39. Nakashima, H., Nishikawa, K., & Ooi, T. (1997). Differences in dinucleotide frequencies of human, yeast, and *Escherichia coli* genes. *DNA Research, 4*, 185–192.

40. Nakashima, H., Ota, M., Nishikawa, K., & Ooi, T. (1998). Genes from nine genomes are separated into their organisms in the dinucleotide composition space. *DNA Research, 5*, 251–259.
41. Kohonen, T. (1990). The self-organizing map. *Proceedings of the IEEE, 78*, 1464–1480.
42. Kanaya, S., Kinouchi, M., Abe, T., Kudo, Y., Yamada, Y., Nishi, T., Mori, H., & Ikemura, T. (2001). Analysis of codon usage diversity of bacterial genes with a self organizing map (SOM): Characterization of horizontally transferred genes with emphasis on the *E. coli* O157 genome. *Gene, 276*, 89–99.
43. Abe, T., Kanaya, S., Kinouchi, M., Ichiba, Y., Kozuki, T., & Ikemura, T. (2003). Informatics for unveiling hidden genome signatures. *Genome Research, 13*, 693–702.
44. Snel, B., Bork, P., & Huynen, M. A. (1999). Genome phylogeny based on gene content. *Nature Genetics, 21*, 108–110.
45. Tekaia, F., Lazcano, A., & Dujon, B. (1999). The genomic tree as revealed from whole proteome comparisons. *Genome Research, 9*, 550–557.
46. Fitz-Gibbon, S. T., & House, C. H. (1999). Whole genome-based phylogenetic analysis of free living microorganisms. *Nucleic Acids Research, 27*, 4218–4222.
47. Bansal, A. K., & Meyer, T. E. (2002). Evolutionary analysis by whole-genome comparisons. *Journal of Bacteriology, 184*, 2260–2272.
48. Gupta, R. S. (1998). Protein phylogenies and signature sequences: A reappraisal of evolutionary relationships among archaebacteria, eubacteria, and eukaryotes. *Microbiology and Molecular Biology Reviews, 62*, 1435–1491.
49. Gupta, R. S. (2001). The branching order and phylogenetic placement of species from completed bacterial genomes, based on conserved indels found in various proteins. *International Microbiology, 4*, 187–202.
50. Dandekar, T., Snel, B., Huynen, M., & Bork, P. (1998). Conservation of gene order: A fingerprint of proteins that physically interact. *Trends in Biochemical Sciences, 23*, 324–328.
51. Huynen, M. A., & Bork, P. (1998). Measuring genome evolution. *Proceedings of the National Academy of Sciences of the United States of America, 95*, 5849–5856.
52. Kunisawa, T. (2001). Gene arrangements and phylogeny in the class Proteobacteria. *Journal of Theoretical Biology, 213*, 9–19.
53. Suyama, M., & Bork, P. (2001). Evolution of prokaryotic gene order: Genome rearrangements in closely related species. *Trends in Genetics, 17*, 10–13.
54. Pride, D. T., Meinersmann, R. J., Wassenaar, T. M., & Blaser, M. J. (2003). Evolutionary implications of microbial genome tetranucleotide frequency biases. *Genome Research, 13*, 145–155.
55. Rzhetsky, A., & Nei, M. (1992). A simple method for estimating and testing minimum-evolution trees. *Molecular Biology and Evolution, 9*, 945–967.
56. Sawada, H., Suzuki, F., Matsuda, I., & Saitou, N. (1999). Phylogenetic analysis of *Pseudomonas syringe* pathovar suggests the horizontal gene transfer of *argK* and the evolutionary stability of hrp gene cluster. *Journal of Molecular Evolution, 49*, 627–644.
57. Heinrichs, D. E., Yethon, J. A., & Whitfield, C. (1998). Molecular basis for structural diversity in the core regions of the lipopolysaccharides of *Escherichia coli* and Salmonella enterica. *Molecular Microbiology, 30*, 221–232.
58. Nakamura, Y., Itoh, T., Matsuda, H., & Gojobori, T. (2004). Biased biological functions of horizontally transferred genes in prokaryotic genomes. *Nature Genetics, 36*, 760–766.
59. Abby, S. S., Tannier, E., Gouy, M., & Daubin, V. (2012). Lateral gene transfer as a support for the tree of life. *Proceedings of the National Academy of Sciences of the United States of America, 109*, 4962–4967.
60. Doolittle, W. (1999). Phylogenetic classification and the universal tree. *Science, 284*, 2124–2129.
61. Saitou, N. (2004). *"Genomu to Shinka" (in Japanese, meaning 'Genome and Evolution' in English)*. Tokyo: Shinyosha.
62. Ikemura, T. (1981). Correlation between the abundance of *Escherichia coli* transfer RNAs and the occurrence of the respective codons in its protein genes: A proposal for a synonymous codon choice that is optimal for the *E. coli* translational system. *Journal of Molecular Biology, 151*, 389–409.

63. Ikemura, T. (1982). Correlation between the abundance of yeast transfer RNAs and the occurrence of the respective codons in protein genes. Differences in synonymous codon choice patterns of yeast and *Escherichia coli* with reference to the abundance of isoaccepting transfer RNAs. *Journal of Molecular Biology, 158*, 573–97.

64. Ikemura, T. (1985). Codon usage and tRNA content in unicellular and multicellular organisms. *Molecular Biology and Evolution, 2*, 13–34.

65. Sharp, P. M., & Li, W. H. (1987). The rate of synonymous substitution in enterobacterial genes is inversely related to codon usage bias. *Molecular Biology and Evolution, 4*, 222–30.

66. Kanaya, S., Yamada, Y., Kinouchi, M., Kudo, Y., & Ikemura, T. (2001). Codon usage and tRNA genes in eukaryotes: Correlation of codon usage diversity with translation efficiency and with CG-dinucleotide usage as assessed by multivariate analysis. *Journal of Molecular Evolution, 53*, 290–298.

67. Venter, J. C., et al. (2004). Environmental genome shotgun sequencing of the Sargasso Sea. *Science, 304*, 66–74.

68. Abe, T., Sugawara, T., Kanaya, S., & Ikemura, T. (2005). Novel phylogenetic studies of genomic sequence fragments derived from uncultured microbe mixtures in environmental and clinical samples. *DNA Research, 12*, 281–290.

69. Rusch, D. B., et al. (2007). The *Sorcerer II* global ocean sampling expedition: Northwest Atlantic through Eastern Tropical Pacific. *PLoS Biology, 5*, e77.

70. Yooseph, S., et al. (2007). The *Sorcerer II* global ocean sampling expedition: Expanding the universe of protein families. *PLoS Biology, 5*, e16.

71. Kurokawa, K., et al. (2007). Comparative metagenomics revealed commonly enriched gene sets in human gut microbiomes. *DNA Research, 14*, 169–181.

72. Yatsunenko, T., et al. (2012). Human gut microbiome viewed across age and geography. *Nature, 486*, 222–227.

Eukaryote Genomes

8

Chapter Summary

General overviews of eukaryote genomes are first discussed, including organelle genomes, introns, and junk DNAs. We then discuss the evolutionary features of eukaryote genomes, such as genome duplication, C-value paradox, and the relationship between genome size and mutation rates. Genomes of multicellular organisms, plants, fungi, and animals are then briefly discussed.

8.1 Major Differences Between Prokaryote and Eukaryote genomes

A eukaryotic cell has a nucleus, surrounded by the nuclear membrane. There are other membrane systems in their cells, such as endoplasmic reticulum, Golgi apparatus, and vacuole. Prokaryotes do not have these membranes nor organella. Therefore, existence of membrane systems and organella, particularly mitochondria, are the two major characteristics of eukaryotes. It should be noted that some parasitic eukaryotes lost mitochondria. Genome sizes of eukaryotes became much bigger than those of prokaryotes. Accordingly, gene numbers are also more abundant in eukaryotes than prokaryotes. It is not clear if the formation of nucleus triggered the increase of the genome size.

There are various differences of genome structures between prokaryotes and eukaryotes, and they are listed in Table 8.1. Most of bacterial genomes are circular, while all eukaryotic genomes so far known are linear (here organelle genomes are not considered). The main reason for a large genome size in eukaryotes is the existence of many repeat sequences, which are minority in prokaryotes. Pseudogenes and introns are also few in prokaryotic genomes, while both are abundant in eukaryotic genomes. High occurrences of gene duplications in eukaryotes prompted production of many pseudogenes. Horizontal gene transfers are known to be quite frequent in prokaryotes, and they are rare in eukaryotes. Finally, genome

N. Saitou, *Introduction to Evolutionary Genomics*, Computational Biology 17,
DOI 10.1007/978-1-4471-5304-7_8, © Springer-Verlag London 2013

Table 8.1 Comparison of prokaryotic and eukaryotic genomes

Category	Prokaryotes	Eukaryotes
Size	Small (1–10 Mb)	Large (3–5,000 Mb)
Gene number	Small (<10,000)	Many (often > 10,000)
Topology	Mostly circular	Linear
Intergenic region	Short (<100 bp)	Long (often >100 kb)
Repeat sequence	Minor component	Major component
Pseudogene	Few	Many
Intron	Few	Usually exit
Complexity	Low	High
Horizontal gene transfer	Frequent	Rare
Gene duplication	Rare	Frequent
Genome duplication	None	Frequent (especially in plants and vertebrates)

Table 8.2 Examples of genes shared among most of eukaryote genomes but nonexisting in prokaryote genomes

DNA polymerase subunit γ 1
DNA topoisomerase 1
Histone H2B
Microtubule binding protein RP/EB family 2
Myosin
Nucleolin
Translation initiating factor
Ubiquitin

duplications sometimes occur in eukaryotes, especially in plants and in vertebrates, but genome duplication is so far not known for prokaryotic genomes.

Because the gene number of typical eukaryotic genomes is much larger than that of prokaryotes, there are many genes shared among most of eukaryote genomes but nonexisting in prokaryote genomes. Some examples are listed in Table 8.2. For example, myosin is located in animal muscle tissues, and its homologous protein exists in cytoskeleton of all eukaryotes, but not found in prokaryotes.

Recently, Kryukov et al. (2012; [1]) constructed a new database on oligonucleotide sequence frequencies and conducted a series of statistical analyses. Frequencies of all possible 1–10 oligonucleotides were counted for each genome, and these observed values were compared with expected values computed under observed oligonucleotide frequencies of length 1–4. Deviations from expected values were much larger for eukaryotes than prokaryotes, except for fungal genomes. Figure 8.1 shows the distribution of the deviation for various organismal groups. The biological reason for this difference is not known.

8.2 Organelle Genomes

There are two major types of organella in eukaryotes: mitochondria and plastids. Figure 8.2 shows schematic views of mitochondria and chloroplasts. These two organella has their independent genomes. This suggests that they were initially

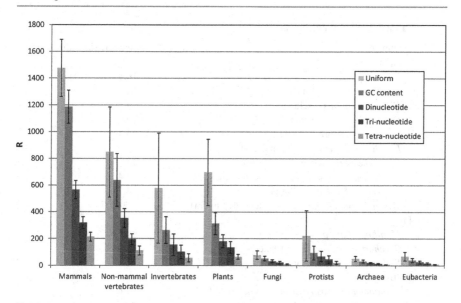

Fig. 8.1 Comparison of genome complexity among eukaryote genomes (From Kryukov et al. 2012; [1])

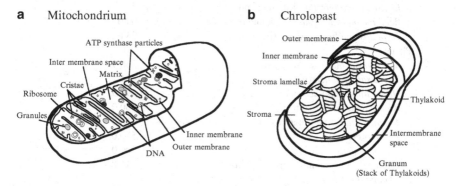

Fig. 8.2 Schematic views of mitochondrion and chloroplast

independent organisms which started intracellular symbiosis with primordial eukaryotic cells. Because most eukaryotes have mitochondria, the ancestral eukaryotes, a lineage that emerged from Archaea, most probably started intracellular symbiosis with mitochondrial ancestor. A parasitic *Rickettsia prowazekii* is so far phylogenetically closest to mitochondria [2], and a rickettsia-like bacterium is the best candidate as the mitochondrial ancestor. However, there is an alternative "hydrogen hypothesis" [3]. Plastids include chloroplasts, leucoplasts, and chromoplasts and exist in land plants, green algae, red algae, glaucophyte algae, and some protists like euglenoids. Mitochondrial genome sizes of some representative eukaryotes are listed in Table 8.3. Most of animal mitochondrial genomes are less than 20 kb, and sizes of protist and fungi mitochondrial genomes are somewhat larger. Mitochondrial genome size of plants is much larger than those of other eukaryotic lineages, yet the size is mostly less than 500 kb.

Table 8.3 Size of mitochondrial genomes

Organism	Genome size (kb)
Animals	
Homo sapiens (human)	16.5
Takifugu rubripes (Torafugu fish)	16.5
Ciona intestinalis (ascidian)	14.8
Drosophila melanogaster (fruit fly)	19.5
Apis mellifera (honey bee)	16.3
Limulus polyphemus (horseshoe crab)	15.0
Caenorhabditis elegans (nematode)	13.8
Schistosoma mansoni (parasitic flatworm)	14.4
Aplysia californica (mollusk)	14.1
Hydra magnipapillata (freshwater polyp hydra)	8.2 + 7.7
Fungi	
Moniliophthora perniciosa	109.1
Saccharomyces cerevisiae (baker's yeast)	75
Suillus grisells (basidiomycete fungus)	121
Protists	
Acanthamoeba castellanii [Acanthamoebidae]	41.6
Paramecium aurelia [Alveolata]	40.5
Plasmodium falciparum [Alveolata]	5.9
Tetrahymena thermophila [Alveolata]	47.6
Phytophthora infestans [Stramenopiles]	39.8
Reclinomonas americana [Jakobida]	69.0
Trypanosoma brucei brucei [Euglenozoa]	23.0
Plants	
Arabidopsis thaliana (Wall cress)	366.9
Oryza sativa indica (indica rice)	434.7
Oryza sativa japonica (japonica rice)	490.5
Brassica oleracea (cabbage)	160.0
Nicotiana tabacum (tobacco)	430.6
Zea mays (corn)	570.0
Cucumis melo (melon)	2,880
Chlamydomonas reinhardtii (green alga)	15.8
Chondrus crispus (red alga)	26

8.2.1 Mitochondria

An ancestral eukaryotic cell, probably an archaean lineage, hosted a bacterial cell, and intracellular symbiosis started. Initially, Archaea and Bacteria shared genes responsible for basic metabolism, and the situation is a sort of gene duplication for many genes, though homologous genes are not identical but already diverged long time ago. In any case, division of labor followed, and only limited metabolic pathways were left in the bacterial system, which eventually became mitochondria.

Animal mitochondrial genomes contain very small number of genes; 13 for peptide subunits, 20 for tRNA, and 2 for rRNA [4]. Figure 8.3 shows gene orders of five

Fig. 8.3 Gene orders of five animal mitochondrial DNA genomes (From Saitou 2007; [103])

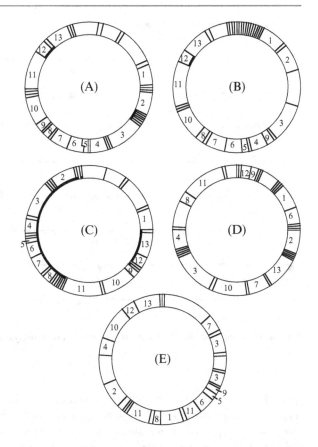

representative animal species mitochondrial DNA genomes. Although most of vertebrate mitochondrial DNA genomes have the same gene order as in human (Fig. 8.3a), gene order may vary from phylum to phylum. Yet the gene content and the genome size are more or less constant among animals. It is not clear why animal mitochondrial genomes are so small. One possibility is that animal individuals are highly integrated compared to fungi and plants, and this might have influenced a drastic reduction of the mitochondrial genome size. Another interesting feature of animal mitochondrial DNA genomes is the heterogeneous rates of gene order change. For example, platyhelminthes exhibit great variability in mitochondrial gene order (Sakai and Sakaizumi, 2012; [5]).

In contrast, plant mitochondrial genomes are much larger (see Table 8.3). Figure 8.4 shows the genome structure of tobacco mitochondrial genome (from Sugiyama et al. 2005; [6]). Horizontal gene transfers are also known to occur in plant mitochondrial DNAs even between remotely related species [7].

The melon (*Cucumis melo*) mitochondrial genome size, ca. 2.9 Mb, is exceptionally large, and recently its draft genome was determined [8]. Interestingly, melon mitochondrial genome looks like the vertebrate nuclear genome in its contents, in

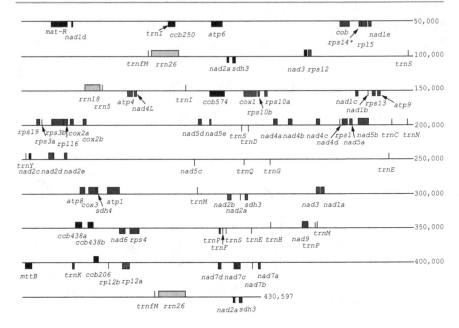

Fig. 8.4 Genome structure of tobacco mitochondria (From Sugiyama et al. 2005; [6])

spite of its genome size being similar to that of bacteria. The protein coding gene region accounted for only 1.7 % of the genome, and about half of the genome is composed of repeats. The remaining part is mostly homologous to melon nuclear DNA, and 1.4 % is homologous to melon chloroplast DNA. Most of the protein coding genes of melon mitochondrial DNAs are highly similar to those of its congeneric species, which are watermelon and squash whose mitochondrial genome sizes are 119 kb and 125 kb, respectively. This indicates that the huge expansion of its genome size occurred only recently. Interestingly, cucumber (*Cucumis sativus*), another congeneric species, also has ~1.8-Mb mitochondrial genome with many repeat sequences [9]. It will be interesting to study whether the increase of mitochondrial genomes of melon and cucumber is independent or not.

8.2.2 Chloroplasts

Chloroplasts exist only in plants, algae, and some protists. It may change to leucoplasts and chromoplasts. Because of this, a generic name "plastids" may also be used. The origin of chloroplast seems to be a cyanobacterium that started intracellular symbiosis as in the case of mitochondria.

A unique but common feature of chloroplast genome is the existence of inverted repeats [10], and they mainly contain rRNA genes. Chloroplast DNA contents may change during the plant growth, and matured leaves are devoid of DNA in their chloroplasts [11].

Chloroplast genomes were determined for more than 340 species as of December 2013 [106]. Their genome sizes range from 59 kb (*Rhizanthella gardneri*) to 521 kb (*Floydiella terrestris*). Although the largest chloroplast genome is still much smaller than atypical bacterial genome, its average intergenic length is 4 kb, much longer than that for bacterial genomes.

8.2.3 Interaction Between Nuclear and Organelle Genomes

Fractions of mitochondrial DNA may sometimes be inserted to nuclear genomes, and they are called "numts." An extensive analysis of the human genome found over 600 numts [12]. Their sequence patterns are random in terms of mitochondrial genome locations. This suggests that mitochondrial DNAs themselves were inserted, not via cDNA reverse-transcribed from mitochondrial mRNA. A possible source is sperm mitochondrial DNA that were fragmented after fertilization [12]. The reverse direction, from nucleus to mitochondria, was observed in melon, as discussed in subsection 8.2.1.

8.3 Intron

Intron is a DNA region of a gene that is eliminated during splicing after transcription of a long premature mRNA molecule. Intron was discovered by Phillip A. Sharp and Richard J. Roberts in 1977 as "intervening sequence" [13], but the name "intron" coined by Walter Gilbert in 1978 [14] is now widely used. It should be noted that some description on intron by Kenmochi [15] was used for writing this section.

8.3.1 Classification of Intron

There are various types of introns, but they can be classified into two: those requiring spliceosomes (spliceosome type) and self-splicing type. Figure 8.5 shows the splicing mechanisms of these two major types. Most of introns in nuclear genomes of eukaryotes are spliceosome type, and there are common GU–AG type and rare AU–AC type, depending on the nucleotide sequences of the intron–exon boundaries [16]. Spliceosomes involving these two types differ [17].

Self-splicing introns are divided into three groups: groups I, II, and III. Group I introns exist in organellar and nuclear rRNA genes of eukaryotes and prokaryotic tRNA genes. Group II are found in organellar and some eubacterial genomes. Cavalier-Smith [18] suggested that spliceosome-type introns originated from group II introns because of their similarity in splicing mechanism and structural similarity between group II introns and spliceosomal RNA. Group III introns exist in organellar genomes, and its splicing system is similar with that of group II intron, though they are smaller and have unique secondary structure.

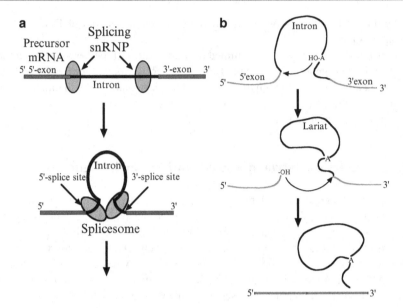

Fig. 8.5 Two major types of introns. (**a**) Spliceosome type. (**b**) Self-splicing type

There is yet another type of introns which exist only in tRNAs of single-cell eukaryotes and Archaea [19]. These introns do not have self-splicing functions, but endonuclease and RNA ligase are involved in splicing. The location of this type of introns is often at a certain position of the tRNA anticodon loop.

8.3.2 Introns Early/Late Controversy

After the discovery of introns, their probable functions and evolutionary origin have long been argued (e.g., [20, 21]). Because self-splicing introns can occur at any time, even in the very early stage of origin of life, we consider only spliceosome-type introns. For brevity, we hereafter call this type of introns as simply "intron." There are mainly two major hypotheses: introns early and introns late. The former claims that exon existed as a functional unit from the common ancestor of prokaryotes and eukaryotes, and "exon shuffling" was proposed for creating new protein functions [14]. Introns which separate exons should also be quite an ancient origin [14, 22]. In contrast, introns are considered to emerge only in the eukaryotic lineage according to the introns-late hypothesis [23, 24].

The protein "module" hypothesis proposed by Go [25] is related to be introns-early hypothesis. Pattern of intron appearance and loss has been estimated by various methods (e.g., [21, 26]). Kenmochi and his colleagues analyzed introns of ribosomal proteins of mitochondrial genomes and eukaryotic nuclear genomes in details [27–29]. These studies supported the introns-late hypothesis, because introns in mitochondrial and cytosolic ribosomal proteins seem to be independent origins and introns seem to emerge in many ribosomal protein genes after eukaryotes appeared.

8.3.3 Functional Regions in Introns

Introns do not code for amino acid sequences by definition. In this sense, most of introns may be classified as junk DNAs (see the next section). There are, however, evolutionarily conserved regions in introns, suggesting the existence of some functional roles in introns.

8.4 Junk DNAs

Ohno (1972; [30]) proclaimed that the most part of mammalian genomes are nonfunctional and coined the term "junk DNA." With the advent of eukaryotic genome sequence data, it is now clear that he was right. There are in fact so much junk DNAs in eukaryotic genomes. Junk DNAs or nonfunctional DNAs can be divided into repeat sequences and unique sequences. Repeat sequences are either dispersed type or tandem type. Unique sequences include pseudogenes that keep homology with functional genes.

8.4.1 Dispersed Repeats

Prokaryote genomes sometimes contain insertion sequences; however, this kind of dispersed repeats constitutes the major portion of many eukaryotic genomes. These interspersed elements are divided into two major categories according to their lengths: short ones (SINEs) and long ones (LINEs).

One well-known example of SINE is Alu elements in primate genomes. It is about 300-bp length, and originated from 7SL ribosomal RNA gene. Let us see the real Alu element sequence from the human genome sequence. If we retrieve the DDBJ/EMBL/GenBank International Sequence Database accession number AP001720 (a part of chromosome 21), there are 128 Alu elements among the 340-kb sequence. The density is 0.38 Alu elements per 1 kb. If we consider the whole human genome of ~3 billion bp, Alu repeats are expected to exist in ~1.13 million copies. One example of Alu sequence is shown below from this entry coordinates from 133600 to 133906:

```
ggcgggagcg atggctcacg cctgtaatgc cagcactttg ggaggccgag
gtgggtggat cacaaggtca ggagatagag accatcctgg ctaacacggt
gaaacactgt ctctactaaa aacacaaaaa actagccagg cgtggtggcg
ggtgcctgta atcccagcta ctcgggaggc tgaggcagga gaatggtgtg
aacccaggaa gtggagcttg cagtgagctc agattgcgcc actgcactcc
agcctgggtg acagagtgag actccatctc aaaaaaaata aaataaataa
aaaaaa
```

Fig. 8.6 An overall pattern
of Alu element evolution
(From Saitou 2007; [103])

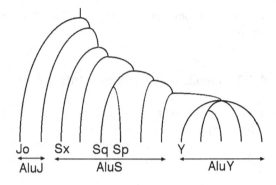

If we do BLAST homology search (see Chap. 14) using DDBJ system (http://
blast.ddbj.nig.ac.jp/blast/blastn) targeted to nonhuman primate sequences (PRI
division of DDBJ database), the best hit was obtained from chimpanzee chromosome 22, which is orthologous to human chromosome 21. I suggest interested readers to do this homology search practice.

Alu elements were first classified into J and S subfamilies [31]. It is not clear
about the reason of selection of two characters (J and S), but probably two authors
(Jurka and Smith) used initials of their surnames. In any case, this division was
based on the distance from Alu consensus sequence; Alu elements which are more
close to the consensus were classified as S and those not as J. Later, a subset of the
S subfamily were found to be highly similar with each other, and they were named
as Y after 'young," for they appeared relatively in young or recent age. Rough estimates of the divergence time of Alu elements are as follows: J subfamily appeared
about 60 million years ago, and S subfamily separated from J at 44 million years
ago, followed by further separation of Y at 32 million years ago [32]. Figure 8.6
shows the overall pattern of Alu element evolution (based on [32]).

8.4.2 Tandem Repeats

Tandemly repeated sequences are also abundant in eukaryotic genomes, and the
representative ones are heterochromatin regions. Heterochromatins are highly
condensed nonfunctional regions in nuclear DNA, in contrast to euchromatins, in
which many genes are actively transcribed. Heterochromatins usually reside at tero-
meres, terminal parts of chromosomes, and at centromeres, internal parts of chromosomes, that connect spindle fibers during cell division. A more than 1-Mb teromeric
regions of *Arabidopsis thaliana* were found to be tandem repeats of ca. 180-bp repeat
unit [33, 34]. The nucleotide sequence below is *Arabidopsis thaliana* tandemly repeated
sequence AR12 (International Sequence Database accession number X06467):

```
aagcttcttc ttgcttctca atgctttgtt ggtttagccg aagtccatat
gagtctttgt ctttgtatct tctaacaagg aaacactact taggctttta
ggataagatt gcggtttaag ttcttatact taatcataca catgccatca
agtcatattc gtactccaaa acaataacc
```

The human genome also has a similar but nonhomologous sequence in centromeres, called "alphoid DNA" with the 171-bp repeat unit [35]. The following is the sequence (International Sequence Database accession number M21746):

```
catcctcaga aacttctttg tgatgtgtgc attcaagtca cagagttgaa
cattcccttt cgtacagcag tttttaaaca ctctttctgt agtatctgga
agtgaacatt aggacagctt tcaggtctat ggtgagaaag gaaatatctt
caaataaaaa ctagacagaa g
```

If we do BLAST homology search (see Chap. 13) targeted to the human genome sequences of the NCBI database, there was no hit with this alphoid sequence. This clearly shows that the human genome sequences currently available are far from complete, for they do not include most of these tandem repeat sequences.

Telomores of the human genome are composed of hundreds of 6-bp repeats, ttaggg. If we search the human genome as 36-bp long 6 tandem repeats of this 6-repeat units as query using the NCBI BLAST, many hits are obtained.

8.4.3 Pseudogenes

As we already discussed in Chap. 4, authentic pseudogenes have no function, and they are genuine members of junk DNAs. When a gene duplication occurs, one of two copies often become a pseudogene. Because gene duplication is prevalent in eukaryote genomes, pseudogenes are also abundant. Pseudogenes are, by definition, homologous to functional genes. However, after a long evolutionary time, many selectively neutral mutations accumulate on pseudogenes, and eventually they will lose sequence homology with their functional counterpart. There are many unique sequences in eukaryote genomes, and majority of them may be this kind of homology-lost pseudogenes.

8.4.4 Junk RNAs and Junk Proteins

A long RNA is initially transcribed from a genomic region having an exon–intron structure, and then RNAs corresponding to introns are spliced out. These leftover RNAs may be called "junk" RNAs, for they will soon be degraded by RNAse. Only a limited set of genes are transcribed in each tissue of multicellular organisms, but leaky expression of some genes may happen in tissues in which these genes should not be expressed. Again these are "junk" RNAs, and they are swiftly decomposed. A series of studies (e.g., [36, 37]) claimed that many noncoding DNA regions are transcribed. However, van Bakel et al. [38] showed that most of them were found to be artifact of chip-chip technologies used in these studies. If nonsense or frameshift mutations occur in a protein coding sequences, that gene cannot make proteins. Yet its mRNA may be produced continuously until the promoter or its enhancer will become nonfunctional. In this case, this sort of mutated genes produces junk RNAs.

If only a small quantity of RNAs are found from cells and when they are not evolutionarily conserved, they are probably some kind of junk RNAs.

As junk DNAs and junk RNAs exist, cells may also have "junk" proteins. If mature mRNAs are not produced in the expected way, various aberrant mRNA molecules will be produced, and ribosomes try to translate them to peptides based on these wrong mRNA information. Proteins produced in this way may be called "junk" proteins, for they often have no or little functions. Even if one protein is correctly translated and is moved to its expected cellular location, it can still be considered as "junk" protein. One good example is the ABCC11 transporter protein of dry-type cerumen (earwax), for one nonsynonymous substitution at this gene caused that protein to be essentially nonfunctional [39].

8.5 Evolution of Eukaryote Genomes

There are various genomic features that are specific to eukaryotes other than existence of introns and junk DNAs, such as genome duplication, RNA editing, C-value paradox, and the relationship between genome size and mutation rates. We will briefly discuss them in this section.

8.5.1 Genome Duplication

The most dramatic and influential change of the genome structure is genome duplications. Genome duplications are also called polyploidization, but this term is tightly linked to karyotypes or chromosome constellation.

Prokaryotes are so far not known to experience genome duplications, which are restricted to eukaryotes. Interestingly, genome duplications are quite frequent in plants, while it is relatively rare in the other two multicellular eukaryotic lineages. An ancient genome duplication was found from the genome analysis of baker's yeast [40], and *Rhizopus oryzae*, a basal lineage fungus, was also found to experience a genome duplication [41]. Among protists, *Paramecium tetraurelia* is known to have experienced at least three genome duplications [42]. Because we human belongs to vertebrates and the two-round genome duplications occurred at the common ancestor of vertebrates (see Chap. 9), we may incline to think that genome duplications often happen in many animal species. It is not the case. So far, only vertebrates and some insects are known to experience genome duplications. The reason for this scattered distribution of genome duplication occurrences is not known.

If we plot the number of synonymous substitutions between duplogs in one genome, it is possible to detect a relatively recent genome duplication. This is because all genes duplicate when a genome duplication occurs, while only a small number of genes duplicate in other modes of gene duplications (see Chap. 3). Figure 8.7 shows the schematic view of two cases: with and without genome duplication. Lynch and Conery (2000; [44]) used this method to various genome sequences and found that the *Arabidopsis thaliana* genome showed a clear peak indicative of relatively recent genome duplication, while the genome sequences of

Fig. 8.7 A schematic view of synonymous distance distribution of duplogs with and without genome duplication (From Saitou 2007; [103])

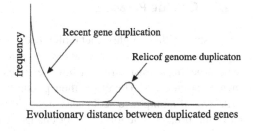

Evolutionary distance between duplicated genes

nematode *Caenorhabditis elegans* and yeast *Saccharomyces cerevisiae* showed the curves of exponential reduction. It is interesting to note that before the genome sequence was determined, the genome duplication was not known for *Arabidopsis thaliana*, while the genome of *Saccharomyces cerevisiae* was later shown to be duplicated long time ago [40].

When genome duplications occurred in some ancient time, the number of synonymous substitutions may become saturated and cannot give appropriate result. In this case, the number of amino acid substitutions may be used, even if each protein may have varied rates of amino acid substitutions. In any case, accumulation of mutations will eventually cause two homologous genes to become not similar with each other. Therefore, although the possibility of genome duplications in prokaryotes are so far rejected [45], it is not possible to infer the remote past events simply by searching sequence similarity. We should be careful to reach the final conclusion.

8.5.2   RNA Editing

Modification of particular RNA molecules after they are produced via transcription is called RNA editing. All three major RNA molecules (mRNA, tRNA, and rRNA) may experience editing [46]. There are various patterns of RNA editing; substitutions, in particular between C and U, and insertions and deletions, particularly U, are mainly found in eukaryote genomes. Guide RNA molecules exist in one of the main RNA editing mechanisms, and they specify the location of editing, but there are some other mechanisms [47].

It is not clear how the RNA editing mechanism evolved. Tillich et al. [47] studied chloroplast RNA editing and concluded that suddenly many nucleotide sites of chloroplast DNA genome started to have RNA editing, but later the sites experiencing RNA editing constantly decreased via mutational changes. They claimed that there was no involvement of RNA editing on gene expression. This result does not give RNA editing a positive significance.

Because there are many types of RNA molecules inside a cell, there also exist many sorts of enzymes that modify RNAs. It may be possible that some of them suddenly started to edit RNAs via a particular mutation. RNA editing which did not cause deleterious effects to the genome may have survived by chance at the initial phase. This view suggests the involvement of neutral evolutionary process in the evolution of RNA editing.

8.5.3 C-Value Paradox

Organisms with complex metabolic pathways have many genes. Multicellular organisms are such examples. Generally speaking, their genome sizes are expected to be large. In contrast, viruses whose genomes contain only a handful of genes have small genome sizes. Therefore, their possibility of genome evolution is rather limited. Even if amino acid sequences are rapidly changing because of high mutation rates, the protein function may not change. Unless the gene number and genome size increase, viruses cannot evolve their genome structures. It is thus clear that the increase of the genome size is crucial to produce the diversity of organisms. However, genomes often contain DNA regions which are not indispensable. Organisms with large genome sizes have many such junk DNA regions. Because of their existence, the genome size and the gene number are not necessarily highly correlated. This phenomenon was historically called C-value paradox (e.g., [48]), after the constancy of the haploid DNA amount for one species was found, yet their values were found to vary considerably among species at around 1950 (e.g., [49–51]). "C-value" is the amount of haploid DNA, and C probably stands as acronym of "constant" or "chromosomes." We now know that the majority of eukaryote genome DNA is junk, and there is no longer a paradox in C-values among species.

8.5.4 Conserved Noncoding Regions

While bacterial genomes are mostly consisting of protein coding genes, a considerable region of eukaryote genomes is noncoding. Most of them are junk DNA and do not have functions. If we find evolutionary conservation, however, these conserved regions should have some function through purifying selection. From the initial stage of molecular evolutionary studies, protein noncoding regions were suspected to be involved in gene regulation (Zuckerkandl and Pauling 1965; Britten and Davidson 1971; King and Wilson 1975). Now it is becoming clear that at least some noncoding regions play important roles in gene regulation (e.g., Carroll 2005; [55]). The functional elements are expected to evolve more slowly than surrounding nonfunctional DNA, as they are under purifying selection. Therefore, conserved noncoding sequences (CNSs) are likely to be important from the functional point of view.

Animal CNSs were discovered by comparison of human and fugu fish genome sequences by How et al. (1996; [52]). CNS analyses have been proved to be powerful for detecting regulatory elements (e.g., Hardison 2000; [53], Levy et al. 2001; [54]). Bejarano et al. (2004; [102]) found highly conserved noncoding sequences through comparison of human, mouse, and rat genomes. Siepel et al. (2005; [56]) found conserved noncoding DNA sequences from insects, nematodes, and yeasts by comparing closely related species. We will discuss more on conserved noncoding sequences of vertebrates in Chap. 9.

As for plants, Kaplinsky et al. (2002; [57]) found six short (<60 bp) CNSs from seven DNA regions related to protein coding gene orthologs between rice and maize. Guo et al. (2003; [58]) identified 20 bp as the minimal criterion for a CNS in grasses. Inada et al. (2003; [59]) examined 3,000 bases upstream and downstream of 52 orthologous protein coding genes of rice and maize and found that most CNSs were less than 20 bases. Thomas et al. (2007; [60]) compared *Arabidopsis thaliana* paralogous sequences, and found 14,944 intronic conserved noncoding sequences, ranging their lengths from 15 to 285 bp. D'Hont et al. (2012; [61]) determined banana genome and found 116 CNSs from genome sequences of commelinid monocot (banana, palm, and grasses). Kristas et al. (2012; [62]) compared genome sequences of Arabidopsis, grape rice, and Brachypodium and found >100 times more abundant CNSs from monocots than dicots. Hettiarachchi and Saitou; [63] compared genome sequences of 15 plant species and searched lineage-specific CNSs. They found 2 and 22 CNSs shared by all vascular plants and angiosperms, respectively, and also confirmed that monocot CNSs are much more abundant than those of dicots.

8.5.5 Mutation Rate and Genome Size

What kind of the relationship exists between the genome size and mutation rates? If all the genetic information contained in the genome of one organism are necessary for survival of that organism, the individual will die even if only one gene of its genome lost its function by a mutation. An organism with a small genome size and hence with a small number of genes, such as viruses, can survive even if the mutation rate is high. In contrast, organisms with many genes may not be able to survive if highly deleterious mutations often happen. Therefore, such organisms must reduce the mutation rate.

Rajic et al. (2005; [43]) compared the rate of synonymous substitutions per year from virus to human and the protein coding region size and found a clear negative correlation, as shown in Fig. 8.8. Sunjan et al. (2010; [64]) compared many studies on viral mutations and found a clear negative correlation between the substitution type mutation rate per nucleotide site per cell infection and viral genome size.

However, when the nucleotide substitution type mutation rate per generation was compared with the whole-genome size, Lynch (2006; [65]) found a positive correlation. More recently, Lynch (2010; [66]) admitted that for organisms with small-sized genomes, these two values were in fact negatively correlated. However, when large-genome-sized eukaryotes are compared, now a positive correlation was observed.

We have to be careful when we discuss these two contradictory reports. One considered the rate using unit as physical year, while the other used one generation as the unit. Another difference is to use either only protein coding gene region DNA sizes or the whole-genome sizes. The relationship between the mutation rate and genome size is not simple. Drake et al. (1998; [67]) examined this problem and found that the mutation rate per genome per replication was approximately 1/300 for bacteria, while mutation rates of multicellular eukaryotes vary between 0.1 and 100 per genome per sexual or individual generation. Table 8.4 shows the list of the mutation rate and the genome size for various organisms. Apparently there is no clear tendency.

Fig. 8.8 A negative correlation between the rate of synonymous substitutions and the protein-coding region size (From Rajic et al. 2005; [43])

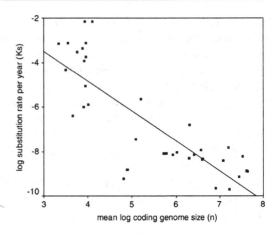

Table 8.4 Mutation rates and genome sizes of various organisms

Organism	Genome	Mutation rate ($\times 10^{-9}$) per		
(Organism group)	Size (bp)	Year	Generation	Reference
By direct method:				
Human	3.2×10^9	~0.4	11–12	1–4
Drosophila	1.7×10^8	35	3.5	5
C. elegans	8.0×10^7	–	2.7	6
Neurospora	4.2×10^7	–	0.072	7
Baker's yeast	1.2×10^7	–	0.22	7
E. coli	4.6×10^6	50	0.5	8
Phage T2,T4	1.7×10^5	–	24	7
Phage λ	4.9×10^4	–	77	7
mtDNA (*C. elegans*)	1.5×10^4	–	160	9
Phage M13	6.4×10^3	–	720	7
By indirect method:				
Human-Chimpanzee	3×10^9	1.0	15	10
Mouse-Rat	3×10^9	5.3	5.3	11
E. coli-Salmonella	4×10^6	4.5	0.04	7
mtDNA (Plants)	4×10^5	0.34	–	12
mtDNA (Mammals)	1.7×10^4	34	–	12
mtDNA (Birds)	1.7×10^4	17	–	12
RNA virus	~10^4	~10^6	~10^4	13

1: Roach, J. C., Glusman, G., Smit, A. F., Huff, C. D., Hubley, R., Shannon, P. T., Rowen, L., Pant, K. P., Goodman, N., Bamshad, M., Shendure, J., Drmanac, R., Jorde, L .B., Hood, L., & Galas, D. J. (2010). *Science, 328,* 636–639

2: Conrad, D. F., et al. (2011). Variation in genome-wide mutation rates within and between human families. *Nature Genetics, 43,* 712–715

3: Kong, A., et al. (2012). Rate of de novo mutations and the importance of father's age to disease risk. *Nature, 488,* 471–475

4: Campbel, C. D. (2012). Estimating the human mutation rate using autozygosity in a founder population. *Nature Genetics, 44,* 1277–1283

5: Keightley, P. D., Trivedi, U., Thomson, M., Oliver, F., Kumar, S., & Blaxter, M. L. (2009). Analysis of the genome sequences of three *Drosophila melanogaster* spontaneous mutation accumulation lines. *Genome Research, 19,* 1195–1201

(continued)

Table 8.4 (continued)

6: Denver, D. R., Dolan, P. C., Wilhelm, L. J., Sung, W., Lucas-Lledó, J. I., Howe, D. K., Lewis, S. C., Okamoto, K., Thomas, W. K., Lynch, M., & Baer, C. F. (2009). A genome-wide view of *Caenorhabditis elegans* base-substitution mutation processes. *Proceedings of the National Academy of Sciences of the United states of America, 106* 16310–16314

7: Drake, J. W., Charlesworth, B., Charlesworth, D., & Crow, J. F. (1998). Rates of spontaneous mutation. *Genetics, 148,* 1667–1686

8: Ochman, H. (2003). Neutral mutations and neutral substitutions in bacterial genomes. *Molecular Biology and Evolution, 20,* 2091–2096

9: Denver, D. R., Morris, K., Lynch, M., & Thomas, W. K. (2004). High mutation rate and predominance of insertions in the *Caenorhabditis elegans* nuclear genome. *Nature, 430,* 679–682

10: Fujiyama, A., Watanabe, H., Toyoda, A., Taylor, T. D., Itoh, T., Tsai, S.-F., Park, H.-S., Yaspo, M.-L., Lehrach, H., Chen, Z., Fu, G., Saitou, N., Osoegawa, K., de Jong, P. J., Suto, Y., Hattori, M., & Sakaki, Y. (2002). Construction and analysis of a human-chimpanzee comparative clone map. *Science, 295*(5552), 131–134

11: Abe, K., Noguchi, H., Tagawa, K., Yuzuriha, M., Toyoda, A., Kojima, T., Ezawa, K., Saitou, N., Hattori, M., Sakaki, Y., Moriwaki, K., & Shiroishi, T. (2004). Contribution of Asian mouse subspecies Mus musculus molossinus to genomic constitution of strain C57BL/6J, as defined by BAC end sequence-SNP analysis. *Genome Research, 14,* 2239–2247

12: Lynch, M., Koskella, B., & Schaack, S. (2006). Mutation pressure and the evolution of organelle genomic architecture. *Science, 311,* 1727–1730

13: Hanada, K., Suzuki, Y., & Gojobori, T. (2004). A large variation in the rates of synonymous substitution for RNA viruses and its relationship to a diversity of viral infection and transmission modes. *Molecular Biology and Evolution, 21*(6), 1074–1080

8.6 Genome of Multicellular Eukaryotes

We will discuss genomes of three multicellular lineages of eukaryotes: plants, fungi, and animals in this section. Unfortunately, there seems to be no common feature of genomes of multicellular organisms, so each lineage is discussed independently.

8.6.1 Plant Genomes

Arabidopsis thaliana was the first plant species whose 125-Mb genome was determined in 2000 [68]. A. *thaliana* is a model organism for flowering plants (angiosperms), with only 2-month generation time. In spite of its small genome size, only 4 % of the human genome, it has 32,500 protein coding genes. The genome sequence of its closely related species, A. *lyrata*, was also recently determined [69].

Angiosperms are divided into monocots and dicots. A. *thaliana* is a dicot, and genome sequences of six more species were determined as of December 2013 (see Table 8.5).

Rice, *Oryza sativa*, is a monocot, and its genome size, 370~410 Mb, is much smaller than that of the wheat genome. Its japonica and indica subspecies genomes were determined [70] and [71], and the origin of rice domestication is currently in great controversy, particularly in single or multiple domestication events (e.g., [72, 73]). The number of protein coding genes in the rice genome is 37,000~40,000 [74].

Table 8.5 List of plant species whose genome sequences were determined

Species name	English common name	Genome size (Mb)	Reference
Dicots:			
Arabidopsis thaliana	Thale Cress	135	1
Brassica rapa	Turnip mustard	273	2
Cucumis sativus	Cucumber	203	3
Ricinus communis	Castor bean	400	4
Populus tricocarpa	Cottonwood	422	5
Vitis vinifera	Grape	487	6
Aquilegia coerulea	Blue columbine	293	Unpublished
Monocots:			
Oryza sativa japonica	Rice (japonica variety)	372	7
Brachypodium distachyon	Purple false brome	272	8
Setaria italica	Foxtail millet	405	9
Sorghum bicolor	Sorghum	697	10
Musa acuminata	Banana	523	11
Phyllostachys heterocycla	Bamboo	2,000	12
Non-seed plants:			
Selaginella moellendorffii	Spikemoss	212	13
Physcomitrella patens	Moss	480	14
Chlamydomonas reinhardtii	Chlamydomonas	120	15

References

1: The Arabidopsis Genome Initiative. (2000). Analysis of the genome sequence of the flowering plant *Arabidopsis thaliana*. *Nature, 408*, 796–815

2: The Brassica rapa Genome Sequencing Project Consortium. (2011). The genome of the meso-polyploid crop species *Brassica rapa*. *Nature Genetics, 43*, 1035–1039

3: Huang, S., et al. (2009). The genome of the cucumber, *Cucumis sativu*s L. *Nature Genetics, 41*(12), 1275–1281

4: Chan, A. P., et al. (2010). Draft genome sequence of the oilseed species *Ricinus communis*. *Nature Biotechnology, 28*, 951–956

5: Tuskan, G., et al. (2006). The Genome of black cottonwood, Populus trichocarpa. *Science, 313*(5793), 1596–1604

6: Jaillon, O., et al. (2007). The grapevine genome sequence suggests ancestral hexaploidization in major angiosperm phyla. *Nature, 449*, 463–467

7: Goff, S. A., et al. (2002). A draft sequence of the rice genome (*Oryza sativa* L. ssp. japonica). *Science, 296*, 92–100

8: Vogel, J. P., et al. (2010). Genome sequencing and analysis of the model grass *Brachypodium distachyon*. *Nature, 463*, 763–768

9: Zhang, G., et al. (2012). Genome sequence of foxtail millet (*Setaria italica*) provides insights into grass evolution and biofuel potential. *Nature Biotechnology, 30*, 549–554

10: Paterson, A. H., et al. (2009). *Nature, 457*, 551–556

11: D'Hont, A., et al. (2012). The banana (*Musa acuminata*) genome and the evolution of mono-cotyledonous plants. *Nature, 488*, 213–217

12: Peng, Z., et al. (2013). The draft genome of the fast-growing non-timber forest species moso bamboo (*Phyllostachys heterocycla*). *Nature Genetics.* doi:10.1038/ng.2569

13: Banks, J. A., et al. (2011). The *Selaginella* genome identifies genetic changes associated with the evolution of vascular plants. *Science, 332*, 960–963

14: Rensing, S. A., et al. (2007). The *Physcomitrella* genome reveals evolutionary insights into the conquest of land by plants. *Science, 319*, 64–69

15: Merchant, S. S., et al. (2007). The *Chlamydomonas* genome reveals the evolution of key animal and plant functions. *Science, 318*, 254–250

Wheat corresponds to genus *Triticum*, and there are many species in this genus. The typical bread wheat is *Triticum aestivum*, and it is a hexaploid with 42 (7×6) chromosomes. Its genome arrangement is conventionally written as AABBDD [75]. Because it is now behaving as diploid, genomic sequencing of 21 chromosomes (A1–A7, B1–B7, and D1–D7) is under way (see http://www.wheatgenome.org/ for the current status). The hexaploid genome structure emerged by hybridization of diploid (DD) cultivated species *T. durum* and tetraploid (AABB) wild species *Aegilops tauschii* [75]. A genome duplication followed hybridization.

Non-seedling land plants are ferns, lycophytes, and bryophytes, in the order of closeness to seed plants (e.g., [76]). A draft genome sequence of a moss, *Physcomitrella patens* was reported in 2008 [77], followed by genome sequencing of a lycophyte, *Selaginella moellendorffii,* in 2011 [78]. These genome sequences of different lineages of plants are deciphering stepwise evolution of land plants.

8.6.2 Fungi Genomes

The genome sequence of baker's yeast (*Saccharomyces cerevisiae*) was determined in 1996, as the first eukaryotic organism [79]. There are 16 chromosomes in *S. cerevisiae,* and its genome size is about 12 Mb. There are a total of 8,000 genes in its genome: 6,600 ORFs and 1,400 other genes. The genome-wide GC content is 38 %, slightly lower than that of the human genome. The proportion of introns is very small compared to that of the human genome, and the average length of one intron is only 20 bp, in contrast to the 1,440-bp average length of exons [80]. As we already discussed, the ancestral genome of baker's yeast experienced a genome-wide duplication [40]. Pseudogenes, which are common in vertebrate genomes, are rather rare in the genome of baker's yeast; they constitute only 3 % of the protein coding genes [80]. The baker's yeast is often considered as the model organisms for all eukaryotes; however, their genome may not be a typical eukaryote genome.

As of December 2013, genome sequences of more than 400 fungi species are available (see NCBI genome list at http://www.ncbi.nlm.nih.gov/genome/browse/ for the present situation). Figure 8.9 shows the relationship between the genome size and gene numbers for 88 genomes. There is a clear positive correlation between them. However, there are some outliers. The Perigord black truffle (*Tuber melanosporum*), shown as A in Fig. 8.9, has the largest genome size (~125 Mb) among the 88 fungi species whose genome sequences were so far determined, yet the number of genes is only ~7,500 [81].

Three other outlier species are *Postia placenta, Ajellomyces dermatitidis,* and *Melampsora laricipopulina,* shown as B, C, and D in Fig. 8.9, respectively. Interestingly, these four outlier species are phylogenetically not clustered well; two are belonging to Pezizomycotina of Ascomycota and the other two are Agaricomycotina and Pucciniomycotina of Basidiomycota. If we exclude these four outlier species, a good linear regression is obtained, as shown in Fig. 8.9. This straight line indicates that in average, one gene size corresponds to 2.9 kb in a typical fungi genome. If we apply this average gene size to the truffle genome, its genome

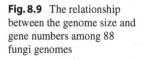

Fig. 8.9 The relationship between the genome size and gene numbers among 88 fungi genomes

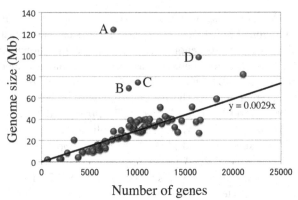

Number of genes

Fig. 8.10 Gain and loss of genes in each branch of the phylogenetic tree for fungi species (Based on [81])

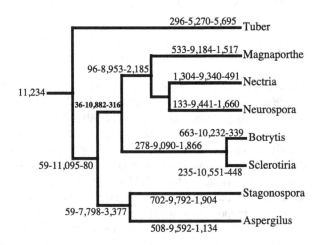

size should be ~22 Mb, but the real size is 103 Mb larger. This suggests that there is unusually large number of junk DNA in this genome. In fact, 58 % of its genome consists of transposable elements [81]. The truffle genome must still have 24 % more junk DNA region. Gain and loss of genes in each branch of the phylogenetic tree for fungi species are shown in Fig. 8.10 (based on [81]). It will be interesting to examine genome sizes of species related to the Perigord black truffle, so as to infer the evolutionary period when the genome size expansion occurred.

8.6.3 Animal Genomes

Animals, or metazoa, are the most integrated multicellular organisms. Genome sequences of four 35 invertebrate species and 32 vertebrate species were determined by end of 2011 according to the GCDB of Kryukov et al. (2012; [1]). As of December 2013, 35 invertebrate and 43 vertebrate species were deteremined according to KEGG database (http://www.genome.jp/kegg/catalog/org_list.html). A major gene

Fig. 8.11 The Hox gene clusters found in each animal phylum (From Saitou 2007; [103])

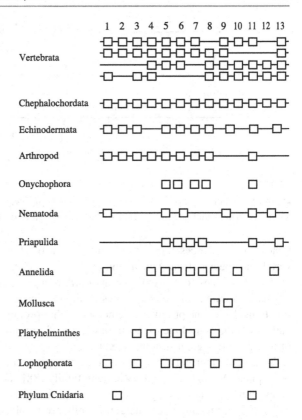

system that is responsible for this is Hox genes. We thus first discuss this gene system in this subsection. The genome of *C. elegans*, first determined genome among animals, will be discussed next, followed by genomes of insects and those of deuterostomes. Because genomes of many vertebrate species were determined, we discuss them in Chap. 9, and in particular, on the human genome in Chap. 10.

Hox Code

Hox genes were initially found through studies of homeotic mutations that dramatically change segmental structure of *Drosophila* by Edward B. Lewis [82]. They code for transcription factors, and a DNA-binding peptide, now called homeobox domain, was later found in almost all animal phyla [83]. Figure 8.11 shows the Hox gene clusters found in 12 animal groups. There are four Hox clusters in mammalian and avian genomes, and they are most probably generated by the two-round genome duplication in the common ancestor of vertebrates (see Chap. 9).

Interestingly, the physical order of Hox genes in chromosomes and the order of gene expression during the development are corresponding, called "collinearity" [84]. This suggests that some sort of cis-regulation is operating in Hox gene clusters, and in fact, many long transcripts are found, and some of their transcription start sites are highly conserved among vertebrates [85]. Figure 8.12 shows highly conserved

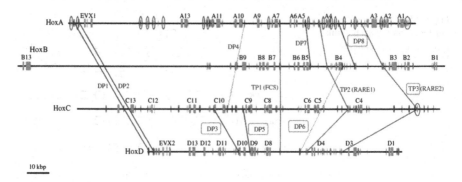

Fig. 8.12 Highly conserved noncoding sequences found from comparison of Hox A cluster regions of many vertebrate species (From Matsunami et al. 2010; [85])

noncoding sequences found from comparison of Hox A cluster regions of many vertebrate species (from Matsunami et al. 2010; [85]).

The Hox genes control expression of different groups of downstream genes, such as transcription factors, elements in signaling pathways, or genes with basic cellular functions. Hox gene products interact with other proteins, in particular, on signaling pathways, and contribute to the modification of homologous structures and creation of new morphological structures [87].

There are other gene families that are thought to be involved in diverse animal body plan. One of them is the Zic gene family [88]. The Zic gene family exists in many animal phyla with high amino acid sequence homology in a zinc-finger domain called ZF, and members of this gene family are involved in neural and neural crest development, skeletal patterning, and left–right axis establishment. This gene family has two additional domains, ZOC and ZF-BC. Interestingly, Cnidaria, Platyhelminthes, and Urochordata lack the ZOC domain, and their ZF-BC domain sequences are quite diverged compared to Arthropoda, Mollusca, Annelida, Echinodermata, and Chordata. This distribution suggests that the Zic family genes with the entire set of the three conserved domains already existed in the common ancestor of bilateralian animals, and some of them may be lost in parallel in the platyhelminthes, nematodes, and urochordates [88]. Interestingly, phyla that lost ZOC domains have quite distinct body plan although they are bilateralian.

Genome of C. *elegans*

Caenorhabditis elegans was the first animal species whose 97-Mb draft genome sequence was determined in 1998 [89]. This organism belongs to the Nematoda phylum which includes a vast number of species [90]. Brenner (1974; [91]) chose this species as model organism to study neuronal system, for its short generation time (~ 4 days) and its size (~1 mm). Figure 3.3 in Chap. 3 shows the cell genealogy of this species.

The following description of this section is based on the information given in online "WormBook" [86]. There are 22,227 protein coding genes in C. *elegans* including 2,575 alternatively spliced forms, with 79 % confirmed to be transcribed

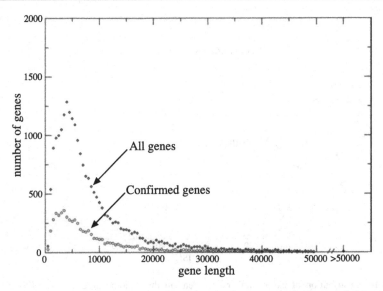

Fig. 8.13 Distribution of the protein coding genes in the genome of *Caenorhabditis elegans* (From [86])

at least partially. The number of tRNA genes is 608, and 274 are located in X chromosome. The three kinds of rRNA genes (18S, 5.8S, and 26S) are located in chromosome I in 100–150 tandem repeats, while ~100 5S rRNA genes are also in tandem form but located in chromosome V. The average protein coding gene length is 3 kb, with the average of 6.4 coding exons per gene. In total, protein coding exons constitute 25.6 % of the whole genome. Figure 8.13 shows the distribution of the protein coding genes, and Fig. 8.14 the distribution of exon numbers per gene. Both distributions have long tails. The median sizes of exons and introns are 123 bp and 65 bp, respectively. Intron lengths of *C. elegans* are quite short compared to these of vertebrate genes (see Chap. 9). The distribution of protein coding genes varies depending on chromosomes, slightly more dense for five autosomes than X chromosome and more dense in the central region than the edge of one chromosome. Processed, i.e., intronless, pseudogenes are rare, and a total of 561 pseudogenes were reported at the Wormbase version WS133. About half of them are homologous to functional chemoreceptor genes.

Genome sequences of four congeneric species of *C. elegans* (*C. brenneri*, *C. briggsae*, *C. japonica*, and *C. remanei*) were determined (http://www.ncbi.nlm.nih.gov/genome/browse/).

Insect Genomes

A fruit fly *Drosophila melanogaster* was used by Thomas Hunt Morgan's group in the early twentieth century and has been used for many genetic studies. Because of this importance, its genome sequence was determined at first among Arthropods in 2000 [92]. Heterochromatin regions of ~50 Mb were excluded from sequencing,

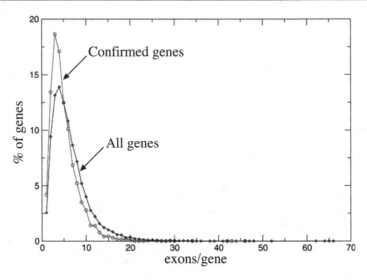

Fig. 8.14 Distribution of exon numbers per gene in the genome of *Caenorhabditis elegans* (From [86])

and only 120-Mb euchromatin regions were determined. Genome sequences of 12 Drosophila species (D. *ananassae*, D. *erecta*, D. *grimshawi*, D. *melanogaster*, D. *mojavensis*, D. *persimilis*, D. *pseudoobscura*, D. *sechellia*, D. *simulans*, D. *virilis*, D. *willistoni*, and D. *yakuba*) were determined in 2007 [93]. Their genome sizes vary from 145 to 258 Mb, and the number of genes is 15,000–18,000. Interestingly, D. *melanogaster* has the largest genome size and the smallest number of genes.

A total of 12 insect species other than *Drosophila* 12 species were sequenced by end of 2011 [1]. As of December 2013, their genome sizes are in the range of 108 Mb and 540 Mb, more than five times difference, and the gene numbers are from 9,000 to 23,000.

Genomes of Deuterostomes

Deuterostomes contain five phyla: Echinodermata, Hemichordata, Chaetognatha, Xenoturbellida, and Chordata. The genome of sea urchin *Strongylocentrotus purpuratus* [94] was determined in 2006. Its genome size is 814 Mb with 23,300 genes. Genomes of another sea urchins, *Lytechinus variegatus* and *Patiria miniata*, are also under sequencing, as well as hemicordate *Saccoglossus kowalevskii*.

Chordata is classified into Urochordata (ascidians), Cephalochordata (lancelets or amphioxus), and Vertebrata (vertebrates). Because we will discuss genomes of vertebrates in Chap. 9, let us discuss genomes of ascidians and lancelets only. The genome of ascidian *Ciona intestinalis* was determined in 2002 [95], and the genome sequence of its congeneric species, *C. savignyi*, was also determined three years later [96]. The genome size of *C. intestinalis* is ~155 Mb with ~16,000 genes. Interestingly it contains a group of cellulose synthesizing enzyme genes, which were probably introduced from some bacterial genomes via horizontal gene transfer [8, 97].

The *C. intestinalis* genome also contains several genes that are considered to be important for heart development ([95]), and this suggests that heart of ascidians and vertebrates may be homologous. Through the superimposition of phylogenetic trees (see Chapter A2) for five genes coding muscle proteins, OOta and Saitou ([98]) estimated that vertebrate heart muscle was phylogenetically closer to vertebrate skeletal muscles. If both results are true, muscles used in heart might have been substituted in the vertebrate lineage. The genome sequences of an amphioxus (Cephalochordate *Branchiostoma floridae*) was determined in by Holland et al. (2008; [104]), and they provide good outgroup sequence data for vertebrates.

8.7 Eukaryote Virus Genomes

Eukaryotic viruses are relying most of metabolic pathways to their eukaryote host species. Therefore, the number of genes in virus genomes is usually very small. For example, influenza A virus has 8 RNA fragments coding for 11 protein genes, and the total genome size is ~13.6 kb.

As in bacteriophages, there are both DNA type and RNA type genomes in eukaryotic viruses. Table 8.6 shows one example of classification of eukaryotic viruses based on their genome structure [99]. Genomes of double-strand DNA genome viruses have four types: circular, simple linear, linear with proteins covalently attached to both ends, and linear but both ends were closed. Genomes of single-strand DNA genome viruses are either circular or linear.

Genomes of RNA genomes are all linear in both single- and double-strand type. Those of single-strand RNA genomes are classified into two types: plus strand and minus strand. A subset of single-plus strand RNA genome type is experiencing

Table 8.6 Classification of eukaryotic viruses based on their genome structure (From Sadaie et al. eds. 2004; [99])

Shape	Example virus
DNA genome:	
Double strand & circular	SV40, polyomavirus
Double strand & linear	T4 bacteriophage, herpes virus
Double strand & linear, proteins attached at both ends	Adenovirus, φ29 bacteriophage
Double strand & linear, both ends are closed	Poxvirus
Single strand & circular	φX174 bacteriophage, M13 bacteriophage
Single strand & linear	Parvovirus
RNA genome:	
Double strand & linear	Reovirus
Single plus strand & linear	Tobacco mosaic virus, poliovirus, coronavirus, norovirus, Japanese encephalitis virus
Single plus strand & linear, including DNA replication intermediate	Retrovius, human T cell leukemia virus
Single minus strand & linear	Rabies virus, measles virus, influenza virus, mumps virus, ebola virus

DNA intermediate during replication, such as retroviruses and human T-cell leukemia virus (HTLV).

Some DNA genome viruses are unusually large and similar to a small bacterial genome. Megavirus, parasitic to amoeba, has 1.26-Mb genome size and there are 1,120 protein coding genes [100]. Megavirus is phylogenetically close to mimivirus [101], a member of nucleoplasmic large DNA viruses, including pox virus. Recently, a larger genome size virus, Pandoravirus, with more than 2.5-Mb genome, was discovered [105]. The phylogenetic status of these large genome size DNA viruses is unknown at this moment.

References

1. Kryukov, K., Sumiyama, K., Ikeo, K., Gojobori, T., & Saitou, N. (2012). A new database (GCD) on genome composition for eukaryote and prokaryote genome sequences and their initial analyses. *Genome Biology and Evolution, 4*, 501–512.
2. Andersson, S. G., et al. (1998). The genome sequence of *Rickettsia prowazekii* and the origin of mitochondria. *Nature, 396*, 133–140.
3. Martin, W., & Muller, M. (1998). The hydrogen hypothesis for the first eukaryote. *Nature, 392*, 37–41.
4. Wolstenholme, D. R., & Jeon, K. W. (Eds.) (1992). *Mitochondrial genome*. San Diego: Academic Press.
5. Sakai, M., & Sakaizumi, M. (2012). The complete mitochondrial genome of *Dugesia japonica* (Platyhelminthes; Order Tricladida). *Zoological Science, 29*, 672–680.
6. Sugiyama, Y., Watase, Y., Nagase, M., Makita, M., Yagura, S., Hirai, A., & Sugiura, M. (2005). The complete nucleotide sequence of the tobacco mitochondrial genome: Comparative analysis of mitochondrial genomes in higher plants and multipartite organization. *Molecular and General Genomics, 272*, 603–615.
7. Bergthorsson, U., Adams, K. L., Thomason, B., & Palmer, J. (2003). Widespread horizontal transfer of mitochondrial genes in flowering plants. *Nature, 424*, 197–201.
8. Rodriguez-Moreno, L., Benjak, A., Marti, M. C., Puigdomenech, P., Aranda, M. A., & Garcia-Mas, J. (2011). Determination of the melon chloroplast and mitochondrial genome sequences reveals that the largest reported mitochondrial genome in plants contains a significant amount of DNA having a nuclear origin. *BMC Genomics, 12*, 424.
9. Lilly, J. W., & Havey, M. J. (2001). Small, repetitive DNAs contribute significantly to the expanded mitochondrial genome of cucumber. *Genetics, 159*, 317–328.
10. Shinozaki, K., Ohme, M., Tanaka, M., Wakasugi, T., Hayashida, N., Matsubayashi, T., Zaita, N., Chunwongse, J., Obokata, J., Yamaguchi-Shinozaki, K., Ohto, C., Torazawa, K., Meng, B. Y., Sugita, M., Deno, H., Kamogashira, T., Yamada, K., Kusuda, J., Takaiwa, F., Kato, A., Tohdoh, N., Shimada, H., & Sugiura, M. (1986). The complete nucleotide sequence of the tobacco chloroplast genome: Its gene organization and expression. *EMBO Journal, 5*, 2043–2049.
11. Oldenburg, D. J., & Bendich, A. J. (2004). Changes in the structure of DNA molecules and the amount of DNA per plastid during chloroplast development in maize. *Journal of Molecular Biology, 344*, 1311–1330.
12. Woischnik, M., & Moraes, C. T. (2002). Pattern of organization of human mitochondrial pseudogenes in the nuclear genome. *Genome Research, 12*, 885–893.
13. The Nobel Prize in Physiology or Medicine. (1993). (http://www.nobelprize.org/nobel_prizes/medicine/laureates/1993/press.html)
14. Gilbert, W. (1978). Why genes in pieces? *Nature, 271*, 501.
15. Kenmochi, N. (2012). Introns. In *Encyclopedia of evolution*. Tokyo: Kyoritsu Shuppan (in Japanese).

16. Sheth, N., Roca, X., Hastings, M. L., Roeder, T., Krainer, A. R., & Sachidanandam, R. (2006). Comprehensive splice-site analysis using comparative genomics. *Nucleic Acids Research, 34*, 3955–3967.
17. Tycowski, K. T., Kolev, N. G., Conrad, N. K., Fok, V., & Steitz, J. A. (2006). The ever-growing world of small nuclear ribonucleoproteins. In R. F. Gesteland, T. R. Cech, & J. F. Atkins (Eds.), *The RNA World*, 3rd ed. (pp. 327–368). Woodbury: Cold Spring Harbor Laboratory Press
18. Cavalier-Smith, T. (1991). Intron phylogeny: A new hypothesis. *Trends in Genetics, 7*, 145–148.
19. Marck, C., & Grosjean, H. (2002). tRNomics: Analysis of tRNA genes from 50 genomes of Eukarya, Archaea, and Bacteria reveals anticodon-sparing strategies and domain-specific features. *RNA, 8*, 1189–1232.
20. Koonin, E. V. (2006). The origin of introns and their role in eukaryogenesis: A compromise solution to the introns-early versus introns-late debate? *Biology Direct, 1*, 22.
21. Roy, S. W., & Gilbert, W. (2006). The evolution of spliceosomal introns: Patterns, puzzles and progress. *Nature Reviews Genetics, 7*, 211–221.
22. Doolittle, W. F. (1978). Genes in pieces: Were they ever together? *Nature, 272*, 581–582.
23. Cavalier-Smith, T. (1978). Nuclear volume control by nucleoskeletal DNA, selection for cell volume and cell growth rate, and the solution of the DNA C-value paradox. *Journal of Cell Science, 34*, 247–278.
24. Logsdon, J. M., Jr. (1998). The recent origins of spliceosomal introns revisited. *Current Opinion in Genetics & Development, 8*, 637–648.
25. Go, M. (1981). Correlation of DNA exonic regions with protein structural units in haemoglobin. *Nature, 291*, 90–92.
26. Rogozin, I. B., Wolf, Y. I., Sorokin, A. V., Mirkin, B. G., & Koonin, E. V. (2003). Remarkable interkingdom conservation of intron positions and massive, lineage-specific intron loss and gain in eukaryotic evolution. *Current Biology, 13*, 1512–1517.
27. Nguyen, D. H., Yoshihama, M., & Kenmochi, N. (2005). New maximum likelihood estimators for eukaryotic intron evolution. *PLoS Computational Biology, 1*, e79.
28. Yoshihama, M., Nakao, A., Nguyen, H. D., & Kenmochi, N. (2006). Analysis of ribosomal protein gene structures: Implications for intron evolution. *PLoS Genetics, 2*, 237–242.
29. Yoshihama, M., Nguyen, H. D., & Kenmochi, N. (2007). Intron dynamics in ribosomal protein genes. *PLoS One, 1*, e141.
30. Ohno, S. (1972). So much "junk" DNA in our genome. *Brookhaven Symposium in Biology, 23*, 366–370.
31. Jurka, J., & Smith, T. (1988). A fundamental division in the Alu family of repeated sequences. *Proceedings of the National Academy of Sciences of the United States of America, 85*, 4775–4778.
32. Price, A. L., Eskin, E., & Pevzner, P. A. (2004). Whole-genome analysis of Alu repeat elements reveals complex evolutionary history. *Genome Research, 14*, 2245–2252.
33. Simoens, C. R., Gielen, J., Van Montagu, M., & Inze, D. (1988). Characterization of highly repetitive sequences of *Arabidopsis thaliana*. *Nucleic Acids Research, 16*, 6753–6766.
34. Murata, M., Ogura, Y., & Mototoshi, F. (1994). Centromeric repetitive sequences in *Arabidopsis thaliana*. *Japanese Journal of Genetics, 69*, 361–370.
35. Wu, J. C., & Manuelidis, L. (1980). Sequence definition and organization of a human repeated DNA. *Journal of Molecular Biology, 142*, 363–386.
36. Yamada, K., et al. (2002). Empirical analysis of transcriptional activity in the *Arabidopsis* genome. *Science, 302*, 842–846.
37. The ENCODE Project Consortium. (2007). Identification and analysis of functional elements in 1% of the human genome by the ENCODE pilot project. *Nature, 447*, 799–816.
38. van Bakel, H., Nislow, C., Blencowe, B. J., & Hughes, T. R. (2010). Most "dark matter" transcripts are associated with known genes. *PLoS Biology, 8*, e1000371.
39. Yoshiura, K., et al. (2006). A SNP in the ABCC11 gene is the determinant of human earwax type. *Nature Genetics, 38*, 324–330.

40. Wolfe, K. H., & Shields, D. C. (1997). Molecular evidence for an ancient duplication of the entire yeast genome. *Nature, 387*, 708–713.

41. Ma, L.-J., et al. (2009). Genomic analysis of the basal lineage fungus *Rhizopus oryzae* reveals a whole-genome duplication. *PLoS Genetics, 5*, e1000549.

42. Aury, J. M., et al. (2006). Global trends of whole-genome duplications revealed by the ciliate *Paramecium tetraurelia. Nature, 444*, 171–178.

43. Rajic, Z. A., Jankovic, G. M., Vidovic, A., Milic, N. M., Skoric, D., Pavlovic, M., & Lazarevic, V. (2005). Size of the protein-coding genome and rate of molecular evolution. *Journal of Human Genetics, 50*, 217–229.

44. Lynch, M., & Conery, J. S. (2000). The evolutionary fate and consequences of duplicated genes. *Science, 302*, 1401–1404.

45. Gregory, T. R., & DeSalle, R. (2005). Comparative genomics in prokaryotes. In T. R. Gregory (Ed.), *The evolution of the genome*, Chapter 10, Burlington: Elsevier Academic Press.

46. Gott, J. M., & Emeson, R. B. (2000). Functions and mechanisms of RNA editing. *Annual Review of Genetics, 34*, 499–531.

47. Tillich, M., Lehwark, P., Morton, B. R., & Maier, U. G. (2006). The evolution of chloroplast RNA editing. *Molecular Biology and Evolution, 23*, 1912–1921.

48. Gall, J. G. (1981). Chromosome structure and the C-value paradox. *Journal of Cell Biology, 91*, 3s–14s.

49. Vendrely, R., & Vendrely, C. (1948). La teneur du noyau cellulaire en acide désoxyribonucléique à travers les organes, les individus et les espèces animales (in French). *Cellular and Molecular Life Sciences, 4*, 434–436.

50. Pollister, A. W., & Ris, H. (1947). Nucleoprotein determination in cytological preparations. *Cold Spring Harbor Symposia on Quantitative Biology, 12*, 147–157.

51. Swift, H. (1950). The constancy of deoxyribose nucleic acid in plant nuclei. *Proceedings of the National Academy of Sciences of the United States of America, 36*, 643–654.

52. How, G. F., Venkatesh, B., & Brenner, S. (1996). Conserved linkage between the puffer fish (Fugu rubripes) and human genes for platelet-derived growth factor receptor and macrophage colony-stimulating factor receptor. *Genome Research, 6*, 1185–1191.

53. Hardison, R. C. (2000). Conserved noncoding sequences are reliable guides to regulatory elements. *Trends in Genetics, 16*, 369–372.

54. Levy, S., Hannenhalli, S., & Workman, C. (2001). Enrichment of regulatory signals in conserved non-coding genomic sequence. *Bioinformatics, 17*, 871–877.

55. Carroll, S. B. (2005). Evolution at two level: On genes and form. *PLoS Biology, 3*, e245.

56. Siepel, A., Bejerano, G., Pedersen, J. S., Hinrichs, A. S., Hou, M., Rosenbloom, K., Clawson, H., Speith, J., Hillier, L. W., Richards, S., Weinstock, G. M., Wilson, R. K., Gibbs, R. A., Kent, W. J., Miller, W., & Haussler, D. (2005). Evolutionarily conserved elements in vertebrate, insect, worm, and yeast genomes. *Genome Research, 15*, 1034–1050.

57. Kaplinsky, N. J., Braun, D. M., Penterman, J., Goff, S. A., & Freeling, M. (2002). Utility and distribution of conserved noncoding sequences in the grasses. *Proceedings of the National Academy of Sciences of the United States of America, 99*, 6147–6151.

58. Guo, H., & Moose, S. P. (2003). Conserved noncoding sequences among cultivated cereal genomes identify candidate regulatory sequence elements and patterns of promoter evolution. *Plant Cell, 15*, 1143–1158.

59. Inada, D. C., et al. (2003). Conserved noncoding sequences in the grasses. *Genome Research, 13*, 2030–2041.

60. Thomas, B. C., Rapaka, L., Lyons, E., Pedersen, B., & Freeling, M. (2007). Arabidopsis intragenomic conserved noncoding sequence. *Proceedings of the National Academy of Sciences of the United States of America, 104*, 3348–3353.

61. D'Hont, A., et al. (2012). The banana (*Musa acuminata*) genome and the evolution of monocotyledonous plants. *Nature, 488*, 213–217.

62. Kritsas, K., Samuel, E., Wuest, S. E., Hupalo, D., Kern, A. D., Wicker, T., & Grossniklaus, U. (2012). Computational analysis and characterization of UCE-like elements (ULEs) in plant genomes. *Genome Research, 22*, 2455–2466.

63. Hettiarachchi, N., & Saitou, N. (2013). Identification and analysis of conserved noncoding sequences in plants. *Genome Biology and Evolution* (in revision).
64. Sanjuan, R., Nebot, M. R., Chirico, N., Mansky, L. M., & Belshaw, R. (2010). Viral mutation rates. *Journal of Virology, 84*, 9733–9748.
65. Lynch, M. (2006). The origins of eukaryotic gene structure. *Molecular Biology and Evolution, 23*, 450–468.
66. Lynch, M. (2010). Evolution of the mutation rate. *Trends in Genetics, 26*, 345–352.
67. Drake, J. W., Charlesworth, B., Charlesworth, D., & Crow, J. F. (1998). Rates of spontaneous mutation. *Genetics, 148*, 1667–1686.
68. Arabidopsis Genome Initiative. (2000). Analysis of the genome sequence of the flowering plant *Arabidopsis thaliana. Nature, 408*, 796–815.
69. Hu, T. T., et al. (2011). The *Arabidopsis lyrata* genome sequence and the basis of rapid genome size change. *Nature Genetics, 43*, 476–481.
70. Goff, S. A., & others. (2002). A draft sequence of the rice genome (*Oryza sativa* L. ssp. japonica). *Science, 296*, 92–100.
71. Yu, J., & others. (2002). A draft sequence of the rice genome (*Oryza sativa* L. ssp. indica). *Science, 296*, 79–92.
72. Londo, J. P., et al. (2006). Phylogeography of Asian wild rice, *Oryza rufipogon*, reveals multiple independent domestications of cultivated rice, *Oryza sativa. Proceedings of the National Academy of Sciences of the United States of America, 103*, 9578–9583.
73. Yang, C.-C., Kawahara, Y., Mizuno, H., Wu, J., Matsumoto, T., & Itoh, T. (2012). Independent domestication of Asian rice followed by gene flow from japonica to indica. *Molecular Biology and Evolution, 29*, 1471–1479.
74. The Rice Annotation Project. (2007). Curated genome annotation of *Oryza sativa* ssp. japonica and comparative genome analysis with *Arabidopsis thaliana. Genome Research, 17*, 175–183.
75. "Chromosomes" of KOMUGI database (http://www.shigen.nig.ac.jp/wheat/komugi/chromosomes/chromosomes.jsp)
76. Nickrent, D. L., et al. (2000). Multigene phylogeny of land plants with special reference to bryophytes and the earliest land plants. *Molecular Biology and Evolution, 17*(12), 1885–1895.
77. Rensing, S. A., et al. (2008). The *Physcomitrella* genome reveals evolutionary insights into the conquest of land by plants. *Science, 319*, 64–69.
78. Banks, J. A., et al. (2011). The *Selaginella* genome identifies genetic changes associated with the evolution of vascular plants. *Science, 332*, 960–963.
79. Mewes, H. W., Albermann, K., Bahr, M., Frishman, D., Gleissner, A., Hani, J., Heumann, K., Kleine, K., Maierl, A., Oliver, S. G., Pfeiffer, F., & Zollner, A. (1997). Overview of the yeast genome. *Nature, 387*, 7–65.
80. Lynch, M. (2007). *Origin of genome architecture*. Sunderland: Sinaur Associates.
81. Martin, F., et al. (2010). Perigord black truffle genome uncovers evolutionary origins and mechanisms of symbiosis. *Nature, 464*, 1033–1038.
82. Biography of E. B. Lewis (http://www.nobelprize.org/nobel_prizes/medicine/laureates/1995/lewis.html)
83. Gehring, W. J. (1999). *Master control genes in development and evolution: The homeobox story*. New Haven: Yale University Press.
84. Carroll, S. B., Grenier, J. K., & Weatherbee, S. D. (2005). *From DNA to diversity*. Malden: Blackwell Publishing.
85. Matsunami, M., Sumiyama, K., & Saitou, N. (2010). Evolution of conserved non-coding sequences within the vertebrate Hox clusters through the two-round whole genome duplications revealed by phylogenetic footprinting analysis. *Journal of Molecular Evolution, 71*, 427–436.
86. Chalfie, M. (Ed.). WormBook – The online review of *C. elegans* biology (http://www.wormbook.org/)
87. Foronda, D., de Navas, L. F., Garaulet, D. L., & Sanchez-Herrero, E. (2009). Function and specificity of Hox genes. *International Journal of Developmental Biology, 53*, 1409–1419.

88. Aruga, J., Kamiya, A., Takahashi, H., Fujimi, T. J., Shimizu, Y., Ohkawa, K., Yazawa, S., Umesono, Y., Noguchi, H., Shimizu, T., Saitou, N., Mikoshiba, K., Sakaki, Y., Agata, K., & Toyoda, A. (2006). A wide-range phylogenetic analysis of Zic proteins: Implications for correlations between protein structure conservation and body plan complexity. *Genomics, 87*, 783–792.

89. C. elegans Sequencing Consortium. (1998). Genome sequence of the nematode *C. elegans*: A platform for investigating biology. *Science, 282*, 2012–2018.

90. Meldal, B. H. M., et al. (2007). An improved molecular phylogeny of the Nematoda with special emphasis on marine taxa. *Molecular Phylogenetics and Evolution, 42*, 622–636.

91. Brenner, S. (1974). The genetics of *Caenorhabditis elegans*. *Genetics, 77*, 71–94.

92. Adams, M. D., & others. (2000). The genome sequence of *Drosophila melanogaster*. *Science, 287*, 2185–2195.

93. Drosophila 12 Genomes Consortium. (2007). Evolution of genes and genomes on the Drosophila phylogeny. *Nature, 450*, 203–218.

94. Sea Urchin Genome Sequencing Consortium. (2006). The genome of the sea urchin *Strongylocentrotus purpuratus*. *Science, 314*, 941–952.

95. Dehal, P., & others. (2002). The draft genome of *Ciona intestinalis*: Insights into chordate and vertebrate origins. *Science, 298*, 2157–2167.

96. Vinson, J. P., Jaffe, D. B., O'Neill, K., Karlsson, E. K., Stange-Thomann, N., Anderson, S., Mesirov, J. P., Satoh, N., Satou, Y., Nusbaum, C., Birren, B., Galagan, J. E., & Lander, E. S. (2005). Assembly of polymorphic genomes: Algorithms and application to Ciona savignyi. *Genome Research, 15*, 1127–1135.

97. Matthysse, A. G., Deschet, K., Williams, M., Marry, M., White, A. R., & Smith, W. C. (2004). A functional cellulose synthase from ascidian epidermis. *Proceedings of the National Academy of Sciences of the United States of America, 101*, 986–991.

98. OOta, S., & Saitou, N. (1999). Phylogenetic relationship of muscle tissues deduced from superimposition of gene trees. *Molecular Biology and Evolution, 16*, 856–867.

99. Sadaie, Y., et al. (Eds.). (2004). *Genome science and microorganismal molecular genetics (in Japanese)*. Tokyo: Baifukan.

100. Arslan, D., Legendre, M., Seltzer, V., Abergel, C., & Claverie, J. M. (2011). Distant Mimivirus relative with a larger genome highlights the fundamental features of Megaviridae. *Proceedings of the National Academy of Sciences of the United States of America, 108*, 17486–17491.

101. Raoult, D., & others. (2004). The 1.2-megabase sequence of mimivirus. *Science, 306*, 1344–1350.

102. Bejerano, G., Pheasant, M., Makunin, I., Stephen, S., Kent, W. J., Mattick, J. S., & Haussler, D. (2004). Ultraconserved elements in the human genome. *Science, 304*, 1321–1325.

103. Saitou, N. (2007). *Genomu Shinkagaku Nyumon (written in Japanese, meaning 'Introduction to evolutionary genomics')*. Tokyo: Kyoritsu Shuppan.

104. Holland, L. G. (2008). The amphioxus genome illuminates vertebrate origins and cephalo-chordate biology. *Genome Research, 18*, 1100–1111.

105. Phillipe, N., et al. (2013). Pandoraviruses: Amoeba viruses with genomes up to 2.5 Mb reaching that of parasitic eukaryotes. *Science, 341*, 281–286.

106. The Chloroplast Genome Database. http://chloroplast.ocean.washington.edu/

Vertebrate Genomes

9

Chapter Summary

We first discuss characteristics of vertebrate genome evolution in this chapter. Special attentions were given to two-round genome duplications and conserved noncoding sequences. We then move to brief description of vertebrate genomes for major phylogenetic lineages: teleost fish, amphibian, and amniotes. Mammals, in which many species genomes were already sequenced, were discussed in more details.

9.1 Characteristics of Vertebrate Genomes

Animals or metazoa were divided into vertebrates and invertebrates ever since Lamarck [1]. Invertebrates are now classified into many phyla in terms of their body plans, and vertebrates are only one group within phylum Chordata. However, from the viewpoint of the genome structure, vertebrates are quite unique among animals. They experienced a series of genome duplications that have not been found in many invertebrates, including vertebrates' close relatives in phylum Chordata: amphioxus and ascidians.

Vertebrates are also unique in their macroscopic characteristics. As their name indicates, they have vertebra and other bones, and a highly differentiated brain structure exists only in vertebrates. Invertebrates also have a central nervous system, but only vertebrates have a segmental structure such as the forebrain, midbrain, hindbrain, and spinal cord. Figure 9.1 shows a schematic view of a typical vertebrate body (based on [2] and [3]). Furthermore, only vertebrates have the acquired immune system with immunoglobulin, T-cell receptor, and MHC gene families.

As we briefly discussed in Chap. 8, the genome complexity in terms of oligonucleotide frequency is higher in most eukaryotes than prokaryotes. Vertebrates, particularly mammals, have the highest complexity among animals. Kryukov et al. (2012; [4]) introduced a measure, R, to represent the complexity of oligonucleotide frequencies in each genome. In short, R is a relative deviation of the observed

N. Saitou, *Introduction to Evolutionary Genomics*, Computational Biology 17, DOI 10.1007/978-1-4471-5304-7_9, © Springer-Verlag London 2013

Fig. 9.1 A schematic view
of vertebrate body (Based on
Refs. [2] and [3])

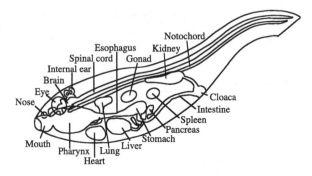

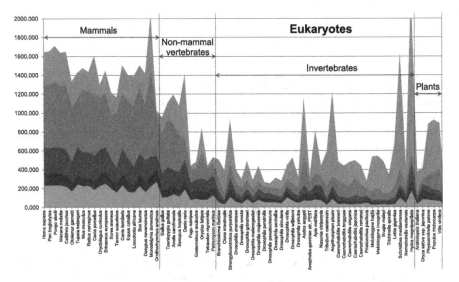

Fig. 9.2 Comparison of R values based on oligonucleotides of 5 bp and all five composition
models R values are shown in vertical axix. Different nucleotide frequency models were used and
they are shown in different gray scales from top to bottom: uniform nucleotide frequency, mononucleotide, dinucleotide, trinucleotide, and tetranucleotide (From Kryukov et al. 2012; [4])

oligonucleotide frequencies from the expected ones under a particular model based
on shorter oligonucleotide length frequencies. Figure 9.2 shows R values for 5-bp
frequencies of animals and plants. In all groups, R values become smaller from the
uniform model (A, C, G, and T are all 25 % frequency) to the tetranucleotide
frequency-based model. Even when the most complex tetranucleotide frequency-
based model was assumed, R is larger than 200 for mammalian genomes, while R
is ~100 for non-mammal vertebrates and ~30 for invertebrates and plants. The reason
for this variation of R values is not clear.

Fig. 9.3 Quartet gene
distribution on the human
genome (From Matsunami
2012; [11])

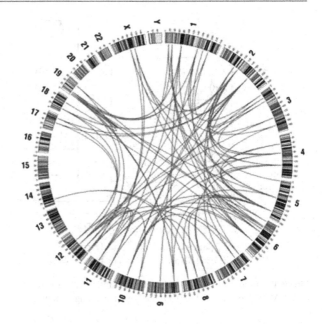

9.2   Two-Round Genome Duplications

Much before any nucleotide sequence data were produced, Susumu Ohno (1970;
[5]) proposed a hypothesis that the vertebrate common ancestor experienced two
rounds of genome duplications (2RGD), mostly based on the DNA amount per cell.

We now know that mammalian genomes often have four homologous genes or
gene clusters in different chromosomes. A notable example is four Hox clusters as
already mentioned in Chap. 8. In the human genome, Hox clusters A, B, C, and D
are located on chromosomes 7, 17, 12, and 2. Another example is the major histocompatibility complex (MHC) genes and their paralogous genes. The MHC is on
chromosome 6 in human, but homologous genes were found in chromosomes 1, 9,
and 19 [6, 7]. Although there was some controversy over the existence of the 2RGD
in the common ancestor of vertebrates (e.g., [8]), it is now established through
genomic comparisons [9, 10]. Figure 9.3 shows the quartet gene distribution on the
human genome [11].

Because amphioxus and ascidian do not seem to have experienced any genome
duplication, 2RGD probably occurred after the vertebrates diverged from ascidian
(see Chap. 6 on the phylogenetic relationship of chordates). Now the problem is the
timing of these two genome duplications with respect to the divergence of gnathostomes (jawed vertebrates, most of us) and cyclostomes (jawless vertebrates, lamprey
and hagfish). There are three possibilities on the timings of 2RGD (Fig. 9.4). Many
examples exist for gene phylogenies in which duplications occurred before the

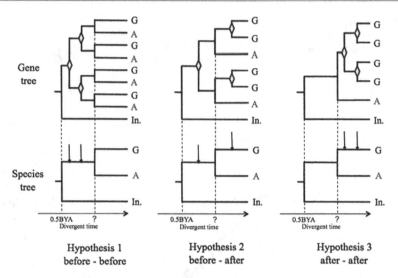

Fig. 9.4 Three possibilities on genome duplications in the common ancestor of vertebrates (From Matsunami 2012; [12]). *G* gnathostomes, *A* agnathans, *In.* invertebrates

Fig. 9.5 An example of gene phylogeny in which duplications occurred before the divergence of gnathostomes and cyclostomes (From Matsunami 2012; [12])

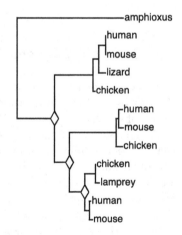

divergence of gnathostomes and cyclostomes. One example is shown in Fig. 9.5 (provided by Dr. Matsunami Masatoshi). Case 3 of Fig. 9.4 is impossible if we accept that the gene phylogeny shown in Fig. 9.5. At least one genome duplication occurred before the divergence of gnathostomes and cyclostomes. The problem is now restricted to case 1 or 2 of Fig. 9.4.

Matsunami [12] analyzed recently released lamprey (agnathan) genome data and newly obtained transcriptome data [13] with many gnathostome genome data. He found that 127 gene families had one lamprey gene (A), two paralogous gnathostome gene (G_1 and G_2), and an out-group species gene (X). Although there are three possible rooted trees for three vertebrate genes, we do not need to distinguish G_1 and G_2, and let us drop their subscripts. There are now only two possible

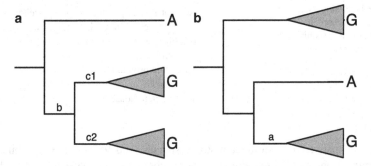

Fig. 9.6 Two possible tree topologies for two gnathostome (G) gene clusters and one agnathan (A) gene. (**a**) ((G,G),A) topology, (**b**) ((A,G),G) topology

tree topologies: ((A,G),G) and ((G,G),A) (Fig. 9.6). There were 69 and 58 gene families showing these two topologies, respectively [12]. These two possible trees superficially correspond to duplication before or after the divergence of gnathostomes and cyclostomes. The ((G,G),A) topology is compatible with the case in which a genome duplication occurred before the lamprey/gnathostome divergence, if we assume that one of the two lamprey genes was lost after the duplication. To examine the possibility of such hidden gene duplication sharing, the internal branch length of each tree was inferred. If genome duplications are shared among all vertebrates, internal branch from the agnathan speciation to the gnathostome divergences will be equal between two gene topologies (a = b + c* in Fig. 9.6, where c* = (c1 + c2)/2). On the contrary, if lamprey genes were lost, the ((G,G),A) topology genes should show longer internal branch length than that of ((A,G),G) topology genes (b + c* > a in Fig. 9.6). In fact, the average internal branch length (0.086 = b + c*) of ((G,G),A) trees was longer than that (0.055 = a) of ((A,G),G) genes with a high statistical significance (P < 0.001, *t*-test) [12]. These results support case 1 of Fig. 9.4 in which both genome duplications occurred before the divergence of gnathostomes and cyclostomes. This result was also supported by Kuraku et al. (2009; [14]) and Smith et al. (2013; [15]).

9.3   Features of Protein Coding Genes

9.3.1   Genes Shared with Other Taxonomic Groups

Kawashima et al. (2008; [16]) compared protein domain combinations of various deuterostome organisms as well as other eukaryotic species using the Pfam database (see Chap. 13). Seven vertebrate species (human, mouse, rat, chicken, *Xenopus tropicalis*, zebra fish, and torafugu), one ascidian species (*Ciona intestinalis*), one amphioxus species (*Branchiostoma floridae*), and one sea urchin species (*Strongylocentrotus purpuratus*) were compared to represent deuterostome lineages. They found that 1,326 domain combinations were found in all deuterostome species compared,

Table 9.1 Examples of proteins with chordate-specific domain combinations (From Ref. [5])	Adaptor-related protein complex 3, delta 1 subunit
	Diacylglycerol kinase, delta 130 kDa
	Insulin-like growth factor-binding protein 4
	Latrophilin 1
	Surfactant protein D
	Tenascin N
	Thyroid peroxidase

as well as shared with non-deuterostome species, and 957 domain combinations were found in most of deuterostome species except for one lineage, as well as shared with non-deuterostome species. If we further include domain combinations that were not found in two deuterostome lineages, in total, about 3,200 domain combinations seem to exist at the common ancestor of deuterostomes. If only the three chordate lineages (vertebrates, ascidian, and amphioxus) were compared, the chordate ancestor was estimated to have 269 chordate-specific domain combinations. Table 9.1 lists examples of vertebrate genes shared with other eukaryotes.

9.3.2   Genes Specific to Vertebrates

We now move to genes specific to vertebrates. Although vertebrates share its ancestry with ascidians and amphioxus at the common ancestor of chordates, there are many vertebrate-specific characteristics, such as cartilage, immune system, complicated craniofacial structures, and well-compartmentalized brain system. Kawashima et al. (2008; [16]) extracted 859 vertebrate-specific domain combinations. One example is aggrecan, the most abundant noncollagenous protein in cartilage. Aggrecan consists of five types of domains, including immunoglobulin and extracellular link domains [16]. Another notable example of vertebrate-specific domain combinations is tectorin-alpha, involved in the vertebrate auditory system. This protein is a major component of the tectorial membrane in the mammalian inner ear [16].

The acquired immune system exists only in vertebrates. A comparison between amphioxus and vertebrate genomes showed that acquired immune system-related genes existing in vertebrates such as VLR, BCR, TCR, and MHC class I and II molecules are absent in amphioxus, while MHC class III genes are involved in the complement system, Toll-like receptor genes, tumor necrosis factor genes, and interleukin-related genes exist in the amphioxus genome [17]. Table 9.2 presents examples of genes specific to vertebrates.

Many characteristics of sex chromosomes are lineage specific in tetrapods. Mammals have the XY system, in which females are XX and males are XY. Birds have an opposite type ZW system; females are WZ and males are WW. The mammalian-specific Sry gene, located in Y chromosome, is responsible for male determination. Recently, DMRT1 gene located in Z chromosome was shown to be responsible for sex determination in birds [18]. Chromosomal involvement in sex determination was not clear in reptiles for a long time, but the mammalian type

Table 9.2 Examples of proteins specific to vertebrates (From Ref. [5])

Proteins used in many kinds of cells
Occludin
SNF2 histone linker PHD RING helicase
Ubiquitin-conjugating enzyme E2B
Vitamin K-dependent plasma glycoprotein
Proteins used for cell-cell communications
Agrin
Attractin
Complement component 4A
Serine protease 7
Serine protease inhibitor (Kuniz type)
Stabilin 1
Von Willebrand factor
Proteins used for intracelluar signal transduction
F-box protein 41
Obscurin, cytoskeletal calmodulin, and titin-interacting RhoGEF
Regulator of G-protein signaling 12
Proteins used for extracellular matrix and cell adhesion
Aggrecan
Cartilage acidic protein 1
Fibronectin 1
Fibulin 1
Nidogen 1

XY chromosomes were found through the genome sequencing of green anole [19]. These sex chromosomes have neither Sry/SOX3 nor DMRT1. Figure 9.7 shows a possible scenario of the evolution of sex determination systems among amniotes, partially based on Graves (2011; [20]).

9.3.3 Olfactory Multigene Families

Distinction between a multigene family and an ordinary gene family is not clear. If there are more than ten homologous genes in one genome, this homologous gene group may be called a multigene family. A good and probably the best example of a multigene family in mammalian genomes is the olfactory receptor family, with more than 100 copies in many vertebrate species genomes. One olfactory receptor gene consists of single exon and is easy to duplicate. Niimura and Nei (2006; [21]) searched for olfactory receptor genes from eight vertebrate genome sequences, and they found 1,391 and 802 olfactory receptor genes from human and

Fig. 9.7 A possible scenario
of the evolution of sex
determination systems among
amniotes (Based on Graves,
2009; [20])

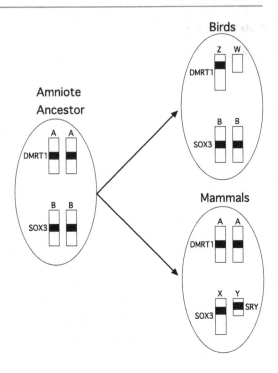

mouse genomes, respectively. Interestingly, 25 % mouse and 52 % human genes
were pseudogenes. Here, functional genes were defined as any gene with a complete
open reading frame and pseudogenes as those with interrupting stop codons,
frameshift mutations, or long deletions [21]. The dog genome also had a low (21 %)
proportion of olfactory receptor pseudogenes among 1,094 genes. The number of
olfactory receptor genes increased after the emergence of tetrapods, for teleost fish
(fugu and zebra fish) genomes contain less than 200 copies [21]. Figure 9.8 shows
a phylogenetic tree of olfactory receptors among five vertebrate genomes. There are
nine major clusters (α-κ), and current fish species retain eight clusters, while only two
clusters were found from mammalian or avian genomes with a few exceptions [21].
Ligand molecules for any olfactory receptor is not yet discovered, but putative ligands
for those in clusters α and γ were airborne molecules, and if this conjecture is true,
ancestral olfactory receptors belonging to these two clusters acquired the ability to
detect airborne odorants when tetrapod lineage landed [21].

9.3.4 Selective Constraints in the Protein Coding Sequences

The primary role of a protein coding gene is to encode amino acids. Therefore,
synonymous sites of codons, which do not change the encoded amino acid, are
regarded as evolving neutrally. However, if a certain region of a protein coding gene

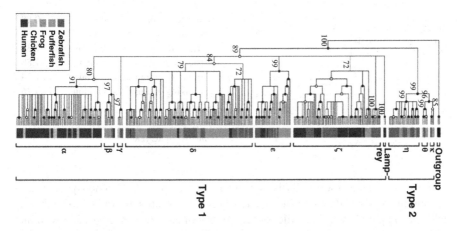

Fig. 9.8   Phylogenetic tree of olfactory receptors (From Ref. [21])

contains a functional nucleotide element such as splicing signals, synonymous sites in the region may have selective pressure. The existence of such elements would be detected by searching regions of low nucleotide substitution. Suzuki and Saitou [22] explored invariant nucleotide sequences in 10,790 orthologous genes of six mammalian species (*Homo sapiens, Macaca mulatta, Mus musculus, Rattus norvegicus, Bos taurus, and Canis familiaris*) and extracted 4,150 sequences whose conservation is significantly stronger than other regions of the gene and named them significantly conserved coding sequences (SCCSs). SCCSs are observed in 2,273 genes. The genes are mainly involved with development, transcriptional regulation, and the neurons and are expressed in the nervous system and the head and neck organs. No strong influence of conventional factors that affect synonymous substitution was observed in SCCSs. These results imply that SCCSs may have double function as nucleotide element and protein coding sequence and retained in the course of mammalian evolution. See also Chap. 5 on this topic.

9.4   Conserved Noncoding Regions

Li et al. (1985; [23]) compared nucleotide sequences of exons and their flanking regions of various mammalian genes available at that time and estimated rates of nucleotide substitutions. Table 9.3 shows these values for each region. Pseudogenes and synonymous codon sites showed similar and highest rates ($\sim 5 \times 10^{-9}$/site/year), which are expected to be equivalent to the mutation rate under the neutral evolution theory (see Chap. 4). The next highest was that for 3'-flanking untranscribed region, suggesting that this region is essentially the intergenic, i.e., junk DNA region. In contrast, the evolutionary rate for 5'-flanking untranscribed region is lower than that, probably because of existence of gene expression control such as promoters. Because they are part of exons, rates ($\sim 2 \times 10^{-9}$/site/year) for 3'- and 5'-flanking

Table 9.3 Rates of
nucleotide substitution for
various genome regions
(From Li et al. 1985; [23])

Nonsynonymous codon sites	0.88
Synonymous codon sites	4.65
5′ franking untranscribed region	2.36
3′ franking untranslated region	1.88
5′ franking untranslated region	1.74
3′ franking untranscribed region	4.46
Intron	3.70
Pseudogene	4.85

Unit: 10^{-9}/site/year

untranslated regions showed further lower values, though the rate of nonsynonymous substitutions is less than half of that. The order of evolutionary rates shown in this table is compatible with the order of importance of each region, consistent with the prediction of the neutral theory of evolution (Chap. 4).

We briefly discussed existence of conserved noncoding sequences in eukaryotic genomes in Chap. 8. These conserved sequences are most abundant in vertebrate genomes, and we discuss some of their features in vertebrate genomes in this section.

9.4.1   Evolutionary Rates of Noncoding Sequences Just Upstream of Protein Coding Regions

Protein coding regions are easy to detect, and they are often well conserved among broad lineages of organisms, as discussed above. In contrast, protein noncoding regions have more lineage-specific characteristics. Kitano and Saitou [24] compared seven untranscribed upstream regions of hominoids and rodents. Human, chimpanzee, gorilla, and orangutan sequences were used as representative of hominoids, while mouse and rat sequences were used as representative of rodents. While all sequences except for that upstream of INMT gene showed evolutionary rates slower than the pure neutrally evolving junk DNA regions, most of them did not have homologous sequences in rodents, except for the POU3F-2 upstream sequence. Because the sequences around transcription start sites of protein coding genes compared were homologous between hominoids and rodents, they compared mouse and rat homologous sequences with the same length just upstream of these homologous regions. Figure 9.9 shows the result of comparison. The diagonal line shows the case in which the same evolutionary constraint for hominoids and rodents. Sequences upstream of POU3F2, NGFB, and MAOA located above this line suggest stronger selective constraint in primates than rodents.

9.4.2   Highly Conserved Noncoding Sequences

Gillipan et al. (2002; [25]) found several highly conserved noncoding sequences from the 85-kb fugu genomic sequence containing 17 protein coding genes through comparison with human draft genome sequences. Bejerano et al. (2004; [26])

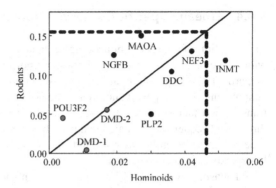

Fig. 9.9 Comparison of evolutionary distances for hominoids and rodents (From Kitano and Saitou, 2005; [24])

searched human, mouse, and rat genome sequences to find out noncoding sequences which are identical among the three genomes with at least 200-bp length, and they found 481 such "ultraconserved elements" (UCEs). Functions of some UCEs were shown to be enhancer of transcription factors [27]; however, mice knockout experiments on four UCEs did not show any clear phenotypic difference from wild mice [28]. It should be noted that even 1 % difference of fertility has a great impact in the long-term evolution, and in fact, Halligan et al. (2011; [29]) showed the selection signals in mouse ultraconserved elements through SNP analyses. Through a global evolutionary analysis of mammalian conserved nonexonic elements, McLean and Bejerano (2008; [30]) also showed that ultraconserved-like elements were over 300-fold more resistant to loss than the neutrally evolving genomic regions during the rodent evolution. It is thus clear that the highly conserved noncoding regions are not really redundant.

9.4.3   Conserved SINEs

Short interspersed elements (SINEs) are representative junk DNAs; however, there was an old hypothesis [31] that some of them are responsible for gene expression control. In fact, some conserved noncoding sequences in vertebrates originated from SINEs [32, 33]. Sasaki et al. [34] demonstrated that one conserved SINE sequence among amniotes, which is located at 178 kbp from the gene FGF8 (fibroblast growth factor 8), was an enhancer that recapitulates *FGF8* expression in two regions of the developing forebrain by using transgenic mice experiments. Because the SINE sequences which were found to have enhancer activities were conserved among mammals, these conserved noncoding sequences may have contributed to mammalian-specific brain formation in the common ancestor of mammals.

Now an old hypothesis was at least partially proven to be true. This drastic change of the role of DNA sequences reminds the author of a famous evolutionary scenario called "hopeful monster" proposed by Richard Goldschmidt in the 1930s [35, 36].

9.4.4 Lineage-Specific Conserved Noncoding Sequences

Identification of conserved noncoding sequences (CNSs), such as UCEs, among a wide range of species genomes is important; however, detection of CNSs restricted to particular lineage of organisms is also important for examining diversity of genome and phenotypes. It should also be noted that a genome-wide occupancy of transcription factors among five vertebrates using immunoprecipitation with high-throughput sequencing revealed that most binding is species specific [37]. Lineage-specific CNSs recently emerged and may have gained new functions to develop lineage-specific characteristics.

Takahashi and Saitou [38] compared human and marmoset genomes for detecting primate-specific CNSs, while mouse and rat genomes were used for detecting rodent-specific CNSs (Fig. 9.10). They identified 8,198 primates and 21,128 specific CNSs that are at least 100-bp length. Figure 9.11 shows nucleotide differences of primate-specific CNSs and their flanking regions. While CNS-flanking regions have very flat nucleotide differences (~0.098) which are expected from the human-marmoset genomic difference, CNSs show very small nucleotide differences. The inset figure shows the distribution near CNSs. They then searched these lineage-specific CNS-flanking protein coding genes (LHF genes). Statistically overrepresented functions of LHF genes were involved in anatomical development as transcriptional regulators, consistent with the characteristics of known mammalian- or vertebrate-shared CNSs. When primate-specific and rodent-specific LHF genes and UCE-flanking protein coding genes were compared (see Fig. 9.12), genes belonging to each of the three categories are often unique. This indicates that independent sets of genes may contribute to develop lineage-specific characteristics during the evolution of primates or rodents through emergence of lineage-specific CNSs. Conversely, the number of LHF genes which were shared by UCE-flanking genes was small yet significantly larger than expected from random coincidence, and many of them were involved in nervous system development as transcriptional regulators, suggesting that certain groups of genes tend to recruit new CNSs in addition to old CNSs which are conserved among vertebrates or mammals. Figure 9.13 shows some examples of primate-specific and rodent-specific CNSs and UCEs with their flanking protein coding genes (from [38]). Babarinde and Saitou (2013; [106]) compared various species genomes of four mammalian orders (primates, rodents, carnivores, and cetartiodactyls) with more relaxed conservation thresholds than those used by Takahashi and Saitou (2012; [38]). The abundance of CNSs varied among lineages, with primates and rodents having highest and lowest number of CNSs, respectively, whereas carnivores and cetartiodactyls had intermediate values. These CNSs cover 1.3–5.5 % of the mammalian genomes and had signatures of selective constraints that are stronger in more ancestral than the recent ones.

9.5 Isochores

Bernardi et al. (1985; [39]) found a high heterogeneity of genomic GC (guanine and cytosine) content among mammals and birds and named one genome block that is seemingly homogeneous in terms of GC content as "isochore." This large-scale

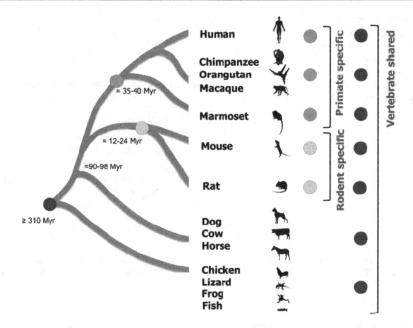

Fig. 9.10 The phylogenetic relationship of human, marmoset, mouse, and rat (From Takahashi and Saitou 2012; [38])

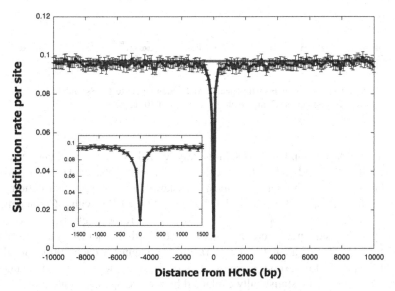

Fig. 9.11 Nucleotide differences of primate-specific CNSs and their flanking regions (From Takahashi and Saitou 2012; [38])

Fig. 9.12 Comparison of primate-specific LHF genes, rodent-specific LHF genes, and UCE-flanking protein coding genes (From Takahashi and Saitou 2012; [38])

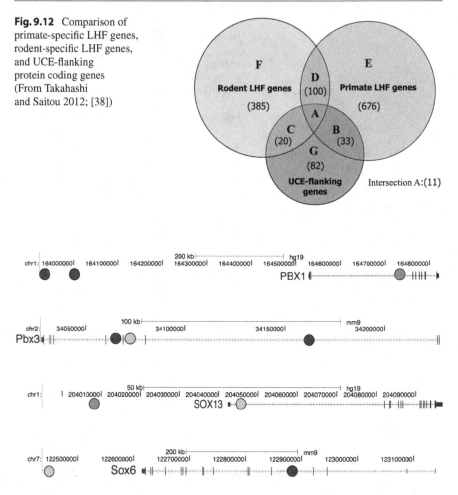

Fig. 9.13 Some examples of primate-specific and rodent-specific CNSs and UCEs with their flanking protein coding genes (From Takahashi and Saitou 2012; [38])

variability of base composition is significantly related to various genomic features: Giemsa bands (G-bands), densities of LINE and Alu elements, level of methylation, recombination rates, and gene densities. The skewed GC content distribution was divided into five types of isochore: L1, L2, H1, H2, and H3, with GC contents of <38 %, 38–42 %, 42–47 %, 47–52 %, and >52 %, respectively [40]. The draft human genome sequences were examined to see whether strict isochores could be identified [41]. They calculated the average GC content for each 300-kb window and 20-kb subwindow. About three quarters of the genome-wide variance among 20-kb windows can be statistically explained by the average GC content of 300-kb windows that contain them, but the residual variance among 20-kb subwindows is far too large to be consistent with a homogeneous distribution. Therefore, the hypothesis of homogeneity was rejected for the draft human genome.

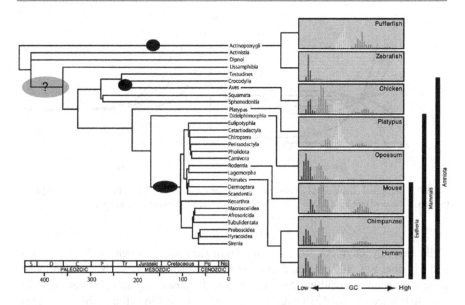

Fig. 9.14 GC content variation among selected vertebrate species (From OOta et al. 2010; [47])

Interestingly, fish and amphibians did not have high GC content heterogeneity, and a hypothesis was proposed that warm-bloodedness in mammals and birds created some kind of selection pressure to create GC content heterogeneity (Bernardi, 2004; [42]). Because mammals and birds belong to amniotes, it is possible that the GC content heterogeneity somehow emerged at the common ancestor of amniotes caused by a change of mutation creating mechanism. If so, reptiles, another group of amniotes, should also have GC content heterogeneity in their genomes. In fact, that was the case [43, 44], though recently determined green anole genome [19] did not have the isochore structure. Ikemura and his collaborators found that a precise switching of DNA replication timing in the GC content transition area exists within the human genome [45, 46]. Heterogeneity of DNA replication timing is widespread among eukaryote genomes. It is thus possible that this heterogeneity came to be connected to the GC content heterogeneity in the amniote common ancestor.

When we consider phylogenies of various organisms whose genome data are recently available, the isochore evolution is quite complex. Figure 9.14 shows the polyphyletic characteristics of isochore families. Human, chimpanzee, mouse, and chicken have clear isochore structures, while opossum and platypus lack GC-rich and GC-poor isochore families, respectively (From OOta et al. 2010; [47]).

One of the difficulties underlying the isochore studies is that there were no sufficient data to analyze the isochore evolution with an appropriate evolutionary model. Major part of isochore resides in noncoding genomic regions, which are in general not alignable between distantly related species like human and mouse. Conventional evolutionary models are empirically based on substitution patterns in coding regions. In other words, we implicitly assume that evolution of alignable genic regions fairly reflects genome-level isochore evolution.

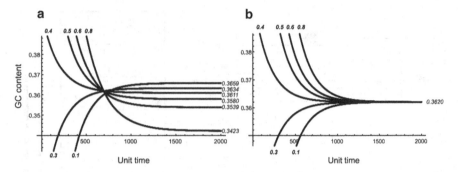

Fig. 9.15 Temporal changes of GC contents of various GC contents (From OOta et al. 2010; [47]). (a) When the constant model was used. (b) When the variable model was used

By using synonymous substitution patterns in coding sequences, Duret and his collaborators [48, 49] predicted that mammalian isochore is vanishing. Although there were some oppositions [50, 51], the vanishing isochore model is now more or less established [47, 52, 53]. OOta et al. (2010; [47]) proposed a time-inhomogeneous Markov process for the GC content evolution, instead of the time-homogeneous Markov process that is usually assumed (e.g., [54]). This new model, called "variable model," and the traditional "constant model" give slightly different predictions on GC content changes, as shown in Fig. 9.15. SNP data of human chromosome 21 were used as the chimpanzee chromosome 22 genome sequence as the out-group. Although there are some differences in equilibrium frequencies under the constant model, all the GC content categories converged on the exactly same equilibrium under the variable model. It should be noted, however, that the estimated convergence time becomes around 350 million years under the following assumptions: (1) the intergenic regions in which most of the SNP loci reside are subject to neutral evolution, (2) the effective population size of the human lineages is 10,000, and (3) the generation time for human is 20 years. Although the variable model has no clear molecular evidence to support this, its nonlinear nature opens a possibility on the transitions of GC contents; the variable model can explain both the gradual and the drastic compositional transitions [47]. If the variable model is more realistic than the constant model, existence of isochore in some lineages is just sporadic and no natural selection is necessary to invoke.

9.6 Genomes of Specific Vertebrate Lineages

Lamprey genome sequences became available in September 2011 [55]. Matsunami et al. [56] recently sequenced cDNAs of Japanese brook lamprey (*Lethenteron reissneri*) using a 454 sequencer (see Chap. 11), and produced short sequence reads were assembled. The average of read length was about 200 bp, and this is much shorter than the expected 454 read length (400–600 bp), probably because of a high GC content in the lamprey genome. The genome sequence of sea lamprey

(*Petromyzon marinus*) was just reported by Smith et al. (2013; [15]). Arctic lamprey (*Lethenteron camtschaticum*) genome sequencing is also under way according to http://www.ncbi.nlm.nih.gov/genome/16905.

Studies on cartilaginous fish genomes are not well progressing. A very low-coverage (~1.4×), incomplete (~70 %) genome of elephant shark, *Callorhinchus milii*, was reported in 2007 [57]. As of December 2013, genome sequencing on skate Leucoraja erinacea, a cartilaginous fish according to http://www.ncbi.nlm.nih.gov/genome/10952. Because cartilaginous fish did not experience the third genome duplication, evolutionary conservation of many proteins with mammalian ones is high [57]. This gives an important status for cartilaginous fish genomes among vertebrates.

9.6.1   Teleost Fish

Teleost fish experienced the third genome duplication in their common ancestor [58]. Additional genome duplications occurred in some teleost lineages. For example, salmons (Oncorhynchus) was known to experience recent genome duplication [59], and there are at least 14 Hox clusters in this group [60]. Their duplication timing was estimated to be about 60 million years ago by comparing two MHC region sequences [61].

There is a huge variety in genome size among teleosts. Fugu (puffer fish) have especially small genome size (~400 Mb), and genome sequences of torafugu (*Takifugu* [or *Fugu*] *rubripes*; [62]) and midorifugu (*Tetraodon nigroviridis*; [63]) were determined. The genome sequence of medaka (*Oryzias latipes*), a small fish native to East Asia, was determined in 2007 by a Japanese group [64]. The genome size of this fish was 700 Mb, and it contained ~20,000 protein coding genes. Interestingly, the genomic divergence between the two geographically different medaka populations was 3.42 %, the highest within-species divergence rate among vertebrates [64]. It should be noted that the divergence time estimate between the two population is controversial, 4 million years ago [64] or 18 million years ago [65].

Sequencing of zebra fish (*Danio rerio*) genome was started much earlier, but it has not been completed as of December 2013 [66]. Meanwhile, the 463-Mb genome of stickleback (*Gasterosteus aculeatus*), which is used for evo-devo studies, was determined in 2012 [67].

9.6.2   Amphibians and Amniotes

Amphibians are basal among the tetrapods or land vertebrates, and comparison of their genomes with those of fish and amniotes may shed lights to identification of genes or genetic changes responsible for landings of vertebrates about 400 million years ago. African clawed frog (*Xenopus laevis*) is often used in developmental biology, and this species was initially chosen as a representative amphibian for genome sequencing [68]. However, *X. laevis* recently experienced a genome duplication and had difficulty in contig formation. A congeneric *X. tropicalis*, or *Silurana tropicalis*, which has not experienced genome duplication, was then chosen. Genome sequencing was started in 2002, and its 1.7-Gb genome was reported in 2010 [69].

Amniotes are monophyletic groups of tetrapods including reptiles, birds, and mammals (see Chap. 6). Genome sequences of many mammalian species are now available (see the next Sect. 9.6.3), while those of reptiles and birds are still limited. The genome sequence of chicken (*Gallus gallus*) was determined in 2004 [70]. There are 38 pairs of autosomal chromosomes and one pair of sex chromosomes in chicken. The chromosome lengths are quite heterogeneous, from 200 Mb to 2 Mb. Chromosomes with less than 20-Mb length are called "microchromosomes," those more than 50 Mb are called "macrochromosomes," and intermediate ones are called "intermediate chromosomes." The GC contents of microchromosomes and macrochromosomes are high and low, respectively. The rate of synonymous substitutions and that of recombinations is higher in microchromosomes than macrochromosomes, and the subtelomeric regions of macrochromosomes have higher rate of synonymous substitutions than the interior regions [70]. It is not clear why such differences exist inside the chicken genome. More recently, the genomes of zebra finch (*Taeniopygia guttata*) and domestic turkey (*Meleagris gallopavo*) were determined in 2010 ([71] and [72], respectively). Green anole (*Anolis carolinensis*) is a lizard often available at pet shops. Its genome sequence was determined in 2011 [19] as the first one among reptiles. The draft genome of python (*Python molurus bivittatus*) soon followed [73], and the genome and transcriptome sequencing of Japanese habu (*Protobothrops flavoviridis*) is under way at Kyushu University (Shibara Hiroki, personal communication).

9.6.3 Mammals

The human genome (see Chap. 10) was first determined as a mammalian species, followed by mouse and rat ([74] and [75], respectively) because of their importance as model organisms, and now genome sequences are available for many other mammalian species (Table 9.4). However, we have to be careful about the credibility of these nonhuman genome sequences, for they were mostly based on whole-genome shotgun sequencing.

Mammals have only 200 million-year history, and mammalian-specific genes are not many, and one of them is Sry for sex determination located at Y chromosome. This gene is a transcription factor and belongs to the Sox gene family [76]. Sry was estimated to appear through a gene duplication from the Sox3 gene located at X chromosome [77]. Masuyama, Ezawa, and Saitou [78] studied mammalian-specific domain combination and found that 34 genes were mammalian specific. As we discussed earlier, expression patterns of orthologous genes may be differentially controlled by their flanking noncoding sequences. Combinations of protein coding gene changes and noncoding sequence changes probably shaped up the mammalian characteristics and diverse mammalian species we see today.

Junk DNAs have many mammalian-specific characteristics. Oligonucleotide frequencies are the most complex in mammals among the vertebrates [3], and there are a series of mammalian-specific SINEs and LINEs [79].

Table 9.4 List of mammalian species whose genomes were sequenced with high coverage (From Ensembl database)

Common name (Latin binomen)
Primates
Human (*Homo sapiens*)
Chimpanzee (*Pan troglodytes*)
Gorilla (*Gorilla gorilla*)
Orangutan (*Pongo abelii*)
White-Cheeked Gibbon (*Nomascus leucogenys*)
Rhesus macaque (*Macaca mulatta*)
Cynomolgous macaque (*Macaca fascicularis*)
Marmoset (*Callithrix jacchus*)
Bushbaby (*Otolemur garnettii*)
Rodents
Mouse (*Mus musculus*)
Rat (*Rattus norvegicus*)
Thirteen-lined ground squirrel (*Spermophilus tridecemlineatus*)
Guineapig (*Cavia porcellus*)
Carnivores
Cat (*Felis catus*)
Dog (*Canis lupus familiaris*)
Ferret (Mustela putorius furo)
Giant Panda (*Ailuropoda melanoleuca*)
Other placental mammals
Rabbit (*Oryctolagus cuniculus*)
Horse (*Equus caballus*)
Microbat (*Myotis lucifugus*)
Cow (*Bos taurus*)
Pig (*Sus scrofa*)
African elefant (*Loxodonta africana*)
Non-eutherian mammals
Opposum (*Monodelphis domestica*)
Tasmanian devil (Sarcophilus harrisii)
Platypus (*Ornithorhynchus anatinus*)

In the following, we will briefly discuss mammalian species whose genome sequences were almost completed.

Nonprimate Mammals

Monotremes are phenotypically the most primitive mammals, and the genome sequences of their important species, platypus (*Ornithorhynchus anatinus*), were determined in 2008 [80]. Opossum (*Monodelphis domestica*) is a marsupial distributing in South and North America. Its genome sequence was shifted to GC content lower than placental mammals, and many CNSs found in placental mammals were missing in the opossum genome [81]. Marsupial genomes are the best out-groups for placental mammals.

Mouse (*Mus musculus*) and rat (*Rattus norvegicus*) are both not only rodent, but also belong to Muridae, and their divergence time is 12–24 million years ago. The rodent lineage was suspected to have a higher evolutionary rate than those of other mammalian lineages for a long time [82]. Although some studies suggested approximate constancy of the evolutionary rates between rodents and nonrodent mammals (e.g., [83]), now it is clear that the evolutionary rate, or the mutation rate per year in rodents, is 2–5 times higher than those of other mammalian lineages [84–86]. However, most of comparisons were made by using mouse sequence data or through mouse and rat sequence comparison, and it is not clear if the high rate applies to non-Muridae rodent species.

Domesticated mammals were targets of genome sequencing because of their economic importance. Genome sequences of dog [87], cow [88], and horse [89] were determined, and those for pig, sheep, goat, and cat were also determined (see http://www.ensembl.org/).

Genome sequences of African elephant [90] were determined, and low-coverage genome sequences of more than 20 mammalian species were determined [91]. Many other mammalian species are planned to be soon sequenced (see Table 9.4), thanks to second-generation sequencing technologies (see Chap. 11).

Nonhuman Primates

We humans belong to primates, but the human genome is discussed separately in Chap. 10, and nonhuman primate genomes are briefly discussed in this section. One important characteristic of primate genomes is the existence of primate-specific SINEs called Alu sequences (see Chap. 8). As we already discussed in Sect. 9.4.4, primate-specific conserved noncoding sequences were found [38, 106], and some of them are expected to be responsible for primate-specific phenotypes.

When the human genome sequencing project was about to complete, it was natural to determine the genome of chimpanzee, the closest organism to human. National Institutes of Health at the USA conducted survey among biologists about the next species whose genome should be sequenced. The majority opinion was chimpanzee, and the draft genome using the whole-genome shotgun method (see Chap. 11) was published in 2005 [92]. Meanwhile, a non-USA international consortium of scientists in Japan, Germany, China, Taiwan, and Korea took a different approach, BAC clone tiling array method (see Chap. 11). They first constructed a chimpanzee BAC clone library and determined more than 110,000 BAC-end sequences in 2002 [93]. A total of 19.8-Mb sequences were obtained, and the difference between human and chimpanzee was 1.23 %. These end-determined BAC clones were mapped to the human genome, and because of limited budgets, the chimpanzee chromosome 22 long arm that is orthologous to the human chromosome 21 long arm was determined in 2004 [94]. Although only 1 % (33.3 Mb) of the chimpanzee genome was determined, the high-resolution sequence comparison produced various interesting results, as already discussed in Chap. 2.

The last 5 years from 2007 to 2012 observed publications of a series of ape and monkey genome sequences: rhesus macaque (*Macaca mulatta*) in 2007 [95], orangutan (*Pongo abelii*) and cynomolgus macaque (*Macaca fascicularis*) in 2011

([96] and [97], respectively), and gorilla in 2012 [98]. Genome sequences of marmoset and bush baby are almost complete, and sequencing is ongoing for many other species.

9.7 Ancient Genomes

Nucleotide sequences of genomes or a particular gene are usually determined from DNAs or RNAs of present-day organisms. Dead cells may leave tiny amount of DNA molecules, depending on their storage conditions. They are generically called "ancient DNA" irrespective of their age, from scores of years old to millions of years old. Study of ancient DNAs were started in 1984 [99] for elucidating the phylogenetic status of extinct quagga, a species in genus *Equus*. Now genome-wide studies for various organisms are conducted. Ancient human genomes will be discussed in Chap. 10, and nonhuman organismal ancient genomes are the targets of this section.

Because mitochondrial DNA (mtDNA) molecules are often hundreds times more abundant than the nuclear DNA molecules per cell, mtDNA was initially the only target of ancient DNA studies. Complete mtDNA genome sequences were first determined for three extinct species of moa in New Zealand [100, 101]. As of February 2013, complete mtDNA genome sequences are available for the following mammals: woolly mammoth (*Mammuthus primigenius*), mastodon (*Mammut americanum*), Columbian mammoth (*Mammuthus columbi*), giant short-faced bear (*Arctodus simus*), cave bear (*Ursus spelaeus*), woolly rhinoceros (*Coelodonta antiquitatis*), black rhinoceros (*Diceros bicornis*), Javan rhinoceros (*Rhinoceros sondaicus*), Tasmanian tiger (*Thylacinus cynocephalus*), and aurochs (*Bos primigenius*) (mostly based on [102]). Recently, complete mtDNA genome sequence of extinct passenger pigeon (*Ectopistes migratorius*) was reported [103].

As for nuclear genomes, main data are for ancient modern humans and more ancient human lineages (Neanderthals and Denisovan) were reported (see Chap. 10), and the only exception as of December 2013 may be ~40-kb sequences from American mastodon and the woolly mammoth compared with extant elephant species [104]. Some fragmental short nuclear DNA sequences were also determined from a limited animal and plant species [105], and ancient genome sequences of a more diverse lineages of organisms are expected to be studied in the near future.

References

1. Lamark, J-B. (1810). *Animal philosophy* (in French).
2. Romer, A. S. (1959). *The vertebrate story*. Chicago: The University of Chicago Press.
3. Radinsky, L. B. (1987). *The evolution of vertebrate design*. Chicago: The University of Chicago Press.
4. Kryukov, K., Sumiyama, K., Ikeo, K., Gojobori, T., & Saitou, N. (2012). A new database (GCD) on genome composition for eukaryote and prokaryote genome sequences and their initial analyses. *Genome Biology and Evolution, 4*, 501–512.
5. Ohno, S. (1970). *Evolution by gene duplication*. New York: Springer-Verlag.

6. Kasahara, M., Hayashi, M., Tanaka, K., Inoko, H., Sugaya, K., Ikemura, T., & Ishibashi, T. (1996). Chromosomal localization of the proteasome Z subunit gene reveals an ancient chromosomal duplication involving the major histocompatibility complex. *Proceedings of the National Academy of Sciences of the United States of America, 93*, 9096–9101.

7. Kasahara, M. (Ed.). (2000). *Major histocompatibility complex – Evolution, structure, and function.* New York: Springer -Verlag.

8. Hughes, A. L. (1998). Phylogenetic tests of the hypothesis of block duplication of homologous genes on human chromosomes 6, 9, and 1. *Molecular Biology and Evolution, 15*, 854–870.

9. Holland, L. Z., et al. (2008). The amphioxus genome illuminates vertebrate origins and cephalochordate biology. *Genome Research, 18*, 1100–1111.

10. Nakatani, Y., Takeda, H., Kohara, Y., & Morishita, S. (2007). Reconstruction of the vertebrate ancestral genome reveals dynamic genome reorganization in early vertebrates. *Genome Research, 17*, 1254–1265.

11. Matsunami, M., & Saitou, N. (2013). Vertebrate paralogous conserved noncoding sequences may be related to gene expressions in brain. *Genome Biology and Evolution, 5*, 140–150.

12. Matsunami, M. (2012). *The evolutionary analysis of the vertebrate two-round whole genome duplications.* Ph.D. dissertation, Graduate University for Advanced Studies.

13. Matsunami, M., Satoh, Y., Saitou, N., et al. (unpublished) Inferring the timing of the 2R WGD from lamprey transcriptome data.

14. Kuraku, S., Meyer, A., & Kuratani, S. (2009). Timing of genome duplications relative to the origin of the vertebrates: Did cyclostomes diverge before or after? *Molecular Biology and Evolution, 26*, 47–59.

15. Smith, J. J., et al. (2013). Sequencing of the sea lamprey (*Petromyzon marinus*) genome provides insights into vertebrate evolution. *Nature Genetics,* published online on 24 Feb 2013.

16. Kawashima, T., et al. (2009). Domain shuffling and the evolution of vertebrates. *Genome Research, 19*, 1393–1403.

17. Huang, S., et al. (2008). Genomic analysis of the immune gene repertoire of amphioxus reveals extraordinary innate complexity and diversity. *Genome Research, 18*, 1112–1126.

18. Smith, C. A., et al. (2009). The avian Z-linked gene DMRT1 is required for male sex determination in the chicken. *Nature, 461*, 267–271.

19. Alfoldi, J., et al. (2011). The genome of the green anole lizard and a comparative analysis with birds and mammals. *Nature, 477*, 587–591.

20. Graves, J. A. M. (2009). Birds do it with a Z gene. *Nature, 461*, 177–178.

21. Niimura, Y., & Nei, M. (2006). Evolutionary dynamics of olfactory and other chemosensory receptor genes in vertebrates. *Journal of Human Genetics, 51*, 505–517.

22. Suzuki, R., & Saitou, N. (2011). Exploration for functional nucleotide sequence candidates within coding regions of mammalian genes. *DNA Research, 18*, 177–187.

23. Li, W.-H., Luo, C.-C., & Wu, C.-I. (1985). Evolution of DNA sequences. In R. J. MacIntyre (Ed.), *Molecular evolutionary genetics,* Chap. 1 (pp. 1–94). New York: Springer.

24. Kitano, T., & Saitou, N. (2005). Evolutionary conservation of 5′ upstream sequence of nine genes between human and great apes. *Genes and Genetic Systems, 80*, 225–232.

25. Gillipan, P., Brenner, S., & Venkatesh, B. (2002). Fugu and human sequence comparison identifies novel human genes and conserved non-coding sequences. *Gene, 294*, 35–44.

26. Bejerano, G., Pheasant, M., Makunin, I., Stephen, S., Kent, W. J., Mattick, J. S., & Haussler, D. (2004). Ultraconserved elements in the human genome. *Science, 304*, 1321–1325.

27. Feng, J., Bi, C., Clark, B. S., Mady, R., Shah, P., & Kohtz, J. D. (2006). The *Evf-2* noncoding RNA is transcribed from the Dlx-5/6 ultraconserved region and functions as a Dlx-2 transcriptional coactivator. *Genes and Development, 20*, 1470–1484.

28. Ahituv, N., Zhu, Y., Visel, A., Holt, A., Afzal, V., Pennacchio, L. A., & Rubin, E. M. (2007). Deletion of ultraconserved elements yields viable mice. *PLoS Biology, 5*, e234.

29. Halligan, D. L., et al. (2011). Positive and negative selection in murine ultraconserved noncoding elements. *Molecular Biology and Evolution, 28*, 2651–2660.

30. McLean, C., & Bejerano, G. (2008). Dispensability of mammalian DNA. *Genome Research, 18*, 1743–1751.

31. Britten, R. J., & Davidson, E. H. (1969). Gene regulation for higher cells: A theory. *Science, 165*, 349–357.

32. Xie, X., Kamal, M., & Lander, E. S. (2006). A family of conserved noncoding elements derived from an ancient transposable element. *Proceedings of the National Academy of Sciences of the United States of America, 103*, 11659–11664.
33. Nishihara, H., Smit, A. F., & Okada, N. (2006). Functional noncoding sequences derived from SINEs in the mammalian genome. *Genome Research, 16*, 864–874.
34. Sasaki, T., Nishihara, H., Hirakawa, M., Fujimura, K., Tanaka, M., Kokubo, N., Kimura-Yoshida, C., Matsuo, I., Sumiyama, K., Saitou, N., Shimogori, T., & Okada, N. (2008). Possible involvement of SINEs in mammalian-specific brain formation. *Proceedings of the National Academy of Sciences of the United States of America, 105*, 4220–4225.
35. Goldschmidt, R. (1933). Some aspects of evolution. *Science, 78*, 539–547.
36. Goldschmidt, R. (1940). *The material basis of evolution.* Yale University Press (Reprint version (1980) Yale University Press).
37. Schmidt, D., et al. (2010). Five-vertebrate ChIP-seq reveals the evolutionary dynamics of transcription factor binding. *Science, 328*, 1036–1040.
38. Takahashi, M., & Saitou, N. (2012). Identification and characterization of lineage-specific highly conserved noncoding sequences in mammalian genomes. *Genome Biology and Evolution, 4*, 641–657.
39. Bernardi, G., et al. (1985). The mosaic genome of warm-blooded vertebrates. *Science, 228*, 953–958.
40. Bernardi, G. (2000). Isochores and the evolutionary genomics of vertebrates. *Gene, 241*, 3–17.
41. International Human Genome Sequencing Consortium. (2001). Initial sequencing and analysis of the human genome. *Nature, 409*, 860–921.
42. Bernardi, G. (2004). *Structural and evolutionary genomics.* Amsterdam: Elsevier Science.
43. Hughes, S., Zelus, D., & Mouchiroud, D. (1999). Warm-blooded isochore structure in Nile crocodile and turtle. *Molecular Biology and Evolution, 16*, 1521–1527.
44. Hamada, K., Horiike, T., Ota, H., Mizuno, K., & Shiozawa, T. (2003). Presence of isochore structures in reptile genomes suggested by the relationship between GC contents of intron regions and those of coding regions. *Genes and Genetic Systems, 78*, 195–198.
45. Tenzen, T., Yamagata, T., Fukagawa, T., Sugaya, K., Ando, A., Inoko, H., Gojobori, T., Fujiyama, A., Okumura, K., & Ikemura, T. (1997). Precise switching of DNA replication timing in the GC content transition area in the human major histocompatibility complex. *Molecular and Cellular Biology, 17*, 4043–4050.
46. Watanabe, Y., Fujiyama, A., Ichiba, Y., Hattori, M., Yada, T., Sakaki, Y., & Ikemura, T. (2002). Chromosome-wide assessment of replication timing for human chromosomes 11q and 21q: Disease-related genes in timing-switch regions. *Human Molecular Genetics, 11*, 13–21.
47. OOta, S., et al. (2010). A new framework for studying the isochore evolution: Estimation of the equilibrium GC content based on the temporal mutation rate model. *Genome Biology and Evolution, 2*, 558–571.
48. Duret, L., Semon, M., Piganeau, G., Mouchiroud, D., & Galtier, N. (2002). Vanishing GC-rich isochores in mammalian genomes. *Genetics, 162*, 1837–1847.
49. Belle, E. M., Duret, L., Galtier, N., & Eyre-Walker, A. (2004). The decline of isochores in mammals: An assessment of the GC content variation along the mammalian phylogeny. *Journal of Molecular Evolution, 58*, 653–660.
50. Alvarez-Valin, F., Clay, O., Cruveiller, S., & Bernardi, G. (2004). Inaccurate reconstruction of ancestral GC levels creates a "vanishing isochores" effect. *Molecular Phylogenetics and Evolution, 31*, 788–793.
51. Gu, J., & Li, W.-H. (2006). Are GC-rich isochores vanishing in mammals? *Gene, 385*, 50–56.
52. Webster, M. T., Smith, N. G., & Ellegren, H. (2003). Compositional evolution of noncoding DNA in the human and chimpanzee genomes. *Molecular Biology and Evolution, 20*, 278–286.
53. Galtier, N., & Duret, L. (2007). Adaptation or biased gene conversion? Extending the null hypothesis of molecular evolution. *Trends in Genetics, 23*, 273–277.
54. Sueoka, N. (1988). Directional mutation pressure and neutral molecular evolution. *Proceedings of the National Academy of Sciences of the United States of America, 85*, 2653–2657.
55. http://asia.ensembl.org/Petromyzon_marinus/Info/Index

56. Matsunami, M., Satoh, Y., Saitou, N., et al. (unpublished) Inferring the timing of the 2R WGD from lamprey transcriptome data.
57. Venkatesh, B., et al. (2007). Survey sequencing and comparative analysis of the elephant shark (*Callorhinchus milii*) genome. *PLoS Biology, 5*, e101.
58. Meyer, A., & de Peer, Y. V. (2005). From 2R to 3R: Evidence for a fish-specific genome duplication (FSGD). *BioEssay, 27*(9), 937–945.
59. Gregory, T. R. (Ed.). (2005). *The evolution of the genome*. Burlington: Elsevier Academic Press.
60. Mooghadam, H. K., et al. (2005). Evidence for Hox gene duplication in rainbow trout (*Oncorhynchus mykiss*): A tetraploid model species. *Journal of Molecular Evolution, 61*, 804–818.
61. Shiina, T., et al. (2005). Interchromosomal duplication of major histocompatibility complex class I regions in rainbow trout (*Oncorhynchus mykiss*), a species with a presumably recent tetraploid ancestry. *Immunogenetics, 56*, 878–893.
62. Aparicio, S., et al. (2002). Whole-genome shotgun assembly and analysis of the genome of *Fugu rubripes*. *Science, 297*, 1301–1310.
63. Jaillon, O., et al. (2004). Genome duplication in the teleost fish *Tetraodon nigroviridis* reveals the early vertebrate proto-karyotype. *Nature, 431*, 946–957.
64. Kasahara, M., et al. (2007). The medaka draft genome and insights into vertebrate genome evolution. *Nature, 447*, 714–719.
65. Setiamarga, D. H., et al. (2009). Divergence time of the two regional medaka populations in Japan as a new time scale for comparative genomics of vertebrates. *Biological Letters, 5*, 812–816.
66. www.sanger.ac.uk/Projects/D_rerio/
67. Jones, F. C., et al. (2012). The genomic basis of adaptive evolution in threespine sticklebacks. *Nature, 484*, 55–61.
68. Pollet, N., & Mazabraud, A. (2005). Insights from Xenopus genomes. *Genome Dynamics, 2*, 138–153.
69. Hellsten, U., et al. (2010). The genome of the Western clawed frog *Xenopus tropicalis*. *Science, 328*, 633–636.
70. International Chicken Genome Consortium. (2004). Sequence and comparative analysis of the chicken genome provide unique perspectives on vertebrate evolution. *Nature, 432*, 695–716.
71. Warren, W. C., et al. (2010). The genome of a songbird. *Nature, 464*, 757–762.
72. Dalloul, R. A., et al. (2010). Multi-platform next-generation sequencing of the domestic turkey (*Meleagris gallopavo*): Genome assembly and analysis. *PLoS Biology, 8*, e1000475.
73. Castoe, T. A., et al. (2011). Sequencing the genome of the Burmese python (*Python molurus bivittatus*) as a model for studying extreme adaptations in snakes. *BMC Genome Biology, 12*, 406.
74. Mouse Genome Sequencing Consortium. (2002). Initial sequencing and comparative analysis of the mouse genome. *Nature, 420*, 520–562.
75. Gibbs, R. A., et al. (2004). Genome sequence of the Brown Norway rat yields insights into mammalian evolution. *Nature, 428*, 493–521.
76. Nagai, K. (2000). Molecular evolution of Sry and Sox gene. *Gene, 270*, 161–169.
77. Kato, K., & Miyata, T. (1999). A heuristic approach of maximum likelihood method for inferring phylogenetic tree and an application to the mammalian SOX-3, origin of the testis-determining gene SRY. *FEBS Letter, 463*, 129–132.
78. Masuyama, W., Ezawa, K., & Saitou, N. (unpublished) Emergence of new domain combinations during the mammalian evolution.
79. Deininger, P. L., et al. (1992). Master genes in mammalian repetitive DNA amplification. *Trends in Genetics, 8*, 307–311.
80. Warren, W. C., et al. (2008). Genome analysis of the platypus reveals unique signatures of evolution. *Nature, 453*, 175–183.
81. Mikkelsen, T. S., et al. (2007). Genome of the marsupial *Monodelphis domestica* reveals innovation in non-coding sequences. *Nature, 447*, 167–177.
82. Wu, C.-I., & Li, W.-H. (1985). Evidence for higher rates of nucleotide substitution in rodents than in man. *Proceedings of the National Academy of Sciences of the United States of America, 82*, 1741–1745.

83. Kumar, S., & Subramanian, S. (2002). Mutation rates in mammalian genomes. *Proceedings of the National Academy of Sciences of the United States of America, 99*, 803–808.
84. Li, W.-H., & Wu, C.-I. (1987). Rates of nucleotide substitution are evidently higher in rodents than in man. *Molecular Biology and Evolution, 4*, 74–82.
85. Kitano, T., OOta, S., & Saitou, N. (1999). Molecular evolutionary analyses of the Rh blood group genes and Rh50 genes in mammals. *Zoological Studies, 38*, 379–386.
86. Abe, K., Noguchi, H., Tagawa, K., Yuzuriha, M., Toyoda, A., Kojima, T., Ezawa, K., Saitou, N., Hattori, M., Sakaki, Y., Moriwaki, K., & Shiroishi, T. (2004). Contribution of Asian mouse subspecies *Mus musculus* molossinus to genomic constitution of strain C57BL/6J, as defined by BAC end sequence-SNP analysis. *Genome Research, 14*, 2239–2247.
87. Lindbald-Toh, K., et al. (2005). Genome sequence, comparative analysis and haplotype structure of the domestic dog. *Nature, 438*, 803–819.
88. The Bovine Genome Sequencing and Analysis Consortium. (2009). The genome sequence of Taurine cattle: A window to ruminant biology and evolution. *Science, 324*, 522–528.
89. Wade, C. S., et al. (2009). Genome sequence, comparative analysis, and population genetics of the domestic horse. *Science, 326*, 865–867.
90. http://www.broadinstitute.org/scientific-community/science/projects/mammals-models/elephant/elephant-genome-project
91. Lindbald-Toh, K., et al. (2011). A high-resolution map of human evolutionary constraint using 29 mammals. *Nature, 478*, 476–482.
92. The Chimpanzee Sequencing and Analysis Consortium. (2005). Initial sequence of the chimpanzee genome and comparison with the human genome. *Nature, 437*, 69–87.
93. Fujiyama, A., Watanabe, H., Toyoda, A., Taylor, T. D., Itoh, T., Tsai, S.-F., Park, H.-S., Yaspo, M.-L., Lehrach, H., Chen, Z., Fu, G., Saitou, N., Osoegawa, K., de Jong, P. J., Suto, Y., Hattori, M., & Sakaki, Y. (2002). Construction and analysis of a human-chimpanzee comparative clone map. *Science, 295*, 131–134.
94. The International Chimpanzee Chromosome 22 Consortium. (2004). DNA sequence and comparative analysis of chimpanzee chromosome 22. *Nature, 429*, 382–388.
95. Rhesus Macaque Genome Sequencing and Analysis Consortium. (2007). Evolutionary and biological insights from the rhesus macaque genome. *Science, 316*, 222–234.
96. Locke, D. P., et al. (2011). Comparative and demographic analysis of orangutan genomes. *Nature, 469*, 529–533.
97. Ebeling, M., et al. (2011). Genome-based analysis of the nonhuman primate *Macaca fascicularis* as a model for drug safety assessment. *Genome Research, 21*, 1746–1756.
98. Scally, A., et al. (2012). Insights into hominid evolution from the gorilla genome sequence. *Nature, 483*, 169–175.
99. Higuchi, R., Bowman, B., Feriberger, M., Ryder, O. A., & Willson, A. (1984). DNA sequences from the quagga, an extinct member of the horse family. *Nature, 312*, 282–284.
100. Cooper, A., et al. (2001). Complete mitochondrial genome sequences of two extinct moas clarify ratite evolution. *Nature, 409*, 704–707.
101. Haddrath, O., & Baker, A. J. (2001). Complete mitochondrial DNA genome sequences of extinct birds: Ratite phylogenetics and the vicariance biogeography hypothesis. *Proceedings of the Royal Society of London Series B, 268*, 939–945.
102. Ho, S. Y. W., & Gilbert, M. T. P. (2010). Ancient mitogenomics. *Mitochondrion, 10*, 1–11.
103. Hung, C. M., et al. (2013). The De Novo assembly of mitochondrial genomes of the extinct passenger pigeon (*Ectopistes migratorius*) with next generation sequencing. *PLoS One, 8*, e56301.
104. Rohland, N., et al. (2010). Genomic DNA sequences from mastodon and woolly mammoth reveal deep speciation of forest and savanna elephants. *PLoS Biology, 8*, e1000564.
105. Paabo, S., et al. (2004). Genetic analyses from ancient DNA. *Annual Review of Genetics, 38*, 645–679.
106. Babarinde, I. A., & Saitou, N. (2013). Heterogeneous tempo and mode of conserved noncoding sequence evolution among four mammalian orders. *Genome Biology and Evolution* (advance publication).

Human Genome

10

Chapter Summary

The human genome can be considered as the representative of mammalian genomes. Basic characteristics of the human genome, such as the overall structure, protein coding genes, and RNA genes, are first discussed. Personal genome sequencing and genomic heterogeneity are described next. We then discuss genetic changes to produce humanness. At the end, ancient human genomes are briefly reviewed.

10.1 Overview of the Human Genome

There are two sets of genomes in one human cell, and they consist of 22 pairs of autosomal chromosomes and one pair of sex chromosomes (X and Y). Females have two X chromosome and males have one X and one Y chromosomes. In total, one human cell usually has 46 chromosomes ($22 \times 2 + 2$). The total DNA amount of one human genome (haploid) was estimated to be about 3.5×10^{-12} g [1]. The molecular weight of one nucleotide pair is 1.08×10^{-21} g (see Chap. 1), and the total number of base pair of one human genome becomes about 3.2×10^9. The X chromosome is much bigger than the Y chromosome; thus, one female autosomal cell has a slightly larger amount of DNA than that of male.

Figure 10.1a, b show microscopic photograph and schematic view of these 46 human chromosomes, respectively. Autosomes were numbered according to their size, and chromosome 1 is the longest. However, the shortest chromosome is not chromosome 22, but chromosome 21, because of a historical reason. There are band patterns in human chromosomes, and these patterns were caused by differential dye concentration depending on the chromosomal locations. The difference is correlated with the GC content (see Chap. 9) of each DNA region [2], but the molecular mechanism for this is still elusive.

Determination of nucleotide sequences of the human genome was often called "Human Genome Project" and was a symbol of the genome sequencing

N. Saitou, *Introduction to Evolutionary Genomics*, Computational Biology 17, DOI 10.1007/978-1-4471-5304-7_10, © Springer-Verlag London 2013

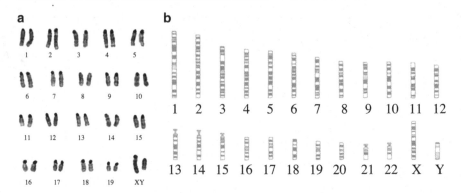

Fig. 10.1 Views of these 46 human chromosomes. (**a**) Microscopic photograph (From Bickmore 2001; [2]). (**b**) Schematic view

efforts of many organisms in the twentieth century. The draft genome sequence was reported in 2001 by the International Consortium [3] and by Celera Genomics [4]. These duplicated determinations were caused by friction between two groups of researchers who used governmental budgets and those who used a private company money. The so-called completion of the human genome sequencing was reported in 2004 [5], but in reality, it was not complete, for only 2.8-Gb (billion base pairs) euchromatins were determined and ~0.4-Gb hetero-chromatins were left unsequenced. Heerochromatins are rich in repeat sequences, and the technology available at that time (see Chap. 11) was not able to produce continuous sequences full of short repeats. This incomplete situation remains now in 2013, even with the advent of second-generation sequencing technologies (see Chap. 11). We should wait for the third- or fourth-generation sequencing technologies to challenge the sequencing mostly short repeat sequences of heterochromatins.

The majority of the human genome is junk DNA (see Chap. 8), and only 1.5 % or 48 Mb is responsible for coding amino acid sequences. Table 10.1 is the content of the human genome [5]. If we include introns and pseudogenes, these gene-related regions consist of 38 % of the human genome, and the remaining 62 % is the inter-genic region. The majority of introns and intergenic regions are junk DNAs. The dispersed repeats cover 1,400 Mb, or the 44 % of the human genome, and they include LINEs and SINEs (see Chap. 8). Microsatellites or short tandem repeats (STRs) are known to have high mutation rates in terms of repeat number changes (see Chap. 2), and they consist of 90 Mb. It should be noted that there are also non-repeat, unique sequences in the intergenic regions, and some of them, about ~3 % of the human genome, are highly conserved (see Chap. 8). If we combine these con-served noncoding DNA regions and also conserved protein coding regions, about 5 % of the human genome inherits the important information to shape up the human, and the rest, 95 %, is mostly junk DNA.

Table 10.1 The content of the human genome (from [5])

Human Genome [3.2 Gb]
Coding regions and related sequences [1.2 Gb]
Exons [48 Mb]
Related sequences (introns, pseudogenes, etc.) [1,152 Mb]
Intergenic sequences [2.0 Gb]
Interspersed repeat sequences [1,400 Mb]
LINEs [640 Mb]
SINEs [420 Mb]
LTR elements [250 Mb]
DNA transposons [90 Mb]
Other intergenic sequences [600 Mb]
Microsatellites [90 Mb]
Various sequences [510 Mb]

10.2 Protein Coding Genes in the Human Genome

A typical protein coding gene in the human genome is as follows (based on description of [3]. It occupies 10–30 kb in the genome with 7–9 exons, and the exon and intron lengths are 120–150 bp and 1–3 kb, respectively. A protein of 350–450 amino acids is translated from a 1.7- to 2.4-kb mRNA. The titin gene has the highest number (234) of exons, while there are thousands of single-exon genes [4].

The number of protein coding genes in the human genome was initially estimated to be 30,000–40,000 based on the draft genome data [3] but was revised to be 20,000–25,000 based on the finished genome data [6]. Venter et al. (2001; [4]) annotated a total of 26,383 genes with the range of 23,000–40,000. Imanishi et al. (2004; [7]) annotated the 21,037 protein coding genes based on cDNA sequences, while Clamp et al. (2007; [8]) estimated the number of protein coding genes to be ~20,500. As of December 2013, OMIM (Online Mendelian Inheritance in Man; http://www.ncbi. nlm.nih.gov/omim) contains 22,129 entries including 3,583 without molecular basis, while PANTHER (Protein Analysis THrough Evolutionary Relationships; [9]; http://www.pantherdb.org/) and GeneCards (http://www.genecards.org/) contain 23,180 and 21,660 protein coding genes, respectively. Therefore, the total number of functional protein coding genes in the human genome may be in the range of 21,000–23,000.

Table 10.2 lists the protein coding genes by 29 categories in the human genome (based on PANTHER). The most frequent gene category is nucleic acid binding (2,806 genes), which occupies 15.3 % of the total protein genes, followed by transcription factors (2,179 genes). Because transcription factors are also DNA binding, ~27 % of the human genome is classified as DNA- or RNA-binding proteins. Genes containing the C2H2-type zinc-finger domain are most frequent (564 genes), followed by those (160 genes) with homeobox domains [5]. When all enzymes (hydrolase, transferase, kinase, oxidoreductase, protease, ligase, phosphatase, lyase, and isomerase) are considered, they cover 6,931 genes, or 30 % of the

Table 10.2 List of the major categories of protein coding genes in the human genome (Based on http://www.pantherdb.org/ as of March 2013)

Class	No.
Nucleic acid binding	2,806
Transcription factor	2,179
Hydrorase[a]	1,884
Transferase[a]	1,571
Receptor	1,904
Enzyme modulator	1,592
Transporter	1,151
Signaling molecule	1,260
Cytoskeletal protein	1,028
Defence/immunity protein	749
Oxidoreductase[a]	742
Protease[a]	717
Cell adhesion molecule	715
Kinase[a]	683
Ligase[a]	613
Extracellular matrix protein	582
Calcium-binding protein	482
Transfer/carrier protein	475
Membrane-traffic protein	425
Structural protein	331
Phosphatase[a]	317
Chaperone	224
Lyase[a]	205
Isomerase[a]	199
Cell junction protein	170
Transmembrane receptor/adaptor protein	85
Surfactant	53
Storage protein	24
Viral protein	14
Total	23,180

[a]Enzyme

total human genes, according to PANTHER annotations. The next frequent class of genes is receptor (1,904 genes), and those for olfactory receptors (~800) are the most dominant but more than half are pseudogenes [10].

10.3 RNA Coding Genes and Gene Expression Control Regions in the Human Genome

When we discuss about genes, they often mean protein coding genes. However, there is another type of genes which code structural RNA molecules vital for the human cell (see Chap. 1). Table 10.3 shows the major RNA coding genes in the human genome. The total number of tRNA genes was estimated to be 625, including 110 predicted pseudogenes [11], though Ensembl BioMart database [12] gives

Table 10.3 List of the RNA coding genes in the human genome

Class	No.	Ref.
tRNA	625	[11]
Pseudogene	110	[11]
rRNA	535	[12]
Pseudogene	179	[12]
snRNA	1,951	[12]
Pseudogene	73	[12]
snoRNA	1,523	[12]
Pseudogene	73	[12]
miRNA	1,809	[12]
Pseudogene	15	[12]
Total	6,732[a]	

[a]Including pseudogenes

a slightly different number (128) for tRNA pseudogenes. Known functional tRNA genes in the human genome are 509 for 20 standard amino acids and selenocysteine (see Chap. 1). The codon GTT corresponding to Asn has the most frequent 32 tRNA genes, followed by GCA for Cys (30 tRNA genes). In contrast, 13 codons do not have their own tRNA genes. There are 535 rRNA genes in the human genome with 179 pseudogenes [12]. More abundant RNA genes in the human genome are those for small nuclear RNA (snRNA) and small nucleolar RNA (snoRNA). They are relatively short (~100 bp) sequences, and 1,951 snRNA and 1,523 snoRNA genes were annotated [12]. Genes for microRNA (miRNA) are much shorter (~22 bp) and are also quite abundant (1,809 genes) in the human genome [12]. There are also genes for small cytoplasmic RNA (scRNA). The total number of these RNA genes is ~7,000, about 1/3 of the protein coding genes.

There also exist DNA regions that control gene expressions. The ENCODE project published a series of papers on the systematic examination of transcription of the human genome in 2012 [13–17]. They previously conducted a pilot study targeted for the 1 % of the human genome [18]. Their result – a high proportion of the human genome is transcribed – was criticized by van Bakel et al. (2010; [19]), who concluded that most "Dark Matter" transcripts are associated with known genes. Although later ENCODE project experiments used the ChIP-seq technology which is more precise than the ChIP-chip technology, there still seem to exist problems in the interpretation of their data. The ENCODE Project Consortium (2012; [13]) assigned biochemical functions for 80 % of the human genome. This particular statement implies that these "functional" regions are not junk DNAs. In fact, some journalist [20] claimed that those ENCODE project papers are "eulogy" for junk DNA. However, Eddy (2012; [21]) ended his commentary as follows:

> Given the C-value paradox, mutational load, and the massive impact of transposons, the data remain consistent with the view that the nonconserved 80–95 % of the human genome is mostly composed of nonfunctional decaying transposons: 'junk'.

Graur et al. (2013; [22]) also condemned the ENCODE project statement as an "evolution-free gospel." After more than 40 years of its proposal by Ohno (1972; [23]), the recognition that our genome has so much junk DNA holds its throne.

10.4   Personal Genomes

The human genome sequences available at genome databases (NCBI, Ensembl, and UCSC) are essentially haploid, and they are mosaic of several individuals [3, 4]. If we consider one particular individual, however, there are two set of genomic sequences because humans are diploids. It is a challenge to determine very similar chromosomes of one human individual. If correctly determined, informations not only on SNPs but also those on insertions and deletions as well as copy numbers (see Chap. 2) can be obtained. Some genetic variations may be associated with important phenotypes such as genetic diseases. The 2010s are thus the decade of personal genomes.

Craig Venter's group first determined the nucleotide sequences of these two genome set for Venter himself in 2007 [24, 25]. The Sanger method (see Chap. 11) was used as before, but the total cost was greatly reduced. The next human individual whose personal genome was more or less determined was James D. Watson [26, 27]. The newly introduced second-generation technology 454 (see Chap. 11) was used for sequencing, and the determined sequence was not really in the "complete" status, though the paper title claimed so. The third and fourth personal genomes were determined for an anonymous Nigerian male [28] and for a Chinese male whose initial was Y. H. [29, 30] using another second-generation technology Solexa (see Chap. 11). At the same time, a personal genome of a patient with acute myeloid leukemia was determined [31]. Two Korean males' personal genomes were also determined independently in 2009 [32, 33], followed by genome sequencing of one Japanese [34].

Craig Venter's diploid genome contained 3.2 million heterozygous sites on autosomal chromosomes, and more than 1/3 of them were newly found SNP loci. A total of 0.85 million insertion–deletion sites with the range of 1-base to 80-kb lengths were found. The inversed regions were also found in 90 locations [24]. James Watson's genome sequence also contained about 3.3 million heterozygous sites, including more than 10,000 nonsynonymous differences [26]. A detailed DNA variation analysis of the genome sequence of Craig Venter [35] showed that 20 % of ~10,000 nonsynonymous differences were rare in the human populations, and 1,500 such sites had possibility of changing protein functions. Table 10.4 shows the decomposition of the SNP data in one Korean male genome sequence (based on [32]. Figure 10.2 presents two Venn diagrams of SNP sharing patterns for five human individual genomes (based on [32]). Although individuals A (Craig Venter) and B (James Watson) are both of European origin, their shared SNP loci (461,281) are smaller than those shared between A and E (a Korean male) and those between B and E. Individual D (a Chinese male) and E are both of East Asian origin, and their shared SNP loci (669,777) are larger than those shared between C (a Nigerian male) and E and those between C and D. The individual-specific SNP loci are highest (1.9 million) for individual C.

Personal genome studies shifted to simultaneous multiple individual sequencing from 2010. Genomes of four Khoisan individuals and one Bantu individual were determined [36], as well as those for 18 Korean individuals [37]. Another aspect of multiple personal genome sequencing is direct estimation of the human mutation rate. Genome sequences of six individuals (two parent-offspring trios) were

Table 10.4 Decomposition of the SNP data in one Korean male genome sequence (from [32])

Total SNPs in SJK genome: 3,439,107
Novel SNP: 420,083
UTR or CDS: 7,172
-5'UTR: 937
-3'UTR: 3,304
-CDS: 2,931
nsSNPs in CDS: 1,348
Others: 1,583
Others: 412,911
Known in dbSNP: 3,019,024
Non-validated: 426,911
UTR or CDS: 4,922
-5'UTR: 820
-3'UTR: 2,232
-CDS: 1,870
nsSNPs in CDS: 638
Others: 1,232
Others: 421,989
Validated: 2,592,113
UTR or CDS: 52,550
-5'UTR: 5,157
-3'UTR: 25,076
-CDS: 22,317
nsSNPs in CDS: 7,348
Others: 14,969
Others: 2,539,563

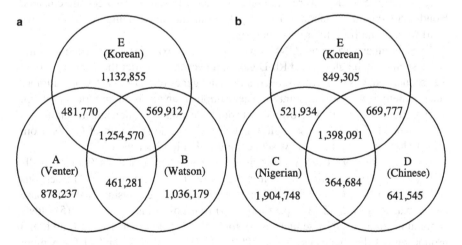

Fig. 10.2 Venn diagrams of SNP sharing patterns for five human individual genomes (Based on [32])

determined by three studies [38–40], and the mutation rate of $\sim 1 \times 10^{-8}$ per nucleotide site per generation was estimated (see Chap. 2). We are now entering personal genome sequencing of 1,000 people; see 1000 genome project homepage (http://www.1000genomes.org/) and Tohoku University's home page (http://www.megabank.tohoku.ac.jp/english/news/detail.php?id=684&c1=4). It is highly likely that genome sequences of more than one million human individuals will be determined by end of 2020, only 7 years later.

10.5 Genomic Heterogeneity of the Human Genome

When the draft human genome sequences were determined [3, 4], multiple human chromosomes were sequenced. This naturally produced individual DNA variation data, and the nucleotide diversity of 0.075 % was estimated from 1.42 million single nucleotide polymorphisms [41]. The human HapMap (haplotype map) project was initiated based on these SNPs. A total of 269 individuals from the three geographically distinct human populations, West Eurasians, East Eurasians, and Africans, were examined for massive SNP typing [42]. The African population from Nigeria was shown to have the highest SNP heterogeneity. A similar scale SNP typing was conducted for samples of 938 individuals from 51 human populations collected under the Human Genome Diversity Project led by L. L. Cavalli-Sforza, and the phylogenetic relationship of populations confirmed previously obtained results based on microsatellite DNA polymorphism data [43]. There are many similar studies on many human populations, and I would like to cite one paper by Yamaguchi-Kabata et al. (2008; [44]), partly because I am Japanese but also for its massive data size; ~273,000 SNP loci were examined for 7,000 Japanese. Two genetic clusters, Hondo ("mainland" in Japanese) and Ryukyu (an alternative name for the Okinawa area) were found from their data analysis.

Copy number variations (CNVs) were known to exist for a long time at some genes such as alpha globin and Rh-D genes; however, only after the human genome was sequenced, CNVs were discovered in many genes (e.g., [45–47]. Copy number heterogeneity is the starting point for a gene duplication and is an important problem in evolutionary genomics. However, currently available sequencing technologies (see Chap. 11) may not be suitable for detecting a wide spectrum of CNVs, from short to long unit lengths. A new sequencing technology is awaited.

Another category of DNA variation is insertions and deletions (indels). Mills et al. (2006; [48]) classified indels into the following five categories: (1) insertions and deletions of single-base pairs, (2) monomeric base pair expansions, (3) multi-base pair expansions of 2–15-bp repeat units, (4) transposon insertions, and (5) indels containing random DNA sequences. A total of ~415,000 indels were found, and proportions of the five categories are 29 %, 18 %, 11 %, 1 %, and 41 %. A more recent study [49] showed that short indels are 1/15 of SNPs. This proportion is in rough agreement with Saitou and Ueda's (1994; [50]) estimate that indel type differences were ~1/10 of nucleotide substitution-type differences from comparison of different primate species sequences.

10.6   Genetic Changes That Made Us Human

Saitou (2005; [51]) estimated the amount of genetic changes responsible for human-specific phenotypes based on data available at that time. Let us follow his arguments using updated values. The overall level of nucleotide substitutional divergence between human and chimpanzee is 1.23 % [52, 53]. Many of those differences are due to mutations that occurred in junk DNAs and most probably have no effect on the level of phenotypic difference between human and chimpanzee. However, some proportion of nucleotide changes, in particular, amino acid replacing nonsynonymous ones, must be responsible for human-specific characters. There are ~21,000 protein coding genes in the human genome as discussed previously in this chapter, and a certain proportion must code proteins that slightly differ in amino acid sequences between human and chimpanzee. Some of those amino acid differences may be responsible for human-specific characters. Amino acid differences are classified into replacement (substitution) and insertions/deletions (indels). Both types can lead to slight-to-considerable changes of the protein 3D structure, which is the source of human-specific phenotypes. DNA regions responsible for the modification of gene expression are also important. The discovery of such regions is difficult and we need to obtain multiple species comparisons of contiguous genomic sequences including both coding and noncoding DNA.

How many significant changes are really responsible for creating humanness or "hominization"? The level of genomic nucleotide difference between human and chimpanzee has been accumulated in both the human and chimpanzee lineage. Therefore, about half (ca. 0.6 %) of this difference can be assumed to have accumulated in the human lineage after divergence from the chimpanzee lineage. All the genetic changes responsible for "humanness" must reside in these differences. The human genome consists of about 3.2 billion nucleotides (see Table 10.1), and thus the 0.6 % difference is tantamount to about 19 million nucleotide changes. Although ~97 % of the human genome may be attributed to be nonfunctional DNAs (see Sect. 10.1), we still have some 570,000 nucleotide changes located in the functional DNA regions, either protein coding or noncoding. Of these, how many changes are really responsible for creating humanness? First, we can safely omit synonymous changes in the coding region. Even if amino acid changes occur, the function of one protein may not drastically change. If we eliminate those nonsignificant nucleotide changes, the really interesting changes may only be ~1 %, or ca. 6,000 nucleotide changes. I would like to add that the genetic changes responsible for the humanness could also be accomplished through the fixation of neutral mutations, as well as that of advantageous mutations.

There are many studies claiming more abundant positive selection in the human lineage compared to the chimpanzee lineage after divergence from the common ancestor of humans and chimpanzees. For example, Clark et al. (2003; [54]) sequenced chimpanzee PCR amplicons using primers designed for exons of ~20,000 human genes and compared them with human and mouse orthologous genes. They estimated that 667 (8.7 %) out of 7,645 genes compared were under positive selection at the 5 % significant level. However, a computer simulation study by Zhang [55] showed that

Table 10.5 Genes showing significantly different nonsynonymous changes between human lineage and ape lineages (from Kitano et al. 2004; [56])

Gene	No. a.a. changes		Results of statistical tests[a]		
	Human	Ape	A.I.[b]	dN×dSc	Dist.[c,d]
BRCA1	17	40	** (NS)	↑*	NS
APOE	6	7	** (NS)	NS	NS
PRM2	8	14	** (**)	NS	↑*
HCR	7	17	* (NS)	NS	NS
FOXP2	2	1	* (**)	↑*	NS
ZFY	2	1	* (NS)	NS	NS
DAF tree A	2	28	NS (NS)	↑*	NS
DAF tree B	5	27	NS (NS)	NS	NS
ACAT2	2	2	NS (NS)	NS	↑*
RNASE3	1	27	NS (NS)	NS	↓*

[a]One asterisk and two asterisks are significant at 5 % level and 1 %, respectively. NS means nonsignificant
[b]Accelelated Index. Results shown in parentheses are when mouse was used as outgroup
[c]Arrows designate direction of selection; up and down for positive and negative, respectively
[d]Result from empirical distribution shown in Fig. 3 of Kitano et al. (2004; [56])

the maximum likelihood method used in this study has tendency to produce many false-positive results. Kitano et al. [56] compared 103 protein coding gene sequences for human, chimpanzee, gorilla, and orangutan and conducted three kinds of statistical tests (see Table 10.5). Nine genes showed statistically significant result in some test; however, none of them was significant for all three tests.

Similar arguments are also popular in the evolution of noncoding regions. Prabhakar et al. (2008; [57]) found one such 239-bp sequence, HACNS1, located neighbor to the GBX2 limb-expressing gene. HACNS1 was shown to act as a limb bud enhancer with enhanced limb enhancer activity specifically in human. This change was caused by 13 human-specific substitutions within an 81-bp region and was interpreted that accumulation of these positively selected substitutions created multiple novel transcription factor binding sites (gain of function) and that the deposition of those facilitated the human-specific enhanced activity [57]. However, loss of function in a repressor element within HACNS1 can be another explanation for it. A GC-biased gene conversion (BGC) may be an alternative explanation for fixation of such mutations without experiencing adaptive evolution (e.g., [58]. Sumiyama and Saitou (2011; [59]) performed transgenic mouse assay of the HACNS1 construct lacking the 81-bp region, and that construct showed similar enhancer activity to the intact human HACNS1. This suggests that the function of the human 81-bp region is not an activating enhancer, but rather a disrupted repressor, supporting the nonselective, neutral BGC hypothesis. We should be careful at jumping to the interpreting data by positive selection, for the default mechanism of evolution is neutral (see Chap. 4).

10.7   Ancient Human Genomes

Neanderthals are one of the closest extinct lineages of humans to modern humans. Although they are often classified as *Homo neanderthalensis*, an alternative classification is *Homo sapiens neanderthalensis*, a subspecies. Under the latter case, reproductive isolation is not assumed. Partial sequences of Neanderthals were reported in 2006 [60] and [61]; however, one of them [61] was found to have a high proportion of contemporary human DNA contaminations [62]. Draft genome sequences of one Neanderthal individual were reported in 2010 [63]. The comparison with various modern human DNA data suggested that the 1–4 % of the modern non-African humans contain the Neanderthal genome sequences, while modern African humans have no signature of the Neanderthal genome. If so, modern humans admixed with Neanderthals after they moved out of Africa. In fact, Yotova et al. (2011; [64]) found a notable presence (9 % overall) of a Neanderthal-derived X chromosome segment among most of contemporary human populations outside sub-Saharan Africa. This supports the admixture model. However, Eriksson and Manica (2012; [65]) suggested that a spatial population structure can explain the patterns similar to those expected under this admixture scenario.

Another ancient DNA draft genome sequences were reported at the end of 2010, based on DNA molecules extracted from a finger bone excavated from Denisova cave in Southern Siberia [66]. Comparison with modern human SNP data suggested that Denisova people admixed only with the common ancestral population of Near Oceanians, Polynesians, Fijians, East Indonesians, and the Philippine Mamanwa [66–68]. Ancient human genomic DNA sequences are shedding lights to the complex migration history of modern humans. Genomic sequences of more recent ancient DNA human samples from Greenland [69], from Australia [70], from Alpes [71], and from Scandinavia [72] are just the first tide of mass ancient human genome sequences to be expected in the near future.

References

1. Gregory, T. R. (Ed.). (2005). *The evolution of the genome*. Burlington: Elsevier Academic.
2. Bickmore, W. A. (2001). *Karyotype analysis and chromosome banding*. Encyclopedia of Life Sciences (http://web.udl.es/usuaris/e4650869/docencia/segoncicle/genclin98/recursos_classe_%28pdf%29/revisionsPDF/bandmethods2.pdf)
3. International Human Genome Sequencing Consortium. (2001). Initial sequencing and analysis of the human genome. *Nature, 409*, 860–921.
4. Venter, J. C., and others (2001). The sequence of the human genome. *Science, 291*, 1304–1351.
5. Brown, T. A. (2007). *Figure 7.13, genomes 3*. New York: Garland Science Publishing.
6. International Human Genome Sequencing Consortium. (2004). Finishing the euchromatic sequence of the human genome. *Nature, 431*, 931–945.
7. Imanishi, T., et al. (2004). Integrative annotation of 21,037 human genes validated by full-length cDNA clones. *PLoS Biology, 2*, 856–875.

8. Clamp, M., et al. (2007). Distinguishing protein-coding and noncoding genes in the human genome. *PNAS, 104*, 19428–19433.
9. Mi, H., Muruganujan, A., Thomas, P. D., et al. (2013). PANTHER in 2013: Modeling the evolution of gene function, and other gene attributes, in the context of phylogenetic trees. *Nucleic Acid Research, 41*, D377–D386.
10. Nozawa, M., Kawahara, Y., & Nei, M. (2007). Genomic drift and copy number variation of sensory receptor genes in humans. *PNAS, 104*, 20421–20426.
11. http://gtrnadb.ucsc.edu
12. www.ensembl.org/biomart/
13. The ENCODE Project Consortium. (2012). An integrated encyclopedia of DNA elements in the human genome. *Nature, 489*, 57–74.
14. Thurman, R. E., et al. (2012). The accessible chromatin landscape of the human genome. *Nature, 489*, 75–82.
15. Neph, S., et al. (2012). An expansive human regulatory lexicon encoded in transcription factor footprints. *Nature, 489*, 83–90.
16. Gestein, M. B. (2012). Architecture of the human regulatory network derived from ENCODE data. *Nature, 489*, 91–100.
17. Djebali, S., et al. (2012). Landscape of transcription in human cells. *Nature, 489*, 101–108.
18. Birney, E., et al. (2007). Identification and analysis of functional elements in 1% of the human genome by the ENCODE pilot project. *Nature, 447*, 799–816.
19. van Bakel, H., Nislow, C., Blencowe, B. J., & Hughes, T. R. (2010). Most "Dark Matter" transcripts are associated with known genes. *Plos Biology, 8*, e1000371.
20. Pennisi, E. (2012). ENCODE project writes eulogy for Junk DNA. *Science, 337*, 1159–1161.
21. Eddy, S. (2012). The C-value paradox, junk DNA and ENCODE. *Current Biology, 22*, R898.
22. Graur, D., Zheng, Y., Price, N., Azevedo, R. B. R., Zufall, R. A., & Elhaik, E. (2013). On the immortality of television sets: function in the human genome according to the evolution-free gospel of ENCODE. *Genome Biology and Evolution*, 5(3), 578–590. (published online on February 20, 2013).
23. Ohno, S. (1972). So much "junk" DNA in our genome. *Brookhaven Symposium in Biology, 23*, 366–370.
24. Levy, S., et al. (2007). The diploid genome sequence of an individual human. *PLoS Biology, 5*, e254.
25. http://huref.jcvi.org/
26. Wheeler, D. A., et al. (2008). The complete genome of an individual by massively parallel DNA sequencing. *Nature, 452*, 872–876.
27. http://jimwastonsequence.cshl.edu/cgi-perl/gbrowse/jwsequence/
28. Bentley, D. R., et al. (2008). Accurate whole human genome sequencing using reversible terminator chemistry. *Nature, 456*, 53–59.
29. Wang, J., et al. (2008). The diploid genome sequence of an Asian individual. *Nature, 456*, 60–66.
30. http://yh.genomics.org.cn/
31. Ley, T. J., et al. (2008). DNA sequencing of a cytogenetically normal acute myeloid leukaemia genome. *Nature, 456*, 66–72.
32. Ahn, S. M., et al. (2009). The first Korean genome sequence and analysis: Full genome sequencing for a socio-ethnic group. *Genome Research, 19*, 1622–1629.
33. Kim, J.-I., et al. (2009). A highly annotated whole-genome sequence of a Korean individual. *Nature, 460*, 1011–1015.
34. Fujimoto, A., et al. (2010). Whole-genome sequencing and comprehensive variant analysis of a Japanese individual using massively parallel sequencing. *Nature Genetics, 42*, 931–936.
35. Ng, P. C., Levy, S., Huang, J., Stockwell, T. B., Walenz, B. P., Li, K., Axelrod, N., Busam, D. A., Strausberg, R. L., & Venter, J. C. (2008). Genetic variation in an individual human exome. *PLoS Genetics, 4*, e1000160.
36. Schster, S. C., et al. (2010). Complete Khoisan and Bantu genomes from southern Africa. *Nature, 463*, 943–947.

37. Ju, Y.-S., et al. (2011). Extensive genomic and transcriptional diversity identified through massively parallel DNA and RNA sequencing of eighteen Korean individuals. *Nature Genetics, 42*, 931–936.

38. Roach, J. C., et al. (2010). Analysis of genetic inheritance in a family quartet by whole-genome sequencing. *Science, 328*, 636–639.

39. The 1000 Genomes Project Consortium. (2010). A map of human genome variation from population-scale sequencing. *Nature, 467*, 1061–1073.

40. Conrad, D. F., et al. (2011). Variation in genome-wide mutation rates within and between human families. *Nature Genetics, 43*, 712–714.

41. The International SNP Map Working Group. (2001). A map of human genome sequence variation containing 1.42 million single nucleotide polymorphisms. *Nature, 409*, 928–933.

42. International HapMap Consortium. (2005). The haplotype map of the human genome. *Nature, 437*, 1299–1320.

43. Rosenberg, N. A., et al. (2002). Genetic structure of human populations. *Science, 298*, 2381–2385.

44. Yamaguchi-Kabata, Y., Nakazono, K., Takahashi, A., Saito, S., Hosono, N., Kubo, M., Nakamura, Y., & Kamatani, N. (2008). Population structure of Japanese based on SNP genotypes from 7,001 individuals in comparison to other ethnic groups: Effects on population-based association studies. *American Journal of Human Genetics, 83*, 445–456.

45. Pinkel, D., et al. (1998). High resolution analysis of DNA copy number variation using comparative genomic hybridization to microarrays. *Nature Genetics, 20*, 207–211.

46. Sebat, J., et al. (2004). Large-scale copy number polymorphism in the human genome. *Science, 305*, 525–528.

47. Redon, R., et al. (2006). Global variation in copy number in the human genome. *Nature, 444*, 444–454.

48. Milles, R. E., et al. (2006). An initial map of insertion and deletion (INDEL) variation in the human genome. *Genome Research, 16*, 1182–1190.

49. The 1000 Genomes Project Consortium. (2010). A map of human genome variation from population-scale sequencing. *Nature, 467*, 1061–1073.

50. Saitou, N., & Ueda, S. (1994). Evolutionary rate of insertions and deletions in non-coding nucleotide sequences of primates. *Molecular Biology and Evolution, 11*, 504–512.

51. Saitou, N. (2005). Evolution of hominoids and the search for a genetic basis for creating humanness. *Cytogenetic and Genome Research, 108*, 16–21.

52. Fujiyama, A., et al. (2002). Construction and analysis of a human-chimpanzee comparative clone map. *Science, 295*, 131–134.

53. The Chimpanzee Sequencing and Analysis Consortium. (2005). Initial sequence of the chimpanzee genome and comparison with the human genome. *Nature, 437*, 69–87.

54. Clark, A. G., et al. (2003). Inferring nonneutral evolution from human-chimp-mouse orthologous gene trios. *Science, 302*, 1960–1963.

55. Zhang, J. (2003). Frequent false detection of positive selection by the likelihood method with branch-site models. *Molecular Biology and Evolution, 21*, 1332–1339.

56. Kitano, T., Liu, Y.-H., Ueda, S., & Saitou, N. (2004). Human specific amino acid changes found in 103 protein coding genes. *Molecular Biology and Evolution, 21*, 936–944.

57. Prabhakar, S., et al. (2008). Human-specific gain of function in a developmental enhancer. *Science, 321*, 1346–1350.

58. Duret, L., & Galtier, N. (2009). Comment on "Human-specific gain of function in a developmental enhancer". *Science, 323*, 714.

59. Sumiyama, K., & Saitou, N. (2011). Loss-of-function mutation in a repressor module of human-specifically activated enhancer HACNS1. *Molecular Biology and Evolution, 28*, 3005–3007.

60. Noonan, J. P., et al. (2006). Sequencing and analysis of Neanderthal genomic DNA. *Science, 314*, 1113–1118.

61. Green, R. E., et al. (2006). Analysis of one million base pairs of Neanderthal DNA. *Nature, 444*, 330–336.

62. Wall, J. D., & Kim, S. K. (2007). Inconsistencies in Neanderthal genomic DNA sequences. *PLoS Genetics, 3,* 1862–1866.
63. Green, R., et al. (2010). A draft sequence of the Neanderthal genome. *Science, 328,* 710–722.
64. Yotova, V., et al. (2011). An X-linked haplotype of Neanderthal origin is present among all non-African populations. *Molecular Biology and Evolution, 28,* 1957–1962.
65. Eriksson, A., & Manica, A. (2012). Effect of ancient population structure on the degree of polymorphism shared between modern human populations and ancient hominins. *Proceedings of the National Academy of Sciences of the United States of America, 109,* 13956–13960.
66. Reich, D., et al. (2010). Genetic history of an archaic hominin group from Denisova Cave in Siberia. *Nature, 468,* 1053–1060.
67. Meyer M. et al. (2012). A high-coverage genome sequence from an archaic Denisovan individual. *Science,* Aug. 31, 2012, (epub ahead of print).
68. Reich, D., et al. (2011). Denisova admixture and the first modern human dispersals into Southeast Asia and Oceania. *American Journal of Human Genetics, 89,* 1–13.
69. Rasmussen, M., et al. (2010). Ancient human genome sequence of an extinct Palaeo-Eskimo. *Nature, 463,* 757–762.
70. Rasmussen, M., et al. (2011). An Aboriginal Australian genome reveals separate human dispersals into Asia. *Science, 334,* 94–98.
71. Keller, A., et al. (2012). New insights into the Tyrolean Iceman's origin and phenotype as inferred by whole-genome sequencing. *Nature Communications, 3,* 698.
72. Skoglund, P., et al. (2012). Origins and genetic legacy of Neolithic farmers and hunter-gatherers in Europe. *Science, 336,* 466–469.
73. Li, J. Z., Absher, D. M., Tang, H., Southwick, A. M., Casto, A. M., Ramachandran, S., Cann, H. M., Barsh, G. S., Feldman, M., Cavalli-Sforza, L. L., & Myers, R. M. (2008). Worldwide human relationships inferred from genome-wide patterns of variation. *Science, 319,* 1100–1104.

Part III

Methods for Evolutionary Genomics

Genome Sequencing

11

Chapter Summary

The wet experimental steps are necessary for sequencing genomes. These steps are summarized in this chapter, starting from DNA sampling to sequence determination with the aid of computational analyses.

11.1 DNA Extraction and Purification

We expect to compare genome sequences of many species in evolutionary genomics. Finding materials for extracting genomic DNA or other resources such as RNA or protein is important, in contrast to studies on model organisms. I focus on genomic DNA extraction in this section.

11.1.1 Special Treatments for Various Samples

We have to take care of individual samples depending on their own characteristics. When one is interested in genome sequencing of an endangered species, it is preferable to obtain samples without killing individuals of this species. Hairs, urines, or feces are often sampled from apes in the wild, as conducted by Shimada et al. [1]. When one wishes to study human individuals for personal genome sequencing, the informed consent should be obtained from these individuals, after approval of the bioethics committee of the institution the researcher belongs. Anonymity is important for human samples to protect individual information of close relatives of sampled individuals. In this regard, disclosures of individual names for some personal genome sequences (e.g., [2–4]) should have not been done. In fact, names of individuals whose genome sequences were determined are no longer open in later studies. General methods for collection and storage of tissue samples are described in [5].

N. Saitou, *Introduction to Evolutionary Genomics*, Computational Biology 17,
DOI 10.1007/978-1-4471-5304-7_11, © Springer-Verlag London 2013

DNA extraction and purification from ancient sample obviously require special care and need special protocols (e.g., [6]). When ancient DNA of modern humans is studied, contamination of investigators should be eliminated through various procedures (e.g., Kanzawa et al. 2013; [7]).

11.1.2 How to Extract Genomic DNA

There are plenty of protocols for extraction of genomic DNAs (e.g., [8, 9]). Because DNA molecules in cell nucleus are usually bound with proteins, we have to eliminate these DNA-bound proteins from the DNA–protein complex. A basic method is to use phenol and chloroform, which was initially devised for extraction of RNAs [10]. Tissues such as blood may be suitable as the starting material from large-sized mammals, but the whole body may be used when the organism is small. Liquid nitrogen is sometimes used to homogenate samples. Tissues are instantaneously frozen and samples can easily be powdered.

The following is the protocol for extracting genomic DNA from human blood (based on [11]). A blood sample is thawed, and after adding the standard citrate buffer, the mixture is centrifuged. The top portion of the supernatant is discarded, and additional buffer is added and again centrifuged. After the supernatant is discarded, the pellet is resuspended in a solution of SDS detergent and proteinase K, and the mixture is incubated. The sample then is phenol extracted once with a phenol/chloroform/isoamyl alcohol solution, and after centrifugation, the aqueous layer is removed. The DNA is ethanol precipitated, resuspended in buffer, and again ethanol precipitated. Once the pellet is dried, buffer is added and the DNA is resuspended by incubation overnight.

Nowadays many types of DNA extraction kits are commercially available, and some of them may be used considering the cost and performance. It should be noted that the initial step of tissue homogenization is often the most important, because it depends on organisms you are interested in.

11.2 Construction of Genomic Library

Any DNA molecules may be used as templates for PCR (polymerase chain reaction) amplification (e.g., [12]). However, a PCR amplicon is only a short fragment of the total genomic DNA. We definitely need a different methodology to access to the whole-genomic DNA in a systematic and exhaustive way. Genomic DNA "library" is the solution to this. As a library in the usual sense contains many books which are accessible for us individually, a genomic DNA library contains many DNA fragments in a special way, called "clone." This word was originated in a Greek word meaning "twig," but in modern biology, it means an entity, such as cell, individual, or DNA, which are genetically identical. DNA cloning (e.g., [8]) is one of the basic technologies of molecular biology, combined with restriction enzymes.

Fig. 11.1 Basic steps for construction of a genomic DNA clone (Based on Fig. 2.15 of [13])

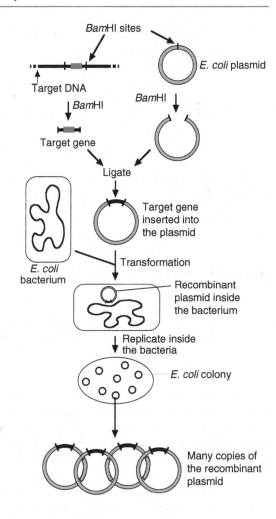

Figure 11.1 shows a strategy for construction of a genomic DNA clones (based on Fig. 2.15 of [13]). The target DNA molecule is digested with a certain restriction enzyme, while the same restriction enzyme was used to digest a bacterial plasmid. They are ligated to produce the plasmid containing a DNA region originated from the target genome. This plasmid is transformed to bacteria, and the recombinant plasmid can clonally replicate within the bacterial cells.

A genomic clone library contains a huge number of such cloning vectors. There are various types of cloning vectors as shown in Table 11.1. The genome of the λ phage is linear 48.5-kb size. Its both end sequences are 12-nucleotide complementary sequences called "cos sites," and they facilitate the linear genome to form a circular molecule. It can contain maximum of 18-kb alien DNA. A fosmid is a plasmid which contains the F plasmid origin of replication and cos sites, and it

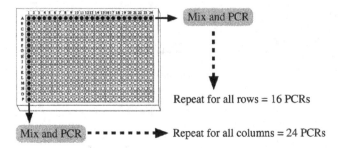

Fig. 11.2 A schematic illustration of our two-step PCR screening system. This shows the second step. After a plate containing a positive clone(s) is found, row-by-row and column-by-column screenings follow by pooling clones from the same row and same column, respectively. (From Kim et al. 2003; [15])

Table 11.1 List of cloning vector systems

Name	Insert size (kb)
λ phage	18
Fosmid	40
Bacterial artificial chromosome (BAC)	100–300
Yeast artificial chromosome (YAC)	100–3000

can contain the maximum of 40-kb insert DNAs. A bacterial artificial chromosome (BAC) is made of *E. coli* F plasmid and can contain up to the 300-kb DNA fragment. BAC is the most popular cloning vector system in genome libraries, and many BAC libraries were made so far (e.g., [14]). A yeast artificial chromosome (YAC), which can contain up to 3-mb alien DNA, is included not in bacteria such as *E. coli* but in yeast, *S. cerevisiae*. YACs have centromeres, telomeres, and origins of replications, and they are similar to natural yeast chromosomes. Unfortunately, YACs are not so stable because of recombinations with other chromosomes. This leads to frequent use of smaller size BACs.

The whole-genomic DNAs are digested with a particular restriction enzyme, and many DNA fragments with the same restriction sites are incorporated into different bacterial cells. These bacteria are cultured to make many colonies, and they are picked, usually by using colony picking robots. A vast number of genomic DNA-containing bacterial clones are arrayed systematically, and they become the genomic DNA resource for that organism. For example, Kim et al. [15] constructed a gorilla fosmid library of 261,120 independent clones, and they are expected to cover the ~3.5 times of the ~3-Gb gorilla genome. Figure 11.2 shows the second step of the screening system using 384 microtiter plates (from [15]). Creation of a BAC library is preferable for large-sized mammalian genomes; however, it requires fresh cells [16]. When only a dead tissue, such as liver tissue kept in ethanol used in [15], is available, a fosmid library may be constructed instead of BAC library. It should be noted that Chap. 2 of Brown [13] was frequently consulted in this section.

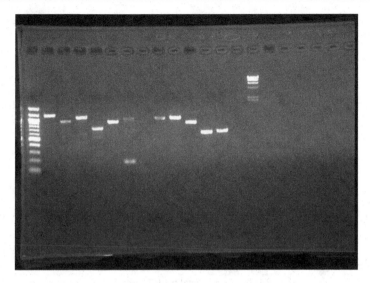

Fig. 11.3   Example of DNA gel electrophoresis

11.3   Determination of Nucleotide Sequences: Chemical Tactics

Determination of nucleotide sequences is essentially chemical process. We thus discuss the two main chemical strategies for DNA sequencing.

11.3.1   General Overview

The first generation of nucleotide sequence determination was based on electrophoresis (see Fig. 11.3). Two types of methods [17, 18] were invented, and Walter Gilbert and Frederic Sanger both received Nobel Chemistry Prize in 1980. Length differences of the DNA sequences are detected through electrophoresis (Fig. 11.3) in both methods. However, there is a clear difference between the two methods: cutting DNA sequences into various lengths in the Maxam–Gilbert method [17] and elongation of DNA with various lengths in Sanger et al.'s method [18]. The latter method became much popular, and most of the automated sequencers used for genome sequencings followed that method, now called Sanger's method or the chain-termination method. We thus explain this method in the next section. Another type of sequencing method used for the second-generation sequencers is detection of DNA replication at each nucleotide addition for relatively short sequences, up to the limit of the first-generation methods. A much longer DNA sequences, say more than 10 kb, are expected to be massively produced when the third-generation sequencers are becoming available soon. Table 11.2 lists the various sequencing technologies.

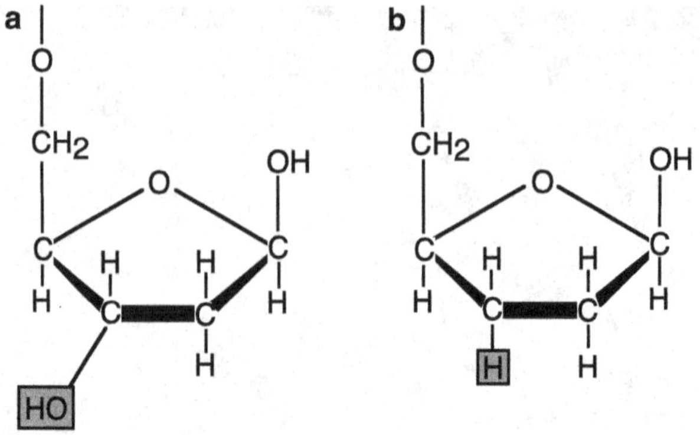

Fig. 11.4 Comparison of molecular structure between deoxyribose (**a**) and dideoxyribose (**b**)

Table 11.2 Various types of nucleotide sequencing methods (as of October 2013)

Type and method	Read length (bp)	Total reads per run	Reference
Length comparison:			
Sanger	500–900	max. 96	http://www.lifetechnologies.com
Sequencing by synthesis:			
Pyrosequencing (454)	~1,000	~1 million	http://www.454.com
Hiseq	~200	~3,000 million	http://www.illumina.com
Ion torrent	~200	~5 million	http://www.lifetechnologies.com
SMRT	~5,000	~50,000	http://www.pacificbiosciences.com
Sequencing by ligation:			
SOLiD	~60	~5,000 million	http://www.appliedbiosystems.com

11.3.2   Sanger's Method

The natural process of DNA polymerization is used in Sanger's method, except for the artificially made dideoxyribose (Fig. 11.4). When DNA polymerase is accessed to natural nucleotide with deoxyribose, DNA polymerization occurs, while the polymerase stops polymerization when it is accessed to dideoxyribose [18]. This produces various lengths of DNAs. For example, a DNA molecule with the sequence 5′-ACCACGTGTA-3′ will produce two kinds of complementary DNAs with lengths 6 and 8, if there are deoxyribose-type and dideoxyribose-type guanine nucleotides. This length difference will be detected by electrophoresis (see Fig. 11.3). Similar procedures can be conducted for the remaining three nucleotides (A, C, and T), and the correct nucleotide sequence can be obtained by combining the results for the four electrophoresis (Fig. 11.5). The automated sequencer is using laser to detect fluorescent-labeled nucleotides, and up to 900 bps can be read

Fig. 11.5 Basic steps of Sanger's method (Based on http://users.rcn.com/jkimball. ma.ultranet/BiologyPages/D/ DNAsequencing.html)

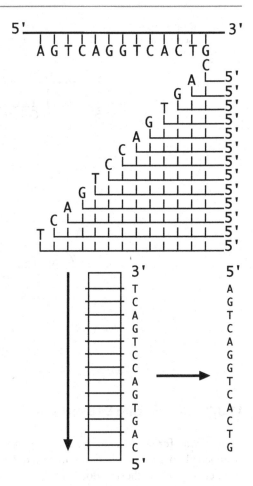

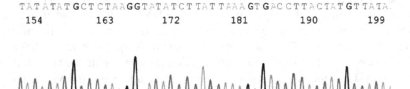

Fig. 11.6 An example of chromatogram

in a standard condition. An example chromatogram produced in my laboratory is shown in Fig. 11.6. A more detailed explanation can be found in Chap. 4 of Brown (2006; [13]).

Fig. 11.7 Basic steps of
the pyrosequencing method
(Based on [19])

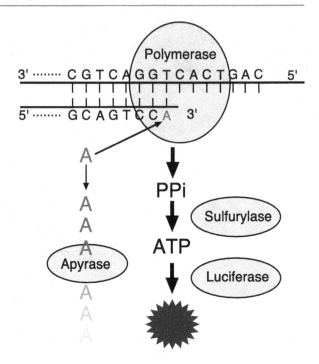

11.3.3 Sequencing by Synthesis

Another way for DNA sequence determination is to detect each nucleotide when it
is attached to the template single-strand DNA molecule. This strategy, called
"sequencing by synthesis," does not depend on DNA lengths, and electrophoresis
is not necessary. One such example is pyrosequencing. As shown in Fig. 11.7
(based on [19]), four kinds of dNTPs (deoxyribonucleotide triphosphates) are
sequentially added to the target single-stranded DNA molecule, and if one, say
ATP, is complementary to the head base (T) of the target, DNA polymerase
covalently bonds this ATP to the new strand. A pyrophosphate molecule, left after
this reaction, is used as trigger to produce light emission via the luciferin–luciferase system. The emitted light is detected by CCD (charge-coupled device) camera.
When there are more than one nucleotide at the target DNA sequence, the amount
of pyrophosphate molecule produced is proportional to the number of homonucleotides, and the height for that nucleotide should be higher. When the number of
homonucleotides is becoming larger, the estimation of number may have higher
error. Human mitochondrial DNA control region is known to have poly-C stretch
in some individuals, and the nucleotide sequences after this region is difficult to
determine if we use Sanger's method (Fig. 11.8a). We therefore used the pyrosequencing method to overcome this problem [20]. Figure 11.8b and 11.8c shows
example pattern of the pyrosequencing method [20].

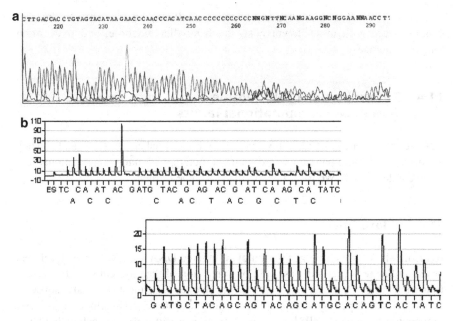

Fig. 11.8 Comparison of Sanger sequencing and pyrosequencing. (**a**) Part of Sanger method sequencing output for human mitochondrial DNA. (**b**) Part of pyrosequencing output for human mitochondrial DNA. (**c**) Enlargement of part of Panel b. (From Shimada et al. 2002; [20])

The pyrosequencing chemistry was later applied to the 454 sequencer [21], one of the first type of the so-called second-generation sequencers, in contrast to the first-generation sequencers based on Sanger's method. Later, the last part of nucleotide detection using luciferin–luciferase system was modified to detect proton ion flux changes [22]. Different techniques on sequencing by synthesis were used in other systems such as Solexa, SOLiD, Helicos, and Ion Torrent [23–26]) and the cost and time of DNA sequencing were drastically decreased during the first decade of the twenty-first century. This dramatic change opened a new window for evolutionary genomics, as briefly described in Part II. Close to one billion US dollars were spent to sequencing one human genome in 2000 A.D., and now, in 2013, the cost is approaching only 1,000 US dollars.

11.3.4 Physical Distinguishing of Each Nucleotide

Seeing is believing. It is ideal if we can literally "see" DNA sequences. Many researchers tried to enhance the electron microscopy (e.g., [27]), but so far not successful. An alternative way of distinguishing four kinds of DNA bases is to physically detect their molecular differences. A promising one is to use specific membrane proteins called "nanopore" (e.g., [28]). Recently, Oxford Nanopore Technology announced a new "strand sequencing" technique based on nanopores [29]. If their sequencers become available, we may shift to the new genomics era with much lower cost, much smaller machine, and with much longer sequences.

11.4 Determination of Nucleotide Sequences: Computational Tactics

DNA sequences are essentially texts with only four alphabets, A, C, G, and T. They are therefore easy to handle using digital computers, and many computational tactics were developed. We would like to discuss some of them.

11.4.1 Base Call

Immediately after the first generation of automated DNA sequencer was developed based on Sanger's method, base call, or translation of the physical signal to one of four nucleotides (A, C, G, or T), was also developed. The most successful software is "phred" developed by Green and his colleagues [30, 31]. The phred reads DNA sequencing trace files, calls bases, and assigns a quality value to each called base. The quality value Q is a log-transformed error probability P_e:

$$Q = -10 \log_{10} P_e \tag{11.1}$$

For example, if P_e is 0.001 or 0.1 %, Q becomes 30. The phred has often been used with two related softwares, phrap and consed, for contig formation in shotgun sequencing [32].

11.4.2 Shotgun Sequencing

DNA sequences determined experimentally from one DNA molecule are much shorter than one chromosome or even one plasmid sequence. It is thus necessary to combine these short sequences into much longer sequences, called "contigs." There are basically two tactics for that: ordered and random. When the initial sequences are short, such as less than 1,000 bp, it is easier to use random tactics. This is the basis of shotgun sequencing, originally proposed by Sanger's group in 1980 [33]. Figure 11.9 depicts its principle. If the final single sequence does not contain many repeat sequences nor unclonable sequences, this method is quite appropriate even for mammalian genome sizes. Although we need to generate short sequences whose total lengths should be several times, often mentioned as "coverage," more than the size of the final sequence, the sequencing cost has been dramatically reduced for these days, and very high coverage is easy to obtain using the second-generation sequencers.

The shotgun sequencing method was successfully applied to the determination of the whole genome of *Haemophilus influenzae* in 1995 [34]. This kind of

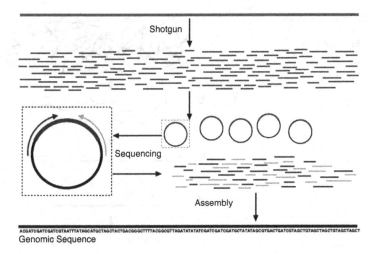

Shotgun

Sequencing

Assembly

ACGATCGATCGATCGTAATTTATAGCATGCTAGCTACTGACGGGCTTTTACGGCGTTAGATATATATCGATCGATCGATGCTATATAGCGTGACTGATCGTAGCTGTAGCTAGCTGTAGCTAGCT
Genomic Sequence

Fig. 11.9 Principle of shotgun sequencing (From https://wiki.cebitec.uni-bielefeld.de/brf-software/index.php/GenDBWiki/IntroductionToGenomics)

application is called "whole-genome shotgun (WGS) sequencing." Most of bacterial genomes are lacking repeat sequences, and the WGS sequencing is appropriate, and more than 1,000 complete bacterial genomes are now available (see Chap. 7). In contrast, there are many short repeat sequences in eukaryote genomes (see Chap. 8), especially in vertebrate genomes (Chap. 9). Therefore, most of the so-called completed eukaryote genomes are often composed of many sequences whose lengths are much shorter than chromosomes. Even the nucleotide sequence of each supercontig (contig of contigs) is questionable, and caution should be made when we try to use these long contiguous genome sequence data. For example, Ezawa et al. (2011; [35]) carefully chose genome sequences of six species (human, mouse, zebra fish, *Caenorhabditis elegans*, *Drosophila melanogaster*, and *D. pseudoobscura*) for the evolutionary study of gene duplications, because they examined physical characteristics of duplogs.

11.4.3 Minimum Tiling Array

A BAC clone contains 150–300-kb DNA of the target organism. Shotgun sequencing can be used to determine its complete nucleotide sequence, although PCR-direct sequencing [36] is often necessary to fill in the gap regions even if a high coverage was attained by shotgun sequencing. In any case, once many BAC sequences (or their restriction site maps) are determined, the next step is to constitute the minimum tiling array of these BACs, as shown in Fig. 11.10 (based on [37]). It is computationally the same as the contig formation in shotgun sequencing, for a BAC clone library can be considered as randomly created clones from one organism. For example, a human genome was covered by 32,855 BAC clones, whose average length

Fig. 11.10 Procedure of the human genome sequencing (Based on Fig. 2 of [37])

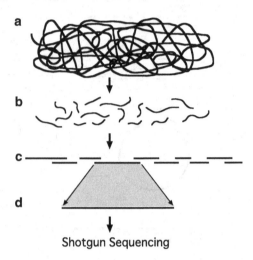

Shotgun Sequencing

Fig. 11.11 Two types of phases. (**a**) Coupling. (**b**) Repulsion

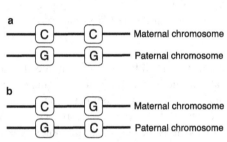

was 176 kb and the average coverage being 1.9X [38]. The minimum tiling array may also be produced when a genomic region is determined based on genome sequence information of the closely related species, as conducted for chimpanzee chromosome 22 genome sequencing [39]. This was achieved by determining BAC-end sequences (BESs) for chimpanzee BAC clone libraries [40]. A total of 77,461 chimpanzee BAC clone BESs were determined, and through BLAST homology searches of these BESs against the human genome sequences, 309 BAC clones were mapped to the human chromosome 21, orthologous to the chimpanzee chromosome 22 [40]. Although the chimpanzee chromosome 22 long arm sequences thus determined consists of only 1 % of the chimpanzee genome, these are much more reliable than the draft chimpanzee genome produced from WGS sequencing [41].

11.4.4   Haplotype Sequence Determination

Autosomes of diploid organisms, such as humans, have both maternally and paternally inherited genes in one locus. Because these nucleotide sequences are usually very similar, both of them are expected to be amplified using the same PCR primer pairs. If we try to determine these sequences by using the PCR-direct sequencing

[36], we may find more than one heterozygous nucleotide site (see Fig. 11.11a) or the sequence itself cannot be obtained because of length variation among the PCR amplicons caused by insertion–deletion polymorphism (see Fig. 11.11b). The latter case often happens for intronic sequences where indel polymorphism is much more frequent than exons.

Although PCR-direct sequencing results may be useful to detect the heterozygous or SNP sites, we cannot determine the maternal and paternal haplotypes, for the phase (see Fig. 11.11) is not known. If there are N heterozygous sites for one PCR amplicon, there are theoretically 2^N possible haplotypes. It is impossible to determine two haplotypes out of these possibilities with only the PCR-direct sequencing result. If some indel polymorphic regions also exist, the result similar to Fig. 11.8a may be obtained. Cloning and sequencing is used to overcome these problems. PCR amplicons are randomly cloned, and nucleotide sequences of many, say 10, clones are determined. If the maternal and paternal haplotype sequences are identical (homozygous), all clone sequences should be identical. However, if the number of sequenced clones is small, say 5, we may obtain the same sequences by chance, even if the two haplotype sequences are different (heterozygous). The simple theoretical expectation to encounter such a case is $(½)^N$ for N clones; however, either maternal or paternal chromosome may be more prone to be amplified (PCR amplification bias). In this case, the real probability of obtaining the biased chromosome may be much higher than ½. If the probability is 0.7, we need to determine at least nine clones to eliminate a chance result of producing identical sequences with less than 5 % error. Multiple clone sequencing also contributes to detect artificial mutations during the PCR process.

Another possible PCR error is artificial recombinants of two alleles. Kitano et al. (2009; [42]) found recombinations in gibbon ABO blood group sequences, and to show that those were not PCR artifacts but genuine natural recombinations, they did the following analysis. They used 23 gibbon individuals and, through multiple cloning techniques described above, determined all haplotype sequences of these individuals. Five individuals were homozygotes with three haplotypes, and these were considered to be authentic sequences. These three haplotypes were also found in nine heterozygotes, and the other haplotype of each heterozygous individual was determined to be authentic. In this way, 15 additional haplotype sequences were fixed. Among the ten remaining haplotypes, only one pair coexisted in a heterozygous individual, and they were in different sequence clusters. Kitano et al. [42] thus sequenced the same region of the ABO blood group gene by using haplotype-specific primers and confirmed that all the haplotype sequences were determined by multiple cloning.

11.4.5 Resequencing

Once the euchromatic nucleotide sequence of the human genome was mostly determined [43], sequencing of various human individuals was carried out [44] to discover DNA variations. This type of sequencing using the reference sequence is called "resequencing." Thanks to the large reduction of the sequencing cost, a large number of human individuals are now resequenced (e.g., [45]). Resequencing is not restricted to individuals of the same species. Because the definition of species is rather artificial, closely related species can be resequenced by using reference sequence of another species. For example, draft chimpanzee genome sequencing was aided by the human reference sequence [41].

It should be noted that the detection of minute duplications and inversions is difficult using the currently available WGS sequencing technology. Discovery of copy number variations needs special care when one is resequencing individuals using a WGS-using technique.

References

1. Shimada, M., Hayakawa, S., Hamle, T., Fujita, S., Hirata, S., Sugiyama, Y., & Saitou, N. (2004). Mitochondrial DNA genealogy of chimpanzees in Nimba mountains and Bossou, West Africa. *American Journal of Primatology, 64*, 261–275.
2. Levy, S., et al. (2007). The diploid genome sequence of an individual human. *PLoS Biology, 5*, e254.
3. Bentley, D. R., et al. (2008). Accurate whole human genome sequencing using reversible terminator chemistry. *Nature, 456*, 53–59.
4. Ahn, S. M., et al. (2009). The first Korean genome sequence and analysis: Full genome sequencing for a socio-ethnic group. *Genome Research, 19*, 1622–1629.
5. Dessauer H. C., Cole H. J., & Hafner, M. S. (1996). Chapter 3. Collection and storage of tissues. In D. M. Hillis, C. Moritz, & B. K. Marble (eds.), *Molecular systematics*, (2nd edn., pp. 29–47). Sunderland: Sinauer Associates.
6. Rohland, N., & Hofreiter, M. (2007). Comparison and optimization of ancient DNA extraction. *BioTechniques, 42*, 343–352.
7. Kanzawa, H., Saso, A., Suwa G., & Saitou, N. (2013). Ancient mitochondrial DNA sequences of Jomon teeth samples from Sanganji, Tohoku district, Japan. *Anthological Science* (in press)
8. Sambrook, J., Fritsch, E. F., & Maniates, T. (1989). *Molecular cloning: A laboratory manual.* Cold Spring Harbor: Cold Spring Harbor Laboratory Press.
9. http://www.protocol-online.org/prot/Molecular_Biology/DNA/DNA_Extraction___ Purification/index.html)
10. Chomczynski, R., & Sacchi, N. (1987). Single-step method of RNA extraction by acid guanidinium thiocyanate-phenol-chloroform extraction. *Analytical Biochemistry, 162*, 156–159.
11. Roe, B. A., Crabtree, J. S., & Khan, A. S. (eds.). (1995). *Protocols for recombinant DNA isolation, cloning, and sequencing. III. Methods for DNA isolation.* http://www.genome.ou.edu/ protocol_book/protocol_partIII.html#III.H
12. Liu, Y.-H., Takahashi, A., Kitano, T., Koide, T., Shiroishi, T., Moriwaki, K., & Saitou, N. (2008). Mosaic genealogy of the Mus musculus genome revealed by 21 nuclear genes from its three subspecies. *Genes and Genetic Systems, 83*, 77–88.
13. Brown, T. A. (2007). *Genomes 3.* New York: Garland Science Publishing.
14. http://bacpac.chori.org/
15. Kim, C.-G., Fujiyama, A., & Saitou, N. (2003). Construction of a gorilla fosmid library and its PCR screening system. *Genomics, 82*, 571–574.
16. Osoegawa, K., & de Jong, P. J. (2004). BAC library construction. *Methods in molecular biology, 255*, 1–46.
17. Maxam, A. M., & Gilbert, W. (1977). A new method for sequencing DNA. *Proceedings of the National Academy of Sciences of the United States of America, 74*, 560–564.
18. Sanger, F., Nicklen, S., & Coulson, A. R. (1977). DNA sequencing with chain-terminating inhibitors. *Proceedings of the National Academy of Sciences of the United States of America, 74*, 5463–5467.

19. http://www.nature.com/app_notes/nmeth/2005/050929/full/nmeth800.html
20. Shimada, M., Kim, C.-G., Takahashi, A., Saitou, N., Ikeo, K., Gojobori, T., & Spitsyn, V. A. (2002). Mitochondrial DNA control region sequences for a Buryats population in Russia (in Japanese). *DNA Takei, 10*, 151–155.
21. http://www.454.com/
22. http://www.iontorrent.com/technology-scalability-simplicity-speed/
23. http://www.illumina.com/technology/solexa_technology.ilmn
24. http://www.appliedbiosystems.com/absite/us/en/home/applications-technologies/solid-next-generation-sequencing/next-generation-systems/solid-sequencing-chemistry.html
25. Pushkarev, D., Neff, N. F., & Quake, S. R. (2009). Single-molecule sequencing of an individual human genome. *Nature Biotechnology, 27*, 847–850.
26. http://www.iontorrent.com/
27. Nagayama, K. (2011). Another 60 years in electron microscopy: Development of phase-plate electron microscopy and biological applications. *Journal of Electron Microscopy, 60*, S43–S62.
28. Maglia, G., Restrepo, M. R., Mikhailova, E., & Bayley, H. (2008). Enhanced translocation of single DNA molecules through α-hemolysin nanopores by manipulation of internal charge. *Proceedings of the National Academy of Sciences of the United States of America, 105*, 19720–19725.
29. http://www.nanoporetech.com/
30. Ewing, B., Hillier, L., Wendl, M. C., & Green, P. (1998). Basecalling of automated sequencer traces using *Phred*. I. Accuracy assessment. *Genome Research, 8*, 175–185.
31. Ewing, B., & Green, P. (1998). Basecalling of automated sequencer traces using *Phred*. II. Error probabilities. *Genome Research, 8*, 186–194.
32. http://www.phrap.org/phredphrapconsed.html
33. Sanger, F., Coulson, A. R., Barrell, B. G., Smith, A. J. H., & Roe, B. A. (1980). Cloning in single-stranded bacteriophage as an aid to rapid DNA sequencing. *Journal of Molecular Biology, 143*, 161–178.
34. Fleischmann, R. D., Adams, M. D., White, O., Clayton, R. A., Kirkness, E. F., Kerlavage, A. R., Bult, C. J., Tomb, J. F., Dougherty, B. A., Merrick, J. M., McKenney, K., Sutton, G., FitzHugh, W., Fields, C., Gocayne, J. D., Scott, J., Shirley, R., Liu, L.-I., Glodek, A., Kelley, J. M., Weidman, J. F., Phillips, C. A., Spriggs, T., Hedblom, E., Cotton, M. D., Utterback, T. R., Hanna, M. C., Nguyen, D. T., Saudek, D. M., Brandon, R. C., Fine, L. D., Fritchman, J. L., Fuhrmann, J. L., Geoghagen, N. S. M., Gnehm, C. L., McDonald, L. A., Small, K. V., Fraser, C. M., Smith, H. O., & Venter, J. C. (1995). Whole-genome random sequencing and assembly of *Haemophilus influenzae* Rd. *Science, 269*, 496–512.
35. Ezawa, K., Ikeo, K., Gojobori, T., & Saitou, N. (2011). Evolutionary patterns of recently emerged animal duplogs. *Genome Biology and Evolution, 3*, 1119–1135.
36. Innis, M. A., Myambo, K. B., Gelfand, D. H., & Brow, M. A. (1988). DNA sequencing with *Thermus aquaticus* DNA polymerase and direct sequencing of polymerase chain reaction-amplified DNA. *Proceedings of the National Academy of Sciences of the United States of America, 85*, 9436–9440.
37. International Human Genome Sequencing Consortium. (2001). Initial sequencing and analysis of the human genome. *Nature, 409*, 860–921.
38. Krzywinski, M., et al. (2004). A set of BAC clones spanning the human genome. *Nucleic Acids Research, 32*, 3651–3660.
39. The International Chimpanzee Chromosome 22 Consortium. (2004). DNA sequence and comparative analysis of chimpanzee chromosome 22. *Nature, 429*, 382–388.
40. Fujiyama, A., Watanabe, H., Toyoda, A., Taylor, T. D., Itoh, T., Tsai, S.-F., Park, H.-S., Yaspo, M.-L., Lehrach, H., Chen, Z., Fu, G., Saitou, N., Osoegawa, K., de Jong, P. J., Suto, Y., Hattori, M., & Sakaki, Y. (2002). Construction and analysis of a human-chimpanzee comparative clone map. *Science, 295*, 131–134.
41. The Chimpanzee Sequencing and Analysis Consortium. (2005). Initial sequence of the chimpanzee genome and comparison with the human genome. *Nature, 437*, 69–87.

42. Kitano, T., Noda, R., Takenaka, O., & Saitou, N. (2009). Relic of ancient recombinations in gibbon ABO blood group genes deciphered through phylogenetic network analysis. *Molecular Phylogenetics and Evolution, 51*, 465–471.
43. International Human Genome Sequencing Consortium. (2004). Finishing the euchromatic sequence of the human genome. *Nature, 431*, 931–945.
44. International HapMap Consortium. (2005). The haplotype map of the human genome. *Nature, 437*, 1299–1320.
45. The 1000 Genomes Project Consortium. (2010). A map of human genome variation from population-scale sequencing. *Nature, 467*, 1061–1073.

Omic Data Collection

12

Chapter Summary

Genome sequences are interacting with molecules inside and outside cells. With the same spirit as genomics to study all genetic informations, there are various categories on studying all transcripts, all proteins, all metabolites, and so on. We briefly discuss these omic worlds including ecome, coined in this book.

12.1 Overview of Omic Worlds

The genome sequences of many organisms are fundamental for the modern biology. We should be able to construct the edifice of biological knowledges on that foundation. In this sense, genome sequencing is only the beginning. We now have many exhaustive data collection, in spirit common with the genomes. There are transcriptomes, proteomes, metabolomes, phenomes, and ecomes. These words all end with "ome," the suffix of "genome." Although this was originally from "chromosome," it now means "every thing." Therefore, "transcriptome" means all the transcripts of one organism or one tissue type, and "proteome" means all the proteins and peptides existing in one organism or one tissue type. All biological activities of life are called metabolism, and "metabolome" means all kinds of metabolites in one cell or in one organism. In this sense, proteome and transcriptome are part of the metabolome, though the research history and the amount of studies of the former two categories are much more abundant than those of metabolome.

12.2 Transcriptome

A transcriptome, the target of transcriptomics, is all the RNA molecules transcribed from DNA. A classic example is Velculescu et al.'s (1997; [1]) study on baker's yeast (*Saccharomyces cerevisiae*). They introduced a new technique called SAGE

N. Saitou, *Introduction to Evolutionary Genomics*, Computational Biology 17,
DOI 10.1007/978-1-4471-5304-7_12, © Springer-Verlag London 2013

(Serial Analysis of Gene Expression). This method is based on two steps: First, a large-scale short (9–11 bp) sequence tags are generated. Each tag should contain sufficient information to identify a unique transcript. Second, many tags are concatenated into a single molecule for sequencing, to identify multiple tags simultaneously. The expression pattern of any population of transcripts is quantitatively evaluated by determining the abundance of individual tags and identifying the gene corresponding to each tag [1]. A total of 60,633 SAGE tags were generated from the three cell cycle phases, log, S, and G2/M, more or less in equal numbers, and 93 % of them uniquely matched to one particular sequence of the yeast genome sequence, corresponding to 4,665 genes [1].

Transcriptome studies are now flourishing for many model organisms, such as mouse (e.g., [2]), human (e.g., [3]), Drosophila (e.g., [4]), and *Arabidopsis thaliana* (e.g., [5]). Because nucleotide sequencing became cheap, direct sequencing of cDNAs is often used. When the reference genome sequence is already available, the complementary tiling microarray hybridization technique may also be used [4].

Diez-Roux et al. (2011; [6]) provided a genome-wide transcriptome atlas of the developing mouse at embryonic day 14.5. RNA in situ hybridization technique was used, and they identified 1,002 tissue-specific genes that are a source of tissue-specific markers for 37 different anatomical structures from the RNA expression profiles of over 18,000 coding genes and over 400 microRNAs. These results can be seen at the Eurexpress atlas (http://www.eurexpress.org).

Transcriptome analysis is also useful for RNA editing study. A widespread adenosine to inosine RNA editing of Alu-containing mRNAs is known. By computational analyses of human mRNA sequences in which Alu elements were embedded, 1.4 % [7] or ~2 % [8] were estimated to be subject to RNA editing. Edited bases are primarily associated with retained introns, with extended UTRs, or with transcripts that have no corresponding known gene. Alu-associated RNA editing may be a mechanism for marking nonstandard transcripts, not destined for translation [8]. This view is consistent with the possibility of neutral evolutionary process in the evolution of RNA editing discussed in Chap. 8.

The ENCODE project systematically mapped regions of transcription factor association, chromatin structure, and histone modification of the human genome in 2012 [9–12] as well as the transcriptome [3]. A series of techniques were used for genome-wide examination. For example, ChIP-seq is a chromatin immunoprecipitation followed by sequencing, RNA-PET is a simultaneous capture of full length RNAs with both a 59 methyl cap and a poly(A) tail, and RRBS (reduced representation bisulphite sequencing) is reducing the genome sequence to those enriched in CpGs followed by sequencing to determine the methylation status of individual cytosines [9]. Various classes of cis-regulatory elements including enhancers, promoters, insulators, silencers, and locus control regions were examined through ~580,000 DNase I hypersensitive sites found from 125 cell and tissue types [10]. If a regulatory factor, such as transcription factor, binds to genomic DNA, this binding protects the underlying sequence from cleavage by DNase I. This DNase I footprinting technique was used for detecting 8.4 million distinct short DNA sequences which are roughly twice the size of the exome [11]. Transcriptional regulatory network of the human genome was also examined for 119 transcription-related factors [12]. Various problems on the interpretation of ENCODE project results were discussed in Chap. 10.

12.3 Proteome

The world of proteins is proteome. Although transcriptome analysis provides useful information at various tissues at various developmental stages, we are not sure if mRNAs are really translated into proteins. Therefore, proteome analysis is definitely necessary.

Two-dimensional protein electrophoresis, developed by O'Farrell (1975; [13]), was used at the initial phase of proteomics. This technique was once widely used. An example of the two-dimensional protein electrophoresis is one conducted by Tsugita et al. (2000; [14]). Later, MALDI (matrix-associated laser desorption/ionization) coupled with mass spectrometer [15] was developed and is now widely used (e.g., [16]). Figure 12.1 is a schematic view of MALDI-TOF based on [15].

One of the initial and important proteome studies was again from baker's yeast (*Saccharomyces cerevisiae*). Gavin et al. (2002; [17]) used tandem affinity purification and mass spectrometry and purified 589 protein assemblies, which showed a massive interactive network. Evolutionary conservation of yeast and human complexes extended from single proteins to their molecular environment [17].

Taniguchi et al. (2010; [18]) carefully compared proteome and transcriptome of *Escherichia coli* at the single-cell level. They found that a single-cell's protein and mRNA copy numbers for any given gene were uncorrelated. This inconsistency was interpreted as a huge difference between mRNA and protein lifetime, within minutes and longer than cell cycle, respectively. Newman et al. (2006; [19]) also compared proteome and transcriptome of single-cell *Saccharomyces cerevisiae* by coupling the high-throughput flow cytometry and a library of GFP-tagged yeast strains for measuring protein levels. Although mRNA levels and protein levels were positively correlated, there were considerable protein-specific differences in noise that were correlated with a protein's mode of transcription and its function [19].

Because of rapid increase of proteome data, there are now several databases such as PRIDE ([20]; http://www.ebi.ac.uk/pride/) and PeptideAtlas ([21]; http://www.peptideatlas.org/). One important biological phenomenon related to proteomics is posttranslational modifications. SysPTM ([22]; www.sysbio.ac.cn/SysPTM) is a systematic resource for proteomic research on posttranslational modifications. There are seven common modification types, and they are phosphorylation, glycosylation, acetylation, methylation, ubiquitination, sumoylation, and S-nitrosylation, in the descending order of number of modified proteins [22]. The number of modified sites per protein can be more than one, and 41 %, 19 %, 11 %, and 8 % of proteins are modified for one, two, three, and four sites, respectively [22].

Another important aspect of the proteome study is inference of the protein 3D structures. Although the total number of its entry is much smaller than UniProt

Fig. 12.1 Scheme of mass spectroscopy for proteomics (From Aebersold and Mann 2003; [15])

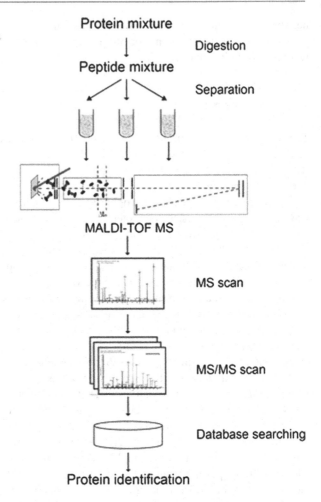

(http://www.uniprot.org/), PDB (http://www.rcsb.org/pdb/home/home.do), the global database of protein 3D structural data, now contains 96,238 structure data as of December 2013. Structural data are useful for inferring a remote homology, which is not easy to detect by amino acid sequence homology.

12.4   Other Omic Data

12.4.1   Metabolome

DNAs, RNAs, and proteins are the three major players in the central dogma of molecular biology (Chap. 1). Genome, transcriptome, and proteome correspond to their own universes, respectively. One cell is, however, composed of much more heterogeneous molecules. A cell is surrounded by cell membrane whose main building unit is lipid. Various types of sugar molecules are important energy source. Any molecule appearing in the metabolism may be called a metabolite, and the metabolome in the wide sense is the world of matters in one organism. One of the earliest papers whose title included the word "metabolite" may be that from Tweeddale et al. (1998; [23]). They examined the metabolome of *Escherichia coli* using the two-dimensional thin-layer chromatography of all 14C-glucose labeled compounds extracted from bacteria. Fu et al. (2011; [24]) used gas chromatography–mass spectrometry to measure concentrations of more than 100 metabolites in the prefrontal and cerebellar cortex in humans, chimpanzees, and rhesus macaques. Although the total number of metabolites measured may not be considered as "metabolome," they found considerable age and species differences on many metabolite concentrations. Metabolome examination may be promising for evolutionary aspects of diverse organisms, as a starting point toward understanding the evolutionary change occurred on some metabolic pathway. At this moment, however, it is still not clear if metabolome study is really complementary to other omic data such as transcriptomes and proteomes.

12.4.2   Phenome

One of the main objectives of genetics is to connect genotypes with phenotypes. If we apply the same view to evolutionary genomics, we should try to connect any changes within the genome with any phenotypic change. As genome sequencing is becoming cheaper and cheaper, now the focus is phenotype data collection. It is then natural to consider the all sorts of phenotypes as "phenome" (e.g., [19]). While nucleotide sequences correspond to objective DNA or RNA molecules, a phenotype is the very product of human recognition. Therefore, as the level of complexity increases, it becomes more difficult to define phenotypes. A transcript and a protein can be considered as phenotypes, but they are directly connected to a certain nucleotide sequence. There are already some phenome projects, especially for human. For example, UK Biobank (http://www.ukbiobank.ac.uk/) is collecting many human phenotype data as well as DNA samples. Another interesting activity is the Matrix of Comparative Anthropogeny (MOCA; http://carta.anthropogeny.org/moca/) organized by the Center for Academic Research and Training in Anthropogeny (CARTA; http://carta.anthropogeny.org/). MOCA is a collection of comparative information regarding humans and our closest evolutionary cousins (chimpanzees, bonobos, gorillas, and orangutans, i.e., "great apes"), with an emphasis on uniquely human features. This sort of human intensive curations is definitely important for phenome study.

Another aspect of phenome study is automatic collection of vast morphological characteristics through imaging using X-ray or NMR CT scanning. Saitou et al. (2011; [25]) used advanced CT images for revealing nonmetric cranial variations in two living human individuals. These bone morphological traits have been studied using bones of dead persons, but thanks to high-resolution CT images, now it is

Table 12.1 List of 23 nonmetric cranial variations observed for the two Japanese males using the high-resolution x-ray CT scanning (Ref. [25])

ID no.	Name of nonmetric characteristic	Subject 1 Right	Left	Subject 2 Right	Left
1	Metopism	−		−	
2	Supraorbital nerve groove	−	−	−	−
3	Supraorbital foramen	+	−	+	−
4	Ossicle at lambda	−		−	
5	Interparietals (inca bone)	+		−	
6	Biasterionic suture vestige	−	−	−	−
7	Asterionic bone	−	−	−	−
8	Occipito-mastoid bone	?	?	−	−
9	Parietal notch bone	−	−	−	−
10	Condylar canal patent	?	?	−	+
11	Precondylar tubercle	−	−	−	−
12	Paracondylar process	−	−	−	−
13	Hypoglossal canal bridging	?	−	−	−
14	Tympanic dehiscence	?	?	?	?
15	Ovale-spinosum confluence	−	−	−	−
16	Pterygo-spinous foramen	−	−	−	−
17	Foramen of Vesalius	−	+	−	−
18	Medial palatine canal	?	?	?	?
19	Transverse zygomatic suture vestige	?	?	−	−
20	Clinoid bridging	?	?	−	−
21	Mylohyoid bridging	−	−	−	−
22	Jugular foramen bridging	−	−	−	−
23	Sagittal groove left	−		−	

+, present; −, absent; ?, unknown

possible to obtain the same nonmetric observations from living humans. Table 12.1 [25] is a list of 23 nonmetric cranial variations observed for the two Japanese males using the high-resolution X-ray CT scanning.

12.4.3   Ecome

Finally, we have ecome. Ecome is all information in one ecosystem, from genome sequences, transcriptomes, proteomes, metabolomes, and phenomes. This wholistic view of life is clearly necessary for the future biology. It should be noted, however, that this approach is definitely a reductionist way, not the classic wholistic one, which produced nothing. *The truth is in the details.* Although we are still far from the ideal ecome data, metagenome studies (e.g., Venter et al. 2004; [26]) are on this trend. If we consider the earth as one ecosystem, one possible future project can be dubbed as EEE, acronym of "Encyclopedia of Earth Ecome."

References

1. Velculescu, V. E., et al. (1997). Characterization of the yeast transcriptome. *Cell, 88,* 243–251.
2. The FANTOM Consortium. (2005). The transcriptional landscape of the mammalian genome. *Science, 309,* 1559–1563.
3. Djebali, S., et al. (2012). Landscape of transcription in human cells. *Nature, 489,* 101–108.
4. Graveley, B. R., et al. (2011). The developmental transcriptome of *Drosophila melanogaster. Nature, 471,* 473–479.
5. Endo, A., et al. (2012). Tissue-specific transcriptome analysis reveals cell wall metabolism, flavonol biosynthesis and defense responses are activated in the endosperm of germinating *Arabidopsis thaliana* seeds. *Plant and Cell Physiology, 53,* 16–27.
6. Diez-Roux, G., et al. (2011). A high-resolution anatomical atlas of the transcriptome in the mouse embryo. *PLoS Biology, 9,* e1000582.
7. Athanasiadis, A., Rich, A., & Maas, S. (2004). Widespread A-to-I RNA editing of Alu-containing mRNAs in the human transcriptome. *PLoS Biology, 2,* e391.
8. Kim, D. D. Y., et al. (2004). Widespread RNA editing of embedded Alu elements in the human transcriptome. *Genome Research, 14,* 1719–1725.
9. The ENCODE Project Consortium. (2012). An integrated encyclopedia of DNA elements in the human genome. *Nature, 489,* 57–74.
10. Thurman, R. E., et al. (2012). The accessible chromatin landscape of the human genome. *Nature, 489,* 75–82.
11. Neph, S., et al. (2012). An expansive human regulatory lexicon encoded in transcription factor footprints. *Nature, 489,* 83–90.
12. Gestein, M. B. (2012). Architecture of the human regulatory network derived from ENCODE data. *Nature, 489,* 91–100.
13. O'Farrell, P. H. (1975). High resolution two-dimensional electrophoresis of proteins. *Journal of Biological Chemistry, 250,* 4007–4021.
14. Tsugita, A., Kawakami, T., Uchida, T., Sakai, T., Kamo, M., Matsui, T., Watanabe, Y., Morimasa, T., Hosokawa, K., & Toda, T. (2000). Proteome analysis of mouse brain: Two-dimensional electrophoresis profiles of tissue proteins during the course of aging. *Electrophoresis, 21,* 1853–1871.
15. Aebersold, R., & Mann, M. (2003). Mass spectrometry-based proteomics. *Nature, 422,* 193–207.
16. Xie, L., et al. (2005). Genomic and proteomic analysis of mammary tumors arising in trans-genic mice. *Journal of Proteome Research, 4,* 2088–2098.
17. Gavin, A.-C., et al. (2002). Functional organization of the yeast proteome by systematic analy-sis of protein complexes. *Nature, 415,* 141–147.
18. Taniguchi, Y., et al. (2010). Quantifying *E. coli* proteome and transcriptome with single-molecule sensitivity in single cells. *Science, 329,* 533–538.
19. Freimer, N., & Sabatti, C. (2003). The human phenome project. *Nature Genetics, 34,* 15–21.
20. Martens, L., et al. (2005). PRIDE: The proteomics identifications database. *Proteomics, 5,* 3537–3545.
21. Deutsch, E. W., Lam, H., & Aebersold, R. (2008). PeptideAtlas: A resource for target selection for emerging targeted proteomics workflows. *EMBO Report, 9,* 429–434.
22. Li, H., Xing, X., Ding, G., Li, Q., Wang, C., Xie, L., Zeng, R., & Li, Y. (2009). SysPTM: A systematic resource for proteomic research on post-translational modifications. *Molecular and Cellular Proteomics, 8,* 1839–1849.
23. Tweeddale, H., Notley-McRobb, L., & Ferenci, T. (1998). Effect of slow growth on metabo-lism of *Escherichia coli,* as revealed by global metabolite pool ("metabolome") analysis. *Journal of Bacteriology, 180,* 5109–5116.
24. Fu, X., et al. (2011). Rapid metabolic evolution in human prefrontal cortex. *Proceedings of the National Academy of Sciences of the United States of America, 108,* 6181–6186.

25. Saitou, N., Kimura, R., Fukase, H., Yogi, A., Murayama, S., & Ishida, H. (2011). Advanced CT images reveal nonmetric cranial variations in living humans. *Anthropological Science, 119*, 231–237.

26. Venter, C. G., et al. (2004). Environmental genome shotgun sequencing of the Sargasso Sea. *Science, 304*, 66–74.

Databases

13

Chapter Summary

Use of databases is essential for evolutionary genomics. Databases for genomes, nucleotide sequences, proteins, literatures, and other categories are discussed.

13.1 Overview of Databases

Genome sequences and their related information are enormous; thus, we definitely need to use databases. Thanks to the rapid development of computers and computational softwares, it is now natural to construct our own databases, as well as using publicly maintained large-scale databases. We would like to summarize these two types of databases.

A database is traditionally considered to have a standardized data structure. An example is a spreadsheet with certain columns and rows. As the exhaustive text search became available, however, any set of digital text can now be considered as a database. It is particularly true for computer files of private use. Any format of text-based files can be the target of character searches, as long as the total volume of the database is not so large. This formatless principle also applies to network searches such as Google [1]. What we need to care in this case is to give appropriate annotation to each information unit, so as to find out that unit with an appropriate keyword.

When a huge-scale information such as the human genome data with annotation is considered, we need a clearly defined data structure. A well-known classic format is that of International Nucleotide Sequence Database entries, as shown in Fig. 13.1. This example entry is based on our study on Rh blood group gene evolution [2]. The meanings of each entry category such as LOCUS, ACCESSION, and VERSION will be explained in a later section of this chapter.

```
LOCUS           AB015189              1257 bp    mRNA    linear   ROD 30-OCT-2008
DEFINITION      Mus musculus mRNA for Rh blood group protein, complete cds,
                strain: C57BL-10SnJ.
ACCESSION       AB015189
VERSION         AB015189.1
KEYWORDS        .
SOURCE          Mus musculus (house mouse)
  ORGANISM      Mus musculus
                Eukaryota; Metazoa; Chordata; Craniata; Vertebrata; Euteleostomi;
                Mammalia; Eutheria; Euarchontoglires; Glires; Rodentia;
                Sciurognathi; Muroidea; Muridae; Murinae; Mus; Mus.
REFERENCE       1  (bases 1 to 1257)
  AUTHORS       Kitano,T. and Saitou,N.
  TITLE         Direct Submission
  JOURNAL       Submitted (04-JUN-1998) to the DDBJ/EMBL/GenBank databases.
                Contact:Takashi Kitano
                National Institute of Genetics, Laboratory of Evolutionary
                Genetics; 1111 Yata, Mishima, Shizuoka 411-8540, Japan
                URL    :http://sayer.lab.nig.ac.jp/
REFERENCE       2
  AUTHORS       Kitano,T., Sumiyama,K., Shiroishi,T. and Saitou,N.
  TITLE         Conserved evolution of the Rh50 gene compared to its homologous Rh
                blood group gene
  JOURNAL       Biochem. Biophys. Res. Commun. 249, 78-85 (1998)
COMMENT
FEATURES             Location/Qualifiers
     source          1..1257
                     /db_xref="taxon:10090"
                     /mol_type="mRNA"
                     /organism="Mus musculus"
                     /strain="C57BL-10SnJ"
                     /sub_species="domesticus"
                     /tissue_type="bone marrow"
     CDS             1..1257
                     /codon_start=1
                     /note="Rh blood group gene"
                     /product="Rh blood group protein"
                     /protein_id="BAA32438.1"
                     /transl_table=1
                     /translation="MGSKYPRSLRCCLPLWALVLQTAFILLSCFFIPHDTAQVDHKFM
                     ESYQVLRNLTLMAALGFGFLSSSFRRHSWSSVAFNLFMLALGVQGTILLDHFLGQVLQ
                     WNKINNLSSIQIATMSTLPVLISAGAVLGKVNLVQLTMMVLMEAMAFGAIRFADEKVF
                     KMTEHIIMMHGHVFGAYFGLTVAWWLSRSLPRRVGENAQTEKVQMATSSSLFAMLGTL
                     FLWIFWPAINSALLEGTKKRNAVFNTYYALAVSAVTATSMSALSHPQGKINMVHIHNA
                     VLAGGVAVGAPGCLISSPWISMVLGLIAGLISIWGAKCPRACLNHMLQNSSGIHYTFG
                     LPGLLGALTYYCLQIVTEPKSSDLWIITQTVTHIGALSFAVAMGMVTGLLTGCLLSVR
                     VWRAPHAAKYFDDQTFWEFPHLAVGF"
BASE COUNT         249 a         351 c         343 g          314 t
ORIGIN
        1 atgggctcta agtacccacg gtccctccgc tgctgcctgc ccctatgggc cttggtgcta
       61 cagacagctt ttattctcct ctcctgtttt ttcatccccc acgacacagc ccaggtggat
      121 cacaagttca tggagagcta tcaagtcctc cggaatttga ccctcatggc agccttgggc
      181 ttcggcttcc tgtcctcgtc ctttcggaga cacagctgga gcagtgtggc cttcaacctc
      241 ttcatgttgg ccctcggggt gcagggggaca atcttgctgg accatttcct gggccaggtc
      301 ctccaatgga acaagatcaa caatctgtcc agcatccaga tagctaccat gagcacctta
      361 cctgtgctga tctcagcggg cgctgtcctg gggaaggtca acctggtgca gctgaccatg
      421 atggtgctga tggaggcaat ggcctttggt gccatcagat ttgccgacga gaaggtcttc
      481 aaaatgacag aacacatcat catgatgcac gggcacgtgt ttggggccta ttttgggcta
      541 actgtggctt ggtggctttc cagatctctg cccaggagag tgggtgagaa cgcccagaca
      601 gagaaggttc aaatggctac gagctccagt ctgtttgcca tgctgggcac cctcttcttg
      661 tggatattct ggccagctat caactctgct ctcctggaag ggacaaagaa aaggaatgct
      721 gtgttcaaca cctactacgc cctggcagtg agcgcagtga cagccacctc catgtcagcc
      781 ctgagtcacc ctcaagggaa gatcaacatg gttcacatcc acaacgcagt gctggcaggg
      841 ggcgtggccg tgggcgcccc gggttgcctg atttcttctc cttggatttc catggtgctg
      901 ggcctcatag ctgggttgat ctccatctgg ggagccaagt gtccacgggc gtgtttgaac
      961 cacatgctgc agaactccag tgggatccac tacaccttcg gcttgccggg tctgctggga
     1021 gcacttacct actactgcct tcagatagtg acagagccca agtcctcgga tctctggatc
     1081 atcacccaga cggtcactca cattggggct ctcagcttcg ctgtggcgat gggtatggtg
     1141 actggactcc tcacaggttg tctcctaagt gtcagagtgt ggagggctcc ccatgcggcc
     1201 aagtattttg atgatcagac tttctgggag ttcccacact tggcggttgg attttaa
//
```

Fig. 13.1 Entry of accession number AB015189 of the International Nucleotide Sequence Database (DDBJ format)

A large-scale database is often constructed by using a database management system, which is based on relational, object-oriented, or some other database model. Computer languages such as SQL and XML are used to construct databases. Interested readers should consult references on these items.

13.2 Genome Sequence Databases

The most important databases for evolutionary genomics are those on genome sequences. There are three comprehensive genome databases in the world: NCBI Genome Biology (http://www.ncbi.nlm.nih.gov/Genomes/), UCSC (University of California, Santa Cruz) Genome Bioinformatics (http://genome.ucsc.edu), and Ensembl (http://www.ensembl.org). The genome database at NCBI is the most systematic among these three databases and contains data for a wide variety of organisms from humans to viruses. The Ensembl database was traditionally focusing on animal genomes; however, it recently started to include genome data of bacteria, protists, fungi, and plants. UCSC database is mostly presenting animal genome informations and is strong on genome comparisons. Figures 13.2, 13.3, and 13.4 are example views of the ABO blood group gene region (chromosome 9, around coordinates 136,130,000–136,150,000) of the human genome at NCBI, UCSC, and Ensembl, respectively. The information shown at UCSC Genome Browser for this genomic region is most abundant, while Ensembl also shows its flanking region. It should be noted that these view formats are often updated and these shown here are as of April 2012.

JGI (Joint Genome Institute: http://www.jgi.doe.gov/) has been sequencing genomes of diverse organisms, and they provide genome sequences for hundreds of fungi species as MycoCosm, plant species as Phytozome, as well as prokaryote species and many metagenomes. There are 11,063 species (including strains) genome projects which finished sequencing, and more than 26,000 genome sequencing projects are underway, according to GOLD (Genomes OnLine Database: http://www.genomesonline.org) as of December 2013. We are now moving to the third-generation sequencing era, and the number of species whose genomes are determined is expected to skyrocket. There are many databases devoted to one model organism, and many of them provide their genomic sequences. Representative databases are listed in Table 13.1.

Many biological journals request or at least recommend authors to submit nucleotide sequences they reported in their papers to one of three INSDC (the International Nucleotide Sequence Database Collaboration) databases: GenBank, ENA, or DDBJ (see the next section). Genome sequences are no exceptions. However, there are now many genome sequencing projects, and unpublished but almost finished genomic sequences are often not deposited to these INSDC databases, although those genomic sequences are available from sequencing center websites.

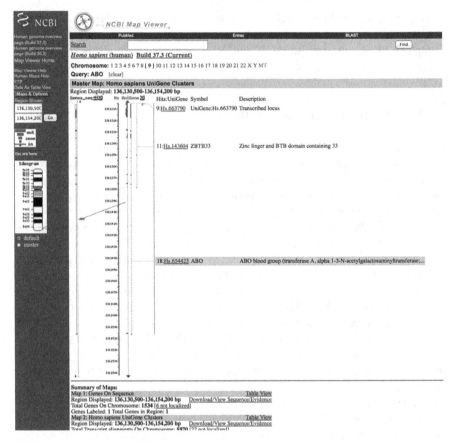

Fig. 13.2 View of the ABO blood group gene region of the human genome at NCBI genome database

13.3 INSDC Databases

There are an immense number of organisms on the earth, and most of their genomes have not been sequenced. In this sense, even their short nucleotide sequences are important for evolutionary studies. GenBank of NCBI (http://www.ncbi.nlm.nih.gov/genbank/), ENA (European Nucleotide Archive) of EMBL (http://www.ebi.ac.uk/ena/), and DDBJ (DNA Data Bank of Japan) of National Institute of Genetics (http://www.ddbj.nig.ac.jp/) formed INSDC (the International Nucleotide Sequence Database Collaboration: http://www.insdc.org/) in the late 1980s and daily exchange nucleotide sequence data submitted to them.

One example of the DDBJ format data entry is shown in Fig. 13.1. Categories of one entry are shown in capital letters at the left side, starting from "LOCUS." Its name is from genetic "locus," and initially the gene-related identification was used for the LOCUS name. However, as the number of entries rapidly increased, the importance of LOCUS names decreased. Because accession number is the

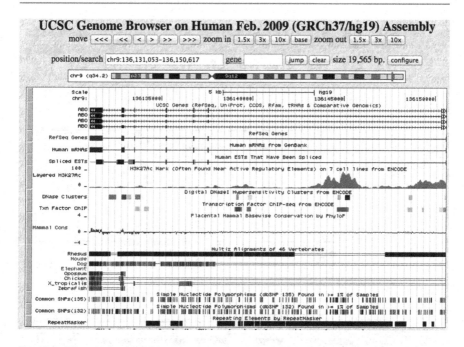

Fig. 13.3 View of the ABO blood group gene region of the human genome at UCSC genome database

unique ID for each entry, LOCUS name is now simply identical with the accession number. The line after "LOCUS" is called LOCUS line, and it includes LOCUS name, the number of nucleotides, type and form of the nucleotide molecule sequenced, division, and the last updated date. In this example, a 1,257-bp DNA sequence complementary to mRNA sequence of a rodent species was submitted. The next line is DEFINITION line, and a succinct explanation of the sequence is given. The third line gives accession number, and its format is two alphabets followed by five digits under the current system. NCBI is using their own accession number system for nucleotide sequences starting with "NM_" followed by six digits, but it is different from the INSDC accession number system. Line 4 is for VERSION, and when the nucleotide sequence was submitted first, version number 1 is given. When any change was made for the original nucleotide sequence later, version number is updated.

The remaining informations may be straightforward, until we encounter FEATURES section. INSDC provides a common format for the three databases (DDBJ, ENA, and GenBank) as "Feature Table Definition" (http://www.insdc.org/documents/feature-table). In the example shown in Fig. 13.1, "source" and "CDS" feature informations are given. Words after slash mark are called "feature key," and they provide carefully annotated information to users. For example, "/transl_table=1" means the standard codon table was used to translate the coding nucleotide sequence to the amino acid sequence given at "/translation" section. Although the DDBJ entry format is very similar to that of GenBank

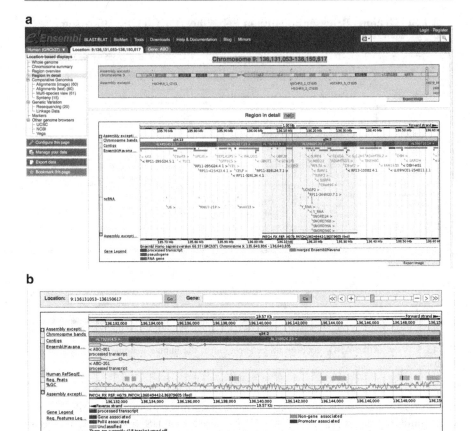

Fig. 13.4 View of the ABO blood group gene region of the human genome at Ensembl genome database. (**a**) Chromosomal view. (**b**) Gene view

Table 13.1 List of genome-related databases for individual organisms

Species	Database name	Home page URL
Mouse	MGI	http://www.informatics.jax.org/
Rat	RGD	http://rgd.mcw.edu/
Zebrafish	ZFIN	http://zfin.org/zf_info/catch/catch.html
Stickleback	Sticklebrowser	http://sticklebrowser.stanford.edu/
Drosophila	FlyBase	http://flybase.org/
Baker's yeast	SGD	http://www.yeastgenome.org/
Fission yeast	PomBase	http://www.pombase.org/
Arabidopsis	ATIDB	http://atidb.org/
Rice	RAPDB	http://rapdb.dna.affrc.go.jp/

format, "BASE COUNT" section, which exists in this example, as well as old entries, no longer exists in GenBank format. The last section, starting with "ORIGIN" is the nucleotide sequence, ending with "//."

13.4 Protein-Related Databases

Databases related to proteins have a long history, and there are many varieties. The most representative one is UniProt (http://www.uniprot.org), which provides amino acid sequence data. Figure 13.5 is an example entry C4B4C0 of UniProt for the partial amino acid sequence of the ABO blood group gene of *Hylobates lar* [3]. Because of synonymous differences among nucleotide sequences, this UniProt entry corresponds to four INSDC entries with accession numbers AB196694-AB196697. As this example shows, most of UniProt entries are translated from the coding regions of nucleotide sequences. Therefore, it is not clear if a "protein" in UniProt database is really produced in the real organism. This is why proteome study (see Chap. 12) is necessary to directly catch really expressed proteins.

PDB (http://www.rcsb.org/pdb/) is specialized in protein 3D structural data, though its name, abbreviation of "Protein Data Bank," sounds a more comprehensive database on proteins. Figure 13.6 is an example PDB entry 1Z98 for spinach aquaporin [4].

Other important protein-related databases are those on domains. Representative ones are Pfam ([5]; http://pfam.sanger.ac.uk/) and InterPro ([6]; http://www.ebi. ac.uk/interpro/). Figure 13.7 is an example of the Pfam search result using human Brain-1 amino acid sequence translated from its mRNA sequence (INSDC accession number AB001835) determined by Sumiyama et al. [7]. The two domains, Hox and POU, were correctly extracted with this search.

13.5 Literature Databases

When the author was a graduate student more than 30 years ago, people had to copy or xerox many papers. Now we download pdf files from journal home pages or via literature databases such as PubMed (http://www.ncbi.nlm.nih.gov/ sites/ entrez?db=pubmed). NCBI has been operating PubMed, a freely available database, as the continuation of the MEDLINE database at National Library of Medicine, USA. As its name suggests, journals collected by PubMed are more abundant in medical and animal oriented studies.

Another freely available journal database is Google Scholar (http://scholar. google.com/). While PubMed is collecting abstracts from the selected set of journals, Google Scholar covers not only journal articles but books, theses, and other categories. If we use "advanced" option of PubMed to search author "Zuckerkandl E," who is the founding editor of Journal of Molecular Evolution, 76 journal articles are found. When we searched Google Scholar using "Advanced Scholar Search" for author "Zuckerkandl E," however, ~280 references were found, both as of March 2013. This large difference is of course books and book chapters are

a

C4B4C0 (C4B4C0_HYLLA) ★ Unreviewed, UniProtKB/TrEMBL
Last modified January 11, 2011. Version 5. 🔊 History...

▞ Clusters with 100%, 90%, 50% identity | 🔲 Third-party data   [text] [xml] [rdf/xml] [gff] [fasta]

🔹 Names · Attributes · Ontologies · Sequence annotation · Sequences · References · Cross-refs · Entry info Customize order

Names and origin

Protein names	*Submitted name:* ABO glycosyltransferase (EMBL BAH58811.1)
Gene names	Name:**ABO** (EMBL BAH58811.1)
Organism	**Hylobates lar (Common gibbon)** (EMBL BAH58811.1)
Taxonomic identifier	9580 [NCBI]
Taxonomic lineage	Eukaryota · Metazoa · Chordata · Craniata · Vertebrata · Euteleostomi · Mammalia · Eutheria · Euarchontoglires · Primates · Haplorrhini · Catarrhini · Hylobatidae · Hylobates

Protein attributes

Sequence length	205 AA.
Sequence status	Fragment.
Protein existence	Predicted

Ontologies

Keywords

Molecular function	Transferase (EMBL BAH58811.1)

Gene Ontology (GO)

Biological process	carbohydrate metabolic process Inferred from electronic annotation. Source: InterPro
Cellular component	membrane Inferred from electronic annotation. Source: InterPro
Molecular function	transferase activity, transferring hexosyl groups Inferred from electronic annotation. Source: InterPro

b

Sequence annotation (Features)

Feature key	Position(s)	Length	Description	Graphical view	Feature identifier
Experimental info					
☐ Non-terminal residue	1	1	(EMBL BAH58811.1)	────────────	
☐ Non-terminal residue	205	1	(EMBL BAH58811.1)	────────────	

Sequences

Sequence		Length	Mass (Da)	Tools
☐ C4B4C0 [UniParc]. Last modified July 7, 2009. Version 1. Checksum: 9255C18778776E3C	FASTA	205	23,927	[Blast ▾] [go]

```
          10         20         30         40         50         60
  PCREDVLVVT PMLAPIVWEG TFNIDILNEQ FRLQNTTIGL TVFAIKKYVA FLKLFLETAE

          70         80         90        100        110        120
  KHFMVGHRVH YYVFTDQPAA VPRVTLGTGR QLMVLEVRAH KRWQDVSMRR MEMISDFCER

         130        140        150        160        170        180
  RFLSEVDYLV CVDVDMEFRD HVGVEILTPL FGTLHPGFYG SSREAFTYER RPQSQAYIPK

         190        200
  DEGDFYYLGG FFGGSVQEVQ RLTRA
```

= Hide

References

[1] "Relic of ancient recombinations in gibbon ABO blood group genes deciphered through phylogenetic network analysis."
Kitano T., Noda R., Takenaka O., Saitou N.
Mol. Phylogenet. Evol. 51:465–471(2009)
Cited for: NUCLEOTIDE SEQUENCE.

Cross-references

Fig. 13.5 An example UniProt entry C4B4C0 for common gibbon ABO blood group protein. (**a**) First part. (**b**) Second part

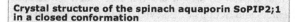

Crystal structure of the spinach aquaporin SoPIP2;1 in a closed conformation **1Z98**

Display Files ▾
Download Files ▾
Share this Page ▾

DOI:10.2210/pdb1z98/pdb

Primary Citation

Structural mechanism of plant aquaporin gating.

Tornroth-Horsefield, S.🔎, Wang, Y.🔎, Hedfalk, K.🔎, Johanson, U.🔎, Karlsson, M.🔎, Tajkhorshid, E.🔎, Neutze, R.🔎, Kjellbom, P.🔎,

Journal: (2006) Nature **439**: 688-694

PubMed: 16340961 ☞
DOI: 10.1038/nature04316 ☞
Search Related Articles in PubMed 🔎

PubMed Abstract:

Plants counteract fluctuations in water supply by regulating all aquaporins in the cell plasma membrane. Channel closure results either from the dephosphorylation of two conserved serine residues under conditions of drought stress, or from the protonation of a conserved histidine... [Read More & Search PubMed Abstracts]

◀ Biological Assembly 1 ❓ ▶

More Images...

🔎 **View in Jmol** | Simple Viewer
Other Viewers ▾ | Protein Workshop

Biological assembly 1 assigned by authors and generated by PISA,PQS (software)

‡ Molecular Description | Hide
Classification: Transport Protein 🔎, Membrane Protein 🔎
Structure Weight: 60084.82

Molecule:	aquaporin
Polymer:	1 **Type:** protein **Length:** 281
Chains:	A, M
Organism	Spinacia oleracea🔎
UniProtKB:	Q41372🔎

‡ MyPDB Personal Annotations Hide

To save personal annotations, please login to your MyPDB account.

‡ Source | Hide
Polymer: 1

‡ Deposition Summary Hide
Authors: Tornroth-Horsefield,

Scientific Name:	Spinacia oleracea🔎	🔎 **Taxonomy** ☞	**Common Name:**	Spinach	**Expression System:**	Pichia pastoris🔎

Fig. 13.6 An example PDB entry 1Z98 for spinach aquaporin

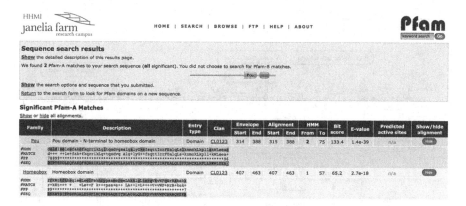

HHMI
janelia farm
research campus

HOME | SEARCH | BROWSE | FTP | HELP | ABOUT

Pfam
keyword search Go

Sequence search results
Show the detailed description of this results page.
We found **2** Pfam-A matches to your search sequence (**all** significant). You did not choose to search for Pfam-B matches.

Pfam

Show the search options and sequence that you submitted.
Return to the search form to look for Pfam domains on a new sequence.

Significant Pfam-A Matches
Show or hide all alignments.

Family	Description	Entry type	Clan	Envelope Start	Envelope End	Alignment Start	Alignment End	HMM From	HMM To	Bit score	E-value	Predicted active sites	Show/hide alignment
Pou	Pou domain - N-terminal to homeobox domain	Domain	CL0123	314	388	315	388	2	75	133.4	1.4e-39	n/a	Hide
Homeobox	Homeobox domain	Domain	CL0123	407	463	407	463	1	57	65.2	2.7e-18	n/a	Hide

Fig. 13.7 An example of PFAM search result for Brain-1 protein

included in Google Scholar, but not in PubMed. Since Google Scholar lists references approximately by the number of citations, celebrated book chapter by Zuckerkandl and Pauling (1965; [8]), which proposed "molecular clock" (see Chap. 4), was listed at the top with 1,151 citations.

There are some other journal databases which are not free, such as Science Citation Index of Thomson Reuters and Scopus of Elsevier. Only institutions which can pay a large amount have accesses to these paid databases, and we will not discuss about these private systems.

13.6   Other Databases

There are many other important databases for evolutionary genomics. KEGG (Kyoto Encyclopedia of Genes and Genomes) pathway database (http://www. genome.jp/kegg/pathway.html) is a collection of manually drawn metabolic pathway maps. Figure 13.8 is its example for purine metabolism. Small circles are metabolites and squares are enzymes with their E.C. numbers. If we crick one of these squares, its information is shown. For example, 2.1.2.2 is phosphoribosylglycinamide formyltransferase, and its orthologous genes and other informations are listed. There are other pathway databases such as REACTOME (http://www.reactome.org/ReactomeGWT/entrypoint.html) and METACYC (http://metacyc.org/).

OMIM (Online Mendelian Inheritance in Man) is a manually curated database for genes with known phenotypes in the human genome. Victor McKusick started to publish "Mendelian Inheritance in Man" from 1966, as a catalog of Mendelian

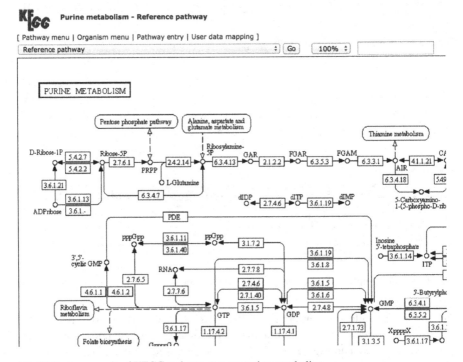

Fig. 13.8 An example of KEGG pathway map on purine metabolism

traits and disorders, until the 12th edition in 1998. OMIM is its online version, now maintained at NCBI. This activity was extended to more than 100 animal species as OMIA (Online Mendelian Inheritance in Animals; http://www.ncbi.nlm.nih.gov/omia). For example, a search of OMIA with keyword "ABO" retrieved nine entries as of December 2013, and one of them, OMIA ID 3221 on agile gibbon, is based on Kitano et al.'s (2009; [3]) paper.

Gene ontology (GO) is an explanation of gene functions with controlled vocabularies and is available from Gene Ontology Consortium web page (http://www.geneontology.org/). For example, Takahashi and Saitou [9] used GO for the analysis of HCNS-flanking protein coding genes.

DNA polymorphism within species is an important evolutionary phenomenon, and it is extensively studied in human populations. MITOMAP (http://www.mitomap.org) is a manually curated database of published and unpublished data on human mitochondrial DNA variation. NCBI hosts dbSNP (http://www.ncbi.nlm.nih.gov/snp), a comprehensive database for short genetic variations. Because it started as a database for SNPs alone, and it became popular, the name of database has not changed even after adding microsatellites and small-scale insertions and deletions.

Omics researches such as transcriptomics and proteomics produce massive data, and there are many databases based on them. We already discussed them in Chap. 12.

References

1. http://www.google.com/
2. Kitano, T., Sumiyama, K., Shiroishi, T., & Saitou, N. (1998). Conserved evolution of the Rh50 gene compared to its homologous Rh blood group gene. *Biochemical and Biophysical Research Communications, 249*, 78–85.
3. Kitano, T., Noda, R., Takenaka, O., & Saitou, N. (2009). Relic of ancient recombinations in gibbon ABO blood group genes deciphered through phylogenetic network analysis. *Molecular Phylogenetics and Evolution, 51*, 465–471.
4. Tornroth-Horsefield, S., et al. (2006). Structural mechanism of plant aquaporin gating. *Nature, 439*, 688–94.
5. Punta, M., et al. (2012). The Pfam protein families database. *Nucleic Acids Research, 40*, D290–D301.
6. Hunter, S., et al. (2012). InterPro in 2011: New developments in the family and domain prediction database. *Nucleic Acids Research, 40*, D306–D312.
7. Sumiyama, K., Washio-Watanabe, K., Saitou, N., Hayakawa, T., & Ueda, S. (1996). Class III POU genes: Generation of homopolymeric amino acid repeats under GC pressure in mammals. *Journal of Molecular Evolution, 43*, 170–178.
8. Zuckerkandl, E., & Pauling, L. (1965). Evolutionary divergence and convergence in proteins. In V. Bryson, & H. J. Vogel (Eds.), *Evolving genes and proteins* (pp. 97–166). New York: Academic Press.
9. Takahashi, M., & Saitou, N. (2012). Identification and characterization of lineage-specific highly conserved noncoding sequences in mammalian genomes. *Genome Biology and Evolution, 4*, 641–657.

Sequence Homology Handling

14

Chapter Summary

How to discover evolutionary homology of nucleotide and amino acid sequences and how to analyze these homologous sequences are discussed. BLAST families are mainly explained in homology search methods. Difference between the biologically true situation and mathematical optimum is stressed in pairwise alignment. CLUSTAL W, MAFFT, and MISHIMA are explained as multiple alignment softwares, and genome-wide homology viewers were finally discussed.

14.1 What Is Homology?

Discovery of homologous sequences is one of the most basic operations in evolutionary genomics. The "homology" concept was initially proposed by Richard Owen (1804–1892), as the contrast to "analogy." He was a British comparative anatomist, and his logic was mainly based on the topological relationship of bones, with no evolutionary thinking. After evolutionary studies were started, homologous bones, say human and cat femurs, were believed to be descended from their common ancestor. Nowadays, two nucleotide sequences, either in the same genome or in different genomes, are called homologous if they go back to a common ancestral sequence. Because amino acid sequences have corresponding genomic sequences, the same logic applies to protein sequences.

While homologous organs or bones are not necessarily easy to find, homologous sequences are easy, because we can have hundreds of nucleotide or amino acid sequences to be compared, and if the two sequences in question have a high similarity, there is almost no possibility for them to be created by chance independently. This is the basis of the methodology for finding sequence homology. For example, let us assume that there are two DNA sequences of 1,000 nucleotides, and their similarity (sequence identity) was 80 %. The probability of obtaining this situation by chance, under the equal frequency of the four nucleotides, is $_{1000}C_{200}[3/4]^{200}[1/4]^{800}$ that is in

N. Saitou, *Introduction to Evolutionary Genomics*, Computational Biology 17, DOI 10.1007/978-1-4471-5304-7_14, © Springer-Verlag London 2013

the order of 10^{-669}. We thus conclude that the two DNA sequences had a common ancestor sometime in the past.

14.2 Homology Search

Any DNA sequence must have their relatives or their homologous sequences connected through evolution. Homology search is to find such homologous sequences with the sequence in question, called the "query" sequence. As genome sequences of many organisms have been determined, there is a high chance to find homologous sequences in those stored in the target database with a query sequence picked up from the newly determined genome sequences.

Homology search methods are developed by considering the balance between the searching time and the homology detection range. Generally speaking, as the more remotely related sequences with the query sequence are searched, the searching time becomes longer. We discuss principles and usages of widely used homology searching softwares in this section, especially BLAST.

14.2.1 BLAST Families

Currently the most frequently used homology search system is BLAST developed by Altschul et al. in 1990 [1, 2]. BLAST is the acronym of "Basic Local Alignment Search Tool" [1], but it is also an English common word meaning a strong gust of wind. As its full name indicates, BLAST produces local or short pairwise alignment (see Sect. 14.3) between the query sequence and one sequence in the target database in a heuristic way. The algorithm of the original BLAST is as follows (based on [2]). Let us assume that the query is a nucleotide sequence with 1,000 bp. BLAST initially searches 11-character sequence by sliding window analysis. If the query sequence starts with "ACGTCCATTGAAAATCTGAG," its search is as follows:

```
                    11111111112
           12345678901234567890
           ACGTCCATTGAAAATCTGAG...
Search  1  ACGTCCATTGA
Search  2   CGTCCATTGAA
Search  3    GTCCATTGAAA
Search  4     TCCATTGAAAA
Search  5      CCATTGAAAAT
Search  6       CATTGAAAATC
Search  7        ATTGAAAATCT
Search  8         TTGAAAATCTG
Search  9          TGAAAATCTGA
Search 10           GAAAATCTGAG
           ...
```

The probability to find the exact 11-nucleotide sequence from a long random sequence with equal frequencies of four nucleotides is 2.38×10^{-7} (= $(\frac{1}{4})^{11}$). When

Table 14.1 Characteristics of query and target databases for five BLAST softwares

Software	Query sequence	Target database
BLASTN	Nucleotides	Nucleotides
BLASTP	Amino acids	Amino acids
BLASTX	Nucleotides[a]	Amino acids
TBLASTN	Amino acids	Nucleotides[a]
TBLASTX	Nucleotides[a]	Nucleotides[a]

[a]Automatically translated into amino acids

the total number of nucleotides of the target database is 10^{10}, there may be more than 2,000 identical 11-mer sequences in the target database. These candidate sequences are used as seeds so as to narrow down the further search by exploring upstream and downstream sequences of each seed. The number of seed nucleotides, called "word size" in BLAST option, can be varied. There are a series of parameters to control if the sequence extension starting from the seed will be continued or stopped. There are the cost to open gap (default is 5), the cost to extend gap (default is 2), the penalty for nucleotide mismatch (default is −3), and the reward for nucleotide match. These penalty or cost concepts were inherited from pairwise alignment methods, as will be discussed in Sect. 14.3.

If the alignment length is short, there is a possibility to erroneously assign a sequence in the target database as "homologous" with the query sequence. BLAST provides an E (expected) value for each pairwise alignment. The E value is the number of BLAST hits one expects to obtain by chance when searching a particular target database. It decreases exponentially as the BLAST score (S) of the match (maximum segment pair) increases, for the relationship between E and S is as follows:

$$E = Kmne^{-\lambda S}, \tag{14.1}$$

where K is search space size in terms of the total nucleotides in the target database; m and n are the number of nucleotides for the query and hit sequences, respectively; and λ is a constant factor determined by the scoring system for evaluating the pairwise alignment between the query sequence and the hit sequence. The E value thus describes the random background noise. For example, an E value of 1 means that in a database of the current size, one might expect to find 1 sequence with a length just by chance. The theoretical basis of the BLAST E value computation was from works of Samuel Karlin and his collaborators (e.g., [3]).

Original version of BLAST [1] found only gapless and thus relatively highly similar sequences. Later, gapped BLAST was developed to find sequences similar to the query sequence but with gaps, as well as PSI-BLAST for finding more remotely related sequences [4]. Web Miller's group then developed a greedy algorithm for pairwise alignment [5], and this was incorporated for BLAST system at NCBI as "megaBLAST."

The BLAST system consists of five softwares depending on the combination of query sequence and the target database: BLASTN, BLASTP, BASTX, TBLASTN, and TBLASTX. The combinations are shown in Table 14.1. It should be noted that query nucleotide sequences are automatically translated into corresponding six possible amino acid sequences in BLASTX and TBLASTX, while the target database

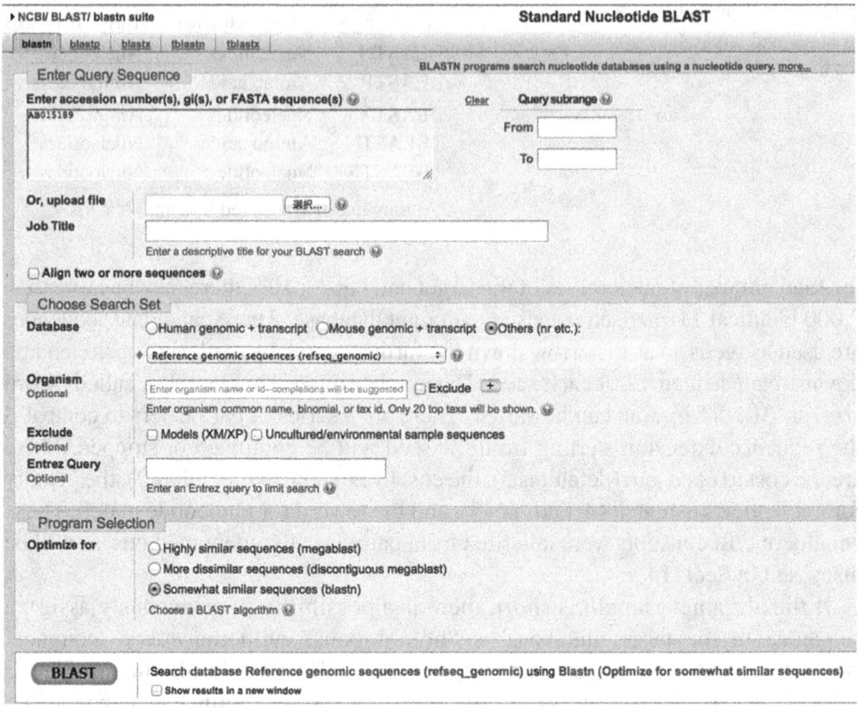

Fig. 14.1 Example of BLAST output using AB015189 as query: initial setting web image

nucleotide sequences were translated into six possible amino acid sequences in TBLASTN and TBLASTX. Why six? This is because there are three possible frames for the query nucleotide sequence and the protein may be coded in the complementary strand of the query sequence. The reason for translating nucleotide sequence to amino acid sequence is that the remote sequence homology can be detected with higher chance with amino acid sequence comparison than nucleotide sequence comparison.

14.2.2 Examples of Nucleotide Sequence Search Using BLASTN

We show how to use the BLAST search using NCBI BLAST system (http://blast. ncbi.nlm.nih.gov/). It should be noted that the examples shown in this section was obtained in March 2013, and it is possible that NCBI will change web page style and various settings. Therefore, even if you visit the same NCBI BLAST page, you may find some differences.

Figures 14.1, 14.2, and 14.3 show example outputs of BLASTN at the NCBI system, using query nucleotide sequence AB015189 shown in Fig. 13.1. Default settings were used in most cases except for two: the program used and the target

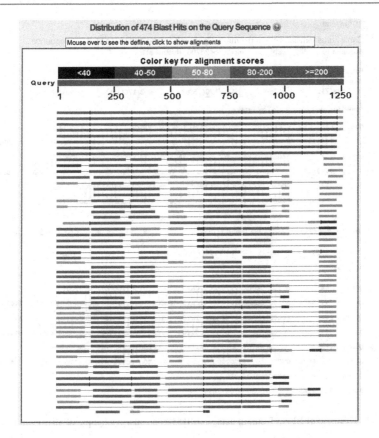

Fig. 14.2 Example of BLAST output using AB015189 as query: color-coded output

database. Because the default program for nucleotide versus nucleotide sequences is megaBLAST [5], the original BLASTN was chosen. The database of "Reference genomic sequences (refsec_genomic)" was chosen instead of the default "Human genomic + transcript." Figure 14.1 is the initial setting web image. Because the author is using a Japanese language system, "Select" button at "Or, upload file" field was written in Japanese. If we click "BLAST" button on the lower left corner, the computation starts. The BLAST search result was shown after a little more than 4 min. It should be noted that physical time depends on various factors such as the query sequence length, the number of total hits, the number of queries, and so on.

Figure 14.2 is the top part of the search result of 474 hits, and the alignment scores of different query sequence regions were shown in five colors. If we click some bar, pairwise alignment between the query sequence and the hit sequence (subject or "Sbjct" in the output) will appear. For example, if we click the tenth bar from the top, we find the output shown in Fig. 14.3. Two ranges of pairwise alignments are shown here, which showed top two E value segments homologous to the query sequence on the same horizontally lined subject sequence in Fig. 14.2. If we

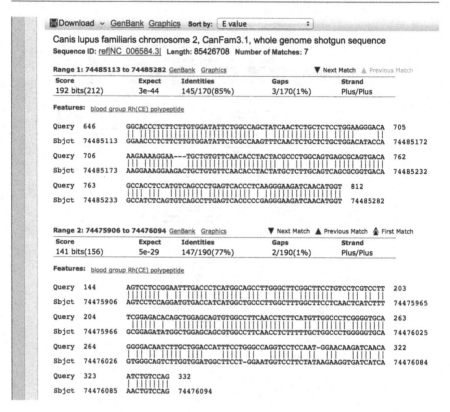

Fig. 14.3 Example of BLAST output using AB015189 as query: example of alignment result

examine the first block, it is the pairwise alignment between nucleotide positions 646–812 of the query sequence (AB015189, mouse Rh blood group gene mRNA) and nucleotide positions 74485113–74485282 of the 85.4-Mb subject sequence, NC_006584.3 on *Canis lupus familiaris* (dog) chromosome 2. Pairwise alignments of two blocks are shown in Fig. 14.3, but in total, there are seven pairwise alignment blocks between these two sequences. We can change the order of these blocks from default sorted by E values to either one of four ways: score, percent identity, query start position, and subject start position. If we choose "query start position," six ranges are as follows:

Range 1: Query = 6–107, S = 50.0 (54), E = 0.23, I & G = 75 & 0/107 (70 % & 0 %)
Range 2: Query = 144–332, S = 141 (156), E = 5e-29, I & G = 147 & 2/190 (77 % & 1 %)
Range 3: Query = 329–448, S = 86.0 (94), E = 3e-12, I & G = 91 & 0/120 (76 % & 0 %)
Range 4: Query = 493–572, S = 59.0 (64), E = 4e-4, I & G = 61 & 0/80 (76 % & 0 %)
Range 5: Query = 646–812, S = 192 (212), E = 3e-44, I & G = 145 & 3/170 (85 % & 1 %)

Range 6: Query=813–949, S=116 (123), E=2e-21, I & G=108 & 0/137 (79 % & 0 %)

Range 7: Query=948–1021, S=55.4 (60), E=0.005, I & G=60 & 6/80 (75 % & 7 %)

S, E, I, and G designate score, expect value, identity, and gap proportion, respectively. These seven ranges can be seen at the top part of the search result as two pink, three green, and one blue blocks (see Fig. 14.2). It should be noted that ranges 2 and 3 are overlapped with four nucleotides (positions 329–332). Range 1 showed the highest E value of 0.23, indicating a low sequence homology between the query and subject sequences. Below is its alignment (note this is different from "Range 1" of Fig. 14.3, for it was sorted by expected values).

```
Query     1      ATGGGCTCTAAGTACCCACGGTCCCTCCGCTGCTGCCTGCCCCTATGGGCCTTGGTGCTA   60
                 ||||||||||||||||| | ||| | || |||||||| ||||| ||| | | | |||
Sbjct  74466988   ATGGGCTCTAAGTACCCGCCGTCTGTGCGGGGCTGCCTCCCCCTGTGGACAATAGCACTA   74467047

Query    61      CAGACAGCTTTTATTCTCCTCTCCTGTTTTTTCATCCCCACGACAC   107
                 || || || | || ||| || |||||||| | || | |||||
Sbjct  74467048   GAGTTGGCCTTCTTGGTCATCTTCTTCTTTTTTCACCTCCTATGACAC   74467094
```

If this pairwise alignment is found alone, this high E value suggests that these two sequences may not be the true homologous sequences. However, this is accompanied with six other pairwise alignment with proximity. Therefore, it is more probable that this block represents an authentic homologous sequence. In fact, its identity (70 %) is quite high.

If we examine the alignment of the last hit, the length is only 46 nucleotides, and E value is 0.23, though the identity was quite high (87 %) with 2 gaps. It looks like just a random hit. However, the annotation of the subject sequence (*Canis lupus familiaris* chromosome 3; NC_006585) is "ammonium transporter Rh type C," and this alignment seems to be really homologous sequence, for the query was mouse Rh blood group gene mRNA. This result is, in a sense, expected, for we chose "Reference genomic sequences (refsec_genomic)" as the target database, and many vertebrate Rh blood group genes are expected to exist as orthologous with the query sequences. When we changed the target database to the default "Human genomic plus transcript (Human G + T)," the BLASTN result is shown in Fig. 14.4. There were seven hits from the human transcripts, and all of their annotations were related to the Rh blood group gene with very low E values except for the last one. In contrast, there were only three hits in the human genomic sequence with low E values, and eight hits had the equal high E values of 5.1. They may be considered to be random hits, for their annotations seem to be not related to the Rh blood group gene. This example clearly shows the importance of the target database when we interpret BLAST E values. Because all organisms are phylogenetically related, simple probability theory assuming a purely random sequence does not apply to the biologically real sequences, for they are products of eons of evolution.

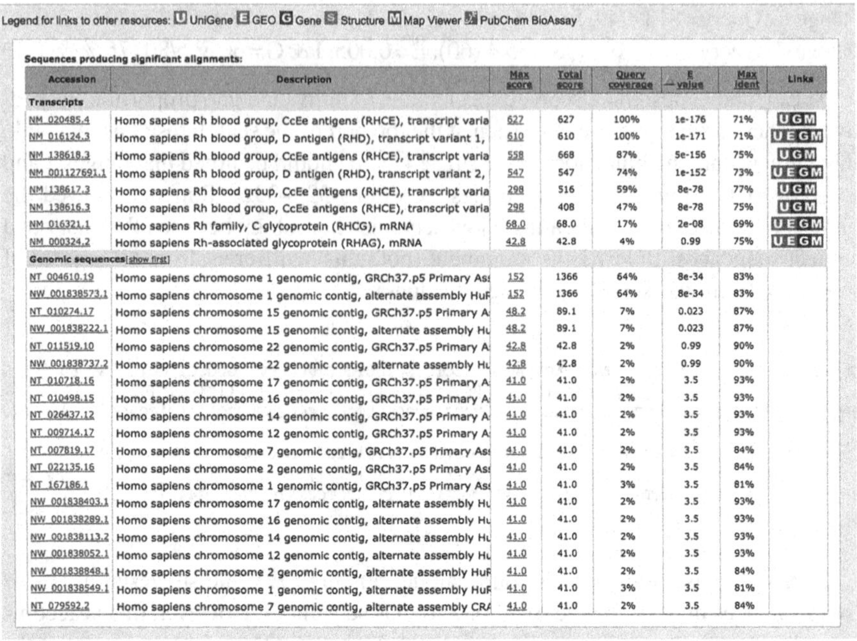

Legend for links to other resources: Ⓤ UniGene Ⓔ GEO Ⓖ Gene Ⓢ Structure Ⓜ Map Viewer 🅑 PubChem BioAssay

Sequences producing significant alignments:

Accession	Description	Max score	Total score	Query coverage	E value	Max ident	Links
Transcripts							
NM_020485.4	Homo sapiens Rh blood group, CcEe antigens (RHCE), transcript varia	627	627	100%	1e-176	71%	U G M
NM_016124.3	Homo sapiens Rh blood group, D antigen (RHD), transcript variant 1,	610	610	100%	1e-171	71%	U E G M
NM_138618.3	Homo sapiens Rh blood group, CcEe antigens (RHCE), transcript varia	558	668	87%	5e-156	75%	U G M
NM_001127691.1	Homo sapiens Rh blood group, D antigen (RHD), transcript variant 2,	547	547	74%	1e-152	73%	U E G M
NM_138617.3	Homo sapiens Rh blood group, CcEe antigens (RHCE), transcript variant	298	516	59%	8e-78	77%	U G M
NM_138616.3	Homo sapiens Rh blood group, CcEe antigens (RHCE), transcript varia	298	408	47%	8e-78	75%	U G M
NM_016321.1	Homo sapiens Rh family, C glycoprotein (RHCG), mRNA	68.0	68.0	17%	2e-08	69%	U E G M
NM_000324.2	Homo sapiens Rh-associated glycoprotein (RHAG), mRNA	42.8	42.8	4%	0.99	75%	U E G M
Genomic sequences[show first]							
NT_004610.19	Homo sapiens chromosome 1 genomic contig, GRCh37.p5 Primary Ass	152	1366	64%	8e-34	83%	
NW_001838573.1	Homo sapiens chromosome 1 genomic contig, alternate assembly Huf	152	1366	64%	8e-34	83%	
NT_010274.17	Homo sapiens chromosome 15 genomic contig, GRCh37.p5 Primary A	48.2	89.1	7%	0.023	87%	
NW_001838222.1	Homo sapiens chromosome 15 genomic contig, alternate assembly Hu	48.2	89.1	7%	0.023	87%	
NT_011519.10	Homo sapiens chromosome 22 genomic contig, GRCh37.p5 Primary A	42.8	42.8	2%	0.99	90%	
NW_001838737.2	Homo sapiens chromosome 22 genomic contig, alternate assembly Hu	42.8	42.8	2%	0.99	90%	
NT_010718.16	Homo sapiens chromosome 17 genomic contig, GRCh37.p5 Primary A	41.0	41.0	2%	3.5	93%	
NT_010498.15	Homo sapiens chromosome 16 genomic contig, GRCh37.p5 Primary A	41.0	41.0	2%	3.5	93%	
NT_026437.12	Homo sapiens chromosome 14 genomic contig, GRCh37.p5 Primary A	41.0	41.0	2%	3.5	93%	
NT_009714.17	Homo sapiens chromosome 12 genomic contig, GRCh37.p5 Primary A	41.0	41.0	2%	3.5	93%	
NT_007819.17	Homo sapiens chromosome 7 genomic contig, GRCh37.p5 Primary Ass	41.0	41.0	2%	3.5	84%	
NT_022135.16	Homo sapiens chromosome 2 genomic contig, GRCh37.p5 Primary Ass	41.0	41.0	2%	3.5	84%	
NT_167186.1	Homo sapiens chromosome 1 genomic contig, GRCh37.p5 Primary Ass	41.0	41.0	3%	3.5	81%	
NW_001838403.1	Homo sapiens chromosome 17 genomic contig, alternate assembly Hu	41.0	41.0	2%	3.5	93%	
NW_001838289.1	Homo sapiens chromosome 16 genomic contig, alternate assembly Hu	41.0	41.0	2%	3.5	93%	
NW_001838113.2	Homo sapiens chromosome 14 genomic contig, alternate assembly Hu	41.0	41.0	2%	3.5	93%	
NW_001838052.1	Homo sapiens chromosome 12 genomic contig, alternate assembly Hu	41.0	41.0	2%	3.5	93%	
NW_001838848.1	Homo sapiens chromosome 2 genomic contig, alternate assembly HuF	41.0	41.0	2%	3.5	84%	
NW_001838549.1	Homo sapiens chromosome 1 genomic contig, alternate assembly HuF	41.0	41.0	3%	3.5	81%	
NT_079592.2	Homo sapiens chromosome 7 genomic contig, alternate assembly CRA	41.0	41.0	2%	3.5	84%	

Fig. 14.4 BLASTN search result for AB015189 using human genome and transcript database

14.2.3 Examples of Amino Acid Sequence Search Using BLAST Families

Let us now consider amino acid sequence searches. We use the translated sequence of AB015189 shown in Fig. 13.1. When the default target database (nonredundant protein sequences) was used with BLASTP, all 100 hits were highly homologous (scores higher than 400 and E values lower than e-141) with the query amino acid sequence, and they are mostly mammalian origin. We are often interested in searching more remotely related sequences. Therefore, by clicking "Algorithm parameters" button at the bottom of the BLASTP page, we find "Max target sequences" option at the top. Let us choose the maximum value (20,000). The last hit in terms of E value (10.0) was a predicted ammonium/methylammonium permease protein of *Blastopirellula marina* (ZP_01090362), as shown below.

```
Score = 39.7 bits (86),  Expect = 10.0, Method: Compositional matrix adjust.
Identities = 35/117 (30%), Positives = 52/117 (44%), Gaps = 5/117 (4%)

Query  193  RVGENAQTEKVQMATSSSLFAMLGTLFLWIFWPAIN--SALLEGTKKRNAVFNTYYALAV  250
            R+G      E  ++  +   A  G LW  W    N  SL     +  + NT  A A+
Sbjct  185  RIGRFDGGEDRPISGHNLTLATFGALVLWFGWFGFNGGSTLSMNSDVPKILVNTNLAGAM  244

Query  251  SAVTATSMSALSHPQGKINMVHIHNAVLAGGVAVGAPGCLISSPWISMVLGLIAGLI     307
            A+T  +S    H +   V  N  +AG V V  A  C I +PW +  ++G  +GLI
Sbjct  245  GALTCLVLSKWVHKRPDVGQVM--NGAIAGLVGVTA-SCHILAPWAAALVGAGSGLI     298
```

Its E value was 10.0; however, the vertebrate Rh blood group gene was known to be homologous to bacterial ammonium transporter [6], and the subject sequence is probably authentic homologous one with the query. This again shows the problem of blindly omitting sequences with high E values.

Let us then try TBLASTN, using the translated sequence of AB015189 as above, but using "Transcriptome Shotgun Assembly" as the target database, with 20,000 Max target sequences. There were 255 hits, and they are mostly animals if we see "Taxonomy reports." However, these taxonomic distributions simply show that animals are more frequently studied in transcriptomics, and they do not show natural distributions.

14.2.4 Other Homology Search Methods

BLAST [1] can be considered as a descendant of FASTP [7] and FASTA [8], because both were developed by David Lipman and his collaborators as in BLAST. Although FASTA is no longer used extensively compared to BLAST, FASTA format for sequence data file was originally proposed for FASTA. The FASTA format is quite simple. It consists of sequence name part and sequence itself. The sequence name part should start with ">" followed by free format text to designate the characteristics of sequence to be shown below. The sequence part can be either nucleotides or amino acids. This is one unit, and the FASTA format can provide multiple sequences as shown below:

```
>AB031368|AB031368.1 Pan troglodytes gene for ABO transferase, partial cds, haplotype:I.
atacgtggctttcctgaagctgttcctggagacggcggagaagcacttcatggtgggcca
ccgtgtccactactatgtcttcaccgaccagccagccgcagtgccccgcgtgacgctggg
gaccggtcggcagctgtcggtgctggaggtgcgcgcctacaagcgctggcaggacgtgtc
catgcgccgcatggagatgatcagtgacttctgccagcggcgcttcctcagcgaggtgga
ttacctggtgtgcgtggacgtggacatggagttccgcgaccacgtgggcgtggagatcct
gactccgctgttcggcaccctgcaccctggcttctacggaagcagccgggaggccttcac
ctacgagcgccggccccagtcccaggcctacatccccaaggatgagggcgatttctacta
cctgggggggttcttcggagggtcggtgcaagaggtgcagcggctcac
>AB031369|AB031369.1 Pan troglodytes gene for ABO transferase, partial cds, haplotype:II.
atacgtggctttcctgaagctgttcctggagacggcggagaagcacttcatggtgggcca
ccgtgtccactactatgtcttcaccgaccagccagccgcagtgccccgcgtgacgctggg
gaccggtcggcagctgtcggtgctggaggtgcgcgcctacaagcgctggcaggacgtgtc
catgcgccgcatggagatgatcagtgacttctgccagcggcgcttcctcagcgaggtgga
ttacctggtgtgcgtggacgtggacatggagttccgcgaccacgtgggcgtggagatcct
gactccgctgttcggcaccctgcaccctggcttctacggaagcagccgggaggccttcac
ctacgagcgccggccccagtcccaggcctacatccccaaggatgagggcgatttctacta
cctgggggggttcttcggagggtcggtgcaagaggtgcagcggctcac
>AB031370|AB031370.1 Pan troglodytes gene for ABO transferase, partial cds, haplotype:III.
atacgtggctttcctgaagctgttcctggagacggcggagaagcacttcatggtgggcca
ccgtgtccactactatgtcttcaccgaccagccagccgcagtgccccgcgtgacgctggg
gaccggtcggcagctgtcggtgctggaggtgcgcgcctacaagcgctggcaggacgtgtc
catgcgccgcatggagatgatcagtgacttctgccagcggcgcttcctcagcgaggtgga
ttacctggtgtgcgtggacgtggacatggagttccgcgaccacgtgggcgtggagatcct
gactccgctgttcggcaccctgcaccctggcttctacggaagcagccgggaggccttcac
ctacgagcgccggccccagtcccaggcctacatccccaaggacgagggcgatttctgcta
cctgggggggttcttcggagggtcggtgcaagaggtgcagcggctcac
```

Blat, developed by Jim Kent in 2002 [9], is an alignment tool similar to BLAST, but it is structured differently. Blat keeps an index of an entire genome in memory in the case of nucleotide sequences as the target database. The index consists of nonoverlapping 11-mers except for those heavily involved in repeats. Blat for nucleotide sequences is designed to quickly find those of 95 % and greater similarity of length 40 bases or more and thus may miss more divergent or short sequence alignments (based on [10]).

PatternHunter [11] is using nonconsecutive sequence pattern as the seed, in contrast to BLAST which uses consecutive sequences for seeds. PatternHunter was used by the Mouse Genome Sequence Consortium, and it finished the mouse and human genome-wide comparison in 20 CPU-days, more than 300 times faster than BLAST with the same condition [12]. Its use is, however, not free.

HMMER [13] is applying a profile hidden Markov models for amino acid sequences. A freely available website is provided ([14]; http://hmmer.janelia.org/) for this method. If we use the translated sequence of AB015189 shown in Fig. 13.1 as we used for BLAST family, 762 hits were obtained using the NR (nonredundant) amino acid sequence data as the target database and the threshold E value of 2. If one is interested in finding amino acid sequences remotely related from the query sequence, it may be better to try both the BLAST system and HMMER.

14.2.5 Problem of Homology Search-Dependent Analyses

It is true that the homology search is a powerful tool to find out evolutionarily related sequences. However, we should be aware that all we can compare through homology searches are those which have a certain degree of sequence similarity with each other. It seems obvious, yet there is a possibility that a pair of sequences with no apparent sequence similarity may share the common ancestral sequence. This can happen if one or both sequence lineages accumulated many mutations and then the sequence similarity was lost. Definition of homology is not sequence similarity but the shared ancestry. We have to be careful about the difference between them.

Archeology has some similarity with evolutionary study, for both are looking for the past. It is a well known caveat in archeology that the reconstruction of the activity of ancient people should not be simply deduced from the remnants found from excavation. Of course, these remnants are precious for archeological studies, yet we have to consider a high heterogeneity on persistence probabilities of various remnants. For example, shells of shell mounds are often kept for thousands of years, while fish bones may be degraded within a few years. The same logic applies to stone tools and wooden tools. We usually find only stone tools from the Paleolithic period, but we should not conclude that people at that time did not use wooden tools, for they were easy to be degraded.

We, who study evolutionary genomics, should also be careful when we speculate the evolutionary history of genomes from homology searches. Highly conserved sequences are easy to detect through homology searches, thanks to Mother Nature called "purifying or negative selection" (see Chap. 4) that protected those sequences from mutation accumulations. If positive selection was very strong in some lineages of genomes, their similarities with other sequences that experienced less magnitude of positive selection may be lost. Mutational hot spot regions may also reduce their sequence similarity more rapidly than the rest of the genome.

Another fundamental problem of genome sequence analyses is ignorance of extinct lineages. Most of organismal lineages are destined to become extinct. Although there are a wide variety of organisms at present, they are a very tiny part of all the organisms that ever existed on the earth. We can use only available genome sequence data, but this methodological problem should always be kept in mind when we consider the evolutionary history of organisms.

14.3 Pairwise Alignment

The theoretical basis of homology searches is pairwise alignment. It also provides a starting point for multiple alignment. One chapter of "Bioinformatics and Molecular Evolution" by Higgs and Atwood (2005; [15]) was devoted to alignment problem meant for biologists, and "Sequence Comparison" by Chao and Zhang (2008; [16]) in Springer Computational Biology Series (this book is part of this series) provides some mathematical details of pairwise alignments. Interested authors may also refer to these two references.

14.3.1 Biologically True Alignment

There are various kinds of mutations as we discussed in Chap. 2. These mutations accumulate as DNA sequences diverge and form a phylogenetic relationship as discussed in Chap. 3. Some lineages may disappear, while other lineages may be kept by chance or by negative (purifying) selection as discussed in Chap. 4. When we compare these homologous sequences, we naturally try to "align" them by considering effects of accumulated mutations. Let us call this "biologically true alignment."

Figure 14.5 shows an example of sequence evolution. Four nucleotide substitutions and two deletions occurred in the two lineages. Figure 14.6a is the biologically true alignment of sequences A and B shown in Fig. 14.5. There are three nucleotide positions in which nucleotides differ between sequences A and B, designated with asterisks. This situation is called "mismatch." There are also two gaps in sequences A and B caused by independent deletions, shown in hyphens.

The real goal of the sequence alignment should be to find this biologically true alignment. Unfortunately, however, it is usually difficult to obtain this, because of our lack of evolutionary histories. It should be noted that even the biologically true alignment (Fig. 14.6a) does not represent evolutionary history shown in Fig. 14.5, for the substituted nucleotide at position 15 in the lineage to sequence A was deleted later. This is why there are only three mismatches in Fig. 14.6a. We therefore usually make mathematically optimum alignment, as we shall see in the next section.

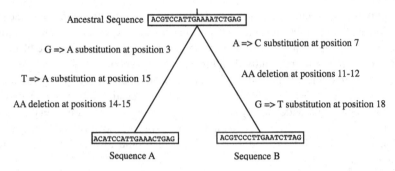

Fig. 14.5 An example of sequence evolution

Fig. 14.6 Two examples of pairwise alignments for sequences shown in Fig. 14.5. (a) The biologically true alignment. (b) An example of mathematically optimum alignment. (c) Another example of mathematically optimum alignment

a
```
Ancestor:    ACGTCCATTGAAAATCTGAG
Sequence A:  ACATCCATTGAAA--CTGAG
Sequence B:  ACGTCCCTTG--AATCTTAG
                 *     *        *
```

b
```
Sequence A:  ACATCCATTGAAACTGAG
Sequence B:  ACGTCCCTTGAATCTTAG
                 *   *    *  *
```

c
```
Sequence A:  ACA-TCC-ATTGAAA-CT-GAG
Sequence B:  AC-GTCCC-TTGAA-TCTT-AG
```

14.3.2   Mathematically Optimum Alignment

As we already discussed in the previous section, two homologous sequences may differ at their alignment with two types: mismatch and gaps. If we align sequences A and B shown in Fig. 14.5, we may end up with an alignment shown in Fig. 14.6b. There are four mismatches and no gap, and it is different from the biologically true alignment (Fig. 14.6a), though the three mismatches were correctly reconstructed. We have to define the mathematical model first to obtain such mathematically optimum alignment. Because there are at least two possible types of differences, mismatch and gap, between two sequences to be aligned, we may define a value, AS, to show the "alignment situation" between two sequences. There are a variety of ways to define AS values. We may count the number (m) of mismatched positions and that (g) of gap positions, and their sum may be considered as AS:

$$AS1 = m + g. \tag{14.2}$$

If we use this measure, $AS1$ becomes 7 and 4 for alignments of Fig. 14.6a, b, respectively. It is natural to choose the alignment that shows the smallest AS value, so we choose alignment B under this mathematical model, even though alignment A is the biologically true alignment.

We implicitly assumed that substitutions causing mismatches and insertions and deletions causing gaps are equally weighted in terms of nucleotides involved in Eq. 14.2. We can slightly modify the definition of g; instead of counting number of positions affected by gaps, we count the number of consecutive gaps. We thus have another definition:

$$AS2 = m + cg, \tag{14.3}$$

where cg means the number of consecutive gaps. Under this new definition, $AS2 = 5$ for Figure 14.6a and $AS2 = 4$ for Fig. 14.6b. Now the difference narrowed, but still Fig. 14.6b is more optimum.

Because nucleotide substitutions are about 10 times more frequent than insertions and deletions in the purely neutrally evolving genomic regions as shown by Saitou and Ueda (1994; [17]), let us give 10 times more weights to one consecutive gap than one substitution. We then have yet another definition:

$$AS3 = m + 10cg. \tag{14.4}$$

We then have 23 and 4 for $AS3$ values of Fig. 14.6a, b, respectively. Because weighting is arbitrary in mathematical definition, we can have a reverse situation as

$$AS4 = 50m + cg. \tag{14.5}$$

We now have 152 and 200 as $AS4$ values of Fig. 14.6a, b, respectively, and the biologically true alignment now shows lower $AS4$ value than that for Fig. 14.6b. However, under this framework, $AS4 = 8$ for another alignment shown in Fig. 14.6c, and this is much more optimum than the two other alignments. It is thus clear that the mathematically optimum alignment is only conditionally obtained under a specific definition, and there are lots of arbitrariness behind the seemingly rigorous mathematical optimum. We should be careful for any alignment result. Another problem is that the relative weights between substitutions and insertions/deletions vary depending on the genomic regions compared. If we compare coding regions, gaps with the length of 3 and its multiple numbers are expected to appear more frequently than those with different lengths, for one amino acid is coded by a set of three nucleotides called codon. It should be always remembered that we are dealing with biological problems here, and mathematical models are only assisting us to solve these biological questions.

14.3.3 Methods Using Dynamic Programming

It is necessary to develop a certain algorithm under a mathematical model as we discussed in a previous section. Needleman and Wunsch (1970; [18]) first developed an algorithm for pairwise alignment based on dynamic programming to compute similarity measures between two sequences. Sellers (1974; [19]) introduced a distance measure for pairwise alignment, and Waterman et al. (1976; [20]) generalized Seller's distance, but it required the computational steps of the order of $M^2 \cdot N$, where M and N are the number of sites (nucleotide or amino acid) for the two sequences to be aligned. Gotoh (1982; [21]) proposed a simplified weighting system for the gap penalty of k characters (nucleotides or amino acids):

$$w[k] = uk + v, \tag{14.6}$$

where u (≥ 0) and v (≥ 0) are gap extension penalty for the length k and gap opening penalty, respectively. This linear weighting system greatly reduced computation to the order of $M \cdot N$. Most of the currently used dynamic programming methods for the pairwise alignment use this weighting system.

Let us explain Gotoh's [21] algorithm. We assume nucleotide sequences, but this algorithm can be applied to any characters, such as amino acids, genes, or certain blocks of characters. Two sequences (A and B) in question are conceptually arranged in horizontal and vertical way (see Table 14.2A). Let us denote ith nucleotide of horizontally arranged sequence A as m_i ($1 \leq i \leq N_A$) and the jth nucleotide of vertically arranged sequence B as n_j ($1 \leq j \leq N_B$). We have a total of $N_A \times N_B$ cells for all possible comparisons of ith and jth nucleotides of the two sequences. We also add 0th row and column from the mathematical requirement to the cells. We then define $D[i, j]$ ($0 \leq i \leq N_A, 0 \leq j \leq N_B$) as the distance in terms of alignment penalty between the partial sequence A from the first to the ith nucleotides and the partial sequence B from the first to the jth nucleotides, and a particular tracing of cells depending on determination of these values defines the optimum pairwise alignment for sequences A and B. These values are computed by using the following procedure called "dynamic programming." Let us assume that $D[i–1, j–1]$, $D[i–1, j]$, and $D[i, j–1]$ are given. We also introduce two kinds of distances, $P[m, n]$ and $Q[m, n]$ ($0 \leq m \leq N_A$, $0 \leq n \leq N_B$). They are defined as follows:

$$P[m, n] = Min\{D[m-k, n] + w[k]\}, \tag{14.7a}$$

$$Q[m, n] = Min\{D[m, n-k] + w[k]\}, \tag{14.7b}$$

where $Min\{\alpha[k]\}$ denotes the minimum value among $\alpha[k]'s$ ($1 \leq k \leq m$ for $P[m, n]$ and $1 \leq k \leq n$ for $Q[m, n]$) [20]. $P[m, n]$ corresponds to the situation below:

Sequence A: [1 2 3 4 … $m-k-2\,m-k-1\,m-k$] $G_1\,G_2\,G_3$ … G_k
Sequence B: [1 2 3 4 … $n-2\,n-1\,n$]

Bracketed parts for sequence A and sequence B represent the pairwise alignment for partial sequences which give the alignment penalty distance $D[m-k, n]$, while $G_i's (1 \leq i \leq k)$ denote a gap with length k at the end of sequence A. The minimum value of $P[m, n]$ is chosen by changing k from 1 to m. $Q[m, n]$ corresponds to the situation by reversing sequences A and B. From these values, $D[i, j]$ is computed by

Table 14.2 Dij, Pi, Qij, and Eij matrices for pairwise alignment of sequences (see text for details)

(A) D[i, j] matrix with the two nucleotide sequences compared

			A	T	C	C	G	C	G	A	T
		0	1	2	3	4	5	6	7	8	9
	0	0	4	5	6	7	8	9	10	11	12
A	1	4	0	4	7	10	11	12	13	10	14
T	2	5	4	0	4	7	10	13	16	14	10
G	3	6	7	4	5	9	7	11	13	17	14
C	4	7	10	7	4	5	9	7	11	14	17
G	5	8	11	10	8	9	5	9	7	11	14
T	6	9	12	11	11	12	9	10	11	12	11
C	7	10	13	15	11	11	12	9	13	16	15
G	8	11	14	17	15	15	11	13	9	13	16
T	9	12	15	14	18	18	15	16	13	14	13
T	10	13	16	15	19	21	18	19	16	18	14

(B) P[i, j] matrix

	0	1	2	3	4	5	6	7	8	9
0	0	3	6	9	10	11	12	13	14	15
1	4	7	4	7	10	13	15	16	17	14
2	5	8	8	4	7	10	13	16	19	18
3	6	9	11	8	9	12	11	14	17	20
4	7	10	13	11	8	9	12	11	14	17
5	8	11	14	14	12	13	9	12	11	14
6	9	12	15	15	15	16	13	14	15	16
7	10	13	16	19	15	15	16	13	16	19
8	11	14	17	20	19	19	15	17	13	16
9	12	15	18	18	21	22	19	20	17	18
10	13	16	19	19	22	25	22	23	20	22

(C) Q[i, j] matrix

	0	1	2	3	4	5	6	7	8	9
0	0	4	5	6	7	8	9	10	11	12
1	3	7	8	9	10	11	12	13	14	15
2	6	4	8	11	13	14	15	16	14	18
3	9	7	4	8	11	14	17	19	17	14
4	10	10	7	9	13	11	15	17	20	17
5	11	13	10	8	9	13	11	15	18	20
6	12	15	13	11	12	9	13	11	15	18
7	13	16	15	14	15	12	14	14	16	15
8	14	17	18	15	15	15	13	17	19	18
9	15	18	21	18	18	15	16	13	17	20
10	16	19	18	21	21	18	19	16	18	17

(D) e[i, j] matrix

	1	2	3	4	5	6	7	8	9
1	1	2	2	6	3	3	3	1	2
2	3	1	2	2	2	2	6	3	1
3	3	3	1	4	1	2	1	6	3
4	6	3	1	1	2	1	2	2	6
5	2	3	3	5	1	2	1	2	2
6	2	1	3	3	3	1	3	1	1
7	2	3	1	1	3	1	2	7	3
8	2	2	3	3	1	3	1	2	2
9	2	1	6	3	3	5	3	1	1
10	2	1	4	3	3	3	3	5	1

choosing the minimum among three alignment penalty values corresponding to the following three paths:

$$Diagonal\ path : D[i-1, j-1] + MP[\mathrm{m}_i, \mathrm{n}_j], \tag{14.8a}$$

$$Horizontal\ path : Min\{D[i, j-1] + w[1], P[i, j-1] + \mathrm{u}\}, \tag{14.8b}$$

$$Vertical\ path : Min\{D[i-1, j] + w[1], Q[i-1, j] + \mathrm{u}\}, \tag{14.8c}$$

where $MP[\mathrm{m}_i, \mathrm{n}_j]$ is the mismatch penalty value between nucleotides m_i and n_j and $Min\{\alpha, \beta\}$ denotes the minimum value between α and β [21]. It should be noted that we do not need to search many $P[m, n]$ and $Q[m, n]$ values as originally proposed by Waterman et al. [20]. Gotoh [21] showed by induction that consideration of $P[i-1, j]$ and $Q[i, j-1]$ is enough under the linear weight system such as that given in Eq. 14.6.

The diagonal, horizontal, and vertical paths correspond to the mismatch, one nucleotide gap at sequence B, and one nucleotide gap at sequence A, respectively. The value for the diagonal path may be easy to understand; we simply add the mismatch penalty value for nucleotide i and j ($MP[\mathrm{m}_i, \mathrm{n}_j]$) to the alignment penalty distance, $D[i-1, j-1]$, for the previous sequence pair on the diagonal line. If we choose the horizontal path, we skip sequence B, namely, one gap is created in sequence B after the $(i-1)$-th nucleotide. We thus add gap penalty with length 1, namely, $w[1]$ ($= \mathrm{u} + \mathrm{v}$) according to Eq. 14.6), to the previously obtained partial alignment with $D[i-1, j]$. It is possible that this new gap is a continuation of the previously existing gap. In this case, only gap extension penalty (u) should be added, and this corresponds to $P[i-1, j] + \mathrm{u}$. We should choose the smaller value between these two situations. Vertical path is just the reversed situation of horizontal path.

The mismatch penalty value in Eq. 14.8a is determined by comparing nucleotide m_i and n_j. If these are identical, there will be no penalty, while some value will be given when mismatch occurred. A variety of mismatch penalty values are possible depending on different situations. For example, in the case of amino acid sequences, chemical difference among 20 amino acids may be considered for mismatch penalty value.

Let us show an example of pairwise alignment using the dynamic programming method for two sequences A and B shown in Table 14.2A:

Sequence A: ATCCGCGAT
Sequence B: ATGCGTCGTT

Let us use gap extension penalty value (u) as 1 and gap opening penalty (v) as 3 in 14.6. The mismatch penalty $MP[\mathrm{m}_i, \mathrm{n}_j]$ ($\mathrm{m}_i \neq \mathrm{n}_j$) is set to be 5. Tables 14.2A, B, C, and D show computation results for $D[i, j]$, $P[i, j]$, $Q[i, j]$, and $e[i, j]$, respectively. These results are output of a perl program align.pl, developed for this textbook, written by Mrs. Mizuguchi Masako.

The matrix $e[i, j]$ of Table 14.2D is showing paths, either diagonal, horizontal, or vertical. If we consider tie situations, we have to consider combination of multiple

a
Sequence A: ATCCG-CGAT
Sequence B: ATGCGTCGTT
            *   *

b
Sequence A: AT-CCG-CG-AT
Sequence B: ATG-CGTCGT-T

Fig. 14.7 Two possible pairwise alignments for the same set of sequences. (**a**) When mismatch penalty = 5, gap opening penalty = 3, and gap extension penalty = 1. (**b**) When mismatch penalty = 10, gap opening penalty = 1, and gap extension penalty = 1

paths: (1) diagonal or horizontal, (2) diagonal or vertical, (3) horizontal or vertical, and (4) diagonal, horizontal, or vertical [22]. Therefore, seven values (1–7) represent the following directions in the $e[i, j]$ matrix:

1 = diagonal
2 = horizontal
3 = vertical
4 = diagonal or horizontal
5 = diagonal or vertical
6 = horizontal or vertical
7 = diagonal, horizontal, or vertical

We should examine this $e[i, j]$ matrix from the end of the sequences, that is, e[9, 10]. Because e[9, 10] = 1, we go back to e[8, 9], which is again 1. There are four consecutive 1 values for $e[i, j]$, and then we found e[6, 5] = 3. We thus go up vertically, and again four consecutive 1 values follow. These underlined $e[i, j]$ values give the pairwise alignment shown in Fig. 14.7a. If we use somewhat different penalty value set (mismatch penalty = 10, gap opening penalty = 1, and gap extension penalty = 1), we obtain the different alignment shown in Fig. 14.7b. The schematically shown alternative alignments in Fig. 14.6b, c do correspond to the difference between Fig. 14.7a, b which are based on dynamic programming results. It is clear that we have to be aware of this arbitrariness on mismatch and gap penalty values.

14.3.4 Dot Matrix Method

Pairwise alignment-producing methods based on dynamic programming generally optimize the alignment by eliminating incongruent sequences while keeping those with high identities. They are thus not expected to obtain good alignments when reversal, relocation, or simple sequence repeats occurred. To circumvent this problem, the dot matrix (also called as "dot plot" or "har plot") method is powerful. This method gives a 2D graph that can be interpreted intuitively by human eye. Fitch (1969; [23]) initiated the use of the dot matrix method. A good explanation of dot matrix methods is given by Schulz et al. [24].

Figure 14.8 shows an example of the dot matrix to explain the basic concept of this method. First half of the two sequences arranged horizontally and vertically is

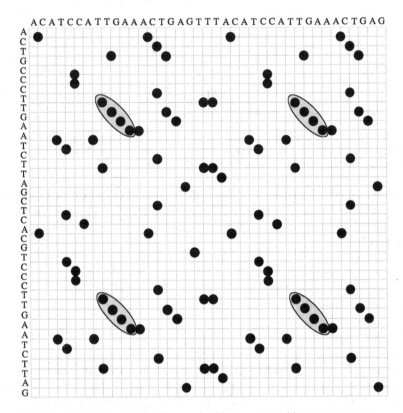

Fig. 14.8 Example of dot matrix for short nucleotide sequences with repeat structures

the same as those presented in Fig. 14.6. A certain length of nucleotides can be used as unit of comparison, called "window size." Because single nucleotide is too short, two nucleotides were used as window size in this example. Thus, a dot was given when the nucleotides (or amino acid in the case of protein sequences) of two sequences compared were identical. If these two sequences are highly similar, a diagonal line is expected to be observed. In fact, two contiguous four dots, marked by shadowed ovals, appeared on the diagonal line. This corresponds to "TTGAA." The same contiguous four dots also appeared off diagonal twice. This indicates the existence of a repeat structure in both sequences.

An example of dot plot for comparison of two long genome sequences is shown in Fig. 14.9. YASS server (http://bioinfo.lifl.fr/yass/index.php) was used, and the human Hox A cluster (DDBJ/EMBL/GenBank accession numberAC004080) is arranged in a vertical way and the chicken Hox A cluster region (coordinates 32472861–32762548 of chromosome 2) in a horizontal way. A clear diagonal line in mostly forward direction (green color) is shown as the main homology, with many short stretches. The latter ones are short repeats in the genome. Figure 14.10 is an application of dot plot to detect a large-scale palindrome structure between human and chimpanzee Y chromosomes (from Kuroki et al. 2006; [25]).

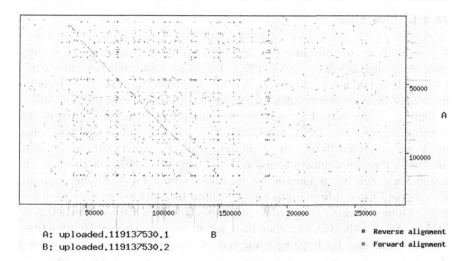

A: uploaded.119137530.1   B   ∎ Reverse alignment
B: uploaded.119137530.2   ∎ Forward alignment

Fig. 14.9 Example of dot matrix for human and chicken Hox A cluster regions

Fig. 14.10 Dot plot
of human-chimpanzee
Y chromosome partial region
(From Kuroki et al. 2006;
[25])

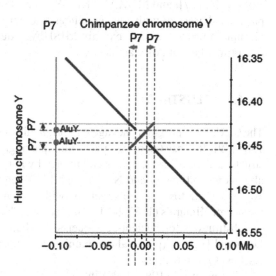

14.4   Multiple Alignment

When more than two sequences are aligned, it is called multiple alignment. One natural way is to apply the methods used for the pairwise alignment. However, the dot matrix for three or more dimensions is almost impossible to visualize. As for the dynamic programming, a method for three sequences was introduced by Murata et al. (1985; [26]), but the generalization to more than three sequences was difficult. When the number of sequences compared becomes large and when the length of sequences becomes long, combinatorial explosion happens.

14.4.1 Overview

Various methods and softwares were developed to overcome this problem, starting from Feng and Doolittle (1987; [27]). We discuss two representative ones: CLUSTAL W [28] and MAFFT [29]. A review by Notredame (2007; [30]) is helpful to compare many other methods for multiple alignment developed until 2007.

Only a certain part of a eukaryote genome is functional, and most of the remaining part is junk DNA (see Chaps. 8, 9, and 10). As a result, functional parts are conserved during the evolution, while the other parts accumulate mutations through neutral evolution (see Chap. 4). Conserved parts of the genome are helpful in identifying the overall homology of long genomic regions, and these informations could be used to assist the process of alignment of long stretches of nucleotide sequences. This idea was used in DIALIGN [31] and MLAGAN [32]. These methods use the pairwise comparison step first. However, this step for large genomic sequences is very time consuming, and it also does not reveal features shared by multiple sequences. A series of heuristic methods such as MUSCLE [33], MAVID [34], MAUVE [35, 36], MISHIMA [37], and MURASAKI [38] were developed for locating conserved patterns shared by multiple sequences that can be directly used in a sequence alignment procedure. We explain MISHIMA developed by the author's group from this class of softwares.

14.4.2 CLUSTAL W

The CLUSTAL program package was originally developed by Higgins and Sharp in 1988 [39]. It consisted of two parts: multiple alignment and phylogenetic tree construction. A progressive approach initiated by Feng and Doolittle [27] was used for alignment, and Saitou and Nei's (1987; [40]) neighbor-joining method (see Chap. 16) was used for tree construction, with Kimura's (1980; [41]) two-parameter method or Kimura's (1983; [42]) equation (see Chap. 15) for computing the number of substitutions for nucleotide sequences and amino acid sequences, respectively. Desmond Higgins (personal communication) chose these two methods because he respects Motoo Kimura.

Higgins et al. (1992; [43]) introduced CLUSTAL V, in which five different programs were integrated. V means five in roman numeral; however, when Thompson et al. (1994; [28]) improved the original multiple alignment algorithm of CLUSTAL, they named it CLUSTAL W. The algorithm consists of three steps: (1) all possible pairwise alignments are obtained and an evolutionary distance matrix is computed, (2) a guide tree is constructed from the distance matrix using the neighbor-joining method [40], and (3) the sequences are progressively aligned according to the branching order of the guide tree. Two options are available for pairwise alignment at step (1): fast but approximate method [44] and slow but rigorous method using a memory-efficient dynamic programming [45]. The unrooted tree produced by using the neighbor-joining method at step (2) is rooted by the midpoint rooting method (see Chap. 16). This rooted tree is used as the guide tree for the progressive alignment at step (3). Each sequence is weighted according to the branch lengths from the root to that sequence, and these weights are used at step (3). At step (3), two gap penalties set by the user are initially used: a gap opening penalty which gives the cost of opening a new gap of any length and a gap extension penalty which gives the cost of every item in a gap. The software then chooses appropriate gap penalties for each sequence alignment, depending on the following three factors: weight matrix, similarity of sequences, and lengths of sequences. This dynamic nature of gap penalty values made CLUSTAL W quite efficient. CLUSTAL W has been used by many researchers. It should be noted, however, that CLUSTAL W is not suitable for aligning long sequences or large number of sequences because of high computational requirements.

User-friendly version with the same algorithms used for CLUSTAL W was later developed and named as CLUSTAL X [46]. Recently, a new version of CLUSTAL series called "CLUSTAL Omega" was developed, and it can produce multiple alignments for huge number of amino acid sequences with short computation time [47]. These softwares are freely downloadable from http://www.clustal.org/.

14.4.3 MAFFT

The fast Fourier transform was introduced for detecting sequence similarity by Felsenstein et al. (1982; [48]). Katoh et al. (2002; [29]) developed a new software called MAFFT (probably the acronym of "Multiple Alignment using Fast Fourier Transform"). MAFFT is applicable both to amino acid sequences and to nucleotide sequences. Here we explain the algorithm of MAFFT for nucleotide sequences. As in CLUSTAL W, all pairwise alignments were obtained first in MAFFT. The correlation of two sequences with k-nucleotide lag (value of k is plus or minus depending on the location of gap in these two sequences) is rapidly computed using the fast Fourier transform. If these two sequences are homologous at some unknown region, the correlation with lag of k nucleotides may become high. The window analysis is then conducted to detect the positions of gaps which caused high correlation values for various k values. Multiple locally homologous sequence segments are expected to be found by this procedure, and these segments are aligned by using a dynamic programming technique. By these series of procedures, the dynamic programming to obtain the nucleotide sequence-based pairwise alignment is conducted with a considerably reduced CPU time while using a newly introduced scoring system [29].

MAFFT can be very fast with small datasets and give alignments comparable to much slower softwares such as T-Coffee [49], but its computation time increases rapidly when aligning larger datasets such as complete bacterial genomes.

a

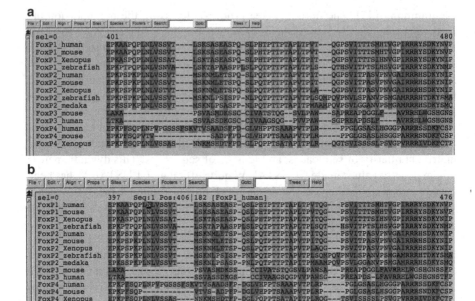

b

Fig. 14.11 Example of amino acid sequence alignment using SeaView system. (**a**) When MUSCLE was used. (**b**) When CLUSTALW2 was used

14.4.4   Example of Amino Acid Sequence Alignment

Let us show an example of amino acid sequence alignment. SeaView [50] is a graphic software for sequence alignment and molecular phylogeny and is freely download-able from http://pbil.univ-lyon1.fr/software/seaview.html. Users can use either MUSCLE [33] or CLUSTAL W2 [46] for multiple alignment. Figure 14.11a, b show a part of SeaView alignment result for 14 FOXP gene family protein sequences using MUSCLE (panel A) or CLUSTAL W2 (panel B). Each amino acid was colored according to its chemical property, and hyphen (–) denotes gaps. Default settings were used for both programs, and slight differences exist between the alignment results. It is difficult to decide which alignment is better.

14.4.5   MISHIMA

MISHIMA, developed by Kryukov and Saitou (2010; [37]), is the acronym of "Method for Inferring Sequence History In terms of Multiple Alignment," as well as the name of the city where the National Institute of Genetics, authors' affiliation, is located. MISHIMA does not depend on pairwise sequence comparison, and a new algorithm is used to quickly find rare oligonucleotide sequences shared by all sequences. The divide-and-conquer approach is then applied to break the sequences into fragments that can be aligned independently by an external alignment program, either CLUSTAL W [28] or MAFFT [29]. These partial alignments are assembled together to form a complete alignment of the original nucleotide sequences.

The core idea of the new algorithm used in MISHIMA is to analyze k-tuples found in the original sequences and to evaluate them based on their frequencies. When a particular nucleotide k-tuple is found exactly once in each sequence, MISHIMA considers it as a likely homology signal. A k-tuple that has its close variants found once in each sequence is considered to be less likely, but still possible homology signal. Analyzing all k-tuples up to certain length allows MISHIMA to select those k-tuples that represent the most probable homology shared by multiple sequences. These k-tuples can then be used to anchor the sequences before employing the divide-and-conquer method to complete the alignment. The basic principle of MISHIMA is as follows:

1. Find potentially useful k-tuples based on the number of their occurrences in the sequence data.
2. Analyze the potentially useful k-tuples, and select those that represent most probable local homology.
3. Use the selected k-tuples as anchors; split the sequences into segments.
4. Align the segments independently from each other.
5. Join partial alignments to complete the final multiple sequence alignment.

MISHIMA counts the number of occurrences of each short k-tuple in the original sequence dataset as a first step. This information is kept in a dictionary structure, indexed by a k-tuple sequence. The number of possible nucleotide k-tuples is 4^k, so the maximum k which MISHIMA can use is limited by the amount of memory it can use for the dictionary. MISHIMA stores k-tuple frequencies as 32-bit numbers, so 4×4^k bytes is required to store the frequencies of all nucleotide k-tuples. This allowed MISHIMA to use k of 13 and 14 on 32-bit machines. If we use 64-bit machines, the maximum k value will become longer.

Knowing the number of occurrences of each k-tuple in the original sequence dataset is not enough to efficiently decide which k-tuples are more likely to represent the local homology. Therefore, MISHIMA stores the number of sequences exhibiting each k-tuple. MISHIMA also stores the index of the last sequence where the k-tuple was found, which allows us to collect all the frequencies using single read through the sequence dataset.

A random sequence of length L is expected to contain $L/4^k$ occurrences of each k-tuple. If we consider N sequences with the average sequence length of M, $M \bullet N$ can be equated as L. If we can find k-tuples with exactly N occurrences once in each sequence, this is an ideal anchor for MISHIMA. In this case, $M \bullet N/4^k = N$, and $M = 4^k$. Thus, to have the best results with MISHIMA, the length of sequences to be analyzed should be comparable to or shorter than 4^k. Of course the real biological sequences are not completely random, and longer sequences can be aligned if they contain significant homology. Therefore, the capability of MISHIMA to align the real sequences should be higher than this estimate. In any case, the length k is the basic parameter affecting the performance of MISHIMA. If k is too small, too many occurrences of each k-tuple will be found. Although this depends on the length of the sequences, it is difficult to predict whether the particular k is appropriate or not. Therefore, MISHIMA uses all k-tuples with k from 1 to Max_k, the upper limit of k. The value of Max_k depends on the memory limitation. Most of the steps of the MISHIMA algorithm are performed in linear time, which makes it practical to use the largest Max_k allowed by the amount of available memory. MISHIMA is using 12 bytes for each k-tuple in a dictionary. The total amount of memory required for dictionary is therefore $12*(4+4^2+\ldots+4^M)$ bytes or less than 300 MB of RAM.

Using identical k-tuples shared by all sequences has a low sensitivity, so MISHIMA also uses inexact matches by updating the dictionary to include the number of inexact copies of a k-tuple in the sequence dataset. MISHIMA allows up to one substitution difference between two k-tuples. Potentially useful k-tuples are defined as those that have one inexact match in every sequence. Such k-tuples, or seeds, are extracted and saved for further analysis. The next step is to find the locations of seeds in sequence data. Locations are extracted in a second read through the sequence data. It should be noted that this step also completes in linear time. In cases where no seeds could be found, the external aligner may be used to align the whole dataset.

The next step is to measure compatibility of two seeds based on their relative coordinates in the sequence data. We consider only seeds that are found no more than once in every sequence. There are two possible orders for such two seeds, A and B: either the coordinate of A is smaller than that of B or vice versa. If all sequences in a dataset exhibit the same order of seeds A and B, these two seeds are compatible. However, if two kinds of orders coexist, we define an incompatibility distance as number of sequences whose order is different from the majority case. A distance defined in this way can be effectively used to evaluate the possibility that two seeds together represent a possible homology signal. This distance is used to construct a maximum nonconflicting set of seeds in the following procedure.

A constant number of best seeds is selected (the maximum number is 600 in MISHIMA because of memory limitations), and the matrix of size T×T is constructed. Each cell of the matrix contains the incompatibility distance between two seeds. The sum of all numbers for every row corresponds to the total amount of incompatibility introduced by one seed, and the seed which introduces the largest amount of incompatibility is removed from the set. The matrix is recalculated and the procedure iterates until no incompatibility is left unresolved. Finally only compatible seeds remain, which are used as alignment anchors.

The sequences are then divided and regions between the anchors are aligned independently from each other. After all partial alignments are complete, they are concatenated to construct a final complete alignment. Alignment of the regions between the seeds is performed using an external aligner provided the sequences are short enough. Otherwise, MISHIMA is used again to divide the sequences into shorter parts. The depth of the recursive operation in MISHIMA can be set with the command line option "-max-depth=x." Since MISHIMA takes a constant time for each step of finding seeds and dividing the sequences, using a limited depth may improve the performance on some datasets. The overall flow of MISHIMA algorithm is shown in Fig. 14.12 (from [37]).

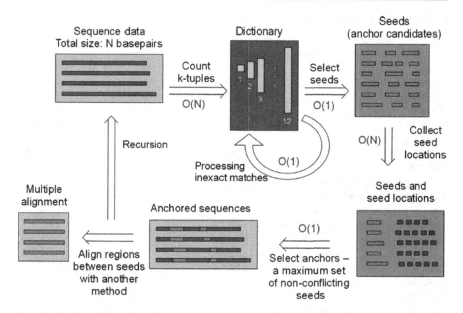

Fig. 14.12 The outline of MISHIMA procedures (From Kryukov and Saitou 2010; [37])

It should be noted that the divide-and-conquer approach used in MISHIMA can-not detect duplications, inversions, nor genomic rearrangements. This approach is valid under the assumption that all sequences can be aligned in a linear fashion to each other to form a multiple alignment.

Table 14.3A shows alignment time comparison of various numbers of human mtDNA genomes. The average nucleotide difference was 0.002~0.003; CLUSTAL W is the slowest among the compared aligners, even with [–quicktree] option, followed by MUSCLE [33]. In contrast, computation time of MISHIMA and MAFFT are much shorter than that of CLUSTAL W and MUSCLE. MUSCLE was unable to complete the alignment of 400 human mtDNA sequences. Two kinds of option sets were used for MAFFT: [–retree 1 –maxiterate 0] and [–retree 2 –maxiterate 1,000]. Both option sets showed much faster results than either MUSCLE or CLUSTAL W. MISHIMA is as fast as MAFFT set at faster option either when CLUSTAL W was used or when MAFFT was used as the external alignment program when 50, 100, and 200 human mtDNA sequences were compared. When 400 human mtDNA sequences are com-pared, MAFFT set at faster option is clearly faster than MISHIMA. However, if we add [–max-depth=1] option to MISHIMA, both aligners finish in similar time. Because human mtDNA sequences are quite similar to each other, good seeds are found in abundance in the first iteration of dictionary analysis, and average block length (length of sequences between two adjacent anchor seeds) is already short enough for CLUSTAL W usage. Therefore, we recommend using this option when aligning closely related sequences, such as sequences from same species.

Table 14.3 Alignment time comparison

(A) Various numbers of human mtDNA genomes

	No. of sequences compared			
Method	50	100	200	400
MISHIMA[a]	43	2:06	6:05	34:19
MISHIMA[b]	1:05	2:23	12:59	1:38:09
MAFFT[c]	2:43	5:44	16:31	38:51
MAFFT[d]	5:47	12:38	34:29	1:32:38
MUSCLE	44:29	1:28:12	2:08:44	_[e]
CLUSTAL W	2:15:32	4:48:59	10:16:22	25:08:23

[a]MISHIMA with CLUSTAL W and only one depth option
[b]MISHIMA with MAFFT and no depth limit option
[c]Options used: retree=1 & maxiterate=0
[d]Options used: retree=2 & maxiterate=1,000
[e]Computation impossible due to memory problem

(B) Various numbers of mammalian mtDNA genomes

	No. of sequences compared		
Method	50	100	200
MISHIMA[a]	6:50	10:24	19:03
MAFFT[b]	12:13	33:36	53:38
MUSCLE	46:13	1:38:59	_[c]
CLUSTAL W	2:15:04	5:05:20	10:40:31

[a]MISHIMA with MAFFT, and no limit of depth option
[b]Options used: retree=1 & maxiterate=0
[c]Computation impossible due to memory problem

Table 14.3B shows alignment time comparison of various numbers of mammalian mtDNA genomes (50, 100, or 200) with the average nucleotide difference of 0.25~0.27. MISHIMA+MAFFT combination is the fastest, followed by MAFFT alone, MISHIMA+CLUSTAL W (not shown here), MUSCLE, and CLUSTAL W. MUSCLE could not complete the alignment of 200 complete mtDNA genomes. If we compare computation times shown in Tables 14.2 and 14.3, those of MISHIMA are certainly much slower for different mammalian species than for human individuals: 23 times and 13 times more for 50 and 100 sequences, respectively. This is because the number of good seeds for anchoring was reduced as sequence divergence increases. MISHIMA greatly accelerates the aligner that it is using: MISHIMA+CLUSTAL W is much faster than CLUSTAL W alone and MISHIMA+MAFFT is faster than MAFFT alone. It should be noted that the improvement in computation time is not achieved at the expense of alignment quality [37].

Figure 14.13 shows a part of an example output of MISHIMA for 20 primate species mtDNA genome data. DDBJ/EMBL/GenBank accession numbers and species names of 20 mtDNA complete genome sequences used are (1) D38112 (*Homo sapiens*), (2) AB286049 (*Propithecus verreauxi coquereli*), (3) AB371085 (*Daubentonia madagascariensis*), (4) AB371087 (*Eulemur fulvus mayottensis*), (5) AB371090 (*Tarsius syrichta*), (6) AB371091 (*Saimiri sciureus*), (7) AB371092

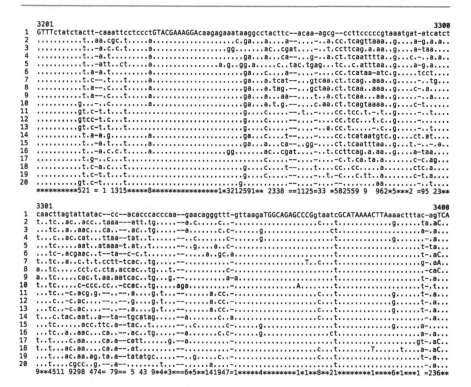

Fig. 14.13 Example output of MISHIMA for 20 primate species mtDNA genome data

(*Galago senegalensis*), (8) AB371093 (*Otolemur crassicaudatus*), (9) AB371094 (*Loris tardigradus*), (10) AB371095 (*Perodicticus potto*), (11) AB504748 (*Hylobates agilis*), (12) AB504750 (*Symphalangus syndactylus*), (13) AB504751 (*Nomascus siki*), (14) AJ309866 (*Cebus albifrons*), (15) AJ421451 (*Lemur catta*), (16) AM905039 (*Daubentonia madagascariensis*), (17) AY612638 (*Macaca mulatta*), (18) AY863426 (*Cercopithecus aethiops*), (19) AY863427 (*Colobus guereza*), and (20) D38113 (*Pan troglodytes*). Capital letter sequences are anchors with only small or no nucleotide differences among all the sequences. At the bottom of each alignment block, asterisk (*) is given when all the nucleotides were identical, following the CLUSTAL W output format. If one, two, …, up to nine sequences were different from the majority nucleotide, 1, 2, …, up to 9 are given at the bottom line of each alignment block. For example, the majority nucleotide at site 3211 was "a," and five sequences (IDs = 10, 11, 12, 13, and 20) were "g," and "5" is shown at the bottom. When more than 9 sequences showed minority nucleotide, the bottom was just blank, and when a gap represented by "-" exists in any sequence, "=" is shown at the bottom.

If sequences are close enough, MISHIMA can align many complete bacterial genomes. Table 14.4 shows alignment time comparison of various numbers of ~2.8 Mb *Staphylococcus aureus* genomes. Because of massive data sizes, only

Table 14.4 Alignment time (hours) comparison of various numbers of bacterial genomes

Method	No. of sequences compared		
	6	10	14
MISHIMA[a]	7	10	12
MLAGAN	14	30	34

[a]MISHIMA with MAFFT, and no limit of depth option

MLAGAN was compared with MISHIMA. When 6, 10, and 14 genome sequences were compared, MISHIMA required the half or less computation times than MLAGAN. It should be noted that MISHIMA gave always higher alignment scores than MLAGAN [37].

If we are interested in a rough result, anchor sequences alone can be obtained with "align seeds only" option. If this option is activated, even large bacterial genomes may be processed within 1 min. In a reverse situation, when there is no clear sequence homology among sequences, it is desirable to know it as soon as possible. MISHIMA can do that. It took only 15 s to return the answer "alignment impossible" for 100 random sequences of 20-kb length, while CLUSTAL W took more than 8 h to return messy results for the same dataset [37]. We may be able to conclude that MISHIMA is quite suitable for aligning closely related relatively short sequences such as animal mtDNA genome sequences and is also suitable for obtaining a rough sketch of bacterial-size closely related genome sequences.

14.4.6 Problem of Evaluation of Multiple Alignment Results

When only a limited number of sequences were compared in molecular evolutionary studies, it is a routine to visually inspect multiple alignments automatically generated by a software. This was because the human pattern recognition ability far surpasses that of typical computer program. As the multiple alignment softwares were improved and as the size of nucleotide sequences quickly increased, it became difficult to manually check a huge amount of multiple alignments. Another problem of visual inspection is its arbitrariness. There is no clear objective measure for improving multiple alignment results when we rely on each human's pattern recognition.

The real problem resides in this point. If the compared sequences are relatively short and highly homologous, a simple mathematical score may be enough. In fact, a series of such alignment scoring systems exist. For example, the sum of pairs alignment scores based on the pairwise sequence identity [51] was used by Kryukov and Saitou (2010; [37]) for comparing various softwares, and MISHIMA was shown to produce high score alignments.

However, when the compared sequences are quite long and highly divergent, it is not clear if a particular scoring formula is better than another one. One practical way to avoid this problem is to confine our interest to align only highly conserved regions of evolutionary related genomes. This is tantamount to abandon global multiple

Table 14.5 Comparison of major nucleotide sequence multiple alignment softwares

Software	Check list[a]			References
	1	2	3	
Class I: both nucleotide and amino acid sequences can be aligned				
CLUSTAL W	Yes	No[b]	No	[28]
CLUSTAL X	Yes	Yes	No	[46]
DIALIGN	Yes	No	No	[31, 52]
FSA	Yes	Yes	No	[53]
MAFFT	Yes	No	No	[29]
MUSCLE	Yes	No	No	[33]
T-COFFEE	Yes	Yes	No	[49]
Class II: only nucleotide sequences can be aligned				
AVID	Yes[c]	Yes[c]	Yes	[54]
MAUVE	No	Yes	Yes	[35, 36]
MAVID	Yes	Yes	Yes	[34]
MISHIMA	Yes	No	Yes	[37]
MLAGAN	Yes[b]	Yes[c]	Yes	[32]
MURASAKI	No	Yes	Yes	[38]
TBA	No	Yes	Yes	[55]

[a]Checklist details are as follows
1 Web server computation service is available?: Yes/No
2 Graphical user interface is available?: Yes/No
3 Bacterial genome-size alignment is possible within reasonable time?: Yes/No
[b]SEAVIEW [47] provides graphical interface for CLUSTAL W
[c]As a part of VISTA server [65]

alignment, although any global search or optimization is mathematically beautiful and desirable. Multiple alignment of genome sequences is the first step for reconstructing the evolutionary history of genomes in question. We do not have a full list of evolutionary mechanisms that shaped up the genome sequence evolution. A mathematically defined model based on a limited knowledge of genome evolution may not lead us to the correct reconstruction of the past genome evolution. We should therefore be modest.

14.4.7 Comparison of Major Multiple Alignment Softwares

There are many other softwares for multiple alignments of nucleotide sequences and amino acid sequences. Table 14.5 lists some major softwares. Computer programs are downloadable in all cases. These are classified into two classes. Both nucleotide and amino acid sequences can be aligned in class I, but bacterial genome size alignment is not possible within reasonable time. Only nucleotide sequences can be aligned in class II, but they are directed to align very long sequences. The author thanks Dr. Kirill Kryukov for precious information on many multiple alignment softwares.

There are many more multiple alignment softwares not listed in this table, such as CHAOS [56], ABA [57], ProbCons [58], Kalign [59], PRANK [60], PSAlign

[61], and MSAProbs [62]; these are ordered according to the time of publications. Some softwares such as GS-Aligner [63], MCALIGN [64], and MUMMER [65], can align only two or three but very long genome-scale sequences.

14.5 Genome-Wide Homology Viewers

PipMaker [66], developed by Web Miller and his collaborators, is a widely used genome-wide homology viewing web server. This server compares two genome-wide sequences using BLAST, and the BLAST outputs are summarized as pip (percent identity plot). Figure 14.14 shows an example, comparing human and chicken Hox A clusters. These two sequences were also used for dot plot shown in Fig. 14.9. Gene annotations are graphically represented in this figure, but we have to provide annotation information to show them. Ensembl database provides such file.

VISTA and multiVISTA [67] produce the result of homology search visually. Figure 14.15 is an example output of Vista (http://genome.lbl.gov/vista/index.shtml) for the Hox A cluster region of human, chicken, fugu, and zebra fish genomes. Unlike PipMaker, VISTA server prepared their own genome sequence data. Red-colored parts are protein noncoding regions showing high evolutionary conservation.

As we mentioned in Chap. 13, there are three major genome databases: NCBI, UCSC, and Ensembl. A genome viewer provided by UCSC Genome Bioinformatics

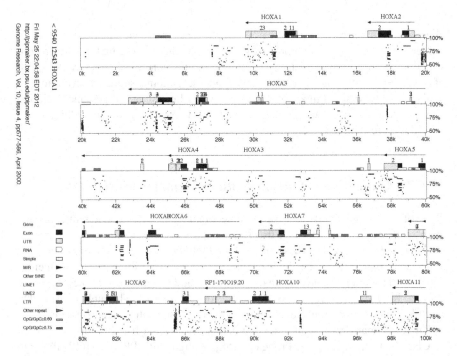

Fig. 14.14 Example of PipMaker output

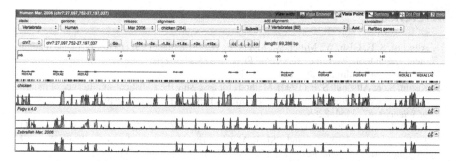

Fig. 14.15 Example of VISTA output

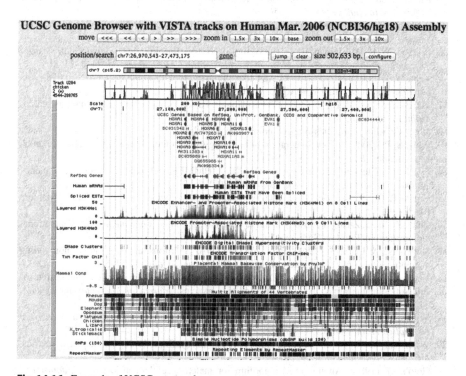

Fig. 14.16 Example of UCSC genome browser output

(http://genome.ucsc.edu) may be most useful for evolutionary genomics studies. Figure 14.16 shows an example of UCSC Genome Browser, linked from VISTA server. The VISTA output, corresponding to Fig. 14.15, is shown immediately below the human genome chromosome-wide view. Bar-code like figures near bottom are highly conserved regions with the human genome for ten vertebrate species, from rhesus macaque to stickleback. Matsunami et al. (2010; [68]) examined vertebrate Hox clusters and found numerous orthologous and paralogous noncoding sequence conservations, as shown in Fig. 8.12. This figure was generated manually.

References

1. Altschul, S. F., Gish, W., Miller, W., Myers, E. W., & Lipman, D. J. (1990). Basic local alignment search tool. *Journal of Molecular Biology, 215*, 403–410.
2. http://www.ncbi.nlm.nih.gov/books/NBK21097/
3. Karlin, S., & Altschul, S. F. (1990). Methods for assessing the statistical significance of molecular sequence features by using general scoring schemes. *Proceedings of the National Academy of Sciences, 87*, 2264–2268.
4. Altschul, S. F., Madden, T. L., Schaffer, A. A., Zhang, J., Zhang, Z., Miller, W., & Lipman, D. J. (1997). Gapped BLAST and PSI-BLAST: A new generation of protein database search programs. *Nucleic Acids Research, 25*, 3389–3402.
5. Zhang, Z., Schwartz, S., Wagner, L., & Miller, W. (2000). A greedy algorithm for aligning DNA sequences. *Journal of Computational Biology, 7*, 203–214.
6. Kitano, T., & Saitou, N. (2000). Evolutionary history of the Rh blood group-related genes in vertebrates. *Immunogenetics, 51*, 856–862.
7. Lipman, D. J., & Pearson, W. R. (1985). Rapid and sensitive protein similarity searches. *Science, 227*, 1435–1441.
8. Pearson, W. R., & Lipman, D. J. (1988). Improved tools for biological sequence comparison. *Proceedings of the National Academy of Sciences of the United States, 85*, 2444–2448.
9. Kent, W. J. (2002). BLAT – the BLAST-like alignment tool. *Genome Research, 12*, 656–664.
10. http://genome.ucsc.edu/FAQ/FAQblat.html
11. Ma, B., Tromp, J., & Li, M. (2002). PatternHunter: Faster and more sensitive homology search. *Bioinformatics, 18*, 440–445.
12. http://www.bioinformaticssolutions.com/all-products/ph
13. Eddy, S. R. (2009). A new generation of homology search tools based on probabilistic inference. *Genome Informatics, 23*, 205–211.
14. Fin, R. D., Clements, J., & Eddy, S. R. (2011). HMMER web server: Interactive sequence similarity searching. *Nucleic Acids Research, 39*, W29–W37.
15. Higgs, P. G., & Atwood, T. K. (2005). *Bioinformatics and molecular evolution*. Malden: Blackwell.
16. Chao, K.-M., & Zhang, L. (2008). *Sequence comparison: Theory and methods* (Computational biology series). London: Springer.
17. Saitou, N., & Ueda, S. (1994). Evolutionary rate of insertions and deletions in non-coding nucleotide sequences of primates. *Molecular Biology and Evolution, 11*, 504–512.
18. Needleman, S. B., & Wunsch, C. D. (1970). A general method applicable to the search for similarities in the amino acid sequence of two proteins. *Journal of Molecular Biology, 48*, 443–453.
19. Sellers, P. H. (1974). On the theory and computation of evolutionary distances. *SIAM Journal on Applied Mathematics, 26*, 787–793.
20. Waterman, M. S., Smith, T. F., & Beyer, W. A. (1976). Some biological sequence metrics. *Advances in Mathematics, 20*, 367–387.
21. Gotoh, O. (1982). An improved algorithm for matching biological sequences. *Journal of Molecular Biology, 162*, 705–708.
22. Altschul, S. F., & Erickson, B. W. (1986). A nonlinear measure of subalignment similarity and its significance levels. *Bulletin of Mathematical Biology, 48*, 603–616.
23. Fitch, W. (1969). Locating gaps in amino acid sequences to optimize the homology between two proteins. *Biochemical Genetics, 3*, 99–108.
24. Schulz, J., Florian Leese, F., & Held, C. (2011). Introduction to dot-plots. Web page available at http://www.code10.info/
25. Kuroki, Y., Toyoda, A., Noguchi, H., Taylor, T. D., Itoh, T., Kim, D. S., Kim, D. W., Choi, S. H., Kim, I. C., Choi, H. H., Kim, Y. S., Satta, Y., Saitou, N., Yamada, T., Morishita, S., Hattori, M., Sakaki, Y., Park, H. S., & Fujiyama, A. (2006). Comparative analysis of chimpanzee and human Y chromosomes unveils complex evolutionary pathway. *Nature Genetics, 38*, 158–167.

26. Murata, M., Richardson, J. S., & Sussman, J. L. (1985). Simultaneous comparison of three protein sequences. *Proceedings of National Academy of Sciences, USA, 82*, 3073–3077.
27. Feng, D.-F., & Doolittle, R. F. (1987). Progressive sequence alignment as a prerequisite to correct phylogenetic trees. *Journal of Molecular Evolution, 25*, 351–360.
28. Thompson, J. D., Higgins, D. G., & Gibson, T. J. (1994). CLUSTAL W: Improving the sensitivity of progressive multiple sequence alignment through sequence weighting, position-specific gap penalties and weight matrix choice. *Nucleic Acids Research, 22*, 4673–4680.
29. Katoh, K., Misawa, K., Kuma, K., & Miyata, T. (2002). MAFFT: a novel method for rapid multiple sequence alignment based on fast Fourier transform. *Nucleic Acids Research, 30*, 3059–3066.
30. Notredame, C. (2007). Recent evolutions of multiple sequence alignment algorithms. *PLoS Computational Biology, 3*, e123.
31. Morgenstern, B., Dress, A., & Werner, T. (1996). Multiple DNA and protein sequence alignment based on segment-to-segment comparison. *Proceedings of National Academy of Sciences, USA, 93*, 12098–12103.
32. Brudno, M., Do, C., Cooper, G., Kim, M. F., Davydov, E., Green, E. D., Sidow, A., & Batzoglou, S. (2003). LAGAN and multi-LAGAN: Efficient tools for large-scale multiple alignment of genomic DNA. *Genome Research, 13*, 721–731.
33. Edgar, R. C. (2004). MUSCLE: Multiple sequence alignment with high accuracy and high throughput. *Nucleic Acids Research, 32*, 1792–1797.
34. Bray, N., & Pachter, L. (2004). MAVID: Constrained ancestral alignment of multiple sequences. *Genome Research, 14*, 693–699.
35. Darling, A. C. E., Mau, B., Blatter, F. R., & Perna, N. T. (2004). Mauve: Multiple alignment of conserved genomic sequence with rearrangements. *Genome Research, 14*, 1394–1403.
36. Darling, A. C. E., Mau, B., & Perna, N. T. (2010). progressiveMauve: Multiple genome alignment with gene gain, loss and rearrangement. *PLoS ONE, 5*, e11147.
37. Kryukov, K., & Saitou, N. (2010). MISHIMA – A new method for high speed multiple alignment of nucleotide sequences of bacterial genome scale data. *BMC Bioinformatics, 11*, 142.
38. Popendorf, K., Tsuyoshi, H., Osana, Y., & Sakakibara, Y. (2010). Murasaki: A fast, parallelizable algorithm to find anchors from multiple genomes. *PLoS ONE, 5*, e12651.
39. Higgins, D. G., & Sharp, P. (1988). CLUSTAL: A package for performing multiple sequence alignment on a microcomputer. *Gene, 73*, 237–244.
40. Saitou, N., & Nei, M. (1987). The neighbor-joining method: A new method for reconstructing phylogenetic trees. *Molecular Biology and Evolution, 4*, 406–425.
41. Kimura, M. (1980). A simple method for estimating evolutionary rates of base substitutions through comparative studies of nucleotide sequences. *Journal of Molecular Evolution, 16*, 111–120.
42. Kimura, M. (1983). *The neutral theory of molecular evolution.* Cambridge: Cambridge University Press.
43. Higgins, D. G., Bleasby, A. J., & Fuchs, R. (1992). CLUSTAL V: Improved software for multiple sequence alignment. *Computational Applied Biosciences, 8*, 189–191.
44. Wilbur, W. J., & Lipman, D. (1984). The context dependent comparison of biological sequences. *SIAM Journal of Applied Mathematics, 44*, 557–567.
45. Myers, E. W., & Miller, W. (1988). Optimal alignments in linear space. *CABIOS, 4*, 11–17.
46. Larkin, M. A., Blackshields, G., Brown, N. P., et al. (13 co-authors) (2007) Clustal W and Clustal X version 2.0. *Bioinformatics, 23*, 2947–2948.
47. Sievers, F., Wilm, A., Dineen, D., Gibson, T. J., Karplus, K., Li, W., Lopez, R., McWilliam, H., Remmert, M., Söding, J., Thompson, J. D., & Higgins, D. G. (2011). Fast, scalable generation of high-quality protein multiple sequence alignments using Clustal Omega. *Molecular Systems Biology, 7*, 539.
48. Felsenstein, J., Sawyer, S., & Kochin, R. (1982). An efficient method for matching nucleotide acid sequences. *Nucleic Acids Research, 10*, 133–139.
49. Notredame, C., Higgins, D. G., & Heringa, J. (2000). T-Coffee: A novel method for fast and accurate multiple sequence alignment. *Journal of Molecular Biology, 302*, 205–217.

50. Galtier, N., Gouy, M., & Gautier, C. (1996). SEA VIEW and PHYLO_WIN: Two graphic tools for sequence alignment and molecular phylogeny. *Computer Applications in the Biosciences, 12*, 543–548.

51. Lipman, D. J., Altschul, S. F., & Kececioglu, J. D. (1989). A tool for multiple sequence alignment. *Proceedings of the National Academy of Sciences of the United States of America, 86*, 4412–4415.

52. Subramanian, A. R., Kaufmann, M., & Morgenstern, B. (2008). DIALIGN-TX: Greedy and progressive approaches for segment-based multiple sequence alignment. *Algorithms for Molecular Biology, 3*, 6.

53. Bradley, R. K., Roberts, A., Smoot, M., Juvekar, S., Do, J., Dewey, C., Holmes, I., & Pachter, L. (2009). Fast statistical alignment. *PLoS Computational Biology, 5*, e1000392.

54. Bray, N., Dubchak, I., & Pachter, L. (2003). AVID: A global alignment program. *Genome Research, 13*, 97–102.

55. Blanchette, M., Kent, W. J., Riemer, C., Elnitski, L., Smit, A. F. A., Roskin, K. M., Baertsch, R., Rosenbloom, K., Clawson, H., Green, E. D., Haussler, D., & Miller, W. (2004). Aligning multiple genomic sequences with the threaded blockset aligner. *Genome Research, 14*, 708–715.

56. Brudno, M., Chapman, M., Gottgens, B., Batzoglou, S., & Morgenstern, B. (2003). Fast and sensitive multiple alignment of long genomic sequences. *BMC Bioinformatics, 4*, 66.

57. Raphael, B., Zhi, D., Tang, H., & Pevzner, P. (2004). A novel method for multiple alignment of sequences with repeated and shuffled elements. *Genome Research, 14*, 2336–2346.

58. Do, C. B., Mahabhashyam, M. S. P., Brudno, M., & Batzoglou, S. (2005). ProbCons: Probabilistic consistency-based multiple sequence alignment. *Genome Research, 15*, 330–340.

59. Lassmann, T., & Sonnhammer, E. L. L. (2005). Kalign—An accurate and fast multiple sequence alignment algorithm. *BMC Bioinformatics, 6*, 298.

60. Lotynoja, A., & Goldman, N. (2005). An algorithm for progressive multiple alignment of sequences with insertions. *Proceedings of the National Academy of Sciences of the United States of America, 102*, 10557–10562.

61. Sze, S.-H., Lu, Y., & Yang, Q. (2006). A polynomial time solvable formulation of multiple sequence alignment. *Journal of Computational Biology, 13*, 309–319.

62. Liu, Y., Schmidt, B., & Maskell, D. L. (2010). MSAProbs: Multiple sequence alignment based on pair hidden Markov models and partition function posterior probabilities. *Bioinformatics, 26*, 1958–1964.

63. Shih, A. C.-C., & Li, W.-H. (2003). GS-Aligner: A novel tool for aligning genomic sequences using bit-level operations. *Molecular Biology and Evolution, 20*, 1299–1309.

64. Keightley, P. D., & Johnson, T. (2004). MCALIGN: Stochastic alignment of noncoding DNA sequences based on an evolutionary model of sequence evolution. *Genome Research, 14*, 442–450.

65. Kurtz, S., Phillippy, A., Delcher, A. L., Smoot, M., Shumway, M., Antonescu, C., & Salzberg, S. L. (2004). Versatile and open software for comparing large genomes. *Genome Biology, 5*, R12.

66. Schwartz, S., Zhang, Z., Frazer, K. A., Smit, A., Riemer, C., Bouck, J., Gibbs, R., Hardison, R., & Miller, W. (2000). PipMaker—A web server for aligning two genomic DNA sequences. *Genome Research, 10*, 577–586.

67. http://genome.lbl.gov/vista/index.shtml

68. Matsunami, M., Sumiyama, K., & Saitou, N. (2010). Evolution of conserved non-coding sequences within the vertebrate Hox clusters through the two-round whole genome duplications revealed by phylogenetic footprinting analysis. *Journal of Molecular Evolution, 71*, 427–436.

Evolutionary Distances

15

Chapter Summary

Definitions and estimation of various types of evolutionary distances are discussed in this chapter, from nucleotide sequences to genomes. Distances based on nucleotide and amino acid substitutions are explained in more detail.

15.1 Overview of Evolutionary Distances

Nucleotide sequences accumulate various types of changes through evolution, and these are measured as evolutionary distances. Because there are a variety of them, we would like to give a general overview in this section.

Mathematically, "distance," $D[A, B]$, is defined as a nonnegative value to represent a certain relationship between two objects A and B. There is no directionality in any distances, namely,

$$D[i,j] = D[j,i] \tag{15.1}$$

for any objects i and j. We also note that

$$D[i, i] = 0 \tag{15.2}$$

for any object i.

If there are more than two objects, all possible pairwise distances may be represented in a matrix form. Examples for five objects are as follows:

$$
\begin{vmatrix}
D_{1,1} & D_{1,2} & D_{1,3} & D_{1,4} & D_{1,5} \\
D_{2,1} & D_{2,2} & D_{2,3} & D_{2,4} & D_{2,5} \\
D_{3,1} & D_{3,2} & D_{3,3} & D_{3,4} & D_{3,5} \\
D_{4,1} & D_{4,2} & D_{4,3} & D_{4,4} & D_{4,5} \\
D_{5,1} & D_{5,2} & D_{5,3} & D_{5,4} & D_{5,5}
\end{vmatrix}
\tag{15.3}
$$

N. Saitou, *Introduction to Evolutionary Genomics*, Computational Biology 17, 335
DOI 10.1007/978-1-4471-5304-7_15, © Springer-Verlag London 2013

Table 15.1 Example of a distance matrix (data from Table 3 of Ishida et al. 1995; [1])

	1	2	3	4	5	6	7
1	0	9	11	6	42	38	35
2	9	0	6	5	45	41	38
3	11	6	0	7	47	43	40
4	6	5	7	0	42	38	35
5	42	45	47	5	0	46	43
6	38	41	43	5	46	0	29
7	35	38	40	5	43	29	0

1: Thoroughbred horse (*Equus caballus*), 2: Przewalskii's wild horse (*E. caballus*), 3: Mongolian native horse (*E. caballus*), 4: Japanese native horse (*E. caballus*), 5: mountain zebra (*E. zebra*), 6: donkey (*E. asinus*), and 7: Grevy's zebra (*E. grevyi*)

Table 15.2 Example of a distance matrix showing only lower-triangle values (data from Table 3 of Ishida et al. 1995; [1])

	1	2	3	4	5	6
2	8.0					
3	11.7	5.4				
4	6.5	5.6	5.4			
5	41.8	49.0	46.5	43.7		
6	35.1	41.3	45.8	35.5	47.9	
7	36.6	34.8	34.8	38.2	41.5	29.1

1–7: Same as those of Table 15.1

$D[i, j]$ represents a particular nonnegative value, and we may give row and column numbers of the matrix as shown in Table 15.1 (data from [1]). Because of the characteristics of distances shown in Eqs. 15.1 and 15.2, we usually show only lower-triangle or upper-triangle values, as shown in Table 15.2 (data from [1]). Although this is no longer a typical matrix with n rows and m columns, we call them "distance matrix" in evolutionary studies. Values in Table 15.1 are all integers, while those in Table 15.2 are non-integer values. If DNA or amino acid sequences are considered, various kinds of mutations are contributing to evolutionary distances, and all mutations are discrete events. Therefore, any evolutionary distances based on mutations should hold ideally integer values. Ishida et al. (1995; [1]) stressed this nature of evolutionary distances, and they estimated the "patristic" distances between a pair of sequences by considering their locations on the estimated phylogenetic tree. In most of the molecular evolutionary studies, however, we use the number of nucleotide or amino acid substitutions per site. In this case, values are non-integer and often less than one. Values in Table 15.2 were estimated numbers of nucleotide substitutions multiplied by the number of nucleotide sites compared. This is why values shown in Tables 15.1 and 15.2 are more or less similar.

In evolutionary genomics, we consider objects such as nucleotide sequences, amino acid sequences, populations, species, or genomes. These objects are generically called "operational taxonomic units" (OTUs) in numerical taxonomy [2].

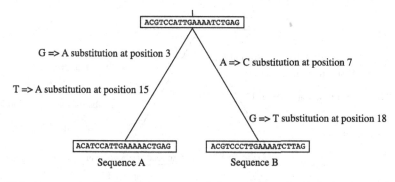

Fig. 15.1 An example of nucleotide sequence evolution only through substitutions

15.2 Nucleotide Substitutions

The number of nucleotide substitutions is the most frequently used distance measure in evolutionary genomics. We often gauge species divergence by genomic average of nucleotide substitutions, for example, 1.23 % for human and chimpanzee [3] and 16–17 % for mouse and rat [4]. We discuss various methods for estimating nucleotide substitutions in this section.

15.2.1 Nucleotide Difference and Nucleotide Substitution

When we compare two homologous nucleotide sequences, we should first align them, as shown in Fig. 14.7a. There are 2 and 1 mismatched and gapped sites, respectively, in that alignment. If we note that the mutational mechanisms (see Chap. 2) and the pattern of natural selection (see Chap. 4) for nucleotide substitutions and insertions/deletions are different, it may be suitable to consider mismatches and gaps independently. Let us consider an example of nucleotide sequence evolution only through substitutions, as shown in Fig. 15.1. Two nucleotide substitutions occurred in the two lineages, and we obtain the alignment below:

$$Sequence \ A : ACATCCATTGAAAAACTGAG$$
$$Sequence \ B : ACGTCCCTTGAAAATCTTAG \qquad (15.4)$$
$$* \qquad * \qquad \qquad * \quad *$$

There are four mismatches or "nucleotide difference" out of 20 nucleotides between sequences A and B. Thus, the proportion of nucleotide difference is 0.2 (=4/20). This value, p, is called "p distance" in molecular phylogenetics, and its range is $0 \leq p \leq 1$. It should be noted that the four nucleotide differences between sequences A and B in this example correspond to four nucleotide substitutions in Fig. 15.1. Because of this nature, often the "nucleotide difference" is called

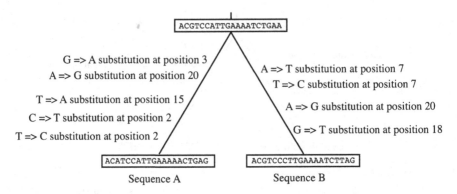

Fig. 15.2 Another example of nucleotide sequence evolution only through substitutions

"nucleotide substitution." It is true that at least m substitutions are required when we observed m differences out of n ($\geq m$) nucleotide sites compared.

However, the real number of nucleotide substitutions may be larger than the nucleotide difference. Figure 15.2 shows another example of nucleotide sequence evolution only through substitutions. Nine nucleotide substitutions occurred, but the current sequences are identical with those of Fig. 15.1, and we obtain the same alignment shown in alignment 15.4. We now have a large discrepancy between the number of nucleotide difference, Nuc_{dif}, and the number of nucleotide substitutions, Nuc_{sub}:

$$\text{Nuc}_{\text{dif}} = 4 \tag{15.5a}$$

$$\text{Nuc}_{\text{sub}} = 9 \tag{15.5b}$$

Therefore, nucleotide differences should be clearly distinguished from nucleotide substitutions. If we examine the nine substitutions that occurred in the two lineages going to sequences A and B in Fig. 15.2, they can be classified into the following four categories:

Category 1: single substitution (positions 3, 15, and 18)
Category 2: one substitution followed by a backward or reverse substitution (position 2)
Category 3: two successive substitutions (position 7)
Category 4: parallel substitutions (position 20)

Substitutions belonging to category 1 contribute to nucleotide difference or mismatch, while those belonging to categories 2 and 4 cancel out, and there is no mismatch. Substitutions in category 3 contribute to single nucleotide difference at that position even if more than one substitution occurred. It is now clear that $\text{Nuc}_{\text{sub}} \geq \text{Nuc}_{\text{dif}}$ and their difference may be large when there are many parallel, reverse, and successive substitutions. It is therefore desirable to estimate Nuc_{sub} from Nuc_{dif}.

15.2.2 Nucleotide Substitution Matrix

We need to introduce a nucleotide substitution matrix **S** for estimating the number of nucleotide substitutions, Nuc_{sub}, from the observed value, Nuc_{dif}:

$$\mathbf{S} = \begin{vmatrix} S_{A,A} & S_{A,C} & S_{A,T} & S_{A,G} \\ S_{C,A} & S_{C,C} & S_{C,T} & S_{C,G} \\ S_{T,A} & S_{T,C} & S_{T,T} & S_{T,G} \\ S_{G,A} & S_{G,C} & S_{G,T} & S_{G,G} \end{vmatrix} \tag{15.6}$$

The diagonal elements $S[i, i]$ (i = A, C, G, or T) are either omitted, or set to have the unity for the row sum, or set to be negative for mathematical requirement. However, essential elements are non-diagonal ones. Let us explain their differences. When we consider the change during an infinitesimally small time, Δt,

$$\text{Freq}_i (t + \Delta t) = \left(1 - \Delta t \cdot \sum_j S[i,j] \right) \cdot \text{Freq}_i (t) + \Delta t \cdot \sum_j S[i,j] \, \text{Freq}_j (t), \tag{15.7}$$

where $\text{Freq}_\alpha(\tau)$ is the frequency of nucleotide α (α = A, C, G, or T) at time τ, and $S[i, j]$ is the transition probability from nucleotide i to j ($i \neq j$). $S[i, i] = 1 - \Sigma_j S[i, j]$. From Eq. 15.7,

$$\frac{\{\text{Freq}_i (t + \Delta t) - \text{Freq}_i (t)\}}{\Delta t} = -\sum_j S[i,j] \cdot \text{Freq}_i (t) + \sum_j S[i,j] \, \text{Freq}_j (t). \tag{15.8}$$

By taking limit ($\Delta t \to 0$), we obtain a differential equation. If we combine the four differential equations for all four nucleotides,

$$\frac{d}{dt} \{\text{Freq}(t)\} = \mathbf{M} \cdot \{\text{Freq}(t)\}, \tag{15.9}$$

where $\{\text{Freq}(t)\}$ is a vector $\{\text{Freq}_A(t), \text{Freq}_C(t), \text{Freq}_G(t), \text{Freq}_T(t)\}$ and **M** is a matrix as follows:

$$\mathbf{M} = \begin{vmatrix} SS_A & S_{A,C} & S_{A,T} & S_{A,G} \\ S_{C,A} & SS_C & S_{C,T} & S_{C,G} \\ S_{T,A} & S_{T,C} & SS_T & S_{T,G} \\ S_{G,A} & S_{G,C} & S_{G,T} & SS_G \end{vmatrix}, \tag{15.10}$$

where $SS[i] = -\Sigma_j S[i, j]$. The differential equation (15.9) can be solved by obtaining the eigenvalue and eigenmatrix of the matrix **M**. Mathematically, matrix M given in Eq. (15.10) may be important, but biologically matrix S given in Eq. (15.6) is called "nucleotide substitution matrix". The author thanks Dr. Yosuke Kawai for his help in writing the content of this section.

Table 15.3 One-parameter and two-parameter models of nucleotide substitution matrix

(A) One-parameter model					
		NEW			
		A	C	T	G
O	A	$1-3\alpha$	α	α	α
L	C	α	$1-3\alpha$	α	α
D	T	α	α	$1-3\alpha$	α
	G	α	α	α	$1-3\alpha$
(B) Two-parameter model					
		NEW			
		A	C	T	G
O	A	$1-\alpha-2\beta$	β	β	α
L	C	β	$1-\alpha-2\beta$	α	β
D	T	β	α	$1-\alpha-2\beta$	β
	G	α	β	β	$1-\alpha-2\beta$

15.2.3 One-Parameter Method

The simplest model for nucleotide substitution matrix is to assume that all 12 types of substitutions occur with the same rate, α (see Table 15.3A). Because only one parameter, α, is used in this model, this is called the one-parameter model, and the method for estimating the number of nucleotide substitutions based on this model is called the one-parameter method. Jukes and Cantor (1969; [5]) first used this method, and it is also called Jukes and Cantor's method.

Let us consider the evolution of one particular nucleotide site from an ancestral nucleotide N_{anc} to a descendant nucleotide N_{des} during time T. We define p_t and q_t as the probabilities of nucleotide difference and identity, respectively, between N_{anc} and the nucleotide at time t ($0 \le t \le T$). Obviously, for any t,

$$p_t + q_t = 1. \tag{15.11}$$

For simplicity and for reality, let us assume N_{anc} was A (adenine). During time t, A may change to either C, G, or T; otherwise, it may remain as A. The first three situations contribute to p_t and the other to q_t. We then consider the situation at time $t+1$ or the unit time after time t. When the nucleotide at time t is still A as N_{anc}, that nucleotide may change to either C, G, or T with the equal rate α, and this event will contribute to the probability of nucleotide difference at time $t+1$. The remaining event, with rate $1-3\alpha$, will contribute to the probability of nucleotide identity at time $t+1$.

When the nucleotide at time t is different from A as N_{anc}, say G, it may further change to either A, C, or T with the equal rate α. Now we have to consider two possible situations: change to A or changes to C or T. If nucleotide G at time t changes to A at time $t+1$, this is backward or reverse substitution, while change to either C or T is successive substitution. The former and latter situations have probabilities α and 2α, respectively.

We thus have five different situations summarized as follows:

Total probability $= q_t$

A $\Longrightarrow$ A $\Rightarrow$ A with probability $1 - 3\alpha$; contribute to q_{t+1}

A $\Longrightarrow$ A $\Rightarrow$ C, G, or T with probability 3α ; contribute to p_{t+1}

Total probability $= p_t$

A $\Longrightarrow$ G $\Rightarrow$ G with probability $1 - 3\alpha$; contribute to p_{t+1}

A $\Longrightarrow$ G $\Rightarrow$ A with probability α ; contribute to q_{t+1}

A $\Longrightarrow$ G $\Rightarrow$ C or T with probability 2α ; contribute to p_{t+1}

By summing all these five situations, we obtain the following recurrent equations:

$$p_{t+1} = 3\alpha q_t + (1-\alpha) p_t \tag{15.12a}$$

$$q_{t+1} = (1-3\alpha) q_t + \alpha p_t. \tag{15.12b}$$

If we note Eq. 15.11,

$$\begin{aligned} q_{t+1} &= (1-3\alpha) q_t + \alpha (1-q_t) \\ &= \alpha + (1-4\alpha) q_t. \end{aligned} \tag{15.13}$$

We thus have

$$q_{t+1} - q_t = \alpha (1-4q_t). \tag{15.14}$$

As the length of unit time is diminished, $q_{t+1} - q_t$ can be equated as the differential, dq/dt, where q is the probability of the nucleotide identity. We thus have the following differential equation:

$$\frac{dq}{dt} = \alpha - 4\alpha q, \tag{15.15}$$

with initial condition as $q = 1$ at $t = 0$. Its solution is given as

$$q = \frac{1}{4} + \frac{3}{4} \exp[-4\alpha t]. \tag{15.16}$$

The value of q starts from 1 at $t = 0$, then monotonically decreases. At the extreme when $t = \infty$, q becomes ¼. The possible range of q is $0 \leq q \leq 1$; however, N_{des} will be either one of four nucleotides with the equal frequency ¼ irrespective of N_{anc}. Therefore, by chance $N_{anc} = N_{des}$ with the probability ¼.

The rate of nucleotide substitution is traditionally written as λ. Under the one-parameter model, this rate is the sum of all three possible changes. Thus,

$$\lambda = 3\alpha. \tag{15.17}$$

We often assume two present-day sequences when we discuss the evolutionary distances. Figure 15.3 shows an evolutionary scheme between two present-day

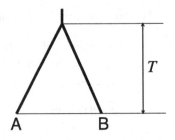

Fig. 15.3 An evolutionary scheme between two present-day sequences

sequences. As both sequences A and B descended from the ancestral sequence after time T, the total time between those two sequences is $2T$. We thus have

$$q = \frac{1}{4} + \frac{3}{4}\exp\left[\frac{-8\lambda T}{3}\right].$$

(15.18)

If we consider λ as the average rate of nucleotide substitution in this time range, the expected number of nucleotide substitution, or the evolutionary distance based on nucleotide substitution, is given by

$$d = 2\lambda T.$$

(15.19)

Putting the relationship shown in Eq. 15.19 to Eq. 15.18 and noting $p = 1 - q$, we have

$$p = \frac{3}{4} - \frac{3}{4}\exp\left[\frac{-4d}{3}\right].$$

(15.20)

It should be noted that $0 \leq p \leq \frac{3}{4}$. Taking logarithms, we finally obtain Jukes–Cantor's [5] formula:

$$d = -\frac{3}{4}\ln\left[1 - \left(\frac{4p}{3}\right)\right].$$

(15.21)

Assuming that p is following a binomial distribution, the standard error of d, SE(d), is given as

$$\text{SE}(d) = \left[\frac{3}{(3-4p)}\right]\sqrt{\left[\frac{p(1-p)}{n}\right]},$$

(15.22)

where n is the number of nucleotide sites compared. This formula was first given by Kimura and Ohta (1972; [6]). Table 15.4 shows some example values of n, p, and d under the one-parameter method. Standard errors (SEs) for p distances were computed assuming the binomial distribution (Nei, 1987; [7]):

$$\text{SE}(p) = \sqrt{\left[\frac{p(1-p)}{n}\right]},$$

(15.23)

When $p = 0.01$, d is almost identical with p, and d becomes only 4 % larger than p when $p = 0.05$. Because $\ln(1+x)$ is approximated as x when $x \ll 1$,

Table 15.4 Some example values of n, p, and d under the one-parameter method

n	p	SE of p	d	SE of d
1,000	0.01	0.003	0.010	0.003
10,000	0.01	0.001	0.010	0.001
1,000	0.05	0.007	0.052	0.007
10,000	0.05	0.002	0.052	0.002
1,000	0.10	0.009	0.107	0.011
10,000	0.10	0.003	0.107	0.003
1,000	0.20	0.013	0.233	0.017
10,000	0.20	0.004	0.233	0.005
1,000	0.40	0.015	0.572	0.033
10,000	0.40	0.005	0.572	0.011
1,000	0.60	0.015	1.207	0.077
10,000	0.60	0.005	1.207	0.024

$$\ln\left[1-\left(\frac{4p}{3}\right)\right] \sim -\left(\frac{4p}{3}\right), \tag{15.24}$$

when $4p/3 \ll 1$ in Eq. 15.17, then $d \sim p$. Therefore, the p distance is a good approximation for d, the number of nucleotide substitutions, when p is small, say less than 0.05.

When the amount of divergence is large and when the compared number of nucleotides is small, the estimates given by using the one-parameter method are not reliable because of huge standard error, as shown in Table 15.4.

Figure 15.4 shows the relationship between p and d. When p, the nucleotide difference, is zero, d is zero, while d approaches ∞ as p approaches ¾, its maximum. It should be noted that the equilibrium frequencies of four nucleotides are all ¼. We thus expect to have the 50 % GC content under the equilibrium. Because many observed nucleotide sequences have GC contents considerably different from 50 %, the one-parameter method is not suitable for estimating the number of substitutions when the amount of divergence is quite large between the two sequences.

Saitou (1990; [8]) showed that the estimate of the nucleotide substitution given by Eq. 15.21 under the one-parameter method is the maximum likelihood estimate. Let us show this. We first consider the probability, $P(i, j, d)$, that two nucleotides happen to be i and j (i and j can be A, C, G, or T) at a particular site of two homologous sequences with the evolutionary distance (nucleotide substitution) d. If $i \neq j$, from Eq. 15.16,

$$P(i,j,d) = U_1 = \frac{3}{4} - \frac{3}{4}\exp\left[\frac{-4d}{3}\right], \tag{15.25}$$

and if $i=j$,

$$P(i,j,d) = U_0 = 1 - U_1 = \frac{1}{4} + \frac{3}{4}\exp\left[\frac{-4d}{3}\right]. \tag{15.26}$$

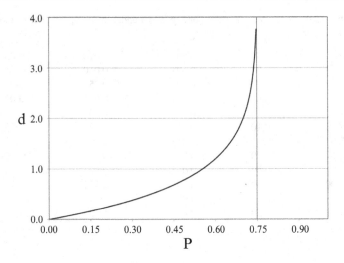

Fig. 15.4 Relationship between p and d under the one-parameter method

U_1 and U_0 are probabilities of configurations 1 (two nucleotides are different) and 0 (two nucleotides are identical) for two sequences. Let us assume that we observed m_0 and m_1 sites for configurations 0 and 1, respectively. The log-likelihood for this observation under the one-parameter model is

$$\ln\left[L\left(m_0,m_1\right)\right] = m_0\,\ln\left[U_0\right] + m_1\,\ln\left[U_1\right] + \ln\left[\frac{\{m_0+m_1\}!}{\{m_0!m_1!\}}\right]$$

$$= m_0\,\ln\left[\frac{1}{4}+\frac{3}{4}x\right] + m_1\,\ln\left[\frac{3}{4}\quad\frac{3}{4}x\right] + \text{constant},$$

(15.27)

where $x = \exp[-4d/3]$. We differentiate log-likelihood in Eq. 15.27 in terms of x and equate it with zero to obtain the maximum likelihood solution. Therefore,

$$\frac{3m_0}{\left(1+3x\right)} - \frac{3m_1}{\left(3-3x\right)} = 0.$$

(15.28)

The solution of Eq. 15.24 becomes

$$\hat{x} = 1 - \frac{4m_1}{3\left(m_0+m_1\right)}$$

(15.29a)

$$= 1 - \frac{4p}{3},$$

(15.29b)

where p is the proportion of nucleotide difference. We thus have

$$\exp\left[\frac{-4d}{3}\right] = 1 - \frac{4p}{3}. \tag{15.30}$$

By taking logarithms of both sides of Eq. 15.30, we obtain Eq. 15.21.

15.2.4 Two-Parameter Method

The assumption of the one-parameter model is too simplistic, because transitions are often more frequent than transversions (see Chap. 2). Kimura (1980; [9]) proposed the two-parameter model shown in Table 15.3B by considering this situation. The number of nucleotide substitutions under this model is the same with Eq. 15.15, with the rate of substitution as

$$\lambda = \alpha + 2\beta, \tag{15.31}$$

from the transition matrix of this model (Table 15.3B). We thus have

$$d = 2(\alpha + 2\beta)T. \tag{15.32}$$

We divided nucleotide sites into identical and different categories for the one-parameter method. Three categories are now necessary for the two-parameter method for each site:

Category 1: two sequences are different with transitional type (frequency $= P$).
Category 2: two sequences are different with transversional type (frequency $= Q$).
Category 3: two nucleotides are identical (frequency $= R = 1 - (P + Q)$).

There are two possible nucleotide pairs in category 1, [A$\Leftrightarrow$G] and [C$\Leftrightarrow$T], while four pairs are possible for category 2, [A$\Leftrightarrow$C], [A$\Leftrightarrow$T], [C$\Leftrightarrow$G], and [G$\Leftrightarrow$T]. It is obvious that there are four pairs for category 3: [A$\Leftrightarrow$A], [C$\Leftrightarrow$C], [G$\Leftrightarrow$G], and [T$\Leftrightarrow$T]. In a similar way as we discussed under the one-parameter model, we consider temporal changes of these three categories. After considering all possible changes, we obtain the following differential equations in a similar manner for the one-parameter method [9]:

$$\frac{dP_t}{dt} = 2\alpha - 4(\alpha + \beta)P_t - 2(\alpha - \beta)Q_t, \tag{15.33a}$$

$$\frac{dQ_t}{dt} = 4\beta - 8\beta Q_t, \tag{15.33b}$$

where P_t and Q_t are the probabilities of transitional and transversional differences at time t, respectively. The solution of this set of equations which satisfies the initial condition

$$P_0 = Q_0 = 0, \tag{15.34}$$

that is, no nucleotide difference when $t=0$, is as follows [9]:

$$P_t = \frac{1}{4} - \frac{1}{2}\exp\left[-4(\alpha+\beta)t\right] + \frac{1}{4}\exp\left[-8\beta t\right] \qquad (15.35a)$$

$$Q_t = \frac{1}{2} - \frac{1}{4}\exp\left[-8\beta t\right]. \qquad (15.35b)$$

Because the number of transitional and transversional substitutions is

$$d_{transition} = 2\alpha t \qquad (15.36a)$$

$$d_{transversion} = 4\beta t, \qquad (15.36b)$$

from Eq. 15.28, respectively,

$$d_{transition} = -\frac{1}{2}\ln\left[\frac{(1-2P-Q)}{\sqrt{(1-2Q)}}\right] \qquad (15.37a)$$

$$d_{transversion} = -\frac{1}{2}\ln(1-2Q). \qquad (15.37b)$$

Suffix t was dropped in the above equations for simplicity. The number of nucleotide substitution per site, d, is thus estimated by using the following equation [9]:

$$d = d_{transition} + d_{transversion}$$

$$= -\frac{1}{2}\ln\left\{(1-2P-Q)\sqrt{(1-2Q)}\right\} \qquad (15.38a)$$

$$= -\frac{1}{2}\ln(1-2P-Q) - \frac{1}{4}\ln(1-2Q) \qquad (15.38b)$$

The temporal changes of P and Q are shown in Fig. 15.5, with $\alpha=6.92\times10^{-8}/$site/year and $\beta=0.13\times10^{-8}/$site/year. The overall rate of substitution, λ, becomes $7.2\times10^{-8}/$site/year from Eq. 15.31 under this situation. This approximates the case for the evolution of human mitochondrial DNA [10]. It should be noted that the equilibrium frequencies of four nucleotides are all ¼ as for the one-parameter model.

The standard error for d is given by

$$SE(d) = \sqrt{\left\{\frac{\left[(a^2P + b^2Q) - (aP + bQ)^2\right]}{n}\right\}}, \qquad (15.39)$$

where

$$a = \frac{1}{(1-2P-Q)}, \qquad (15.40a)$$

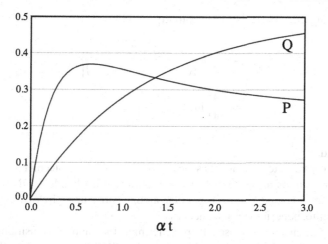

Fig. 15.5 Temporal changes of P and Q under the two-parameter model with $\alpha = 10\beta$

$$b = \frac{1}{2}\left[\frac{1}{(1-2P-Q)} + \frac{1}{(1-2Q)}\right], \tag{15.40b}$$

and n is the number of nucleotide sites compared [9].

If the transitional rate (α) and the transversional rate (β) are the same, the two-parameter model reduces to the one-parameter model, and Eq. 15.38 should reduce to Eq. 15.21. Let us show this based on Kimura (1980; [9]). Because $\alpha = \beta$ in Eqs. 15.35a and 15.35b,

$$P_t = \frac{1}{4} - \frac{1}{4}\exp\left[-8\alpha t\right] \tag{15.41a}$$

$$Q_t = \frac{1}{2} - \frac{1}{2}\exp\left[-8\alpha t\right]. \tag{15.41b}$$

Therefore, dropping suffix t,

$$P = \frac{1}{2}Q. \tag{15.42}$$

Because the proportion of nucleotide difference $p = P + Q$,

$$P = \frac{p}{3} \tag{15.43a}$$

$$Q = \frac{2p}{3}. \tag{15.43b}$$

By putting these two relationships to Eq. 15.38b,

$$d = -\frac{1}{2}\ln\left\{\left(\frac{1-4p}{3}\right)\sqrt{\left(\frac{1-4p}{3}\right)}\right\} \qquad (15.44a)$$

$$= -\frac{3}{4}\left[\ln\left(1-\left(\frac{4p}{3}\right)\right)\right], \qquad (15.44b)$$

as expected.

The two-parameter method has been widely used in molecular phylogenetic studies. Transitions are especially high in animal mitochondrial DNAs, and the two-parameter method has been popular among molecular phylogeneticists who often use mitochondrial DNA sequences.

Joseph Felsenstein proposed to predetermine the ratio of transitions (α) and transversions (β) for the two-parameter model in his PHYLIP package (http://evolution. genetics.washington.edu/phylip.html). Although this modified method is called "two-parameter method" [11], this is different from the original two-parameter method proposed by Kimura [9].

Saitou (1990; [8]) showed that the estimate of the nucleotide substitution given by Eq. 15.38 under the two-parameter method is the maximum likelihood estimate. We have to consider probabilities to obtain three kinds of nucleotide configurations: U_0, U_1, and U_2, corresponding to no difference, transitional difference, and transversional difference. From Eqs. 15.35a and 15.35b,

$$U_1 = \frac{1}{4} - \frac{1}{2}ab + \frac{1}{4}b^2, \qquad (15.45a)$$

$$U_2 = \frac{1}{2} - \frac{1}{2}b^2, \qquad (15.45b)$$

$$U_0 = 1 - (U_1 + U_2) = \frac{1}{4} + \frac{1}{2}ab + \frac{1}{4}b^2, \qquad (15.45c)$$

where $a = \exp[-4\alpha t]$ and $b = \exp[-4\beta t]$. The log-likelihood for this observation under the two-parameter model then becomes

$$\ln[L(Pn,Qn,Rn)] = Pn\ln[U_1] + Qn\ln[U_2] + Rn\ln[U_0] + \text{constant}, \qquad (15.46)$$

where P, Q, and R are frequencies of transitional difference, transversional difference, and no difference, and n is the number of nucleotides compared between the two sequences. Pn, Qn, and Rn are thus observed numbers of three configurations with probabilities U_1, U_2, and U_0, respectively. We differentiate the log-likelihood in Eq. 15.46 in terms of a and b and equate them with zero and obtain the solutions below:

$$\hat{a} = \frac{(1 - 2P - Q)}{\hat{b}} \qquad (15.47a)$$

Table 15.5 Pattern of nucleotide substitutions for the human mitochondrial genome (from [10])

		NEW			
		A	C	T	G
O	A	–	0.0031	0.0030	0.0901
L	C	0.0070	–	0.0593	0.0028
D	T	0.0042	0.2566	–	0.0048
	G	0.5574	0.0080	0.0037	–

$$\hat{b} = (1 - 2Q)^{\frac{1}{2}} \tag{15.47b}$$

Thus, considering Eq. 15.32, the number of nucleotide substitutions becomes

$$
\begin{aligned}
d &= 2\alpha t + 4\beta t \\
&= -\frac{1}{2}\ln\left(ab^2\right) \\
&= -\frac{1}{2}\ln\left[(1 - 2P - Q)(1 - 2Q)^{\frac{1}{2}}\right]
\end{aligned}
\tag{15.48}
$$

Equation 15.48 is equivalent with Eq. 15.38.

One may think that there is only one two-parameter model, distinguishing transitions and transversions. There are, however, 2047 $(= 2^{11} - 1)$ possible ways to divide 12 directions of nucleotide substitutions. If we find some organism which shows an enigmatic nucleotide substitution patterns, an alternative two-parameter model may be necessary. In fact, the nucleotide substitution pattern of the human mitochondrial DNA may be of this kind. Kawai, Kikuchi, and Saitou (2012; [10]) estimated that pattern from 7,264 complete sequences, as shown in Table 15.5. The 12 values are relative frequencies of nucleotide substitutions, and they sum up to the unity. It is true that four transitional substitutions are much higher than eight transversional ones; however, $G \Rightarrow A$ and $T \Rightarrow C$ transitions outnumber the remaining two, and only these two types occupy more than 80 % of the total nucleotide substitutions. Therefore, one may divide all possible 12 substitutions into two classes ($G \Rightarrow A$ and $T \Rightarrow C$ types and all others) for a new two-parameter model.

15.2.5 Methods Incorporating Observed Nucleotide Frequencies

As we saw, both one- and two-parameter methods give equilibrium nucleotide frequencies of ¼, and they may not be suitable for sequences with heterogeneous nucleotide frequencies. We can use the observed nucleotide frequencies for estimating the numbers of nucleotide substitutions to alleviate this problem. Felsenstein (1981; [12]) used the following transition probabilities from nucleotide i to j during

Table 15.6 Models of nucleotide substitution matrices incorporating nucleotide frequencies

(A) Equal input model

| | NEW | | | |
	A	C	T	G
O A	$1-\Sigma\lambda_A\bullet$	$\pi_C\alpha$	$\pi_T\alpha$	$\pi_G\alpha$
L C	$\pi_A\alpha$	$1-\Sigma\lambda_C\bullet$	$\pi_T\alpha$	$\pi_G\alpha$
D T	$\pi_A\alpha$	$\pi_C\alpha$	$1-\Sigma\lambda_T\bullet$	$\pi_G\alpha$
G	$\pi_A\alpha$	$\pi_C\alpha$	$\pi_T\alpha$	$1-\Sigma\lambda_G\bullet$

(B) Equal output model

| | NEW | | | |
	A	C	T	G
O A	$1-\Sigma\lambda_A\bullet$	$\pi_A\alpha$	$\pi_A\alpha$	$\pi_A\alpha$
L C	$\pi_C\alpha$	$1-\Sigma\lambda_C\bullet$	$\pi_C\alpha$	$\pi_C\alpha$
D T	$\pi_T\alpha$	$\pi_T\alpha$	$1-\Sigma\lambda_T\bullet$	$\pi_T\alpha$
G	$\pi_G\alpha$	$\pi_G\alpha$	$\pi_G\alpha$	$1-\Sigma\lambda_G\bullet$

(C) Hasegawa-Kishino-Yano model

| | NEW | | | |
	A	C	T	G
O A	$1-\Sigma\lambda_A\bullet$	$\pi_C\beta$	$\pi_T\beta$	$\pi_G\alpha$
L C	$\pi_A\beta$	$1-\Sigma\lambda_C\bullet$	$\pi_T\alpha$	$\pi_G\beta$
D T	$\pi_A\beta$	$\pi_C\alpha$	$1-\Sigma\lambda_T\bullet$	$\pi_G\beta$
G	$\pi_A\alpha$	$\pi_C\beta$	$\pi_T\beta$	$1-\Sigma\lambda_G\bullet$

(D) Tamura-Nei model

| | NEW | | | |
	A	C	T	G
O A	$1-\Sigma\lambda_A\bullet$	$\pi_C\beta$	$\pi_T\beta$	$\pi_G\alpha1$
L C	$\pi_A\beta$	$1-\Sigma\lambda_C\bullet$	$\pi_T\alpha2$	$\pi_G\beta$
D T	$\pi_A\beta$	$\pi_C\alpha2$	$1-\Sigma\lambda_T\bullet$	$\pi_G\beta$
G	$\pi_A\alpha1$	$\pi_C\beta$	$\pi_T\beta$	$1-\Sigma\lambda_G\bullet$

(E) General time reversible model

| | NEW | | | |
	A	C	T	G
O A	$1-\Sigma\lambda_A\bullet$	$\pi_C\beta1$	$\pi_T\beta2$	$\pi_G\alpha1$
L C	$\pi_A\beta1$	$1-\Sigma\lambda_C\bullet$	$\pi_T\alpha2$	$\pi_G\beta3$
D T	$\pi_A\beta2$	$\pi_C\alpha2$	$1-\Sigma\lambda_T\bullet$	$\pi_G\beta4$
G	$\pi_A\alpha1$	$\pi_C\beta3$	$\pi_T\beta4$	$1-\Sigma\lambda_G\bullet$

an infinitesimal time when he introduced the maximum likelihood method for inferring gene trees:

$$\lambda_{ij} = u\pi_j, \tag{15.49}$$

where u is the total rate of nucleotide substitution per unit time and π_j is the frequency of nucleotide j. Its nucleotide substitution matrix is shown in Table 15.6A. When Tajima and Nei (1982; [13]) introduced the equal input model of nucleotide substitutions, they were not aware that Felsenstein's model was identical with their equal input model. Tajima and Nei [13] also proposed the equal output model of nucleotide substitutions (Table 15.6B). These two models reduce to the one-parameter model if all four π_i (i = A, C, G, or T) frequencies become identical. Tajima and Nei (1984; [14]) later derived the analytical formula for estimating the number of substitutions from empirical nucleotide frequencies based on the equal input model:

$$d = -b \ln\left[\frac{1-p}{b}\right], \tag{15.50}$$

where $b = 1 - \Sigma\pi_j^2$, and these observed nucleotide frequencies are assumed to be the equilibrium frequencies. When all four π_i's are ¼, b becomes ¾, and the Eq. 15.50 becomes Eq. 15.21 for the one-parameter model. Tajima and Nei [14] provided a slightly more complicated equation by considering situations that the real substitution matrix does not follow the equal input model.

Hasegawa, Kishino, and Yano (1985; [15]) distinguished transitions and transversions in the equal output model, as shown in Table 15.6C. Because of the two additional parameters, there are five parameters in this model. This model (often abbreviated as the HKY model) was later incorporated to MOLPHY package (http://www.ism.ac.jp/ismlib/softother.html#molphy) developed by Adachi and Hasegawa [16] and was widely used. It may be noted that Felsenstein independently considered a model similar to the HKY model and implemented it in his PHYLIP package in 1984.

Tamura and Nei (1993; [17]) further divided transitions into $A \Leftrightarrow G$ and $T \Leftrightarrow C$ types (Table 15.6D) and proposed a six-parameter model. They gave the analytical formula for the estimation of nucleotide substitutions from observed frequencies of $A \Leftrightarrow G$ differences (P_{AG}), $T \Leftrightarrow C$ differences (P_{TC}), and transversional differences (Q) as well as observed nucleotide frequencies (g_A, g_G, g_T, and g_C):

$$d = -AG \ln \left[1 - \left(\frac{1}{AG} \right) P_{AG} - \left(\frac{1}{2g_R} \right) Q \right] - TC \ln \left[1 - \left(\frac{1}{TC} \right) P_{TC} - \left(\frac{1}{2g_Y} \right) Q \right]$$
$$- VER \ln \left[1 - \left(\frac{1}{2g_R g_Y} \right) Q \right] \tag{15.51}$$

where $AG = 2g_A g_G / g_R$, $TC = 2g_T g_C / g_Y$, $VER = 2(g_R g_Y - g_A g_G g_Y / g_R - g_T g_C g_R / g_Y)$, $g_R = g_A + g_G$, and $g_Y = g_T + g_C$. Equation 15.51 is also the maximum likelihood estimate [17]. If we do not distinguish $A \Leftrightarrow G$ and $T \Leftrightarrow C$ substitutions, the Tamura–Nei model (Table 15.6D) reduces to the HKY model (Table 15.6C). However, a possible analytical formula for the HKY model has two solutions for transitional parameters, as shown by Rzhetsky and Nei (1995; [59]).

We can further divide transversions into four classes by distinguishing $A \Leftrightarrow C$, $A \Leftrightarrow T$, $C \Leftrightarrow G$, and $G \Leftrightarrow T$ from the Tamura–Nei model. This model is usually called "general time reversible" (GTR) or REV (Table 15.6E) and was first proposed by Tavaré (1986; [18]). Yang (1994; [19]) applied this model for estimation of the nucleotide substitution patterns within the maximum likelihood framework (Yang 2006; [20]). Without knowing these developments, Zharkikh (1994; [21]) called GTR as "the equal input-related model."

15.2.6 Other Methods

There are many other methods for estimating the number of nucleotide substitutions, and we briefly mention representative ones in the order of publication years. Kimura (1981; [22]) proposed a method based on a three-parameter model (Table 15.7A) by distinguishing complementary bond forming transversions ($A \Leftrightarrow T$ and $G \Leftrightarrow C$) and the remaining transversions ($A \Leftrightarrow C$ and $G \Leftrightarrow T$). Takahata and Kimura (1981; [23]) proposed two methods using four and five parameters, respectively (Table 15.7B, C). In both models, different parameters were given to $A \Rightarrow G$ and its complementary $T \Rightarrow C$ transitions and the remaining two transitions ($G \Rightarrow A$ and $C \Rightarrow T$). Gojobori, Ishii, and Nei (1982; [24]) and Blaisdell (1985; [25]) proposed a six-parameter model (Table 15.7D) and a four-parameter model (Table 15.7E), respectively, by dividing the two types of transitions following Takahata and Kimura [23]. Tamura (1992; [26]) proposed an alternative 3-parameter model (Table 15.6F). We can also have the symmetric 6-parameter model (Table 15.7G; [21]) and the strand complementarity

Table 15.7 Various models of nucleotide substitutions

(A) Kimura 3P [21]					(B) Takahata-Kimura 4P [22]			
NEW					NEW			
	A	C	T	G		A	C	T	G
O A	$1-\Sigma\lambda_A\bullet$	γ	β	α	O A	$1-\Sigma\lambda_A\bullet$	$\theta\alpha$	β	α
L C	γ	$1-\Sigma\lambda_C\bullet$	α	β	L C	$\theta\gamma$	$1-\Sigma\lambda_C\bullet$	γ	β
D T	β	α	$1-\Sigma\lambda_T\bullet$	γ	D T	β	α	$1-\Sigma\lambda_T\bullet$	$\theta\alpha$
G	α	β	γ	$1-\Sigma\lambda_G\bullet$	G	γ	β	$\theta\gamma$	$1-\Sigma\lambda_G\bullet$

(C) Takahata-Kimura 5P [22]					(D) Gojobori-Ishii-Nei 6P [23]			
NEW					NEW			
	A	C	T	G		A	C	T	G
O A	$1-\Sigma\lambda_A\bullet$	δ	β	α	O A	$1-\Sigma\lambda_A\bullet$	α	α^1	α
L C	ε	$1-\Sigma\lambda_C\bullet$	γ	β	L C	β	$1-\Sigma\lambda_C\bullet$	β	α^2
D T	β	α	$1-\Sigma\lambda_T\bullet$	δ	D T	β_1	α	$1-\Sigma\lambda_T\bullet$	α
G	γ	β	ε	$1-\Sigma\lambda_G\bullet$	G	β	β_2	β	$1-\Sigma\lambda_G\bullet$

(E) Blaisdell 4P [24]					(F) Tamura 3P [25]			
NEW					NEW			
	A	C	T	G		A	C	T	G
O A	$1-\Sigma\lambda_A\bullet$	β	β	α	O A	$1-\Sigma\lambda_A\bullet$	$\theta\beta$	$(1-\theta)\beta$	$\theta\alpha$
L C	δ	$1-\Sigma\lambda_C\bullet$	γ	δ	L C	$(1-\theta)\beta$	$1-\Sigma\lambda_C\bullet$	$(1-\theta)\alpha$	$\theta\beta$
D T	δ	α	$1-\Sigma\lambda_T\bullet$	δ	D T	$(1-\theta)\beta$	$\theta\alpha$	$1-\Sigma\lambda_T\bullet$	$\theta\beta$
G	γ	β	β	$1-\Sigma\lambda_G\bullet$	G	$(1-\theta)\alpha$	$\theta\beta$	$(1-\theta)\beta$	$1-\Sigma\lambda_G\bullet$

(G) Symmetric 6P [26]					(H) Strand complementarity 6P [18]			
NEW					NEW			
	A	C	T	G		A	C	T	G
O A	$1-\Sigma\lambda_A\bullet$	γ_1	β_1	α_1	O A	$1-\Sigma\lambda_A\bullet$	β_1	β_2	α_1
L C	γ_1	$1-\Sigma\lambda_C\bullet$	α_2	β_2	L C	β_3	$1-\Sigma\lambda_C\bullet$	α_2	β_4
D T	β_1	α_2	$1-\Sigma\lambda_T\bullet$	γ_2	D T	β_2	α_1	$1-\Sigma\lambda_T\bullet$	β_1
G	α_1	β_2	γ_2	$1-\Sigma\lambda_G\bullet$	G	α_2	β_4	β_3	$1-\Sigma\lambda_G\bullet$

(I) Rzhetsky-Nei 8P					(J) General 12P			
NEW					NEW			
	A	C	T	G		A	C	T	G
O A	$1-\Sigma\lambda_A\bullet$	β_2	β_3	α_4	O A	$1-\Sigma\lambda_A\bullet$	λ_{AC}	λ_{AT}	λ_{AG}
L C	β_1	$1-\Sigma\lambda_C\bullet$	α_3	β_4	L C	λ_{CA}	$1-\Sigma\lambda_C\bullet$	λ_{CT}	λ_{CG}
D T	β_1	α_2	$1-\Sigma\lambda_T\bullet$	β_4	D T	λ_{TA}	λ_{TC}	$1-\Sigma\lambda_T\bullet$	λ_{TG}
G	α_1	β_2	β_3	$1-\Sigma\lambda_G\bullet$	G	λ_{GA}	λ_{GC}	λ_{GT}	$1-\Sigma\lambda_G\bullet$

6-parameter model (Table 15.7H) first proposed by Sueoka (1995; [27]). Rzhetsky and Nei (1995; [59]) proposed 8-parameter model (Table 15.7I). The final model is the general 12-parameter model (Table 15.7J), which has no limitation or maximum freedom, where all the twelve substitutions can have different parameters.

Although there is only one pattern for the 1-parameter model, there can be many 2-parameter models, as we discussed in Sect. 15.2.4. However, it is clear that Kimura's [9] 2-parameter model in which different parameters are given transitions and transversions is biologically most reasonable as the 2-parameter model. How about models with three parameters? So far, only two models by Kimura [22] and by Tamura [26] were proposed. In both models, transversions are expected to

Table 15.8 Two new models of nucleotide substitutions

| (A) 2-Transition 3P (this book) | | | | | (B) 4-Transition 5P (this book) | | | |
| NEW | | | | | NEW | | | |
	A	C	T	G		A	C	T	G
O A	$1-\Sigma\lambda_A\bullet$	β	β	α_1	O A	$1-\Sigma\lambda_A\bullet$	β	β	α_1
L C	β	$1-\Sigma\lambda_C\bullet$	α_2	β	L C	β	$1-\Sigma\lambda_C\bullet$	α_2	β
D T	β	α_1	$1-\Sigma\lambda_T\bullet$	β	D T	β	α_3	$1-\Sigma\lambda_T\bullet$	β
G	α_2	β	β	$1-\Sigma\lambda_G\bullet$	G	α_4	β	β	$1-\Sigma\lambda_G\bullet$

be heterogeneous. Transitions are uniform in Kimura's 3-parameter model, while there are two groups of transitions in Tamura's model. If we examine the observed nucleotide substitution patterns in the human nuclear genome shown in Table 2.2, $C \Rightarrow T$ and $G \Rightarrow A$ transitions are both ~20 %, while $T \Rightarrow C$ and $A \Rightarrow G$ transitions are both ~15 %, and transversions are much lower than all four transitions. Therefore, we can consider a new 3-parameter model in which $C \Rightarrow T$ and $G \Rightarrow A$ transitions are rate α_1, and $T \Rightarrow C$ and $A \Rightarrow G$ transitions are rate α_2, and all the eight transversions are rate β (Table 15.8A). This new three-parameter model can be considered as the reduced 5-parameter model of Takahata and Kimura [23] by equating β, δ, and ϵ in Table 15.7C. We also obtain this new 3-parameter model if we equate parameters β and δ in Blaisdell's [25] 4-parameter model (Table 15.7E). If we equate α_1 and α_2, this new 3-parameter model reduces to Kimura's [9] 2-parameter model.

Let us consider the nucleotide substitution patterns for human mitochondrial genome (Table 15.5). All four types of transitions vary greatly, from 0.5574 $(G \Rightarrow A)$ to 0.0593 $(C \Rightarrow T)$, while all transversions are much rarer (between 0.0028 and 0.0080). We can therefore propose a new 5-parameter model in which all four transitions have different parameters $\alpha_1 - \alpha_4$, while all eight transversions have single parameter β (see Table 15.8B).

Figure 15.6 shows the relationship of various nucleotide substitution models so far discussed in this chapter, hinted by Zharkikh's (1994; [21]) Fig. 1. Some connections are different from his figure and from description in [59], for the nested structure of parameters was respected in Fig. 15.6. Each box represents one model, and the top value of each column is the number of parameters for the model(s) in that column.

15.2.7 Handling of Heterogeneity Among Sites

So far, we implicitly assumed that all nucleotide sites are evolving under the same rates. In reality, however, mutation rates vary from site to site, as first pointed out by Benzer (1961; [28]). It is also well known that the substitution rates among the first, the second, and the third codon positions are different because of heterogeneous selective constraints (Chap. 4). When we compare very long nucleotide sequences, substitution rates may vary depending on the genomic region. Golding (1983; [29]) considered a gamma distribution and a lognormal distribution to fit to varying rates.

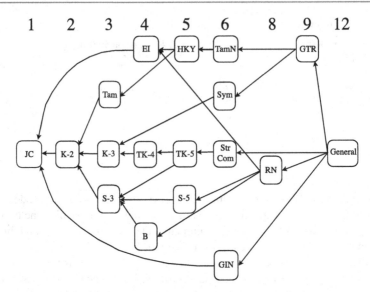

Fig. 15.6 Relationship of various nucleotide substitution models. *JC* Jukes-Cantor 1-parameter model [5], *2K-2* Kimura's 2-parameter model [6], *K-3* Kimura's 3-parameter model [9], *Tam* Tamura's 3-parameter model [26], *S-3* Saitou's 3-parameter model (this book), *TK-4* Takahata-Kimura's 4-parameter model [16, 17, 21–23], *EI* Equal input model [12–14], *B* Blaisdel's 4-parameter model [25], *TK-5* Takahata-Kimura's 5-parameter model [23], *HKY* Hasegawa-Kishino-Yano model [15], *S-5* Saitou's 5-parameter model (this book), *TamN* Tamura-Nei's model [17], *Sym* Symmetric model [21], *KS* Kawai and Saitou's 6-parameter model (this book), *GIN* Gojobori-Ishii-Nei model [24], *EIr* Zharkikh's EI-related model [21], *RN* Rzhetsky and Nei's 8-parameter model [59], *General* General 12-parameter model

Kocher and Wilson (1991; [30]) and Tamura and Nei (1993; [17]) analyzed the pattern of mutation accumulations in the control region of human mitochondrial DNA sequences and showed that the distribution of nucleotide substitution numbers per site fits much better with the negative binomial distribution than the Poisson distribution. This indicates that the substitution rate varies considerably among nucleotide sites. The negative binomial distribution is generated when the Poisson parameter λ (the mean) follows a gamma distribution,

$$f(\lambda) = \frac{b^a}{\Gamma(a)} \bullet e^{-b\lambda} \lambda^{a-1}, \tag{15.52}$$

where $a = M^2/V$ and $b = M/V$, and M and V are mean and variance of λ, respectively, and

$$\Gamma(a) = \Sigma e^{-t} t^{a-1} dt, \tag{15.53}$$

is a gamma function [17].

Gamma distribution is certainly convenient; however, we need to estimate the appropriate shape parameter beforehand. We need many sequences to estimate it, then pairwise distances can be obtained.

15.3 Synonymous and Nonsynonymous Substitutions

Nucleotide substitutions that occurred in amino acid coding nucleotide sequences can be classified into synonymous and nonsynonymous substitutions (Chap. 2). Synonymous substitutions are approximately following pure neutral evolution (Chap. 4), while nonsynonymous substitutions may face negative (purifying) and positive selection (Chap. 5). It is thus appropriate if we can separate these two types of substitutions. We discuss this problem in this section. For simplicity, only the standard codon table (see Chap. 1) is considered in this section.

15.3.1 Estimations of the Numbers of Synonymous and Nonsynonymous Sites

It is necessary to estimate the number of synonymous ($Nsite_S$) and nonsynonymous ($Nsite_N$) sites for estimating the numbers of synonymous and nonsynonymous substitutions "per site," as initially proposed by Kafatos et al. (1977; [31]); see also Miyata and Yasunaga (1980; [32]). Synonymous sites can be divided into fourfold degenerate, threefold degenerate, and twofold degenerate sites. All nonsynonymous sites are, by definition, nondegenerate. Let us explain how to estimate these two kinds of sites, following Miyata and Yasunaga [32]. Nei and Gojobori (1986; [33]) also used this scheme. We denote by f_i the fraction of synonymous changes at the ith position of a given codon ($i = 1, 2$, or 3). The number (n_S) of synonymous site and that (n_N) of nonsynonymous site for this codon are then given by

$$n_S = f_1 + f_2 + f_3,$$ (15.54a)

$$n_N = 3 - n_S.$$ (15.54b)

For example, $n_S = 2/3$ and $n_N = 7/3$ for codon TTA for Leu. This is because $f_1 = 1/3$ (only T $\Rightarrow$ C), $f_2 = 0$ (all changes are nonsynonymous), and $f_3 = 1/3$ (only A $\Rightarrow$ G) for this codon. When there are r codons for the coding sequence in question, we simply sum n_S and n_N values for all codons for obtaining the total numbers of synonymous ($Nsite_S$) and nonsynonymous ($Nsite_N$) sites for one protein coding sequence:

$$Nsite_S = \sum_{i=1,r} n_{Si},$$ (15.55a)

$$Nsite_N = \sum_{i=1,r} n_{Ni},$$ (15.55b)

where n_{Si} and n_{Ni} are n_S and n_N values for the ith codon.

David Lipman (cited as personal communication in Nei and Gojobori [33]) proposed a much simpler method to divide synonymous and nonsynonymous sites as follows. If we assume that four nucleotides are equally frequent and the substitutions among them are random, following the 1-parameter model (see Sect. 15.2.3), 5 % and 72 % of the substitutions in the first and third positions of a codon are synonymous, and all others are nonsynonymous. Thus, $Nsite_S = (0.05 + 0.72)r$ and $Nsite_N = (3 - 0.77)r$. Nei and Gojobori [33] showed, through computer simulations, that these very simple estimates, only relying on r, were quite unreliable under realistic situations. Nevertheless, these two proportions, namely, ~0.26 (=0.77/3) and ~0.74 (2.23/3), or more simply, ¼ and ¾, give us a rough idea on the proportions of synonymous and nonsynonymous sites.

If one codon changes to stop codon, how should we treat such changes? For example, codon TCA for Ser becomes stop codon TAA or TGA if the second position nucleotide C changes to A or G, respectively. Miyata and Yasunaga [32] mentioned that termination codons were considered as the twenty-first amino acid, while there was no mention about this problem in Kafatos et al. [31] nor in Nei and Gojobori [33]. Yang and Nielsen (1998; [34]) suggested to disallow mutations to stop codons and lose some sites so that $n_S + n_N < 3$. This is not a simple problem. Suzuki (2007; [35]) studied the detection of natural selection at one codon by computer simulations, and he found that the proportion of false detection rate of positive selection was decreased dramatically when he eliminated stop codons.

Li et al. (1985; [36]) proposed a different classification for protein coding nucleotide sequences. They classified sites into 4-fold, 2-fold, and 0-fold degenerate (nondegenerate in their original terminology) classes. The 4-fold degenerate sites are those in the third codon position in which any nucleotide change is synonymous, and 2-fold degenerate sites are those in the third codon position in which only transitional changes are synonymous. All the second codon positions are 0-fold degenerate or nondegenerate sites. Many of the first codon positions are also classified as 0-fold degenerate sites, and exceptions are those for four Arg codons and four Leu codons, which are classified as 2-fold degenerate. The third positions of Met and Trp codons are also 0-fold degenerate sites. Although the third positions of three Ile codons are 3-fold degenerate sites, they are treated as 2-fold degenerate for simplicity.

Another problem of synonymous and nonsynonymous sites is their estimation for pairwise sequences. It may be possible to adequately assign all nucleotide sites into synonymous and nonsynonymous sites for one nucleotide sequence. However, we need two homologous sequences for estimating evolutionary distances, and these two sequences are expected to have slightly different synonymous and nonsynonymous sites. Miyata and Yasunaga [32], Li et al. [36], and Nei and Gojobori [33] all used averaged values of these two sequences. When the amount of divergence is small, the numbers of synonymous and nonsynonymous sites for two sequences may be similar, and this averaging may not cause a serious problem. However, as the amount of divergence becomes larger and larger, a synonymous site may frequently change to a nonsynonymous site and vice versa. Therefore, we should be careful in this kind of comparisons.

One way to overcome this problem is to estimate the numbers of nucleotide substitutions for the each codon position separately. Since the mean proportions of synonymous and nonsynonymous substitutions are known for each codon position, we can have a rough idea about them from the estimates on different codon positions. For example, Gojobori and Yokoyama (1985; [37]) and Saitou and Nei (1986; [38]) estimated evolutionary rates of the first, second, and third positions of codons of retroviral oncogenes and influenza A virus genes, respectively. When the amount of divergence is large, it may be more suitable to use these ways rather than blindly estimating numbers of synonymous and nonsynonymous substitutions by using some software package.

15.3.2 Weightings for Multiple Paths

Another problem inherent in the synonymous and nonsynonymous substitutions is multiple paths for codon changes. Let us consider one homologous codon in two sequences, and let us assume that the codon of one sequence is TTT and that of the other sequence is GTA as an example. We need at least two substitutions between these two codons, and there are two possible paths:

$$\text{Path 1: TTT (Phe)} \Leftrightarrow \text{GTT (Val)} \Leftrightarrow \text{GTA (Val)}$$

$$\text{Path 2: TTT (Phe)} \Leftrightarrow \text{TTA (Leu)} \Leftrightarrow \text{GTA (Val)}$$

There are one nonsynonymous and one synonymous substitutions for path 1, while both substitutions are nonsynonymous for path 2. Miyata and Yasunaga [32] proposed to give different weights to synonymous and nonsynonymous substitutions, and further, depending on chemical distances of amino acids defined by Miyata et al. (1979; [39]), they gave different weights to nonsynonymous substitutions. According to their weighting scheme, a nonsynonymous substitution between chemically quite different amino acids, such as Gly and Trp, has the weight of only 1/100 of that for a synonymous substitution. Li et al. [36] adopted this weighting scheme. Nei and Gojobori [33], through computer simulations, showed that the equal weighting on different pathways did not considerably change the final results compared to the results based on the original method of Miyata and Yasunaga [32]. The biological reason for this may be as follows: amino acid substitutions which really occurred during evolution are often selectively neutral and did not alter protein function. Therefore, their behaviors were similar to synonymous changes.

15.3.3 Correction of Multiple Hits

As we discussed so far in Sect. 15.2, there can be many parallel, backward, or successive substitutions in nucleotide sequence evolution. They are often collectively called "multiple hits." Kafatos et al. [31] did not pay attention to this, and Miyata and Yasunaga [32] considered their estimates as minimum substitution numbers because of the parsimonious nature of their method.

Perler et al. (1980; [40]) proposed a method to estimate numbers of synonymous and nonsynonymous substitutions by considering multiple hits, probably for the first time. They cited Kimura and Ohta (1972; [6]) instead of Jukes and Cantor (1969; [5]) and proposed the following equations for estimating the number of synonymous substitutions by summing distances for three categories of sites:

$$Dsyn[1] = 3 \times \left\{ -\left(\frac{1}{2}\right) \ln\left(1 - 2p_1\right) \right\},$$ (15.56a)

$$Dsyn[2] = \frac{3}{2} \times \left\{ -\left(\frac{2}{3}\right) \ln\left(1 - \frac{3p_2}{2}\right) \right\},$$ (15.56b)

$$Dsyn[3] = 1 \times \left\{ -\left(\frac{3}{4}\right) \ln\left(1 - \frac{4p_3}{3}\right) \right\},$$ (15.56c)

where $Dsyn[i]$ is the number of synonymous substitutions at category i site (i = 1, 2, or 3) and p_i is the proportion of nucleotide difference for category i which consists of sites that can afford i changes. $Dsyn[3]$ in Eq. 15.56c is the same as Eq. 15.21. Equations for $Dsyn[1]$ and $Dsyn[2]$ for categories 1 and 2 sites are conjectures based on the one-parameter model in which all four nucleotides can change to any one of the remaining three with the uniform rate. These equations may be the causes of the saturation effect when the rate of mutation is not equal as Gojobori (1983; [41]) showed through computer simulations.

Nei and Gojobori [33] also proposed to use Eq. 15.21 based on the one-parameter model, for estimating the number of synonymous (d_S) and nonsynonymous (d_N) substitutions both from synonymous (p_S) and nonsynonymous (p_S) proportions of nucleotide differences:

$$d_S = -\left(\frac{3}{4}\right) \ln\left(1 - \frac{4p_S}{3}\right),$$ (15.57a)

$$d_N = -\left(\frac{3}{4}\right) \ln\left(1 - \frac{4p_N}{3}\right).$$ (15.57b)

Synonymous (p_S) and nonsynonymous (p_S) proportions of nucleotide differences are estimated as

$$p_S = \frac{Dif_S}{Nsite_S},$$ (15.58a)

$$p_N = \frac{Dif_N}{Nsite_N},$$ (15.58b)

where Dif_S and Dif_N are numbers of synonymous and nonsynonymous differences. An unweighted version of Miyata and Yasunaga [32] was used for estimating Dif_S and Dif_N in method 1 of Nei and Gojobori [33], while David Lipman's (cited as personal communication) suggestion was used in their method 2. Lipman's idea was to apply information from the single-base-change codons to the multiple-base-change codons. We first consider homologous codons of the two protein coding sequences in which only one nucleotide difference which is synonymous difference is observed. Depending on the position (1, 2, or 3) of the different nucleotides, we denote the number and fraction of such sites as $dif_{S-Sin}[i]$ and $p_{S-Sin}[i]$, respectively (i = 1, 2, or 3). The corresponding numbers and fractions of nonsynonymous changes are $dif_{N-Sin}[i]$ and $p_{N-Sin}[i]$, respectively (i = 1, 2, or 3). When more than one nucleotide difference (either two or three) exist between the two homologous codons, the number of such sites is denoted as $dif_{Mul}[i]$ (i = 1, 2, or 3). Dif_S and Dif_N are then estimated as

$$Dif_S = dif_{S-Sin}[1] + dif_{Mul}[1] \cdot p_{S-Sin}[1] + dif_{S-Sin}[3] + dif_{Mul}[3] \cdot p_{S-Sin}[3], \quad (15.59a)$$

$$Dif_N = dif_{N-Sin}[1] + dif_{Mul}[1] \cdot p_{N-Sin}[1] + dif_{N-Sin}[2]$$
$$+ dif_{Mul}[2] + dif_{N-Sin}[3] + dif_{Mul}[3] \cdot p_{N-Sin}[3] \quad (15.59b)$$

It should be noted that the one-parameter model does not apply to some sites, as Nei and Gojobori [33] admitted. Therefore, some modifications were made later (see [42]).

Li et al. [36] applied Kimura's [9] two-parameter model, and the transitional-type and transversional-type synonymous and nonsynonymous substitutions were independently estimated as follows:

$$TSi = \frac{1}{2}\ln\left(\frac{1}{[1-2P_i-Q_i]}\right) - \frac{1}{4}\ln\left(\frac{1}{[1-2Q_i]}\right), \quad (15.60a)$$

$$TVi = \frac{1}{2}\ln\left(\frac{1}{[1-2Q_i]}\right), \quad (15.60b)$$

where P_i and Q_i are proportions of transitional and transversional differences at the ith class of nucleotide sites. The numbers of synonymous (dS) and nonsynonymous (dA) substitutions per site are estimated by

$$dS = \frac{3[L_2 TS_2 + L_4(TS_4 + TV_4)]}{[L_2 + 3L_4]}, \quad (15.61a)$$

$$dN = \frac{3[L_2 TV_2 + L_0(TS_0 + TV_0)]}{[2L_2 + 3L_0]}, \quad (15.61b)$$

where L_i (i = 0, 2, or 4) is the total number of ith-fold degenerate sites.

Pamilo and Bianchi (1993; [43]) and Li (1993; [44]) modified Li et al.'s method, and Ina (1995; [45]) further modified previous methods. Interested readers should refer to Nei and Kumar [42].

15.4 Amino Acid Substitutions

We now consider methods for estimating the numbers of amino acid substitutions. There are 20 amino acids mainly used for protein sequences, and mathematically it seems simple to extend the idea used for estimation of nucleotide substitutions. However, a simpler method has been widely used, and we first explain this classic method. It should be noted that in some earlier papers, "replacement" was used instead of "substitution" for amino acid changes, following Fitch's (1976; [46]) suggestion.

15.4.1 Poisson Correction

Let us assume that the rate of amino acid substitution is equal in any amino acid site and at any time period. The unit of this rate, λ, is usually per amino acid site per year. The expected number of amino acid substitution during t years then becomes λt at one amino acid site. If amino acid substitutions are random events, they should follow a Poisson distribution with mean λt. We then have the probability of having r amino acid substitution during t years at one site as

$$\mathrm{Prob}(r) = \frac{\mathrm{e}^{-\lambda t}\left(\lambda t\right)^{r}}{r!}. \tag{15.62}$$

When $r=0$, $\mathrm{Prob}(0)=\mathrm{e}^{-\lambda t}$. If we ignore parallel and backward changes, this value can be equated with the proportion of identical amino acid sites between the ancestral sequence and its descendant one after t years. Because there are 20 amino acids, the possibility of experiencing parallel or backward substitutions at one site is expected to be not so high, compared to nucleotide sequences with only four kinds of bases. We often compare amino acid sequences of two present-day organisms which started to diverge t years ago. Since these two sequences are $2t$ years apart, the evolutionary distance between these two sequences is thus given by

$$d = 2\lambda t. \tag{15.63}$$

Therefore, when the proportion of identical amino acids for two homologous present-day amino acid sequences is q and that of different ones is p,

$$q = 1 - p = \mathrm{e}^{-2\lambda t} = \mathrm{e}^{-d}. \tag{15.64}$$

By taking logarithms in the above equation, we have

$$d = -\ln(1 - p).$$ (15.65)

This equation was first used by Zuckerkandl and Pauling (1965; [47]). When $p \ll 1$, $d \sim p$, in a similar manner with the nucleotide substitutions. The sampling variance of d is given by considering a binomial (two states: identical or different) distribution:

$$V(d) = \frac{p}{\left[(1-p)n\right]},$$ (15.66)

where n is the number of amino acids compared. It should be noted that rate heterogeneity can be handled as we discussed for nucleotide substitutions.

15.4.2 Dayhoff Matrix and Its Descendants

When the amount of divergence becomes large between the two compared amino acid sequences, the Poisson correction is expected to give underestimates of amino acid substitutions. Dayhoff et al. (1972; [48]) collected many amino acid sequences in the 1960s and initiated the protein sequence database (see Chap. 13). She and her colleagues constructed phylogenetic trees of many proteins and estimated the average pattern of amino acid substitutions from these results. This pattern is now called Dayhoff matrix, which is associated with PAM, acronym of "point accepted mutations." For example, the Ile ⇒ Val change is three times higher than Ile ⇒ Leu. Kimura (1983; [49]) showed that the following equation is a good approximation of the number of amino acid substitutions based on Dayhoff's matrix:

$$d = -\ln\left(\frac{1 - p - p^2}{5}\right),$$ (15.67)

where p is the proportion of amino acid difference for the homologous amino acid sequence pair.

Later, similar estimations of amino acid substitutions were conducted. Popular matrices are JTT matrix by Jones et al. (1992; [50]) and BLOSUM matrix by Henikoff and Henikoff (1992; [51]). Adachi and Hasegawa (1996; [52]) estimated a transition probability matrix of the general reversible Markov model of amino acid substitution for mtDNA-encoded proteins from the complete sequence data of mtDNA from 20 vertebrate species. Table 15.9 shows their matrix. Now Ile ⇒ Val change is almost the same with Ile ⇒ Leu. With huge genomic sequences now available, it may be better to estimate lineage-specific and protein family-specific amino acid substitution matrices if we are interested in a long-term evolution of proteins.

Table 15.9 Amino acid substitution matrix for vertebrate mitochondrial DNA coded proteins

	Ala	Arg	Asn	Asp	Cys	Gln	Glu	Gly	His	Ile	Leu	Lys	Met	Phe	Pro	Ser	Thr	Trp	Tyr	Val
Ala	***	2	9	2	3	1	2	64	4	84	35	0	63	5	26	243	380	0	2	67
Arg	7	***	4	0	7	51	0	13	38	0	23	29	0	3	14	5	2	6	0	3
Asn	17	4	***	134	2	31	17	25	114	22	30	102	17	4	39	327	161	3	44	4
Asp	7	0	279	***	0	14	135	32	34	6	1	3	0	4	4	42	24	2	3	0
Cys	35	20	14	0	***	9	0	18	41	42	60	0	0	40	8	200	140	102	75	0
Gln	2	38	49	10	2	***	73	4	148	10	57	98	25	16	68	43	78	0	11	5
Glu	7	0	28	105	0	75	***	11	12	0	0	59	0	0	4	39	12	0	6	9
Gly	82	4	18	11	2	2	5	***	0	6	2	3	1	0	0	81	7	2	0	1
His	10	25	159	23	9	132	10	10	***	13	17	15	0	21	21	40	44	2	198	0
Ile	70	3	10	1	3	3	0	4	4	***	465	2	216	40	6	23	285	0	9	461
Leu	15	24	7	0	2	9	0	1	3	241	***	2	244	118	20	49	90	8	12	33
Lys	0	0	173	2	0	105	61	8	18	8	12	***	34	5	22	60	102	8	15	1
Met	86	1	13	0	0	12	0	1	0	354	772	15	***	42	8	73	408	6	9	155
Phe	6	5	3	1	4	7	0	0	10	57	328	2	37	***	8	42	23	2	124	2
Pro	35	5	28	2	1	31	2	0	11	10	62	10	7	9	***	100	96	1	4	3
Ser	246	1	179	11	18	15	13	64	16	28	116	20	54	36	76	***	452	8	16	0
Thr	315	1	72	5	10	22	3	5	14	283	172	27	247	16	60	369	***	4	8	80
Trp	0	4	4	2	2	0	0	4	2	0	48	6	12	5	3	21	12	***	7	2
Tyr	4	0	53	2	14	8	4	0	169	23	62	10	14	227	7	34	22	6	***	2
Val	112	1	4	0	0	3	5	2	0	918	128	0	118	3	4	0	160	2	2	***

15.5 Evolutionary Distances Not Based on Substitutions

We considered nucleotide substitutions in the previous three sections as the main mechanism of nucleotide sequence changes. Insertions and deletions are another important mutations (see Chap. 2). We therefore discuss evolutionary distances based on insertions and deletions.

Tajima and Nei (1984; [14]) proposed a simple method for estimating the number (γ) of nucleotide changes due to insertions and deletions between sequences X and Y:

$$\gamma = -2 \ln \left\{ \frac{n_{XY}}{\sqrt{(n_X + n_Y)}} \right\}, \qquad (15.68)$$

where n_{XY}, n_X, and n_Y are the total number of shared nucleotide sites between X and Y, the number of nucleotides for X and that for Y, respectively. Saitou (1992; [53]) applied this method to the two sets of noncoding sequences of primates and found that the species tree topologies were correctly reconstructed, while the branch lengths varied more than two times in some cases. When the mutational events were used as units of evolution [54], a rough constancy of the evolutionary rate was observed. This suggests that Eq. 15.68 is somewhat biased by large deletions and insertions.

Evolutionary distances based on gene order changes or genome rearrangements are also studied for a long time (e.g., [55]). There are many types in genome rearrangements. Duplication, inversion, transposition, block exchange, and fission can occur on a single chromosome, while translocation and fusion require two chromosomes. Yancopoulos et al. (2005; [56]) proposed the "double-cut-and-join" (DCJ) operation which accounts for inversions, translocations, fissions, and fusions. Although this operation was devised from an algorithmic point of view, its possible molecular mechanism reminds me of the "drift duplication" model we recently proposed (Ezawa et al., 2011; [57]). It should also be noted that Luo et al. (2012; [58]) applied the DCJ operation-based genomic distances for mammalian phylogeny, and the tree thus obtained was identical with those constructed by using various tree-making methods from nucleotide sequence data.

References

1. Ishida, N., Oyunsuren, T., Mashima, S., Mukoyama, H., & Saitou, N. (1995). Mitochondrial DNA sequences of various species of the genus Equus with a special reference to the phylogenetic relationship between Przewalskii's wild horse and domestic horse. *Journal of Molecular Evolution, 41*, 180–188.
2. Sneath, P. A. P., & Sokal, R. (1970). *Numerical taxonomy*. San Francisco: W. H. Freeman.
3. Fujiyama, A., Watanabe, H., Toyoda, A., Taylor, T. D., Itoh, T., Tsai, S.-F., Park, H.-S., Yaspo, M.-L., Lehrach, H., Chen, Z., Fu, G., Saitou, N., Osoegawa, K., de Jong, P. J., Suto, Y., Hattori, M., & Sakaki, Y. (2002). Construction and analysis of a human-chimpanzee comparative clone map. *Science, 295*, 131–134.

4. Abe, K., Noguchi, H., Tagawa, K., Yuzuriha, M., Toyoda, A., Kojima, T., Ezawa, K., Saitou, N., Hattori, M., Sakaki, Y., Moriwaki, K., & Shiroishi, T. (2004). Contribution of Asian mouse subspecies Mus musculus molossinus to genomic constitution of strain C57BL/6J, as defined by BAC end sequence-SNP analysis. *Genome Research, 14*, 2239–2247.

5. Jukes, T. H., & Cantor, C. R. (1969). Evolution of protein molecules. In H. N. Munro (ed.), Mammalian protein metabolism (pp. 21–132). New York: Academic.

6. Kimura, M., & Ohta, T. (1972). Title of paper. *Journal of Molecular Evolution, 2*, 87–90.

7. Nei, M. (1987). *Molecular evolutionary genetics*. New York: Columbia University Press.

8. Saitou N. (1990) Maximum likelihood methods. In R. Doolittle (ed.), Methods in enzymology, Vol 183: Computer analysis of protein and nucleic acid sequences (pp. 584–598). San Diego: Academic.

9. Kimura, M. (1980). A simple method for estimating evolutionary rates of base substitutions through comparative studies of nucleotide sequences. *Journal of Molecular Evolution, 16*, 111–120.

10. Kawai, Y., Kikuchi, T., & Saitou N. (unpublished) Evolutionary dynamics of nucleotide composition of primate mitochondrial DNA inferred from human SNP data and nuclear pseudogenes.

11. Felsenstein, J. (2004). *Inferring phylogenies*. Sunderland: Sinauer Associates.

12. Felsenstein, J. (1981). Evolutionary trees from DNA sequences: A maximum likelihood approach. *Journal of Molecular Evolution, 17*, 368–376.

13. Tajima, F., & Nei, M. (1982). Biases of the estimates of DNA divergence obtained by the restriction enzyme technique. *Journal of Molecular Evolution, 18*, 115–120.

14. Tajima, F., & Nei, M. (1984). Estimation of evolutionary distance between nucleotide sequences. *Molecular Biology and Evolution, 1*, 269–285.

15. Hasegawa, M., Kishino, H., & Yano, T. (1985). Dating of the human-ape splitting by a molecular clock of mitochondrial DNA. *Journal of Molecular Evolution, 22*, 160–174.

16. Adachi, J., & Hasegawa, M. (1996). *MOLPHY version 2.3: Programs for molecular phylogenetics based on maximum likelihood* (Computer science monographs, Vol. 28). Tokyo: Institute of Statistical Mathematics.

17. Tamura, K., & Nei, M. (1993). Estimation of the number of nucleotide substitutions in the control region of mitochondrial DNA in humans and chimpanzees. *Molecular Biology and Evolution, 10*, 512–526.

18. Tavaré, S. (1986). Some probabilistic and statistical problems on the analysis of DNA sequences. *Lectures on Mathematics in the Life Sciences, 17*, 57–86.

19. Yang, Z. (1994). Estimating the pattern of nucleotide substitution. *Journal of Molecular Evolution, 39*, 105–111.

20. Yang, Z. (2006). *Computational molecular evolution* (Oxford series in ecology and evolution). Oxford/New York: Oxford University Press.

21. Zharkikh, A. (1994). Estimation of evolutionary distances between nucleotide sequences. *Journal of Molecular Evolution, 39*, 315–329.

22. Kimura, M. (1981). Estimation of evolutionary distances between homologous nucleotide sequences. *Proceedings of National Academy of Sciences of the United States of America, 78*, 454–458.

23. Takahata, N., & Kimura, M. (1981). A model of evolutionary base substitutions and its application with special reference to rapid change of pseudogenes. *Genetics, 98*, 641–657.

24. Gojobori, T., Ishii, K., & Nei, M. (1982). Estimation of average number of nucleotide substitutions when the rate of substitution varies with nucleotide. *Journal of Molecular Evolution, 18*, 414–423.

25. Blaisdell, B. E. (1985). A method of estimating from two aligned present-day DNA sequences their ancestral composition and subsequent rates of substitution, possibly different in the two lineages, corrected for multiple and parallel substitutions at the same site. *Journal of Molecular Evolution, 22*, 69–81.

26. Tamura, K. (1992). The rate and pattern of nucleotide substitution in Drosophila mitochondrial DNA. *Molecular Biology and Evolution, 9*, 814–825.
27. Sueoka, N. (1995). Intrastrand parity rules of DNA base composition and usage biases of synonymous codons. *Journal of Molecular Evolution, 40*, 318–325.
28. Benzer, S. (1961). On the topography of genetic fine structure. *Proceedings of National Academy of Sciences of the United States of America, 47*, 403–415.
29. Golding, B. (1983). Estimates of DNA and protein sequence divergence: An examination of some assumptions. *Molecular Biology and Evolution, 1*, 125–142.
30. Kocher, T. D., & Wilson, A. C. (1991). Sequence evolution of mitochondrial DNA in humans and chimpanzees: Control region and a protein-coding region. In S. Osawa & T. Honjo (Eds.), *Evolution of life* (pp. 391–413). Tokyo/New York: Springer-Verlag.
31. Kafatos, F. C., Efstratiadis, A., Forget, B. G., & Weissman, S. M. (1977). Molecular evolution of human and rabbit beta-globin mRNAs. *Proceedings of National Academy of Sciences of the United States of America, 74*, 5618–5622.
32. Miyata, T., & Yasunaga, T. (1980). Molecular evolution of mRNA: A method for estimating evolutionary rates of synonymous and amino acid substitutions from homologous nucleotide sequences and its application. *Journal of Molecular Evolution, 16*, 23–36.
33. Nei, M., & Gojobori, T. (1986). Simple methods for estimating the numbers of synonymous and nonsynonymous nucleotide substitutions. *Molecular Biology and Evolution, 3*, 418–426.
34. Yang, Z., & Nielsen, R. (1998). Synonymous and nonsynonymous rate variation in nuclear genes of mammals. *Journal of Molecular Evolution, 46*, 409–418.
35. Suzuki, Y. (2007). Inferring natural selection operating on conservative and radical substitution at single amino acid sites. *Genes and Genetic Systems, 82*, 341–360.
36. Li, W.-H., Luo, C.-C., & Wu, C.-I. (1985). A new method for estimating synonymous and nonsynonymous rates of nucleotide substitution considering the relative likelihood of nucleotide and codon changes. *Molecular Biology and Evolution, 2*, 150–174.
37. Gojobori, T., & Yokoyama, S. (1985). *Proceedings of the National Academy of Sciences of the United States of America, 82*, 4198–4201.
38. Saitou, N., & Nei, M. (1986). Polymorphism and evolution of influenza A virus genes. *Molecular Biology and Evolution, 3*, 57–74.
39. Miyata, T., Miyazawa, S., & Yasunaga, T. (1979). Two types of amino acid substitutions in protein evolution. *Journal of Molecular Evolution, 12*, 219–236.
40. Perler, F., Efstratiadis, A., Lomedico, P., Gilbert, W., Kolodner, R., & Dodgson, R. (1980). The evolution of genes: The chicken preproinsulin gene. *Cell, 20*, 555–566.
41. Gojobori, T. (1983). Codon substitution in evolution and the "saturation" of synonymous changes. *Genetics, 105*, 1011–1027.
42. Nei, M., & Kumar, S. (2000). *Molecular evolution and phylogenetics*. Oxford/New York: Oxford University Press.
43. Pamilo, P., & Bianchi, O. (1993). Evolution of the Zfx and Zfy genes: Rates and interdependence between the genes. *Molecular Biology and Evolution, 19*, 271–281.
44. Li, W.-H. (1993). Unbiased estimation of the rates of synonymous and nonsynonymous substitution. *Journal of Molecular Evolution, 36*, 96–99.
45. Ina, Y. (1995). New methods for estimating the numbers of synonymous and nonsynonymous substitutions. *Journal of Molecular Evolution, 40*, 190–226.
46. Fitch, W. M. (1976). Molecular evolutionary clocks. In F. J. Ayala (Ed.), *Molecular evolution* (pp. 160–178). Sunderland: Sinauer Associates.
47. Zuckerkandl, E., & Pauling, L. (1965). Evolutionary divergence and convergence in proteins. In V. Bryson & H. J. Vogel (Eds.), *Evolving genes and proteins* (pp. 97–166). New York/London: Academic.
48. Dayhoff, M., Eck, R. V., & Park, C. M. (1972). A model of evolutionary change in proteins. In M. O. Dayhoff (Ed.), *Atlas of protein sequence and structure* (Vol. 5, pp. 89–99). Silver Spring: National Biomedical Research Foundation.

49. Kimura, M. (1983). *The neutral theory of molecular evolution*. Cambridge: Cambridge University Press.
50. Jones, D. T., Taylor, W. R., & Thornton, J. M. (1992). The rapid generation of mutation data matrices from protein sequences. *Computer Applications in the Biosciences: CABIOS, 8*, 275–282.
51. Henikoff, S., & Henikoff, J. G. (1992). Amino acid substitution matrices from protein blocks. *Proceedings of National Academy of Sciences of the United States of America, 101*, 10915–10919.
52. Adachi, J., & Hasegawa, M. (1996). Model of amino acid substitution in proteins encoded by mitochondrial DNA. *Journal of Molecular Evolution, 42*, 459–468.
53. Saitou, N. (1992). Evolutionary rate of insertions and deletions. In N. Takahata (Ed.), *Population paleo-genetics* (pp. 335–343). Tokyo: Japan Science Society Press.
54. Saitou, N., & Ueda, S. (1994). Evolutionary rate of insertions and deletions in non-coding nucleotide sequences of primates. *Molecular Biology and Evolution, 11*, 504–512.
55. Sankoff, D., & Nadeau, J. H. (2003). Chromosome rearrangements in evolution: From gene order to genome sequence and back. *Proceedings of National Academy of Sciences of the United States of America, 100*, 11188–11189.
56. Yancopoulos, S., Attie, O., & Friedberg, R. (2005). Efficient sorting of genomic permutations by translocation, inversion and block interchange. *Bioinformatics, 21*, 3340–3346.
57. Ezawa, K., Ikeo, K., Gojobori, T., & Saitou, N. (2011). Evolutionary patterns of recently emerged animal duplogs. *Genome Biology and Evolution, 3*, 1119–1135.
58. Luo, H., Arndt, W., Zhang, Y., Shi, G., Akekseyev, M., Tang, J., Huges, A. L., & Friedman, R. (2012). Phylogenetic analysis of genome rearrangements among five mammalian orders. *Molecular Phylogenetics and Evolution, 65*, 871–882.
59. Rzhetsky, A., & Nei, M. (1995). Tests of applicability of several substitution models for DNA sequence data. *Molecular Biology and Evolution, 12*, 131–151.

Tree and Network Building

16

Chapter Summary

Construction of phylogenetic trees from nucleotide or amino acid sequence data is one of the important areas of evolutionary genomics. We start from classification of tree-building methods, both by type of data and by type of tree search algorithm. Various distance matrix methods including UPGMA, minimum deviation methods, minimum evolution methods, transformed distance methods, and neighbor-joining method are explained. Among character-state methods, maximum parsimony methods, maximum likelihood methods, and Bayesian method are explained. These many phylogenetic tree-making methods were compared mainly based on computer simulation studies. Phylogenetic network constructions from distance matrix and from multiply aligned sequences are also discussed as well as phylogeny construction without multiple alignments.

16.1 Classification of Tree-Building Methods

Many methods have been proposed for finding the phylogenetic tree from observed data. To clarify the nature of each method, it is useful to classify these methods from various aspects. Two aspects of classifications are discussed in this section, by type of data and by type of tree search algorithm.

16.1.1 Classification by Type of Data

Tree-building methods can be divided into two types in terms of the type of data they use: distance matrix methods and character-state methods. A distance matrix consists of a set of $n(n-1)/2$ distance values (see Chap. 15) for n OTUs (operational taxonomic units), whereas an array of character states is used for the character-state methods.

N. Saitou, *Introduction to Evolutionary Genomics*, Computational Biology 17, DOI 10.1007/978-1-4471-5304-7_16, © Springer-Verlag London 2013

Table 16.1   Examples of distance matrices

(A) When the evolutionary rate is constant

	A	B	C	D	E	F
A	0	2	6	4	16	16
B	2	0	6	4	16	16
C	6	6	0	6	16	16
D	4	4	6	0	16	16
E	16	16	16	16	0	10
F	16	16	16	16	10	0

(B) When all distances are purely additive (from [1])

	1	2	3	4	5	6	7	8
1	0	7	8	11	13	16	13	17
2	7	0	5	8	10	13	10	14
3	8	5	0	5	7	10	7	11
4	11	8	5	0	8	11	8	12
5	13	10	7	8	0	6	6	10
6	16	13	10	11	5	0	9	13
7	13	10	7	8	6	9	0	8
8	17	14	11	12	10	13	8	0

(C) Distance matrix estimated from real nucleotide sequence data (from [2])

	1	2	3	4	5	6	7	8	9
2	0.0516								
3	0.0550	0.0031							
4	0.0483	0.0221	0.0253						
5	0.0582	0.0651	0.0685	0.0549					
6	0.0094	0.0416	0.0450	0.0384	0.0549				
7	0.0125	0.0584	0.0619	0.0551	0.0651	0.0157			
8	0.0284	0.0687	0.0722	0.0654	0.0754	0.0317	0.0285		
9	0.0925	0.1221	0.1259	0.1185	0.1370	0.0820	0.0786	0.0927	
10	0.1921	0.2183	0.2228	0.2054	0.2309	0.1798	0.1795	0.1833	0.1860

OTU ID; 1=*M. m. domesticus* functional gene, 2=*M. m. domesticus* pseudogene, 3=*M. m. castaneus* pseudogene, 4=*M. spicilegus* pseudogene, 5=*M. leggada* pseudogene, 6=*M. m. domesticus* functional gene, 7=*M. leggada* functional gene, 8=*M. platythrix* functional gene, 9=*Rattus norvegicus* functional gene, 10=*Homo sapiens* functional gene

In distance matrix methods such as the neighbor-joining method of Saitou and Nei (1987; [1]), a phylogenetic tree is constructed by considering the relationship among the distance values of a distance matrix in which a pairwise distance, D_{ij} between OTUs i and j, is element of the matrix. By definition, $D_{ii}=0$ and $D_{ij}=D_{ji}$. Various examples of distance matrices are presented in Table 16.1. Table 16.1A is an artificial example when the evolutionary rate is strictly constant. Some distance values are identical because of the hierarchical structure of this tree (see Fig. 16.1a). The distance matrix shown in Table 16.2B satisfies the condition that the distance between any OTU pair is the sum of all branches connecting these two OTUs in the phylogenetic tree (see Fig. 16.1b). This is also an artificial example, while Table 16.1C is based on a real nucleotide sequence data: functional p53 genes of mouse, rat, and human and their mouse pseudogenes [2]. Kimura's (1980; [3]) 2-parameter method

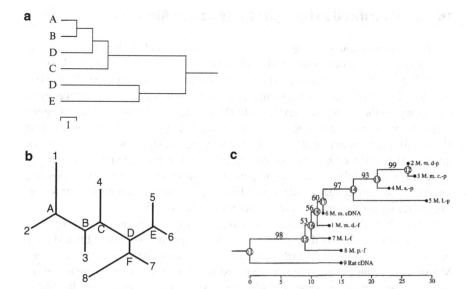

Fig. 16.1 Phylogenetic trees corresponding to distance matrices shown in Table 16.1

Table 16.2 Operation of UPGMA for distance matrix of Table 16.1A

(A) After OTUs A and B are clustered						(B) After OTUs AB and D are clustered			
	AB	C	D	E	F		ABD	C	E	F
AB	0	6	4	16	16	ABD	0	6	16	16
C	6	0	6	16	16	C	6	0	16	16
D	4	6	0	16	16	E	16	16	0	10
E	16	16	16	0	10	F	16	16	10	0
F	16	16	16	10	0					

(C) After OTUs ABD and C are clustered				(D) After OTUs E and F are clustered		
	ABDC	E	F		ABDC	EF
ABDC	0	16	16	ABDC	0	16
E	16	0	10	EF	16	0
F	16	10	0			

was used for estimating the number of nucleotide substitutions (see Chap. 15). It should be noted that only distances below the diagonal are shown in Table 16.1B, C.

A character state is either A, C, G, T, or gap in the case of multiple alignment of nucleotide sequences (see Chap. 14). One of 20 amino acids or gap is a character state for one position in the case of multiple alignment of protein sequences. These multiply aligned character-state data are used in character-state methods such as maximum parsimony methods and maximum likelihood methods. Broader kinds of character states, such as presence/absence of a certain molecular markers, may also be used for maximum parsimony method.

16.1.2   Classification by Type of Tree Search Algorithm

Another classification is by the strategy of a method to find the best tree. One way is to examine all or a large number of possible tree topologies and choose the best one according to a certain criterion. This type is called the "exhaustive search method" (e.g., [4]). Fitch and Margoliash's method, minimum evolution methods, the maximum parsimony method, and the maximum likelihood method belong to this category. When the number of sequences compared to construct phylogenetic trees is less than 10, an exhaustive search of all possible tree topologies may be possible. Nowadays, however, we often construct a phylogenetic tree for more than 100 sequences. In this case, any method with fast computation is impossible to examine all possible topologies within a reasonable time, because of the combinatorial explosion. Therefore, we no longer search tree topologies exhaustively.

Saitou (2007; [5]) therefore proposed to call this type of method as "completely bifurcating tree search method." Various measures are computed for a limited number of topologies to find the best tree in methods belonging to this category. Measures or criterions used are the required number of minimum changes for the maximum parsimony method, the likelihood value for the maximum likelihood method, the sum of branch lengths for the minimum evolution method, difference between observed and estimated distances in the minimum deviation method, and so on. There are plenty of algorithms to search completely bifurcating trees, as we shall see in later sections of this chapter.

The other strategy is to examine a local topological relationship of OTUs and find the best one, then apply the same procedure after reducing the search space. This type of methods is called the "stepwise clustering method" [6]. Majority of the distance matrix methods such as the neighbor-joining method [1] are stepwise clustering methods, but this approach can also be used for character-state methods.

16.1.3   Character-State Data

An array of character states is used for the character-state methods. In evolutionary genomics, nucleotide sequences and amino acid sequences are the two representative character states used for phylogeny construction. These sequences are expected to be multiply aligned (see Chap. 14), though there are some methods which do not require multiple alignment process, as we will see in Sect. 16.9.

It is useful to consider the concept of "configuration" in multiply aligned character-state data. For simplicity, let us consider nucleotide sequences. If there are multiply aligned nucleotide sequences, we can have some specific site patterns. When we do not distinguish nucleotides and only focus on patterns, we call them configurations. For example, if the list of nucleotides for seven sequences at one site is AAGGAAC, this pattern has the identical nucleotide configuration as TTAATTG; sequence 7 has a unique nucleotide, while sequences 3 and 4 have the same nucleotide, and the remaining four sequences have a different nucleotide. A "nucleotide configuration" is a distribution pattern of nucleotides for a given number of sequences. Saitou and Nei [7] showed that the possible number (C[N]) of configurations for N sequences is given by

$$C[N] = \frac{\left(4^{N-1} + 3 \cdot 2^{N-1} + 2\right)}{6}. \tag{16.1}$$

For example, there are 15 and 51 possible nucleotide configurations for 4 and 5 nucleotide sequences, respectively. Table 16.6 shows all possible 15 nucleotide configurations for four nucleotide sequences. Unfortunately, C[N] suffers from the combinatorial explosion, for there are 43,819 configurations when we have mere 10 OTUs, and C[N] becomes ~6.7×10^{58} when $N = 100$. Therefore, it is not appropriate to consider all possible configurations for tree construction. We should consider only observed configurations.

16.2 Distance Matrix Methods

There are rooted and unrooted trees (see Chap. 3), and some distance matrix methods such as UPGMA and WPGMA produce rooted trees, for they assume a constancy of the evolutionary rate. However, majority of distance matrix methods produce only unrooted trees. We first discuss UPGMA and WPGMA in this section and then minimum deviation method, minimum evolution method, distance parsimony method, and other distance matrix methods are discussed, before introducing the neighbor-joining method developed by Saitou and Nei (1987; [1]).

Tamura et al. (2004; [8]) proposed a likelihood method for the simultaneous estimation of all pairwise distances by using biologically realistic models of nucleotide substitution. They showed that this modification corrects up to 60 % of tree topology errors when the neighbor-joining method was used for conventional pairwise distances.

16.2.1 UPGMA and WPGMA

When the constancy of the evolutionary rate, or molecular clock, is assumed, we can reconstruct rooted trees. There are many ways to obtain such rooted trees from a distance matrix. Interested readers may read Sneath and Sokal (1973; [9]); see also Sokal and Sneath (1968; [10]). In this section, only UPGMA (unweighted pair group method with arithmetic mean) and very similar WPGMA (weighted pair group method with arithmetic mean) are discussed. UPGMA was originally proposed by Sokal and Michener (1958; [11]) and is a stepwise clustering method. UPGMA was later independently proposed by Nei (1975; [12]).

Let us explain the UPGMA algorithm using the distance matrix of Table 16.1A. We first choose the smallest distance, D_{AB} (= $D_{BA} = 2$). Then OTUs 1 and 2 are combined and the distances between the combined OTU [AB] and the remaining four OTUs

Fig. 16.2 A phylogenetic
tree constructed by using
UPGMA from distance
matrix of Table 16.1B

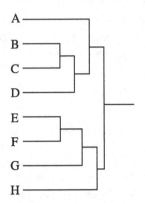

are computed by taking arithmetic means. A new distance matrix (see Table 16.2A)
for five OTUs is produced after this operation. At the next step, the same procedure
is used, that is, the smallest distance ($D_{[AB]D}=D_{D[AB]}=4$) is chosen from the new
distance matrix. Then the OTU [AB] and OTU D are further combined into OTU
[ABD]. This process is continued until all the OTUs are finally clustered into a
single one (see Table 16.2). The resultant tree is shown in Fig. 16.2a. Because an
exact constancy of evolutionary rate was satisfied for the distance matrix of Table
6.1A, this tree is the correct one. There is no difference between trees constructed
by using UPGMA and WPGMA in this case.

   It should be noted that UPGMA will construct a correct rooted tree when an
approximate constancy of the evolutionary rate holds. However, UPGMA is
expected to construct an erroneous tree if the evolutionary rate heterogeneity is
strong. The correct unrooted tree for the distance matrix of Table 16.1B is that
shown in Fig. 16.1b, while the tree shown in Fig. 16.2a is constructed if we use
UPGMA. Let us explain how to construct a UPGMA tree from this distance matrix.
We face a tie situation, for three distances ($D_{BC}=D_{CD}=D_{EF}=5$) show the same value.
In many distance matrix methods, the tie situation is simply handled either by
choosing the first one or the last one found in the distance value search. If tied pairs
are compatible, there are not much problem; however, tie of incompatible pair, such
as B-C and C-D in Table 16.1B, produces different tree topology. In any case, let us
choose the B-C pair at step 1. The E-F pair is then clustered at step 2. The BC-D pair
is clustered at step 3. When distances are averaged after step 3, operations of
UPGMA and WPGMA now differ with each other. In UPGMA, distances between
the new OTU BCD and the remaining four OTUs (A, EF, G, and H) are computed
as simply averaging three distances involving original three OTUs, B, C, and D,
while these distances are weighted according to the topological relationship in
WPGMA. For example,

$$D_{A(BCD)} = \frac{\left(D_{AB}+D_{AC}+D_{AD}\right)}{3} = \frac{\left(7+8+11\right)}{3} = 8.67 \qquad (16.2)$$

in UPGMA, while in WPGMA,

$$D_{A(BCD)} = \frac{\dfrac{D_{AB} + D_{AC}}{2} + D_{AD}}{2} = \frac{\dfrac{7+8}{2} + 11}{2} = 9.25. \tag{16.3}$$

There is a slight difference in the averaging. Because of this, UPGMA is called unweighted, while WPGMA is weighted.

UPGMA has been criticized by many researchers including the author (e.g., [1]) because it implicitly assumed a constancy of the evolutionary rate. In reality, however, an approximate constancy of the evolutionary rate often holds, and it may be recommended to use UPGMA tree when its tree topology is the same with the ones determined by using some other methods which do not assume rate constancy. It should also be noted that the branch lengths estimated by using UPGMA are least-squares estimates (Chakraborty 1977; [13]).

16.2.2 Minimum Deviation Method and Related Methods

Fitch and Margoliash (1967; [14]) proposed a completely bifurcating tree search method from distance matrix data. The criterion of choosing the best topology is "percent standard deviation" (PSD), defined as

$$PSD = \left[\sum_{i<j} \frac{\left\{ \dfrac{(Dij - Eij)}{Dij} \right\}^2}{\left\{ \dfrac{n(n-1)}{2} \right\}} \right]^{\frac{1}{2}} \times 100, \tag{16.4}$$

where Eij is the estimated distance between OTUs i and j out of n OTUs for a given completely bifurcating tree topology. Although this method is usually called as Fitch and Margoliash's method, they proposed various ideas in their paper, and additional modifications were also made for their method later. Therefore, Saitou [5] proposed to call this and all related methods as "minimum deviation methods."

Tateno et al. (1982; [15]) proposed a related measure for comparison between observed and expected distances:

$$So = \left[\sum_{i<j} \frac{\left\{ (Dij - Eij)^2 \right\}}{\left\{ \dfrac{n(n-1)}{2} \right\}} \right]^{\frac{1}{2}}. \tag{16.5}$$

Fig. 16.3 An unrooted
tree for three OTUs

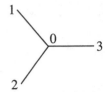

The estimated distance between a pair of OTUs for a given tree topology is
obtained by summing all branches connecting these two OTUs. Each branch is
estimated as follows. When there are only three OTUs 1, 2, and 3, there is a unique
unrooted tree topology with one interior node 0, and the problem is just to estimate
three branch lengths, BL_{10}, BL_{20}, and BL_{30}, where BL_{ij} is the branch length between
nodes i and j (see Fig. 16.3). If we assume the additivity of distances,

$$D_{12} = BL_{10} + BL_{20}, \tag{16.6a}$$

$$D_{13} = BL_{10} + BL_{30}, \tag{16.6b}$$

$$D_{23} = BL_{20} + BL_{30}. \tag{16.6c}$$

We then obtain the following formulas for branch lengths:

$$BL_{10} = \frac{\left[D_{12} + D_{13} - D_{23}\right]}{2}, \tag{16.7a}$$

$$BL_{20} = \frac{\left[D_{12} + D_{23} - D_{13}\right]}{2}, \tag{16.7b}$$

$$BL_{30} = \frac{\left[D_{13} + D_{23} - D_{12}\right]}{2}. \tag{16.7c}$$

When there are more than three OTUs, we first choose two OTUs whose distance
is the smallest, and consider the remaining ones as the composite third OTU. We can
then apply 16.7a, b, c for branch length estimation. Saitou and Nei (1986; [7])
showed the detailed way of computing all branch lengths when the number of
OTUs is either 4 or 5.

Algorithm of program FITCH of PHYLIP (http://evolution.genetics.washington.
edu/phylip.html) is based on this method, though the estimated distance is obtained
after several cycles of optimization that was not included in the original Fitch and
Margoliash's method.

Fitch and Margoliash's method [14] can be considered as an approximation of
the least-squares method. Cavalli-Sforza and Edwards (1967; [16]) proposed the
"additive tree" method in which the residual sum of squares obtained by applying
the least-squares method is used for choosing the best tree. It should be noted that
this is different from the use of the least-squares method only for estimating branch
lengths in the minimum evolution method of Rzhetsky and Nei [17].

16.2.3 Minimum Evolution Method and Related Methods

The tree topology which has the smallest sum of branch lengths is searched in the minimum evolution method. The principle of minimum evolution is in spirit similar to the maximum parsimony method (see Sect. 16.5); however, the former is a distance matrix method and the latter is a character-state method. The principle of minimum evolution is also used for the neighbor-joining method (see Sect. 16.3), but this is a stepwise clustering method, while the minimum evolution method is a completely bifurcating tree search method.

The minimum evolution method was first proposed by Edwards and Cavalli-Sforza (1964; [18]); see also Cavalli-Sforza and Edwards (1967; [16]). They were interested in the phylogenetic relationship of human populations and used the genetic distances between populations estimated from allele frequency data of several blood group loci. N populations to be compared are located within the Euclidean space with $N-1$ dimensions, and the tree whose sum of branch lengths is the minimum is chosen. This is Steiner's minimum spanning tree problem in mathematics (e.g., [19]).

Saitou and Imanishi [6] proposed a minimum evolution method by using the branch estimating method of Fitch and Margoliash [14]. They showed through computer simulations that the efficiency of this minimum evolution method was almost identical with that of the neighbor-joining method. Later, Rzhetsky and Nei (1992; [20]) used the least-squares method for estimating the sum of branch lengths.

The minimum evolution method is a completely bifurcating tree search method, and we need a certain algorithm for searching tree topologies. Rzhetsky and Nei [20] used the neighbor-joining method to obtain the initial tree and compared its surrounding tree topologies in terms of tree topology distances of Robinson and Folds [21]. If a tree with smaller sum of branches is found compared to the initial tree, that tree is used as the new initial tree, and surrounding tree topologies are searched. Computer simulations showed that the tree thus found to have the smallest sum of branch lengths was often not statistically different from the initial neighbor-joining tree when the number of sequences is not large [20]. Nei et al. (1998; [22]) also showed that the optimization principle used in minimum evolution method tends to give incorrect topologies when the number of nucleotides or amino acids used is small. This clearly indicates that the mathematically optimum tree after using a huge computation time may not be the biologically correct tree.

Pauplin (2000; [23]) proposed a new algorithm (direct calculation) to estimate the total length of any bifurcating trees, and Semple and Steel (2004; [24]) generalized this algorithm to multifurcating trees (Gascuel and Steel 2006; [25]). Mihaescu and Pachter (2008; [26]) considered a phylogenetically desirable property that weighted least-squares methods should satisfy, and they provided a complete characterization of methods that satisfy the property. Price et al. (2009; [27]) proposed a new method called FastTree that produces a minimum evolution tree not from distance matrix but from sequence profiles of internal nodes in the tree.

Another algorithm for finding the minimum evolution tree is the distance parsimony methods. Farris (1972; [28]) proposed a stepwise clustering method for distance

Fig. 16.4 Explanation of the distance Wagner method

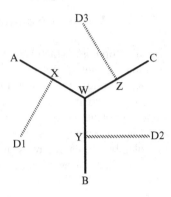

matrix data and named it "distance Wagner" method. Its algorithm is as follows. The two OTUs whose distance is the smallest are clustered first. We then compute average distance between these two OTUs, say, A and B, and the remaining OTUs, and choose the OTU, say, C, which shows the smallest average distance. At this moment, the tree is like the one in Fig. 16.3 except for OTU IDs. When the next OTU, say, D, is found as having the smallest distance with OTUs A, B, and C, there are three possibilities, D1, D2, and D3, for this OTU to be attached, as shown in Fig. 16.4. We choose the location in which the external branch for OTU D becomes the shortest. Three possible external branch lengths are estimated as

$$BL_{D1X} = \frac{\left[BL_{DW} + D_{AD} - BL_{AW}\right]}{2} \tag{16.8a}$$

$$BL_{D2X} = \frac{\left[BL_{DW} + D_{BD} - BL_{BW}\right]}{2} \tag{16.8b}$$

$$BL_{D3X} = \frac{\left[BL_{DW} + D_{CD} - BL_{CW}\right]}{2} \tag{16.8c}$$

where

$$BL_{AW} = \frac{\left[D_{AB} + D_{AC} - D_{BC}\right]}{2}, \tag{16.9a}$$

$$BL_{BW} = \frac{\left[D_{AB} + D_{BC} - D_{AC}\right]}{2}, \tag{16.9b}$$

$$BL_{CW} = \frac{\left[D_{AC} + D_{BC} - D_{AB}\right]}{2}, \tag{16.9c}$$

$$BL_{DW} = \text{Max.}\left(D_{AD} - BL_{AW}, D_{BD} - BL_{BW}, D_{CD} - BL_{CW}\right). \tag{16.9d}$$

Table 16.3 A transformed distance matrix of Table 16.1B using OTU H as the outgroup

	A	B	C	D	E	F
B	12					
C	10	10				
D	9	9	9			
E	7	7	7	7		
F	7	7	7	7	9	
G	6	6	6	6	6	6

Max. (α, β, γ) denotes the maximum value among α, β, and γ, and Eq. 16.9d came from consideration of discrete character changes which were hypothesized to be behind the distance values [28]. This procedure is continued until all OTUs are connected to the single unrooted tree.

A slight modification ("modified Farris" method) of its algorithm was proposed by Tateno et al. [15], and Faith [29] also proposed yet another modification.

16.2.4   Transformed Distance Methods

Farris et al. (1970; [30]) pointed out that a transformation of original pairwise distance with using distances from a reference OTU gives a good indicator of the relative location of the pair in the phylogenetic tree. Klotz and Blanken (1981; [31]) independently devised "the present-day ancestor method" in which distances were transformed using one arbitrarily chosen OTU as a reference. Let us assume that a phylogenetic tree contains one reference OTU 0 and N other OTUs 1 to N. If we use 16.6a, 16.6b, and 16.6c, any pairwise distance Dij ($1 \leq i < j \leq N$) can be transformed to the transformed distance (TD$_{ij}$), that is, the branch length between the reference OTU 0 and the interior node A which is connecting OTUs i and j:

$$\text{TD}_{ij} = \text{L}_{0A} = \frac{\left[D_{i0} + D_{j0} - D_{ij} \right]}{2}. \tag{16.10}$$

Let us use the distance matrix shown in Table 16.1B and consider OTU H as the reference group. The transformed distance matrix is shown in Table 16.3. We can now apply an UPGMA-like method to construct the phylogenetic tree; instead of connecting the OTU pair which shows the minimum distance, we connect the pair having the largest distance. The rationale for this is that transformed distances are in fact branch lengths between the reference OTU and the interior node that connects two OTUs in question. Therefore, the OTU pair with the largest transformed distance is neighbors (see the next section). In any case, we first cluster OTUs A and B that have the largest transformed distance, 12. Then the combined OTU AB and OTU C are clustered, followed by ABC-D and E-F simultaneously for having the same distance, then ABCD-EF, and finally G-ABIDE. The resultant tree topology is the correct one (see Fig. 16.1b). Branch lengths are also correctly reproduced by applying 16.6a, 16.6b, and 16.6c. It should be noted that this method does not produce a rooted tree. The putative root between OTU G and the combined OTU ABCDEF is simply the interior node connecting OTUs G and H, which are neighbors.

Li (1981; [32]) proposed yet another method using distance transformation. UPGMA is first used for estimating the root of the tree, and then all OTUs are divided into two clusters. Each cluster is used as combined reference OTU for estimating the phylogenetic relationship of the remaining cluster and vice versa. If the root position is correctly estimated using UPGMA, this method is expected to be quite efficient.

OOta (1998a; [33], 1998b; [34]) proposed a new method for phylogenetic tree construction called ThreeTree. This is also yet another application of 6.4a, 6.4b, and 6.4c. If we apply these equations to the three distances among any three OTUs of the distance matrix in Table 16.1B, we notice some constancy only at some cases. When two OTUs, i and j, are neighbors among the three OTUs chosen, BL_{io} and BL_{jo} are faithfully reproducing the exterior branch lengths for OTUs i and j. For example, if we consider OTUs 1 and 2 in Fig. 16.1b, they are neighbors, and we always obtain $BL_{1A}=5$ and $BL_{2A}=2$ from the distance matrix of Table 16.1B. This is because node 0 in 6.4a, 6.4b, and 6.4c corresponds to node A in Fig. 16.1b irrespective of the third OTU chosen (from 3 to 8). In contrast, if we choose non-neighboring pair of OTUs such as OTUs 1 and 4 in Fig. 16.1b, $BL_{1A}=5$ only when the third OTU is OTU 2, and $BL_{1A}=7$ for OTU 3 and $BL_{1A}=8$ for OTUs 5–8. OOta [33, 34] thus proposed to use this invariance of external branch length estimations as the indicator of neighbors. If the distance matrix is purely additive, this ThreeTree method was proven to reconstruct the correct phylogenetic tree by simultaneously finding neighboring pairs at each step [34]. The problem is, however, the real data often do not satisfy the pure additivity, and the heterogeneity of external branches may be obtained even when the OTU pair in question is neighbors.

16.2.5 Use of Quartet OTUs for Tree Construction

When there are only four OTUs 1–4 or quartets, there are three possible tree topologies: OTUs 1 and 2 (or OTUs 3 and 4) are neighbors, OTUs 1 and 3 (or OTUs 2 and 4) are neighbors, or OTUs 1 and 4 (or OTUs 2 and 3) are neighbors (see Fig. 3.25). Buneman (1971; [35]) showed that for one particular unrooted tree, say, OTUs 1 and 2 are neighbors, to be true, the distances should satisfy the following condition:

$$D_{12} + D_{34} < D_{13} + D_{24} \quad and \quad D_{12} + D_{34} < D_{14} + D_{23}. \quad (16.11)$$

These inequalities are often called "four-point metric." Let us show its justification. If we examine the tree shown in Fig. 16.5 with four OTUs 1–4, five branch lengths are estimated as

$$BL_{1X} = \frac{\left[D_{12} + D_{13} - D_{23}\right]}{2} = \frac{\left[D_{12} + D_{14} - D_{24}\right]}{2}, \quad (16.12a)$$

$$BL_{2X} = \frac{\left[D_{12} + D_{23} - D_{13}\right]}{2} = \frac{\left[D_{12} + D_{24} - D_{14}\right]}{2}, \quad (16.12b)$$

Fig. 16.5 An unrooted
tree for four OTUs

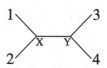

Table 16.4 Matrix of neighboring pair for all quartet trees

	A	B	C	D	E	F	G
B	15						
C	10	10					
D	6	6	6				
E	1	1	2	4			
F	1	1	2	4	15		
G	1	1	2	4	6	6	
H	1	1	2	4	6	6	15

$$BL_{3Y} = \frac{\left[D_{34} + D_{13} - D_{14}\right]}{2} = \frac{\left[D_{34} + D_{23} - D_{24}\right]}{2}, \qquad (16.12c)$$

$$BL_{4Y} = \frac{\left[D_{34} + D_{14} - D_{13}\right]}{2} = \frac{\left[D_{34} + D_{24} - D_{23}\right]}{2}, \qquad (16.12d)$$

$$BL_{XY} = \frac{\left[D_{13} + D_{24} - D_{12} - D_{34}\right]}{2} = \frac{\left[D_{14} + D_{23} - D_{12} - D_{34}\right]}{2}, \qquad (16.12e)$$

when the pure additivity is assumed. If the tree shown in Fig. 16.5 is the correct one, BL_{XY} should be positive. Therefore, we obtain two simultaneous inequalities in 16.11. Under the pure additivity, inequalities (16.11) become more strong as

$$D_{12} + D_{34} < D_{13} + D_{24} = D_{14} + D_{23}. \qquad (16.13)$$

Therefore, inequalities (16.11), called relaxed additivity (Fitch 1981; [36]), may be used when the distances are not purely additive because of parallel or backward changes.

Saitou and Nei (1986; [7]) showed that inequalities (16.11) are also required for the transformed distance method and the distance Wagner method when the number of OTUs is four. Saitou and Nei (1987; [1]) also showed that the neighbor-joining method requires inequalities (16.11) to choose the tree shown in Fig. 16.5. Therefore, if a distance matrix method does not reduce its tree choice criterion to the four-point metric shown in (16.11) for the four OTU situation, that criterion may not be appropriate. Unfortunately, the minimum deviation method of Fitch and Margoliash [14] does not show this characteristic, as shown by Saitou and Nei [7].

Sattath and Tversky [37] proposed a method for examining all possible quartet trees out of N OTUs. Therefore, we have to search $_NC_4$ (=N(N−1)(N−2)(N−3)/ 4•3•2•1) quartets. Each quartet should have two neighbors after checking 4-point metric. If these two are really neighbors in the N-OTU tree, they become neighbors $_{N-2}C_2$ times. If we consider the distance matrix of Table 6.1B, N=8, we should expect 15 (=$_6C_2$) neighbors in all quartet trees if that pair of OTUs is really neighbors. Table 16.4 shows values of neighbors for all possible pairs, and in fact, there are three pairs of OTUs (A-B, E-F, and G-H) whose neighbor numbers are 15. We then join them as neighbors and compute average distances between these two and the remaining ones. The same procedure will be conducted for 5 OTUs at the next step, and finally the correct unrooted tree shown in Fig. 16.2b can be obtained. The problem of this method is combinatorial explosion as N becomes large.

Fitch (1981; [36]) proposed "neighborliness method" in which the relaxed additivity (16.11) is also applied. Although that method was meant to find the minimum length tree, Saitou and Nei [1] showed that their neighbor-joining method found more minimum tree than the neighborliness method in one empirical data. However, all neighbors are expected to be found in the neighborliness method as in the case of Sattath and Tversky's method [37], and the use of quartets is a promising way of finding the tree structure.

16.3   Neighbor-Joining Method

The author is proud of the proposal of the neighbor-joining method as one chapter of his Ph.D. dissertation (Saitou, 1986; [38]), which was later published in *Molecular Biology and Evolution* (Saitou and Nei 1987; [1]). Partly because of this personal connection, but also because of its popularity even a quarter century after its publication, we will discuss this method in detail.

16.3.1   What Are Neighbors?

A pair of OTUs is called "neighbors" when these are connected through a single internal node in an unrooted bifurcating tree [1]. Essentially the same definition was first given by Sattath and Tversky [37]; see also Fitch [36]. For example, OTUs A and B of Fig. 16.1b are a pair of neighbors. If we combine these OTUs, these combined OTU AB and OTU C become a new pair of neighbors. It is thus possible to define the topology of a tree by successively joining pairs of neighbors and producing new pairs of neighbors as mentioned in Chap. 2. In general, $N-3$ pairs of neighbors are necessary to define the topology of an unrooted tree with N OTUs. It should be noted that the "neighbors" concept can also be applied to rooted trees; OTU pairs A-B and D-E in Fig. 16.1a are neighbors.

16.3.2   Algorithm of the Neighbor-Joining Method

The neighbor-joining method is a distance matrix method as well as a stepwise clustering method. A unique final unrooted tree is produced by sequentially finding pairs of neighbors by examining a distance matrix if we apply the neighbor-joining method to a distance matrix data. The principle of minimum evolution is used for a local tree with multifurcation. When the phylogenetic tree is purely additive, the neighbor-joining method is always producing the correct tree from the corresponding

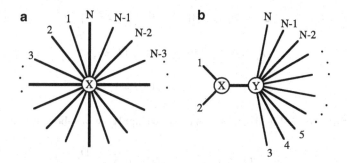

Fig. 16.6 Two types of multifurcating trees. (**a**) Star phylogeny with no interior branch. (**b**) Multifurcating tree with only one interior branch

distance matrix [1]. Rzhetsky and Nei [17] proved that the expected value of the sum of branch lengths is smallest for the tree with the true branching pattern. Because of the simple algorithm, a large number of OTUs can be handled within a relatively short computer time by using the neighbor-joining method.

The algorithm of the neighbor-joining method is as follows. We start from a starlike tree, which is produced under the assumption of no clustering among all the N OTUs compared (see Fig. 16.6a). Under this tree, all branches are external ones, and

$$D_{ij} = BL_{iX} + BL_{jX}. \tag{16.14}$$

Thus, the sum (So) of N external branch lengths can be shown to be

$$So = \sum_{i=1}^{i=N} BL_{iX} = \frac{Q}{(N-1)}, \tag{16.15}$$

where

$$Q = \sum_{i<j} D_{ij}. \tag{16.16}$$

This is because BL_{iX} ($i = 1 \sim N$) is covered $N-1$ times when all lower-diagonal (or upper-diagonal) pairwise distances are added.

In practice, some pairs of OTUs are more closely related to each other than other pairs are. Among all the possible pairs of OTUs ($N[N-1]/2$ pairs for N OTUs), we choose the one that gives the smallest sum of branch lengths. Let us consider the tree of Fig. 16.6b, where OTUs i and j are assumed to be neighbors. The sum of branch lengths is defined by

$$S_{ij} = \left(BL_{iX} + BL_{jX}\right) + BL_{XY} + \sum_{k \neq i,j} BL_{kY} \tag{16.17}$$

where $BL_{\alpha\beta}$ is branch length between nodes α and β. There are the following relationships between distances and branch lengths:

$$D_{ij} = BL_{iX} + BL_{jX}, \tag{16.18a}$$

$$D_{ik} = BL_{iX} + BL_{XY} + BL_{kY} \left(k \neq i, j \right) \tag{16.18b}$$

$$D_{jk} = BL_{jX} + BL_{XY} + BL_{kY} \left(k \neq i, j \right) \tag{16.18c}$$

$$D_{kl} = BL_{iY} + BL_{jY} \left(k, l \neq i, j \right). \tag{16.18d}$$

Under the tree of Fig. 16.6b, it can be shown applying the above relationship:

$$BL_{XY} = \frac{\left[Q - (N-1)D_{ij} - (N-1) \sum_{k,l \neq i,j} \frac{D_{kl}}{(N-3)} \right]}{2(N-2)}. \tag{16.19}$$

If we neglect OTUs i and j in Fig. 16.6b, the remaining $N-2$ OTUs form a starlike tree. We thus apply 16.15 and obtain

$$\sum_{k \neq i,j} B_{kY} = \sum_{k,l \neq i,j} \frac{D_{kl}}{(N-3)}. \tag{16.20}$$

We also note that

$$\sum_{k,l \neq i,j} D_{kl} = Q - \left(R_i + R_j - D_{ij} \right), \tag{16.21}$$

where

$$R_i = \sum_j D_{ij}, \tag{16.22a}$$

$$R_j = \sum_i D_{ij}. \tag{16.22b}$$

Putting 16.18a, 16.19, and 16.20 into 16.17 with considering 16.21, we obtain

$$S_{ij} = \frac{D_{ij}}{2} + \frac{\left[2Q - R_i - R_j \right]}{2(N-2)}. \tag{16.23}$$

The author modified the FORTRAN program algorithm based on Saitou and Nei (1987; [1]) to much simplified version corresponding to 16.23 in 1986 following a suggestion by Dr. Clay Stephens who pointed out some redundant DO loops in the original program. Equation 16.23 was also shown by Studier and Keppler (1988; [39]).

In any case, this S_{ij} value is computed for $N(N-1)/2$ pairs of OTUs, and the pair that has the smallest S_{ij} value is chosen as neighbors. This pair of OTUs is then regarded as a single OTU, and the new distances between the combined OTU and the remaining ones are computed by averaging. This procedure is continued until all pairs of neighbors are found. This derivation was given by Saitou (1996; [4]).

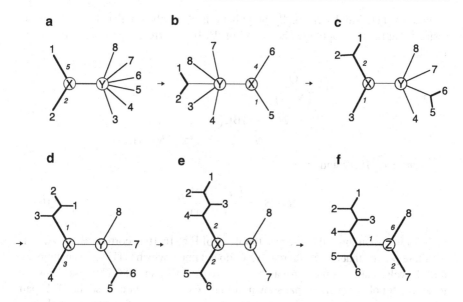

Fig. 16.7 Construction of the neighbor-joining tree for eight OTUs (From Saitou and Nei 1987; [1])

If OTUs i and j are chosen as neighbors as shown in Fig. 16.6b, the branch lengths are estimated as

$$BL_{iX} = \frac{D_{ij}}{2} + \frac{\left(R_i - R_j\right)}{2\left(N-2\right)},$$ (16.24a)

$$BL_{jX} = D_{ij} - BL_{iX}.$$ (16.24b)

Therefore, all the branch lengths as well as the tree topology will be determined after $N-2$ steps for N OTUs. Saitou and Nei (1987; [1]) showed that branch length estimates by using the neighbor-joining method are least-squares estimates. The stepwise clustering algorithm used in the neighbor-joining method is the same with that of UPGMA/WPGMA, and the branch length estimation method is the same with WPGMA.

When discrete character data such as nucleotide or amino acid sequences are used for estimating distance matrices, it is more informative to estimate the number of mutational events occurred at each branch of the phylogenetic tree. Ishida et al. (1995; [40]) thus proposed a method to transform decimal values to integer values for each branch length by multiplying the number of nucleotides and rounding (see also Chap. 15 on this point).

The unrooted tree shown in Fig. 16.1b corresponds to the distance matrix shown in Table 16.1B. Figure 16.7 shows how this tree is constructed from that distance matrix using the neighbor-joining method [1].

When OTUs i and j are really neighbors, is S_{ij} really smaller than S_O? Let us assume that the tree topology shown in Fig. 16.1b is correct. Then, from 16.15 and 16.18a,

$$So = \frac{Q}{(N-1)}$$

$$= D_{ij} + \frac{2(N-2)BL_{XY}}{(N-1)} + \sum_{k,l \neq i,j} \frac{D_{kl}}{(N-3)}.$$

(16.25)

Then, from 16.17 and 16.6,

$$S_O - S_{ij} = \frac{(N-3)BL_{XY}}{(N-1)}.$$

(16.26)

Because we assume that the tree topology of Fig. 16.1b is correct, $BL_{XY}>0$. It is therefore proven that $S_{ij}<S_O$. In reality, we do not know which OTU pair is neighbors; thus, S_{ij} values are computed for all possible $N(N-1)/2$ pairs of OTUs, assuming the local tree topology such as that shown in Fig. 16.6b. We then choose the OTU pair whose S_{ij} value is the minimum, by applying the principle of minimum evolution. The OTU pair which shows the smallest S_{ij} value may not be the neighbors. However, the neighbor-joining method was shown to give correct neighbors if distances are purely additive. It should be noted that the "proof" of this proposition given by Saitou and Nei [1] was incorrect, and the real proof was given by Studier and Keppler [39].

After the OTU pair which shows the smallest S_{ij} value is found, these two OTUs are joined, and the average distance between D_{ik} and D_{jk} $(k \neq i, j)$ is computed:

$$D_{(ij)k} = \frac{(D_{ik} + D_{jk})}{2}.$$

(16.27)

It should be noted that this distance averaging is the same with that of WPGMA, not UPGMA. After that, the same search is conducted for $N-1$ pairs of OTUs. Repeats of this procedure for up to $N-3$ times will finally produce an unrooted tree with all branch lengths [1].

16.3.3   A Worked-Out Example of Tree Construction Using the Neighbor-Joining Method

Let us use the distance matrix of Table 16.1C estimated from the real nucleotide sequence data [2]. The resulting neighbor-joining tree is shown in Fig. 16.2c with bootstrap probabilities [117] (see also website of this book). The human sequence was used as out-group to root this tree (see Chap. 3 for out-group rooting). The integer branch lengths were estimated by using Ishida et al.'s [39] method. The four pseudogene sequences formed a tight cluster with bootstrap probability of 97 %,

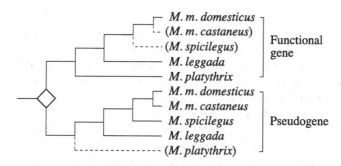

Fig. 16.8 Expected tree for 10 sequences which produced the distance matrix of Table 16.1C

and their evolutionary rates are much larger than those for functional genes. These pseudogenes are processed type and are considered to be "dead on arrival." Because of this, they became free of selective constraints on their functional homologous genes, and the evolutionary rate was increased, probably equivalent to the mutation rate (see Chap. 4). This pseudogene cluster contains sequences from *Mus musculus*, *M. spicilegus*, and *M. leggada*, suggesting that the pseudogene was created at the common ancestor of these three species. Therefore, the expected tree (corresponding to tree 11 of Ohtsuka et al. [2]) is the one shown in Fig. 16.8. Saitou [4] showed the UPGMA tree for this distance matrix, in which four pseudogenes did not form a cluster. This is apparently because their evolutionary rates are much higher than those for functional genes, and the violation of constancy of the evolutionary rate produced an erroneous tree when UPGMA was applied.

16.3.4 Methods Related to the Neighbor-Joining Method

There are four ancestors for the neighbor-joining method [1]. First of all, the minimum evolution principle was from Edwards and Cavalli-Sforza's original minimum evolution method [16, 18], and the branch length estimation is from that originally proposed by Fitch and Margoliash [14], distance averaging is following the WPGMA algorithm [9], and finally its name was hinted from the neighborliness method of Fitch [36].

The simultaneous partitioning method by Saitou (1986; [38]) may be considered as the predecessor of the neighbor-joining method. While partitions of N OTUs into 2 and $N-2$ OTUs are considered in the neighbor-joining method, any possible partitions (splits) of N OTUs into the two clusters A and B, with m and n OTUs ($m+n=N$; $m, n \geq 2$), respectively, are considered in the simultaneous partitioning method, and the sum of branch lengths, S_{AB}, is computed as

$$S_{AB} = \sum_{A,B} \frac{D_{ij}}{mn} + \sum_{A} \frac{D_{ij}}{m} + \sum_{B} \frac{D_{ij}}{n}, \tag{16.28}$$

where the first, second, and third summation are for every pair of i and j belonging to both set A and set B, to set A, and to set B, respectively [38]. We compute S_{AB} values for all possible partitions and first select the partition that is the smallest. We then check the partition compatibility with second smallest with that of the first one, and if they are mutually compatible, the second partition is chosen. If not, we examine the third and later ones until the compatible pair is found. If we apply this algorithm to the distance data of Table 16.1B, S_{AB} is smallest for partition 1234–5678, followed by 123–45678 and 12–345678. These three partitions are the correct ones (see Fig. 16.1b); however, the fourth one, 12348–567, is not a correct partition, although this partition is compatible with the former three partitions [38]. The distance matrix shown in Table 16.1B is purely additive and ideal one for any good distance matrix method. Therefore, the simultaneous partitioning method is not guaranteed to find the true tree. Another problem is the huge possibility of partitions; the total number, N_{AB}, to separate N OTUs into two sets is

$$S_{AB} = 2^{N-1} - N - 1. \tag{16.29}$$

As N becomes large, say, 100, S_{AB} becomes $\sim 6.34 \times 10^{29}$. In this regard, any method to try to separate the whole OTUs into two, or so-called top-down method, has the same problem of combinatorial explosion.

Gascuel (1997; [41]) proposed the BIONJ method in which a simple first-order model of the variances and covariances among distances are used when the new distances are computed after finding the neighbors. Molecular sequences such as nucleotide or amino acid sequences are required to estimate variances and covariances. When the number of sites compared is S, approximate variances (V_{ij}) and covariances ($COV_{ij,kl}$) are estimated as

$$V_{ij} \sim \frac{D_{ij}}{S} \tag{16.30a}$$

$$COV_{ij} \sim \frac{D_{uv}}{S} \tag{16.30b}$$

where u and v represent common ancestral nodes for sequences i and j and for k and l, respectively. Interestingly, as a branch length is estimated from three involved distances in Eq. (16.7), the covariance of $D_{ik,jk}$ is given as

$$COV_{ik,jk} \sim \frac{1}{2}\left(V_{ik} + V_{jk} - V_{ij}\right). \tag{16.31}$$

In any case, these variances and covariances are used for estimating the weighting factor λ in the following equation for estimating the averaged distance between the newly found neighboring sequences i and j and sequence k:

$$D_{(ij)k} = \lambda\left(D_{ik} - BL_{iu}\right) + \left(1 - \lambda\right)\left(D_{jk} - BL_{ju}\right). \tag{16.32}$$

λ is always half for the neighbor-joining method. It should be noted that the new averaged distance is not the simple average between the original distances as in the neighbor-joining method as given in Eq. 16.27, but the external branches are subtracted. Computer simulations showed that when the evolutionary rates are quite heterogeneous and the amount of divergence was large, the BIONJ method performed better than the neighbor-joining method, while results of the two methods were essentially the same when the amount of divergence is small or when the approximate constancy of the evolutionary rate is held [41]. It should be noted that the reliability of estimated distances is not high when the amount of divergence is large and any method may not perform well to reconstruct correct phylogenetic trees.

Bruno et al. (2000; [42]) proposed Weighbor, another version of the weighted neighbor-joining method, as well as introducing two criteria to find out neighbors. One criterion is called additivity, where the pair of OTUs i and j which shows constancy of $D_{ik} - D_{jk}$ is chosen as neighbors. This logic is the same with that of ThreeTree method [33, 34]. Another criterion of neighbors is called positivity, where $D_{ik} + D_{jl} - D_{ik} - D_{kl}$ is always nonnegative for any OTUs k and j. This is mathematically identical with the relaxed additivity relation, as we discussed in Sect. 16.2.5.

The single tree is selected at each step in the original neighbor-joining method, and there are some modifications to consider multiple candidate trees [43, 44]. Gascuel and Steel (2006; [25]) gave a review on theoretical development of branch length and tree topology estimation related to the neighbor-joining method.

16.4 Phylogenetic Network Construction from Distance Matrix Data

A tree is a special case of networks in the graph theory (Chap. 3). Dress (1984; [45]) pioneered in this field, and Dress et al. (2012; [46]) presented the most updated explanations on phylogenetic networks. Let us discuss the network construction from distance matrix data. The simplest case is quartets. There are three possible phylogenetically informative splits for OTUs 1, 2, 3, and 4: split 1 (12–34), split 2 (13–24), and split 3 (14–23). However, there are only six pairwise distances for 4 OTUs, and we can consider only two-dimensional networks in this case, composed of two phylogenetically informative splits as well as four phylogenetically noninformative splits: 1–234, 2–134, 3–124, and 4–123. If the distance matrix follows relation 16.13, we have only split 1, and a nonreticulated unrooted tree (Fig. 16.5) is obtained. Let us assume that the following relaxed additivity relation holds for the given distance matrix:

$$D_{12} + D_{34} < D_{13} + D_{24} < D_{14} + D_{23}. \tag{16.33}$$

Under this condition, we have the network shown in Fig. 16.9. Lengths of four external branches (corresponding to phylogenetically noninformative splits) and two phylogenetically informative splits (α and β) are estimated by solving the following equations:

$$D_{12} = BL_{15} + BL_{26} + \beta, \tag{16.34a}$$

Fig. 16.9 A phylogenetic
network with two splits for
four OTUs

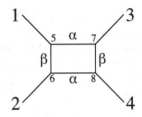

Table 16.5 Partial distance matrix for four OTUs of
Table 16.1C (data from [2])

	1	2	3
2	0.0582		
3	0.0754	0.0284	
4	0.0651	0.0125	0.0285

OTU IDs; 1 = *M. leggada* pseudogene, 2 = *M. m. domesticus* functional gene, 3 = *M. platythrix* functional gene,
4= *M. leggada* functional gene. Please note that OTU ID
numbers are different from those of Table 16.1C

$$D_{34} = BL_{37} + BL_{48} + \beta, \tag{16.34b}$$

$$D_{13} = BL_{15} + BL_{37} + \alpha, \tag{16.34c}$$

$$D_{24} = BL_{26} + BL_{48} + \alpha, \tag{16.34d}$$

$$D_{14} = BL_{15} + BL_{48} + \alpha + \beta, \tag{16.34e}$$

$$D_{23} = BL_{26} + BL_{37} + \alpha + \beta. \tag{16.34f}$$

If we use the distance matrix shown in Table 16.5, $D_{12}+D_{34}=0.0867$, $D_{13}+D_{24}=0.0879$, and $D_{14}+D_{23}=0.0935$. Therefore, condition 16.33 is satisfied, and sequences 1–4 should have the phylogenetic network shown in Fig. 16.9. From 16.34a, 16.34b, 16.34c, 16.34d, 16.34e, and 16.34f,

$$BL_{15} = \frac{\left\{\left(D_{12} + D_{13}\right) - D_{23}\right\}}{2}, \tag{16.35a}$$

$$BL_{26} = \frac{\left\{\left(D_{12} + D_{24}\right) - D_{14}\right\}}{2}, \tag{16.35b}$$

$$BL_{37} = \frac{\left\{\left(D_{13} + D_{34}\right) - D_{14}\right\}}{2}, \tag{16.35c}$$

$$BL_{48} = \frac{\left\{\left(D_{24} + D_{34}\right) - D_{23}\right\}}{2}, \tag{16.35d}$$

$$\alpha = D_{13} - BL_{15} - BL_{37} = \frac{\{(D_{23} + D_{14}) - (D_{12} + D_{34})\}}{2}, \quad (16.35e)$$

$$\beta = D_{12} - BL_{15} - BL_{26} = \frac{\{(D_{23} + D_{14}) - (D_{13} + D_{24})\}}{2}. \quad (16.35f)$$

It is interesting to note that 16.35a, 16.35b, 16.35c, and 16.35d are the application of 16.7a, 16.7b, and 16.7c. Thus, using the real data shown in Table 16.6, $BL_{15} = 0.0526$, $BL_{26} = 0.0028$, $BL_{37} = 0.0194$, $BL_{48} = 0.0063$, $\alpha = 0.0034$, and $\beta = 0.0028$. Clearly, a reticulation was observed.

Bandelt and Dress (1992; [47]) proposed the Split Decomposition method for distance matrix data. All possible quartets are first examined as in Sattath and Tversky's [37] method, and an unrooted tree for four OTUs is constructed for each quartet. The interior branch length ("isolation index" in their terminology) is computed for each tree using 16.12e. Because a phylogenetic network is a combination of certain splits, quartets with the same interior branches are searched to find appropriate splits. Fig. 16.10a shows a phylogenetic network constructed by using the Split Decomposition method from a distance matrix estimated by Kitano et al. (2009; [48]).

Bryant and Moulton (2004; [49]) further developed the Neighbor-Net method by extracting more information from distance matrix data. Its name came from some similarity on search space procedure with the neighbor-joining method [1], yet it produces not trees but networks. The final set of splits obtained by using the Neighbor-Net method form a circular link. Interested readers are advised to read the original paper on the mathematical details.

Huson and Bryant (2006; [50]) developed a software for constructing and drawing phylogenetic networks from distance matrix data using Neighbor-Net. Fig. 16.10b shows a phylogenetic network by using the Neighbor-Net method for the same dataset except for adding the human sequence used for Fig. 16.10a. It is clear that much more splits were extracted in the Neighbor-Net network compared to the Split Decomposition network.

16.5   Maximum Parsimony Method

There are several kinds of maximum parsimony methods based on various assumptions, but the maximum parsimony principle is used for any parsimony method. We will only discuss the parsimony method that is frequently used for molecular data. This type of parsimony method produces unrooted trees as in the case of the neighbor-joining method [1] but is for character-state data, and it typically takes a completely bifurcating tree search.

The maximum parsimony principle is the minimization (or maximize parsimony) of the changes on character states, such as nucleotide or amino acid sequences, on the given tree topology (usually perfectly bifurcating type), and is related to the principle used in minimum evolution methods. However, the performance of these

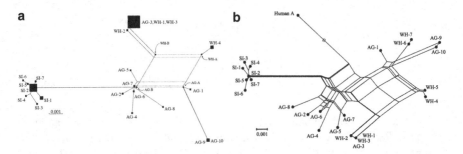

Fig. 16.10 Two phylogenetic networks constructed from a distance matrix data of 24 gibbon ABO blood group gene partial sequences (From Kitano et al. 2009; [48]). (**a**) When the Split Decomposition method [47] was used. (**b**) When the Neighbor-Net method [49, 50] was used (human A allele sequence is also added)

two methods in choosing the best topology can be quite different. There was a very long and often misleading controversy between phenetics and cladistics in the 1970s–1980s. The maximum parsimony principle was the basis of cladistics and was originally conceptualized for morphological data (e.g., Camin and Sokal (1965; [51])). However, the maximum parsimony method that produces unrooted tree was also proposed at the same time in molecular evolutionary studies by Eck and Dayhoff (1966; [52]). We will assume nucleotide sequences in this section. A more comprehensive description on various kinds of maximum parsimony methods is given in Felsenstein (2004; [53]).

16.5.1   The Basic Algorithm of the Maximum Parsimony Method

The essence of the maximum parsimony method is to minimize the changes. Because each nucleotide site is assumed to evolve independently, we consider only one site at one time, and the number of necessary change for each site is summed for all sites compared. Nucleotide sites that do not contribute to the selection of tree topology are called "noninformative" for the maximum parsimony method. Because invariant sites with only one type of nucleotide require no change, they are noninformative. This corresponds to configuration 1 of Table 16.6. When there are two types of nucleotides at one site, we need at least one change in any tree topology. Configurations 2–8 of Table 16.6 are of this type. When one nucleotide is found only in one sequence (configurations 2–5), this type of configuration or site is called "singleton." Any tree topology requires one change for singleton sites, and they are also phylogenetically noninformative. This is because the change is assigned to the exterior branch connecting to the sequence with the unique nucleotide. A substitution $\alpha1$ ($A \Rightarrow G$) at site 1 in Fig. 16.11 is this type of change. Of course, these changes are important when we try to estimate external branch lengths, but they are by definition not involved in tree topology determination.

If three kinds of nucleotides exist in one site, such as configurations 9–14 of Table 16.6, we need at least two changes for this site. If one nucleotide, say, A, is

Table 16.6 Fifteen nucleotide configurations for four nucleotide sequences

	A	B	C	D
1	°	°	°	°
2	°	°	°	X
3	°	°	X	°
4	°	X	°	°
5	X	°	°	°
6	°	°	X	X
7	°	X	°	X
8	°	X	X	°
9	°	°	X	Y
10	°	X	°	Y
11	°	X	Y	°
12	X	°	°	Y
13	X	°	Y	°
14	X	Y	°	°
15	°	X	Y	Z

Note: °, X, Y, and Z are different nucleotides with each other

Fig. 16.11 The maximum parsimony tree (*left*) for the sequence data (*right*)

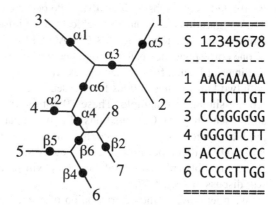

```
==========
S  12345678
----------
1  AAGAAAAA
2  TTTCTTGT
3  CCGGGGGG
4  GGGGTCTT
5  ACCCACCC
6  CCCGTTGG
==========
```

the majority and the other two nucleotides, say, C and T, exist only in one sequence each, this site is also noninformative. This is because two independent changes in exterior branches are assigned for this case. Substitutions $\alpha2$ (T$\Rightarrow$C) and $\beta2$ (T$\Rightarrow$G) at site 2 in Fig. 16.11 correspond to this situation. Because there are only four sequences in Table 16.6, configurations 9–14 are all noninformative. The same logic applies to sites in which four kinds of nucleotides exist, and configuration 15 of Table 16.6 is also noninformative. Gaps in multiple alignments can be considered as the fifth nucleotide.

An informative nucleotide configuration or site should have more than one kind of nucleotides, and at least two of them are observed at more than one sequence. Sites 3–6 in Fig. 16.11 are all informative. Sequences 1 and 2 share nucleotide C at

Table 16.7 Ten informative configurations for five nucleotide sequences

	A	B	C	D	E
1	X	X	*	*	*
2	X	*	X	*	*
3	X	*	*	X	*
4	X	*	*	*	X
5	*	X	X	*	*
6	*	X	*	X	*
7	*	X	*	*	X
8	*	*	X	X	*
9	*	*	X	*	X
10	*	*	*	X	X

Note: * and X are different nucleotides with each other

site 3, and substitution α3 (G⇔C) is responsible for this pattern. There are three types of nucleotides in site 4, but nucleotide C is observed only at sequence 6; thus, substitution β4 and substitution α4 are assumed for the tree shown in Fig. 16.11. Although sequences 1 and 5 share nucleotide A at site 5, they are outcomes of two parallel substitutions (α5 and β5). The nucleotide distribution at site 6 can be interpreted as two substitutions α6 and β6 occurred at two interior branches.

Configurations 6, 7, and 8 of Table 16.6 are informative, and they correspond to three possible tree topologies AB-CD, AC-DB, and AD-BC, respectively. This one-to-one correspondence is the unique feature for quartet sequences, for there is only one interior branch for four sequences. Table 16.7 shows nucleotide configurations with two kinds of nucleotides for five sequences. The remaining 41 configurations are not shown in this table. These 10 configurations correspond to 10 possible splits (see Chap. 3). Combinations of two compatible configurations or splits will produce a tree. For example, configurations 1 and 10 will produce tree topology AB-C-DE. Fifteen possible tree topologies for five sequences are thus produced. If incompatible splits are combined, they will produce reticulated network, as we will discuss in Sect. 16.7.

We now move to the determination of a tree topology. Let us use the multiple alignment of 8 sequences in Fig. 16.11 as an example. Because noninformative configurations do not contribute to the determination of the best topology, we consider only informative sites 3 to 6. Sites 3 and 5 support splits 12–345678 and 15–234678, respectively. There are three nucleotides at sites 4 and 6, and two splits are embedded in each site: 1234–5678 and 12346–578 for site 4 and 123–45678 and 12356–478 for site 6. There are 10,395 completely bifurcating tree topologies for 8 sequences. Obviously, we do not need to examine all these topologies and should consider tree topologies which are compatible with some of these six splits. For example, let us consider tree topology $(((1,2),(3,4)),((5,6),(7,8)))$. The necessary numbers of changes for this tree for sites 3–6 are 1, 2, 2, and 3, respectively. Thus, the total number of required changes for this tree topology becomes 11, with changes for sites 1 and 2. If we consider tree topology $((((1,2),3),4),((5,6),(7,8)))$, shown in Fig. 16.11, the total number of required changes becomes 10. In fact, this tree is a maximum parsimony tree. Alternative trees ((((1,2),3),4),(((5,6),7),8)) and ((((1,2),3),4),(((5,6),8),7)) are also maximum parsimony trees. These three trees are called "equally parsimonious trees." Fitch (1977; [54]) devised an algorithm to count the total number of required changes for a given tree topology. As we will discuss in Sect. 16.8, various tree topology searching methods are available.

It may be useful to consider the maximum (T_{max}) and the minimum (T_{min}) numbers of changes for N sequences compared in the maximum parsimony method (nomenclature is from Saitou 1986; [38]):

$$T_{max} = \sum_i \left(N - N_{majority}[i]\right), \tag{16.36a}$$

$$T_{min} = \sum_i \left(K_{nuc}[i] - 1\right), \tag{16.36b}$$

where $N_{majority}[i]$ is the number of sequences with the majority nucleotide at site i and $K_{nuc}[i]$ is the number of kinds of nucleotides observed at site i. The number of required changes for the maximum parsimony tree should be between these two values. For example, T_{max} and T_{min} for the sequence data shown in Fig. 16.11 are 15 and 9, respectively.

16.5.2 Stepwise Clustering Algorithms for Constructing Maximum Parsimony Trees

Although the popular way to find the maximum parsimony tree(s) from a given multiply aligned data is to compare perfectly bifurcating trees, there are a series of stepwise clustering algorithms. Hartigan (1973; [55]) proposed a stepwise algorithm for constructing maximum parsimony trees. The tree construction algorithm is similar to that of the neighbor-joining method [1]. Saitou (1986; [38]) independently devised a similar algorithm when he proposed the neighbor-joining method. Let us explain this character-state version of the neighbor-joining method following Saitou [38]. We start from the starlike tree which has no interior branches (Fig. 16.6a). If there are no informative sites in the multiply aligned sequences, this is the maximum parsimony tree. When the sequence length is short, this kind of tree may be obtained from the real data. If there are phylogenetically informative sites, we should expect to have some interior branches which minimize the total number of changes. As in the local topology search used for the neighbor-joining method, we consider a multifurcating tree such as one shown in Fig. 16.6b and count the required number (T_{ij}) of changes for that tree assuming sequences i and j as neighbors. The counting algorithm is a straightforward extension of Fitch's [54] algorithm. When the required number of changes at one site becomes identical with T_{min} given in 16.30b, this site is eliminated from the later steps.

Zharkikh (1977; [56]; see also [57]) proposed an algorithm for finding the maximum parsimony tree in a stepwise way. Saitou (1998; [58]) independently discovered the algorithm mathematically identical with that of Zharkikh [56] and named it SSJ

Table 16.8 Application of the SSJ algorithm to an ideal sequence data

Step 0: Initial sequences after elimination of invariant sites and sequence identity check	Step 1: After elimination of non-informative sites	Step 2: After joining identical sequences	Step 3:
`00000000011111111112222`	`0000000001`	`0000000001`	
`12345678901234567890123`	`1234567890`	`1234567890`	`2345679`
A aacgtttcattgagatacgtgca	A aacgtttcat	AB aacgtttcat	AB acgttta
B ....gc...........	B	C c.........	C ....
C c........g.g..........	C c.........	D cta.......	D ta.....
D cta.......g..tt........	D cta.......	E ctacga....	E tacga..
E ctacga....g....cg......	E ctacga....	FG ctacgagt..	FG tacgag.
F ctacgagt..g......a.....	F ctacgagt..	H ctacgag.c.	H tacgagc
G ctacgagt..g.......ca...	G ctacgagt..	IJ ctacgag.cg	IJ tacgagc
H ctacgag.c.g.........t..	H ctacgag.c.		
I ctacgag.cgg..........g.	I ctacgag.cg		
J ctacgag.cgg...........c	J ctacgag.cg		

Step 4:	Step 5:	Step 6:
`2345679`	`4567`	`4567`
		`-------p`
ABC acgttta	ABC gttt	ABCD gttt
D ta.....	D	E cga.
E tacga..	E cga.	FGHIJ cgag
FG tacgag.	FG cgag	
HIJ tacgagc	HIJ cgag	

(Simultaneous Sequence Joining) for preprocessing multiply aligned sequences to produce phylogenetic networks (see Sect. 16.7). If all informative sites are mutually compatible, or if there are no parallel nor backward changes, this method is very quick to construct maximum parsimony trees. Algorithm of SSJ consists of sequence-checking operation and site-checking operation. We first eliminate invariant sites. We then examine identity of sequences, and identical ones are joined. The next step is detection of noninformative sites, and these are eliminated. We then return to sequence identity check. If all the noninformative sites are eliminated, neighbors should have the identical sequences, and they are joined. This joining transforms informative sites to noninformative sites in some cases. Thus, we again detect noninformative sites. These recursive sequence-checking and site-checking operations find the tree structure in a stepwise way. Table 16.8 shows the application of SSJ algorithm to an ideal sequence data with 10 sequences. Step 0 shows the initial sequence alignments. Invariant sites were omitted, and for simplicity, sites are sorted depending on their configurations. When we eliminate noninformative sites 11–23 at step 1, three pairs of sequences (A-B, F-G, and I-J) become identical, and they are joined as neighbors at step 2. Three new noninformative sites (1, 8, and 10) are found, and they are eliminated at step 3. Two identical pairs of sequences (AB-C and H-IJ) are found and they are joined at step 4. Three noninformative sites (2, 3, and 9) are eliminated at step 5, and two pairs of sequences (ABC-D and FG-HIJ) are joined at step 6. All three remaining sites become noninformative

Fig. 16.12 The maximum
parsimony tree for sequence
data shown in Table 16.8

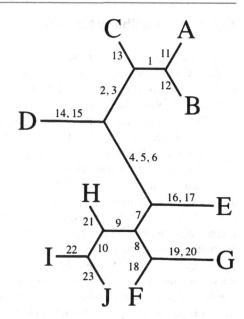

at this step, *and then there were none*, if we move to the next step. The resulting tree
shown in Fig. 16.12 is the maximum parsimony tree.

If the informative sites are mutually compatible, we do not need such recursion.
Classification of sites into 1-ton, 2-ton, etc., is enough. Naming of sites is from
"singleton," which is now called 1-ton. When there are N sequences whose nucleotides are different from the remaining ones with more than N sequences at one site,
let us call this as "N-ton" site. Invariant sites may also be called 0-ton. Each N-ton
($N > 0$) site supports a particular branch of the tree. I would like to call this algorithm
as "N-ton sorting."

Tateno (1990; [59]) proposed "the stepwise ancestral sequence (SAS) method,"
for nucleotide sequence data. The algorithm of the SAS method is similar to that of
the distance Wagner method [28]. We should first define similarity score matrix
among the four nucleotides. Any score can be used, but for simplicity, let us give
score 1 when two sequences at the homologous site have the same nucleotide and
give score 0 when they are different. The sum (W_{ij}) of scores for all homologous
sites for sequences i and j is computed, and this counting is done for all possible
pairs for N sequences. We then examine all possible three sequences in question,
and the trio of sequences which have the smallest sum of W_{ij} is chosen. At this
moment, the sequence at the interior node connecting the three sequences is determined by a majority rule. When three nucleotides at one site are all different, IUPAC
code designating nucleotide ambiguity is given. We then examine the remaining
$N - 3$ sequences, and for each sequence, three possible connections (see Fig. 16.4)
are compared. Therefore, we choose the best tree which has the smallest total branch
length. We then proceed to connect the fifth sequence out of $N - 4$ sequences in the
same manner, and finally the tree topology is determined.

16.5.3   Compatibility Method

All the phylogenetically informative sites are used in the maximum parsimony method. Configurations of some sites may not support the final tree topology. As we already discussed in the simultaneous partitioning method [38], we may select partitions or splits which are mutually compatible for constructing a phylogenetic tree. The compatibility method is based on this simple idea. Wilson (1965; [60]) already proposed a test of consistency (same as compatibility) for a set of characters for constructing phylogenetic trees. Le Quesne (1969; [61]) proposed to choose the tree that maximizes the number of mutually compatible characters. Later, this set of sites is called "maximum clique." A clique is a set of sites whose configurations are all pairwise compatible, and the maximum clique has the largest number of sites [53]. It should be noted that the tree constructed by applying the compatibility method is identical with that applying the maximum parsimony method, if all sites are mutually compatible.

In any case, the compatibility method is closely related to the maximum parsimony method. Its problem is that sites which are not compatible with the tree corresponding to the maximum clique do not contribute to the final tree. Because of this problem, the compatibility method is not so popular compared to the maximum parsimony method.

16.5.4   Theoretical Problems of the Maximum Parsimony Method

The principle of maximum parsimony attracted many people because of its simplicity and logical clarity. However, there are some problems in this method when molecular data are used. First of all, because of parsimony principle, the total number of changes is expected to be underestimated. Saitou (1989; [62]) showed that the gross underestimation of the branch lengths occurs when the divergence (number of nucleotide substitutions per site) among sequences is larger than 0.2. Another problem is uncertainty of branch length estimation. Let us consider a four-OTU tree shown in Fig. 16.5. Let us assume that this tree was chosen as the most parsimonious tree from a long sequence data. One site may have nucleotides A, G, A, and G, for OTUs 1–4, respectively. In this case, nucleotides at the two interior nodes X and Y may be both either A or G. If we assign A for nodes X and Y, we have to assume parallel substitutions at branches X-2 and Y-4, while the alternative branches (X-1 and Y-3) are expected to have substitutions if we assign G for nodes X and Y. If we estimate a distance matrix with a certain method, this kind of uncertainty does not exist, though we have no information on nucleotides at interior nodes. These branch length problems can be avoided if we use the maximum parsimony method only for determining tree topology. In fact, many researchers nowadays only show tree topologies when they used the maximum parsimony method.

More serious problem is its efficiency. Felsenstein (1978; [63]) analytically showed that the maximum parsimony method may be positively misleading when the rate of evolution is grossly different among lineages of four sequences. When the expected number of required substitutions for the true tree is larger than that for a wrong one, the maximum parsimony method will give more and more wrong answer as the number of compared nucleotides is increased (problem of efficiency). The same problem was found even when the constancy of the evolutionary rate is assumed [64, 65]. Therefore, we should be careful in using the maximum parsimony method for the tree topology determination.

The parsimony principle may be more effective for estimating changes that occurred at each branch after the tree topology is determined. Tamura and Nei [66] estimated human mitochondrial DNA nucleotide substitution patterns using the parsimony principle after constructing the phylogenetic tree using the neighbor-joining method. Saitou and Ueda [67] also used the parsimony principle for mapping nucleotide insertions and deletions on the established primate phylogeny. However, underestimation of events can occur when we compare sequences with large divergence. Therefore, comparison using the parsimony principle should be limited to only closely related sequences where parallel and backward changes are not frequent.

16.6 The Maximum Likelihood Method

The maximum likelihood method has been widely used in molecular phylogenetic tree construction for nucleotide sequences as well as amino acid sequences. We will discuss this method to some details. Interested readers are suggested to also read Felsenstein (2004; [53]).

16.6.1 The Principle of the Maximum Likelihood Method

The maximum likelihood method, originally proposed by R. A. Fisher in the early twentieth century, is often used for parameter estimation in statistics. Let us explain the maximum likelihood principle. We consider the ABO blood group locus with three alleles (A, B, and O) and four phenotypes (A, B, AB, and O). A and B alleles are codominant, while O is recessive both to A and to B. The problem to be solved is the estimation of population frequencies p_A, p_B, and p_O for alleles A, B, and O, respectively, from the observed numbers n_A, n_B, n_{AB}, and n_O for phenotypes A, B, AB, and O, respectively. If we assume the Hardy–Weinberg equilibrium (see Chap. 5) for this locus,

$$n_A = n\left(p_A^2 + 2p_A p_O\right), \tag{16.37a}$$

$$n_B = n\left(p_B^2 + 2p_B p_O\right), \tag{16.37b}$$

$$n_{AB} = 2np_A p_B, \tag{16.37c}$$

$$n_O = np_O^2, \tag{16.37d}$$

where n $(=n_A+n_B+n_{AB}+n_O)$ is the sample size (number of individuals typed). Bernstein (1925; [68]) gave the following equations:

$$\hat{p}_A = \sqrt{\left\{\frac{(n_A+n_O)}{n}\right\}} - \hat{p}_O, \qquad (16.38a)$$

$$\hat{p}_B = \sqrt{\left\{\frac{(n_B+n_O)}{n}\right\}} - \hat{p}_O, \qquad (16.38b)$$

$$\hat{p}_O = \sqrt{\left\{\frac{n_O}{n}\right\}}. \qquad (16.38c)$$

It should be noted that n_{AB} is not used in the above equations and $\hat{p}_A+\hat{p}_B+\hat{p}_O$ may not sum up to one. Let us now consider the maximum likelihood estimation. The probability of jointly observing n_A, n_B, n_{AB}, and n_O in a randomly mating population with p_A, p_B, and p_O is

$$P[n_A,n_B,n_{AB},n_O] = C^* \bullet \left(p_A^2 + 2p_Ap_O\right)n_A \bullet \left(p_B^2 + 2p_Bp_O\right)n_B$$
$$\bullet \left(2np_Ap_B\right)n_{AB} \bullet \left(p_O^2\right)n_O, \qquad (16.39)$$

where C^* is a constant (polynomial coefficient) and is not necessary for the later discussion. Derivation of Eq. 16.39 is rather straightforward, but the next step is both statistically and technically somewhat difficult. We introduce a concept called "likelihood." A likelihood is identical with the probability of obtaining the observed value, and this is given by Eq. 16.39 in the current example. But we now treat the unknown yet fixed values in the population (three allele frequencies in the current example) as variables and known but mere observed data from a random sample (four phenotype numbers in the current example) as the given, constant values. In the conventional view, the world of population is the truth or the idea in Plato's analogy of the cave, and we try to find that truth from mere samples. In the maximum likelihood method, samples or observations are placed as if they are at the center of the world, and the truth is searched just as to fit best to the observations. The word "likelihood" comes from this reversed thinking, for we look for the most likely population value(s) from the observations.

Estimation of the maximum likelihood value(s) is theoretically simple; we should find the population parameters which maximize the probability of obtaining the given sample distribution. In the case of 16.39, this search is to find the highest probability within the two-dimensional space spanned by population allele frequencies p_A and p_B. Please note that $p_O = 1 - (p_A+p_B)$, for $p_A+p_B+p_O$ should be the unity. Estimation of these two maximum likelihood values is technically not necessarily simple. Because a likelihood or a probability value is often very small, we usually take logarithm and compute log-likelihood. From 16.39,

$$\text{Log}\ \ L[p_A, p_B] = \log C^* + n_A \log\left(p_A{}^2 + 2p_A\left\{1 - (p_A + p_B)\right\}\right)$$
$$+ n_B \log\left(\left(p_B{}^2 + 2p_B\left\{1 - (p_A + p_B)\right\}\right)\right)$$
$$+ n_{AB} \log\left(2np_A p_B\right) + n_O \log\left\{1 - (p_A + p_B)\right\}^2. \tag{16.40}$$

We should differentiate the above log-likelihood function regarding p_A and p_B, but its solution is not easy to be obtained analytically. Instead, numerical methods such as the EM algorithm or the Newton–Raphson method are usually used to obtain the maximum likelihood solutions. Yasuda and Kimura (1968; [69]) developed a counting method for the maximum likelihood estimation of the ABO blood group system. This is essentially an application of the EM algorithm.

Let us consider an example data, in which the numbers of individuals with A, B, AB, and O phenotypes were 186, 38, 13, and 284 (the sum was 521). Application of 16.38a, 16.38b, and 16.38c gives estimations of p_A, p_B, and p_O as 0.212, 0.048, and 0.738, respectively, while the corresponding maximum likelihood estimations are 0.214, 0.050, and 0.736 (based on http://galton.uchicago.edu/~eichler/stat24600/Handouts/s04.pdf).

The maximum likelihood method was first applied to building phylogenetic trees by Cavalli-Sforza and Edwards (1967; [16]) for allele frequency data. Later, various maximum likelihood methods were developed [70–75], but the foundation of the currently popular algorithms was developed by Felsenstein (1981; [76]).

16.6.2   The Basic Algorithm of the Maximum Likelihood Method

Let us now explain the maximum likelihood estimation of the phylogenetic tree for nucleotide sequence data. Both the completely bifurcating tree search and the stepwise clustering are possible, and we first discuss the former one.

We consider a certain completely bifurcating tree, and let us pay attention to one particular branch connecting nodes A and B. Because each nucleotide site is assumed to evolve independently, we focus on one particular nucleotide site. Although a maximum likelihood tree is usually unrooted, let us assume that node A is ancestral and node B is descendant. We then define the probability Prob[$Nuc_A \Rightarrow Nuc_B$; t] that nucleotide N_A will change to nucleotide N_B during time t. This probability can be determined if we assume a particular nucleotide substitution model. For example, if we assume the one-parameter model (see Chap. 15), Eq. 15.16 applies when $Nuc_A \neq Nuc_B$:

$$Prob\left[Nuc_A \Rightarrow Nuc_B; t\right] = \frac{3}{4} - \frac{3}{4}\exp\left[-4\alpha t\right], \tag{16.41}$$

where α is the rate of nucleotide substitution per unit time. Because we cannot separate α and t in any nucleotide substitution model, the product of the evolutionary rate and the evolutionary time is usually considered. This product is tantamount to the number of nucleotide substitutions expected to occur at this particular branch

Fig. 16.13 An unrooted
tree of six sequences

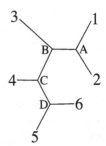

of the tree. Let us denote Prob[$Nuc_A \Rightarrow Nuc_B$; t] as P_{AB} for simplicity. Because most
of the mathematical models used in the maximum likelihood methods for tree
construction are time reversal, we also use this characteristic. We consider the
unrooted tree for six sequences shown in Fig. 16.13, and the interior node D was
assumed as the imaginary root for mathematical convenience. The log-likelihood
L[i] for site i thus becomes

$$L[i] = \sum_D \left[g P_{D5} P_{D6} \left\{ \sum_C P_{DC} P_{C4} \left\{ \sum_B P_{CB} P_{B3} \left\{ \sum_A P_{BA} P_{A1} P_{A2} \right\} \right\} \right\} \right], \quad (16.42)$$

where $\Sigma\alpha$ denote the summation of likelihood values for four possible nucleotides
at node α and g is the nucleotide frequency at node D, the imaginary root. The final
log-likelihood is summation of L[i]'s for all nucleotide sites compared.

This log-likelihood is maximized in terms of all branch lengths (product of the
evolutionary rate and evolutionary time) to obtain the maximum likelihood solution
for that given tree topology. This procedure is repeated for many possible tree
topologies, and the one with the highest maximum likelihood value is chosen as
the best one. More than 30 years have passed after Felsenstein's paper (1981; [76]),
and many improvements were made (e.g., [77]). However, they are mostly on
searching completely bifurcating trees (see Sect. 16.5) or computational speedup,
and the essential framework has not changed.

Fig. 16.14 is likelihood values for one interior node of a four-sequence tree,
shown by Saitou (1988; [78]). He developed a rather primitive computer program
written in FORTRAN (see Appendix of [38]) for the maximum likelihood method.
When there are only four nucleotide sequences, there are 15 nucleotide configurations as shown in Table 16.6. The maximum likelihood values are computed based
on the observed numbers for these 15 possible configurations via a simulation
assuming tree 1 shown in Fig. 16.5. All the five branches were optimized for each
case, but only the interior branch length (corresponding to branch XY of Fig. 16.5)
is shown as the horizontal axis, and the vertical axis gives maximum likelihood
values. The true tree (tree 1) does show the maximum likelihood value, while the
highest likelihood values were found at the quadrifurcating tree with zero interior
branch length for the two erroneous trees.

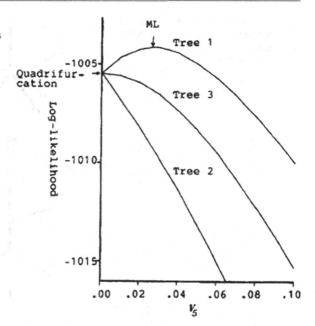

Fig. 16.14 Likelihood values for three possible trees with four sequences (From [76])

16.6.3   Stepwise Clustering Algorithm for the Maximum Likelihood Method

Although the completely bifurcating tree search is popular for the maximum likelihood method, there are a series of stepwise clustering searches for the maximum likelihood method.

One way is to sequentially add sequences, as in the distance Wagner method [28]. The maximum likelihood tree is kept for the initial n sequences, then $(n+1)$-th sequence is added for that tree. This is one of the chief methods used to obtain the initial tree to start searching completely bifurcating trees [53].

Saitou (1988; [78]; see also Saitou 1990; [79]) proposed a maximum likelihood method whose tree topology search is reminiscent of the neighbor-joining method [1]. We start from the star phylogeny. We then search local trees with only one interior branch, separating two sequences with the remaining ones. As we will discuss later in Sect. 16.6.5, this nesting structure does not violate the theoretical problem when the completely bifurcating tree search is used for the maximum likelihood method.

Fig. 16.15 (data from [79]) is an application of this method to Hixson et al. (1982; [80]) mitochondrial sequence data for human, bonobo, chimpanzee, gorilla, and orangutan. Interestingly, tree 3, in which human and gorilla are clustered, was chosen as the maximum likelihood tree. PhyML 3.0 [77] implemented in SeaView [81] also chose tree 3 as the maximum likelihood tree. Tree 2, in which human and chimpanzee are clustered, was chosen by using the neighbor-joining method [79],

Fig. 16.15 Application of Saitou's NJ-like stepwise clustering search – using ML method (From [79])

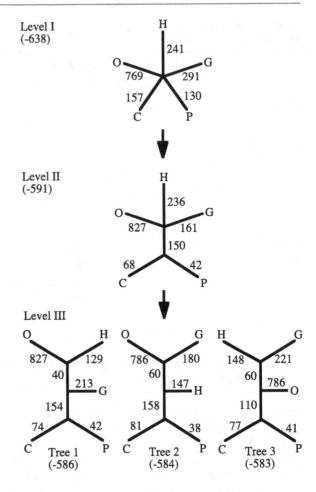

while trees 2 and 3 were equally parsimonious [80]). We now know that the true gene tree for those five species is tree 2 (Horai et al. 1995; [82]). The erroneous tree 3 was chosen by using the maximum likelihood method apparently because of insufficient lengths of nucleotides, as Saitou and Nei (1986; [7]) showed through computer simulations.

Adachi and Hasegawa (1996; [83]) implemented Saitou's NJ-like ML method as "star-decomposition" option in their MOLPHY package program. The same option was also included in PAML developed by Yang (1997; [84]).

Strimmer and von Haeseler (1996; [85]) developed the Quartet-Puzzle method, whose stepwise clustering procedure is quite similar to Sattath and Tversky's [37] quartet examination method. The difference is that they used the maximum likelihood method, not the distance matrix operation.

Ota and Li (2000; [86]) developed the NJ–ML method. The neighbor-joining method [1] is first used for constructing the initial tree. When some branches have low statistical reliability, surrounding trees are searched by using the maximum likelihood method. They later expanded this method for amino acid sequences [87]. Interestingly, their computer simulations showed that the star-decomposition option of MOLPHY [83] was less reliable than their NJ–ML method [87]. This suggests that the maximum likelihood criterion is not necessarily good for finding the correct neighbors when multifurcations are involved.

16.6.4 Algorithm of the Bayesian Method

The origin of the Bayesian method goes back to Thomas Bayes, who was active in the eighteenth century. The Bayesian theorem is treating the relationship between prior and posterior conditional probabilities, and this is closely related to the maximum likelihood method. A historical perspective on the Bayesian approach to the phylogenetic tree construction is given in Felsenstein (2004; [53]). If we apply the Bayesian method for tree construction,

$$Prob(Tree \mid Data) = Prob(Data \mid Tree) \bullet Prob(Tree) \mid Prob(Data), \qquad (16.43)$$

where $Prob(\alpha|\beta)$ is the probability of observing α given (conditional with) β. The problem is to find the tree which has the highest posterior probability, Prob (Tree|Data), for the given sequence data. Prob (Data|Tree) is essentially the likelihood value considered in the maximum likelihood method. Prob (Tree) is the prior probability of the tree in question. Cavalli-Sforza and Edwards (1967; [16]) proposed to use the probability of tree topology assuming the Yule process, introduced by Yule (1924; [88]), when they proposed the maximum likelihood method. However, speciation or gene differentiation may be quite complex, and it is not clear whether this approach is valid. When Felsenstein (1981; [76]) proposed his maximum likelihood method, this tree topology problem was not discussed, and the tree probability was implicitly assumed to be equal. Rannala and Yang (1996; [89]) used a birth–death process for the prior tree probability, while Huelsenbeck et al. (2001; [90]) assumed an equal prior probability. Prob(Data) sounds a bit funny, but it is the sum of all possible Prob(Data|Tree) •Prob (Tree) so as to normalize different probabilities to add up to the unity. When we compare Prob(Tree|Data) for different trees, Prob(Data) cancel out, and it is not necessary for computing it. Therefore, the main problem is Prob (Tree); see also Felsenstein [53]. We will discuss this problem with the case of the maximum likelihood method in the next section.

16.6.5 Theoretical Problems of ML and Bayesian Methods

The standard maximum likelihood method is to find the set of parameters that maximizes the likelihood in a certain likelihood function under the given data, as we discussed in Sect. 16.6.1 using the ABO blood group allele frequency estimation as an example. Because of its consistency (approach the true value as the sample size increases) and efficiency (requirement of small sample size to achieve a certain statistical power), the maximum likelihood method is widely used for statistical inferences of many phenomena.

When we use the maximum likelihood method for phylogenetic tree construction, it is fine as long as we discuss only one particular tree topology. As shown in Eq. 16.42, we have single likelihood function for that tree topology, and we can find the set of branch lengths which give the highest likelihood. However, when we try to compare different tree topologies, we have to use different likelihood functions. It is clear that Eq. 16.42 is only for the completely bifurcating tree shown in Fig. 16.13, shown as ((((1,2),3),4),(5,6)), and it is not applicable for a different tree topology such as (((1,5),3),(2,6),4). When we force some branch length to be zero, we have a multifurcating tree, and comparison of this tree and the original completely bifurcating tree is no problem, for we are using the same likelihood function.

Some researchers questioned the property of comparison of multiple likelihood functions in phylogenetic tree estimation, such as Nei (1987; [91]), Saitou (1988; [78]), Li and Guoy (1991; [92]), and Yang (1996; [93]). For example, Nei [91] argued that the likelihood computed in this method is conditional for each topology, so that it is not clear whether or not the topology showing the highest likelihood has the highest probability of being the true topology when a relatively small number of nucleotides are examined. Felsenstein (1984; [94]) tried to justify his algorithm [76] by using Bayes' theorem. Saitou [78] commented that Felsenstein's argument did not seem to be justified, because different likelihood functions require different probability spaces and we usually do not know the prior probability of each topology. This argument seems to be not applicable to the Bayesian method. However, prior probability of a tree is again the big problem for this method. It should be emphasized that the prior probability is not usually controversial when we consider a simple probability space, as originally considered by Bayes. In contrast, the phylogenetic tree space is quite complicated, and we need to be very cautious to any approach to simply applying a standard statistical method to tree construction.

If we confine our discussion only to the pragmatic side, though, any method which produces the correct tree with a high probability is useful. There are many simulation studies, including ours [6], to show that the maximum likelihood method is good in this sense.

Another theoretical problem with the maximum likelihood and Bayesian methods is the branch length estimation. When a gene tree is considered, we usually try to estimate the realized gene tree, in which branch lengths are the accumulated numbers of mutations. However, as we discussed on Eq. 16.41, the expected number of changes is estimated for each branch, not the observed one, in the maximum likelihood method. The situation is the same for the Bayesian method. This is as if we are considering an expected gene tree (see Chap. 3). We should remember that there are three layers in gene trees: the expected or true tree, the realized tree, and the estimated tree. In usual statistical inferences, we have only two layers: the population and the sample. The former has true values, while the latter has the estimated values. Phylogenetic trees are not simple. As I stressed in the Introduction, everything in this universe has its own history, and what we can estimate is this historical event, not a timeless "true" situation in an imaginary population, which is assumed in the standard statistics theory. We can see a clear limit of modern statistics when we discuss the estimation of phylogenies which are indeed evidences of history of organisms.

16.7   Phylogenetic Network Construction from Character-State Data

We would like to consider construction of phylogenetic networks from character-state data. As we already discussed in the maximum parsimony method, a combination of mutually compatible splits will produce trees, while incompatible splits will produce reticulations. In this sense, the phylogenetic network method can be considered as an extension of the compatibility method for tree construction.

When there are n kinds of splits, we may need $n-1$ dimensions as the maximum to describe the relationship of these splits, if all of them are mutually incompatible. We have only two characters, such as $+$ and $-$ for describing one split, while there are four kinds of nucleotides, and if we also consider the gap caused by insertions or deletions, we need a maximum of four dimensions just for one nucleotide position. When there are three nucleotides at one site, we need a triangle, and a tetrahedron for four nucleotides (see Fig. 16.16).

When a distance matrix for 4 OTUs is given, we have a phylogenetic network shown in Fig. 16.9. We now have a 3D structure for character states (Fig. 16.17). Three edges of the cube correspond to three informative configurations 6, 7, and 8 in Table 16.6, and four edges connected to four sequences are corresponding to singleton configurations 2–5 in Table 16.6.

Some heuristic algorithms were developed for constructing phylogenetic networks from character-state data. Bandelt et al. (1999; [95]) developed the median-joining

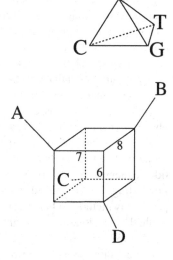

Fig. 16.16 A tetrahedron for four nucleotides

Fig. 16.17 A phylogenetic network with three splits for four sequences

method for recombination-free sequences sampled from a population. Kruskal's (1956; [96]) algorithm for finding a minimum spanning tree and the sequential addition of new sequences [59, 97] were combined. The median-joining method is suitable for relatively short mitochondrial DNA sequences and was widely used especially for human data. As determination of complete mitochondrial DNA sequences became a standard way, conventional nonreticulated trees are now produced (e.g., [98]). This is because the evolutionary history of recombination-free sequences should be a tree.

Kryukov and Saitou (2003; [99]) developed a program called NetView, in which all nucleotide sites incompatible with the neighbor-joining tree produced from nucleotide sequence data are detected, and the split is visualized as overlay to the neighbor-joining tree. This program is useful for checking local nontree structures.

Grunewald et al. (2007; [100]) proposed a new method for constructing phylogenetic network called Qnet. The algorithm of Qnet is similar to Neighbor-Net [49]; however, character-state data are used in Qnet, in contrast to Neighbor-Net which uses distance matrix data.

There are some other methods for constructing phylogenetic networks from sequence data; however, as Wooley et al. (2007; [101]) showed by simulations, all of them failed to construct the correct phylogenetic networks when recombinations are frequently occurring.

16.8   Tree-Searching Algorithms

As we saw in this chapter, there are various algorithms for stepwise clustering methods. It is also true for completely bifurcating tree search methods, for the exhaustive tree search is no longer practical with so many sequence data. One way is to search trees surrounding the initial tree, if the initial tree is expected be rather close to the most optimum tree topology. Rzhetsky and Nei (1992; [20]) used the neighbor-joining method [1] to produce the initial tree and then searched tree topologies whose topological distances [21] are small. There is a problem of local optimum for this kind of limited search strategy, and various techniques were proposed for trying to find the global optimum, such as nearest-neighbor interchanges, subtree pruning and grafting, and tree bisection and reconnection. Interested readers may refer to Felsenstein (2004; [53]) for these techniques. Takahashi and Nei (2000; [102]) compared various search strategies and found that the neighbor-joining method with p distance often showed a better performance than ML methods with various searching methods.

As we already discussed in Sect. 16.5, it is important to consider the list of splits observed in the sequence data in character-state methods. Even if the number of sequences is large, there is no need to search a vast topology space for the short sequence length data. For example, there are only six variant sites for eight nucleotide sequences in Fig. 16.11. We need to examine only a handful of completely bifurcating trees produced from combinations of splits if we apply the maximum parsimony method. When we apply the maximum likelihood method or the Bayesian method, the situation is slightly more complex. However, some split such as 17–234568 is not supported by any sites, and completely bifurcating tree topologies which contain this split can be safely ignored from comparison. Development of new tree topology search algorithms including this is left to future studies.

16.9 Comparison of Phylogenetic Tree-Making Methods

It is desirable if we can use a single tree-making method for any data, because there are so many methods developed so far. There are two major ways to compare different methods: using real data and using simulated data. Real data are preferable for discovering previously unrecognized evolutionary patterns, while the true phylogenetic tree behind them is often unknown. Simulated data have advantages over real data, for all the evolutionary mechanisms and evolutionary histories which produced them are known, while the only restricted parameters can be used for generating them. Because of the clarity of the result, computer simulation is the dominant method for comparing different tree-making methods. In some cases, however, numerical computations are possible for comparing various tree-making methods. Saitou and Nei (1986; [7]) computed the probabilities of obtaining the correct tree topology for three sequences with 100 nucleotides using UPGMA by directly computing all possible configurations. DeBry (1992; [103]) compared the consistency of UPGMA, the neighbor-joining method, modified Farris method, and the maximum parsimony method through the numerical computations. This was possible because the infinite lengths of nucleotides were assumed.

Tateno et al. (1982; [15]) and Nei et al. (1983; [104]) conducted computer simulations for nucleotide sequence substitutions and random genetic drift of allele frequencies, respectively. Saitou and Nei (1987; [1]) conducted an extensive computer simulations at that time by modifying model trees used by Tateno et al. (1982; [15]). Three kinds of nucleotide sequences for eight-sequence trees were generated. Both constant and varying evolutionary rates were considered, and the two types of tree topologies with the maximum number (4) and minimum number (2) of neighbors were considered for various amounts of divergences. In total, 20 datasets with 100 replications each were generated. They compared UPGMA [11], the distance Wagner method [28], Sattath and Tversky's method [37], the modified Farris method [15], Li's method [32], and the neighbor-joining method [1]. Two measures were used to quantify the efficiency of a tree-making method in recovering the topology of the model tree. One was the proportion of the correct tree and the other is the distortion index [15] based on the metric of Robinson and Foulds [21].

UPGMA performed the worst even when the constancy of the evolutionary rate was assumed. This is apparently because of the stochastic fluctuation of the numbers of nucleotide substitutions which really happened in each branch, for the realized gene tree has varying branch lengths (see Chap. 3). Another important message from their simulation is that the longer sequences give better performance irrespective of the methods used. If we can produce very long sequence data, any method may produce the correct tree, except for UPGMA in which the constancy of the evolutionary rate is assumed. If we consider all the results from 20 simulated datasets, the neighbor-joining method and Sattath and Tversky's method were equally better than the other four methods. Because all possible quartets should be examined for Sattath and Tversky's method, the computation time will become much longer than the neighbor-joining method as the number of nucleotide sequences becomes larger. Another advantage of the neighbor-joining method was estimation of the branch lengths as well as the tree topology.

Saitou and Imanishi (1989; [6]) compared the following five tree-making methods: the maximum parsimony method [52], Fitch and Margoliash's minimum deviation method [14] and its modification [15], the minimum evolution method proposed by themselves [6], the maximum likelihood method [76], and the neighbor-joining method [1]. Eight types of phylogenetic trees only for six nucleotide sequences were considered, but all the 105 possible bifurcating trees were examined for the five completely bifurcating tree search methods. The general conclusions are as follows. The efficiency of obtaining the correct tree for the minimum deviation method and its modification was considerably lower than the remaining four methods. The results using the neighbor-joining method were similar with those using the minimum evolution method, and both showed a high performance in obtaining the correct tree. When the constancy of the evolutionary rate was assumed, the results of the four distance matrix methods when uncorrected distance (p distances, proportion of nucleotide difference) was used were slightly better than those for corrected distances using the 1-parameter method (d distances; see Chap. 15); however, results using p distances were very poor compared to those using d distances when the varying rate of evolution was assumed. The latter result may be somewhat related to the problem of the maximum parsimony method (see Sect. 16.5), but the performance of the maximum parsimony method was better than those based on p distances. Results for the neighbor-joining method, the minimum evolution method, and the maximum likelihood method are more or less the same and performed best, followed by the maximum parsimony method.

There are many computer simulation studies in later years (e.g., [102, 105–107]). Nei et al. (1998; [107]) compared the maximum parsimony method, the minimum evolution method, and the maximum likelihood method. These three methods have three different criteria for choosing the best tree topology; the minimum number of mutational changes required, the smallest sum of branch lengths, and the highest likelihood, respectively. They showed that these optimization principles tend to give incorrect topologies when the number of compared sites is small. When the sequence lengths are long, any tree-making method may produce very similar trees. Therefore, in my view, a fast method such as the neighbor-joining method [1] should be used first. More sophisticated methods may be used not for tree topology determination but for estimating branch lengths or inferring specific mutations occurred at each branch.

Although computer simulations are powerful for comparing various phylogenetic tree-making methods, it is not clear whether models of molecular changes used in simulations are good approximations of real changes which occurred in evolution. A different strategy is to use real sequence data with known phylogenetic relationship. For example, Russo et al. (1996; [108]) compared 13 nucleotide and amino acid sequences of vertebrate mitochondrial DNAs using known phylogeny and used four tree-making methods.

Another important aspect is consistency between a multiple alignment (see Chap. 14) and a constructed tree. If the tree is a good representation of the multiple alignment of nucleotide or amino acid sequences, they should have a high consistency with each other. Nguyen et al. (2011; [109]) developed a software called MISFITS which introduces a minimum number of extra substitutions on the inferred tree to provide an explanation why the alignment may deviate from the expectation given by a tree. Using a similar concept, Nguyen et al. (2012; [110]) introduced another software called ImOSM which creates extra substitutions occurring randomly on branches of a tree as model violation into a multiple alignment. They then compared the robustness of three tree-making methods (maximum likelihood, maximum parsimony, and distance matrix method) when ImOSM was used. Interestingly, a distance matrix method (BIONJ [41] was used) performed much better than the two other methods when the assumed model was violated. This indicates that tree-making methods using summary statistics such as pairwise distances may be robust compared to methods directly using multiple alignments of sequences.

16.10   Phylogeny Construction Without Multiple Alignment

Phylogenetic relationship of organisms is usually estimated by comparing homologous genes. As an alternative method to whole bacterial genome comparison, many studies have shown that dinucleotide frequencies within DNA sequences exhibit species-specific signals (e.g., [111, 112]). Species-specific signals for oligomers up to a length of four nucleotides have also been detected [113, 114]. Phylogenetic analysis using trees based on tetranucleotide frequencies demonstrates a level of congruence with trees based on single genes, such as 16S rRNA [115]. These studies have been revealing the effectiveness of tetranucleotide frequencies. However, the evolutionary significance of oligonucleotides longer than tetranucleotide was not studied well. Takahashi et al. (2009; [116]) performed phylogenetic analysis by

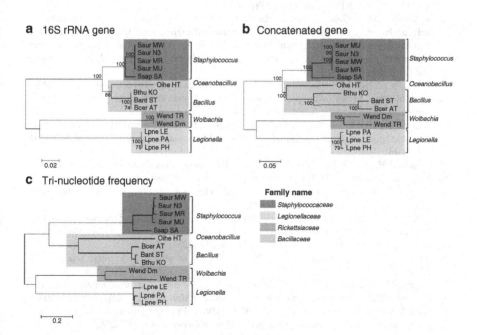

Fig. 16.18 Phylogenetic trees of bacterial species with GC content 32–38 %. (**a**) 16S rRNA gene-based tree. (**b**) Concatenated gene-based tree. (**c**) Tri-nucleotide frequency-based tree

using the Euclidean distances calculated from the di- to deca-nucleotide frequencies in bacterial genomes and compared these oligonucleotide frequency-based tree topologies with those for 16S rRNA gene and concatenated seven genes. When oligonucleotide frequency-based trees were constructed for bacterial species with similar GC content, their topologies at genus and family level were congruent with those based on homologous genes (see Fig. 16.18). Their results suggest that oligonucleotide frequency is useful not only for classification of bacteria but also for estimation of their phylogenetic relationships for closely related species. Because of rapid increase of bacterial genome and metagenome sequences (see Chap. 7), this sort of methods which does not rely on multiple alignment may become more and more useful in the near future.

References

1. Saitou, N., & Nei, M. (1987). The neighbor-joining method: A new method for reconstructing phylogenetic trees. *Molecular Biology and Evolution, 4*, 406–425.
2. Ohtsuka, H., Oyanagi, M., Mafune, Y., Miyashita, N., Shiroishi, T., Moriwaki, K., Kominami, R., & Saitou, N. (1996). The presence/absence polymorphism and evolution of p53 pseudogene within the genus Mus. *Molecular Phylogenetics and Evolution, 5*, 548–556.
3. Kimura, M. (1980). A simple method for estimating evolutionary rates of base substitutions through comparative studies of nucleotide sequences. *Journal of Molecular Evolution, 16*, 111–120.
4. Saitou, N. (1996). Reconstruction of gene trees from sequence data. In R. Doolittle (Ed.), *Methods in enzymology, 266: Computer methods for macromolecular sequence analysis* (pp. 427–449). San Diego: Academic Press.
5. Saitou, N. (2007). *Genomu Shinkagaku Nyumon*. Tokyo: Kyoritsu-Shuppan (in Japanese).
6. Saitou, N., & Imanishi, T. (1989). Relative efficiencies of the Fitch-Margoliash, maximum-parsimony, maximum-likelihood, minimum-evolution, and neighbor-joining methods of phylogenetic tree construction in obtaining the correct tree. *Molecular Biology and Evolution, 6*, 514–525.
7. Saitou, N., & Nei, M. (1986). The number of nucleotides required to determine the branching order of three species, with special reference to the human-chimpanzee-gorilla divergence. *Journal of Molecular Evolution, 24*, 189–204.
8. Tamura, K., Nei, M., & Kumar, S. (2004). Prospects for inferring very large phylogenies by using the neighbor-joining method. *Proceedings of National Academy of Sciences, USA, 101*, 11030–11035.
9. Sneath, P. H. P., & Sokal, R. (1973). *Numerical taxonomy*. San Francisco: W. H. Freeman.
10. Sokal, R., & Sneath, P. H. P. (1968) *Principles of numerical taxonomy*.
11. Sokal, R., & Michener, C. D. (1958). A statistical method for evaluating systematic relationship. *University of Kansas Science Bulletin, 38*, 1409–1438.
12. Nei, M. (1975). *Molecular population genetics and evolution*. Amsterdam: North-Holland.
13. Chakraborty, R. (1977). Estimation of the time of divergence from phylogenetic studies. *Canadian Journal of Genetics and Cytology, 19*, 217–223.
14. Fitch, W. M., & Margoliash, E. (1967). Construction of phylogenetic trees. *Science, 155*, 279–284.
15. Tateno, Y., Nei, M., & Tajima, F. (1982). Accuracy of estimated phylogenetic trees from molecular data. I. Distantly related species. *Journal of Molecular Evolution, 18*, 387–404.
16. Cavalli-Sforza, L. L., & Edwards, A. W. F. (1967). Phylogenetic analysis: Models and estimation procedures. *American Journal of Human Genetics, 19*, 233–257.

17. Rzhetsky, A., & Nei, M. (1992). Statistical properties of the ordinary least-squares, generalized least-squares, and minimum-evolution methods of phylogenetic inference. *Journal of Molecular Evolution, 35*, 367–375.

18. Edwards, A. W. F., & Cavalli-Sforza, L. L. (1964). A method for cluster analysis. *Biometrics, 21*, 362–375.

19. Courant, R., Robbins, H., & Stewart, I. (1996). *What is mathematics?* Second edition: Oxford University Press.

20. Rzhetsky, A., & Nei, M. (1992). A simple method for estimating and testing minimum-evolution trees. *Molecular Biology and Evolution, 9*, 945–967.

21. Robinson, D. F., & Foulds, L. R. (1981). Comparison of phylogenetic trees. *Mathematical Bioscience, 53*, 131–147.

22. Nei, M., Kumar, S., & Takahashi, K. (1998). The optimization principle in phylogenetic analysis tends to give incorrect topologies when the number of nucleotides or amino acids used is small. *Proceedings of National Academy of Sciences, USA, 95*, 12390–12397.

23. Pauplin, Y. (2000). Direct calculation of a tree length using a distance matrix. *Journal of Molecular Evolution, 51*, 41–47.

24. Semple, C., & Steel, M. (2004). Cyclic permutations and evolutionary trees. *Advances in Applied Mathematics, 32*, 669–680.

25. Gascuel, O., & Steel, M. (2006). Neighbor-joining revealed. *Molecular Biology and Evolution, 23*, 1997–2000.

26. Mihaescu, R., & Pachter, L. (2008). Combinatorics of least-squares trees. *Proceedings of the National Academy of Sciences of the United States of America, 105*, 13206–13211.

27. Price, M., Dehal, P. S., & Arkin, A. P. (2009). FastTree: Computing large minimum evolution trees with profiles instead of a distance matrix. *Molecular Biology and Evolution, 26*, 1641–1650.

28. Farris, J. S. (1972). Estimating phylogenetic trees from distance matrices. *American Naturalist, 106*, 645–668.

29. Faith, D. P. (1985). Distance methods and the approximation of most-parsimonious trees. *Systematic Zoology, 34*, 312–325.

30. Farris, J. S., Kluge, A. G., & Exkardt, M. J. (1970). A numerical approach to phylogenetic systematics. *Systematic Zoology, 19*, 172–191.

31. Klotz, L. C., & Blanken, R. L. (1981). A practical method for calculating evolutionary trees from sequence data. *Journal of Theoretical Biology, 91*, 261–272.

32. Li, W.-H. (1981). Simple method for constructing phylogenetic trees from distance matrices. *Proceedings of National Academy of Sciences, USA, 78*, 1085–1089.

33. OOta, S. (1998). ThreeTree: A new method to reconstruct phylogenetic trees. *Genome Informatics, 9*, 340–341.

34. OOta S. (1998b). Ph.D. dissertation.

35. Buneman, P. (1971). The recovery of trees from measurements of dissimilarity. In F. R. Hodson, D. G. Kendall, & P. Tautu (Eds.), *Mathematics in the archeological and historical sciences* (pp. 387–395). Edinburgh: Edinburgh University Press.

36. Fitch, W. M. (1981). A non-sequential method for constructing trees and hierarchical classifications. *Journal of Molecular Evolution, 18*, 30–37.

37. Sattath, S., & Tversky, A. (1977). Additive similarity trees. *Psychometrika, 42*, 319–345.

38. Saitou N. (1986). *Theoretical studies on the methods of reconstructing phylogenetic trees from DNA sequence data*. Ph.D. dissertation. Graduate University of Biomedical Sciences, University of Texas Health Science Center at Houston.

39. Studier, J. A., & Keppler, K. J. (1988). A note on the neighbor-joining algorithm of Saitou and Nei. *Molecular Biology and Evolution, 5*, 729–731.

40. Ishida, N., Oyunsuren, T., Mashima, S., Mukoyama, H., & Saitou, N. (1995). Mitochondrial DNA sequences of various species of the genus Equus with a special reference to the phylogenetic relationship between Przewalskii's wild horse and domestic horse. *Journal of Molecular Evolution, 41*, 180–188.

41. Gascuel, O. (1997). BIONJ: an improved version of the NJ algorithm based on a simple model of sequence data. *Molecular Biology and Evolution, 14*, 685–695.

42. Bruno, W. J., Socci, N. D., & Halpern, A. L. (2000). Weighted neighbor joining: A likelihood-based approach to distance-based phylogeny reconstruction. *Molecular Biology and Evolution, 17*, 189–197.
43. Kumar, S. (1996). A stepwise algorithm for finding minimum evolution trees. *Molecular Biology and Evolution, 13*, 584–593.
44. Pearson, W. R., Robins, G., & Zhang, T. (1999). Generalized neighbor-joining: More reliable phylogenetic tree reconstruction. *Molecular Biology and Evolution, 16*, 806–816.
45. Dress, A. (1984). Trees, tight extensions of metric spaces, and the cohomological dimension of certain groups: A note on combinatorial properties of metric spaces. *Advances in Mathematics, 53*, 321–402.
46. Dress, A., Huber, K. H., Koolen, J., Moulton, V., & Spillner, A. (2012). *Basic phylogenetic combinatorics.* Cambridge: Cambridge University Press.
47. Bandelt, H. J., & Dress, A. W. (1992). Split decomposition: A new and useful approach to phylogenetic analysis of distance data. *Molecular Phylogenetics and Evolution, 1*, 242–252.
48. Kitano, T., Noda, R., Takenaka, O., & Saitou, N. (2009). Relic of ancient recombinations in gibbon ABO blood group genes deciphered through phylogenetic network analysis. *Molecular Phylogenetics and Evolution, 51*, 465–471.
49. Bryant, D., & Moulton, V. (2004). Neighbor-Net: An agglomerative method for the construction of phylogenetic networks. *Molecular Biology and Evolution, 21*, 255–265.
50. Huson, D. H., & Bryant, D. (2006). Application of phylogenetic networks in evolutionary studies. *Molecular Biology and Evolution, 23*, 254–267.
51. Camin, J. H., & Sokal, R. R. (1965). A method for deducing branching sequences in phylogeny. *Evolution, 19*, 311–326.
52. Eck, R. V., & Dayhoff, M. (1966). *Atlas of protein sequence and structure.* Silver Spring: National Biomedical Research Foundation.
53. Felsenstein, J. (2004). *Inferring phylogenies.* Sunderland: Sinauer Associates.
54. Fitch, W. M. (1977). On the problem of discovering the most parsimonious tree. *American Naturalist, 111*, 223–257.
55. Hartigan, J. A. (1973). Minimum mutation fits to a given tree. *Biometrics, 29*, 53–65.
56. Zharkikh, A. A. (1977). Algorithm for constructing phylogenetic trees from amino acid sequences. In V. A. Ratner (Ed.), *Mathematical models of evolution and selection* (pp. 5–52). Novosibirsk: Institute of Cytology and Genetics (in Russian).
57. Zharkikh, A. A., & Ratner, V. A. (1996). Methods for studying the evolution of macromolecules. In V. A. Ratner et al. (Eds.), *Molecular evolution* (pp. 71–91). Berlin/New York: Springer-Verlag.
58. Saitou, N. (1998). Simultaneous sequence joining (SSJ): A new method for reconstruction of phylogenetic networks of closely related sequences (Abstract). *Anthropological Science, 106*, 141–142.
59. Tateno, Y. (1990). A method for molecular phylogeny construction by direct use of nucleotide sequence data. *Journal of Molecular Evolution, 30*, 85–93.
60. Wilson, A. O. (1965). A consistency test for phylogenies based on contemporaneous species. *Systematic Zoology, 14*, 214–220.
61. Le Quesne, W. J. (1969). A method of selection of characters in numerical taxonomy. *Systematic Zoology, 18*, 201–205.
62. Saitou, N. (1989). A theoretical study of the underestimation of branch lengths by the maximum parsimony principle. *Systematic Zoology, 38*, 1–5.
63. Felsenstein, J. (1978). Cases in which parsimony or compatibility methods will be positively misleading. *Systematic Zoology, 27*, 401–410.
64. Zharkikh, A., & Li, W.-H. (1993). Inconsistency of the maximum parsimony method: The case of five taxa with a molecular clock. *Systematic Biology, 42*, 113–125.
65. Takezaki, N., & Nei, M. (1994). Inconsistency of the maximum parsimony method when the rate of nucleotide substitution is constant. *Journal of Molecular Evolution, 39*, 210–218.
66. Tamura, K., & Nei, M. (1993). Estimation of the number of nucleotide substitutions in the control region of mitochondrial DNA in humans and chimpanzees. *Molecular Biology and Evolution, 10*, 512–526.

67. Saitou, N., & Ueda, S. (1994). Evolutionary rate of insertions and deletions in non-coding nucleotide sequences of primates. *Molecular Biology and Evolution, 11*, 504–512.
68. Bernstein, F. (1925). Zusammenfassende betrachtungen uber die erblichen blutstrukturen des menschen. *Molecular and General Genetics, 37*, 237–370.
69. Yasuda, N., & Kimura, M. (1968). A gene-counting method of maximum likelihood for estimating gene frequencies in ABO and ABO-like systems. *Annals of Human Genetics, 31*, 409–420.
70. Neyman, J. (1971). Molecular studies of evolution: A source of novel statistical problems. In S. S. Gupta & J. Yackel (Eds.), *Statistical decision theory and related topics* (pp. 1–27). New York: Academic Press.
71. Felsenstein, J. (1973). Maximum-likelihood estimation of evolutionary trees from continuous characters. *American Journal of Human Genetics, 25*, 471–492.
72. Felsenstein, J. (1973). Maximum-likelihood and minimum-steps methods for estimating evolutionary trees from data on discrete characters. *Systematic Zoology, 22*, 240–249.
73. Kashap, R. L., & Subas, S. (1974). Statistical estimation of parameters in a phylogenetic tree using a dynamic model of the substitutional process. *Journal of Theoretical Biology, 47*, 75–101.
74. Langley, C., & Fitch, W. M. (1974). An examination of the constancy of the rate of molecular evolution. *Journal of Molecular Evolution, 3*, 161–177.
75. Thompson, E. A. (1975). *Human evolutionary trees*. Cambridge/New York: Cambridge University Press.
76. Felsenstein, J. (1981). Evolutionary trees from DNA sequences: A maximum likelihood approach. *Journal of Molecular Evolution, 17*, 368–376.
77. Guindon, S., Dufayard, J. F., Lefort, V., Anisimova, M., Hordijk, W., & Gascuel, O. (2010). New algorithms and methods to estimate maximum-likelihood phylogenies: Assessing the performance of PhyML 3.0. *Systematic Biology, 59*, 307–321.
78. Saitou, N. (1988). Property and efficiency of the maximum likelihood method for molecular phylogeny. *Journal of Molecular Evolution, 27*, 261–273.
79. Saitou, N. (1990). Maximum likelihood methods. *Methods in Enzymology, 183*, 584–598.
80. Hixson, J., & Brown, W. M. (1986). A comparison of the small ribosomal RNA genes from the mitochondrial DNA of the great apes and humans: Sequence, structure, evolution, and phylogenetic implications. *Molecular Biology and Evolution, 3*, 1–18.
81. Gouy, M., Guindon, S., & Gascuel, O. (2010). SeaView version 4: A multiplatform graphical user interface for sequence alignment and phylogenetic tree building. *Molecular Biology and Evolution, 27*, 221–224.
82. Horai, S., Hayasaka, K., Kondo, R., Tsugane, K., & Takahata, N. (1995). Recent African origin of modern humans revealed by complete sequences of hominoid mitochondrial DNAs. *Proceedings of the National Academy of Sciences of the United States of America, 92*, 532–536.
83. Adachi, J., & Hasegawa, M. (1996). MOLPHY version 2.3: Programs for molecular phylogenetics based on maximum likelihood. *Computer Science Monographs, 28*, 1–150.
84. Yang, Z. (1997). PAML: A program package for phylogenetic analysis by maximum likelihood. *CABIOS Applications Note, 13*, 555–556.
85. Strimmer, K., & von Haeseler, A. (1996). Quartet puzzling: A quartet maximum-likelihood method for constructing phylogenetic trees. *Molecular Biology and Evolution, 13*, 1401–1409.
86. Ota, S., & Li, W.-H. (2000). NJML: A hybrid algorithm for the neighbor-joining and maximum-likelihood methods. *Molecular Biology and Evolution, 17*, 1401–1409.
87. Ota, S., & Li, W.-H. (2001). NJML+: An extension of the NJML method to handle protein sequence data and computer software implementation. *Molecular Biology and Evolution, 18*, 1983–1992.
88. Yule, G. U. (1924). *A mathematical theory of evolution, based on the conclusions of Dr. J. C. Willis, F.R.S* (Philosophical transaction of royal society of London, series B, Vol. 213, pp. 21–87). London: Harrison and Sons.
89. Rannala, B., & Yang, Z. (1996). Probability distribution of molecular evolutionary trees: A new method of phylogenetic inference. *Journal of Molecular Evolution, 17*, 368–376.
90. Huelsenbeck, J. P., Ronquist, F., Nielsen, R., & Bollback, J. P. (2001). Bayesian inference of phylogenetic trees and its impact on evolutionary biology. *Science, 294*, 2310–2314.

91. Nei, M. (1987). *Molecular evolutionary genetics.* New York: Columbia University Press.
92. Li, W.-H., & Guoy, M. (1991). Statistical methods for testing molecular phylogenies. In M. M. Miyamoto & J. Cracraft (Eds.), *Phylogenetic analysis of DNA sequences* (pp. 249–277). New York: Oxford University Press.
93. Yang, Z. H. (1996). Phylogenetic analysis using parsimony and likelihood methods. *Journal of Molecular Evolution, 42,* 294–307.
94. Felsenstein, J. (1984). The statistical approach to inferring evolutionary trees and what it tells us about parsimony and compatibility. In T. Duncan & T. F. Steussy (Eds.), *Cladistics: Perspectives on the reconstruction of evolutionary history* (pp. 169–191). New York: Columbia University Press.
95. Bandelt, H. J., Forster, P., & Rohl, A. (1999). Median-joining networks for inferring intraspecific phylogenies. *Molecular Biology and Evolution, 16,* 37–48.
96. Kruskal, J. B. (1956). On the shortest spanning subtree of the graph and the travelling salesman problem. *Proceedings of the American Mathematical Society, 7,* 48–57.
97. FarrisJ, S. (1970). Methods for computing Wagner trees. *Systematic Zoology, 19,* 83–92.
98. Jinam, T. A., Hong, L. -C., Phipps, M. E., Stoneking, M., Ameen, M., Edo, J., HUGO Pan-Asian SNP Consortium, & Saitou, N. (2012). Evolutionary history of Continental Southeast Asians: "Early train" hypothesis based on genetic analysis of mitochondrial and autosomal DNA data. *Molecular Biology and Evolution, 29,* 3513–3527.
99. Kryukov, K., & Saitou, N. (2003). Netview: Application software for constructing and visually exploring phylogenetic networks. *Genome Informatics, 14,* 280–281.
100. Grunewald, S., Farslund, K., Dress, A., & Moulton, V. (2007). QNet: An agglomerative method for the construction of phylogenetic networks from weighted quartets. *Molecular Biology and Evolution, 24,* 532–538.
101. Wooley, S., Posada, D., & Crandall, K. A. (2007). A comparison of phylogenetic network methods using computer simulation. *PLoS One, 3,* e1913.
102. Takahashi, K., & Nei, M. (2000). Efficiencies of fast algorithms of phylogenetic inference under the criteria of maximum parsimony, minimum evolution, and maximum likelihood when a large number of sequences are used. *Molecular Biology and Evolution, 17,* 1251–1258.
103. DeBry, R. W. (1992). The consistency of several phylogeny-inference methods under varying evolutionary rates. *Molecular Biology and Evolution, 9,* 537–551.
104. Nei, M., Tajima, F., & Tateno, Y. (1983). Accuracy of estimated phylogenetic trees from molecular data. II. Gene frequency data. *Journal of Molecular Evolution, 19,* 153–170.
105. Tateno, Y., Takezaki, N., & Nei, M. (1994). Relative efficiencies of the maximum likelihood, neighbor-joining, and maximum parsimony methods when substitution rate varies with site. *Molecular Biology and Evolution, 11,* 261–277.
106. Kuhner, M. K., & Felsenstein, J. (1994). A simulation comparison of phylogeny algorithms under equal and unequal evolutionary rates. *Molecular Biology and Evolution, 11,* 459–468. Erratum in: Molecular Biology and Evolution, 12, p. 525.
107. Nei, M., Kumar, S., & Takahashi, K. (1998). The optimization principle in phylogenetic analysis tends to give incorrect topologies when the number of nucleotides or amino acids used is small. *Proceedings of the National Academy of Sciences of the United States of America, 95,* 12390–12397.
108. Russo, C., Takezaki, N., & Nei, M. (1996). Efficiencies of different genes and different tree-making methods in recovering a known vertebrate phylogeny. *Molecular Biology and Evolution, 13,* 525–536.
109. Nguyen, M. A. H., Klaere, S., & von Haeseler, A. (2011). MISFITS: Evaluating the goodness of fit between a phylogenetic model and an alignment. *Molecular Biology and Evolution, 28,* 143–152.
110. Nguyen, M. A. H., Gesell, T., & von Haeseler, A. (2012). ImOSM: Intermittent evolution and robustness of phylogenetic methods. *Molecular Biology and Evolution, 29,* 663–673.
111. Karlin, S., & Ladunga, I. (1994). Comparisons of eukaryotic genomic sequences. *Proceedings of the National Academy of Sciences of the United States of America, 91,* 12832–12836.
112. Nakashima, H., Nishikawa, K., & Ooi, T. (1997). Differences in dinucleotide frequencies of human, yeast, and *Escherichia coli* genes. *DNA Research, 4,* 185–192.

113. Karlin, S., Mrazek, J., & Campbell, A. (1997). Compositional biases of bacterial genomes and evolutionary implications. *Journal of Bacteriology, 179,* 3899–3913.
114. Abe, T., et al. (2003). Informatics for unveiling hidden genome signatures. *Genome Research, 13,* 693–702.
115. Pride, D. T., Meinersmann, R. J., Wassenaar, T. M., & Blaser, M. J. (2003). Evolutionary implications of microbial genome tetranucleotide frequency biases. *Genome Research, 13,* 145–155.
116. Takahashi, M., Kryukov, K., & Saitou, N. (2009). Estimation of bacterial species phylogeny through oligonucleotide frequency distances. *Genomics, 93,* 525–533.
117. Felsenstein, J. (1985). Confidence limits on phylogenies: An approach using the bootstrap. *Evolution, 39,* 783–791.

Population Genomics

Chapter Summary

Population genetics is a part of evolutionary studies. Now with genome sequences, population genomics emerged, starting from the analysis of multiple human mitochondrial DNA genome sequences. It was extended to nuclear DNA of human individuals, and genome-wide SNP data comparison is now flourishing, slowly followed by comparisons of personal genomes. This is also true in bacteria in which different strains of the same species are now determined and compared. As the genome sequencing cost is becoming drastically reduced, population genomics will definitely expand to many other organisms. We discuss both methods and examples of population genomics in this chapter.

17.1 Evolutionary Distances Between Populations

We discussed population (and species) trees in Chap. 3. Branch lengths of expected population trees are evolutionary times, while those of estimated population trees may also be genetic differences. They are often estimated from genetic distances between populations. How to estimate these distances is introduced in this section. Because data are very abundant for human populations, examples are from human data.

17.1.1 Distance Between Populations Based on Gene Genealogy

Let us consider the evolutionary history of two populations, α and β, which started to differentiate some time ago (Fig. 17.1). We focus on one small piece of DNA sequence and assume that the sequence and the history are both short enough to ignore recombinations. A constant population size, N, is assumed, and we also assume that the gene flow does not occur between the two populations after the

Fig. 17.1 A schematic gene
genealogy of two populations
which differentiated long
time ago

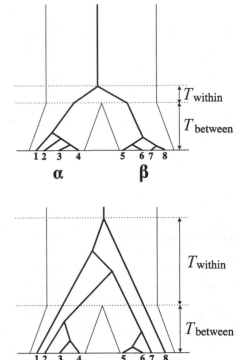

Fig. 17.2 A schematic gene
genealogy of two populations
which differentiated recently

differentiation started. It should be noted that the expected coalescent time for the
constant population size of N individuals is ~$4N$ generations for diploid autosomal
genes and ~N for maternally inherited haploid mitochondrial DNAs (see Chap. 4).

There are four genes (1–4 and 5–8 for populations α and β, respectively) in each
population. If the divergence time ($T_{between}$) between the two populations is much
larger than the coalescent time (T_{within}) within the common ancestral population,
these four genes may find their common ancestor within the population, as shown
in Fig. 17.1. When $T_{between}$ is smaller than the expected coalescent time, there is a
possibility that some genes sampled from the same population do not coalesce, and
the two coalescences are never expected if $T_{between}$ is much smaller than the expected
coalescent time. This situation is shown in Fig. 17.2.

If we assume that the population size has been constant among the common
ancestral population (γ) of α and β, population α, and population β, we can use the
averaged nucleotide diversity (D_α and D_β) observed in the current populations α
and β, respectively, for estimating the nucleotide diversity (D_γ) of the common
ancestral population of α and β:

$$D_\alpha = \left[\frac{n_\alpha}{(n_\alpha - 1)} \right] \sum d_{ij}, \qquad (17.1a)$$

$$D_\beta = \left[\frac{n_\beta}{(n_\beta - 1)}\right]\sum d_{kl}, \tag{17.1b}$$

$$D_\gamma = \frac{1}{2}\left[D_\alpha + D_\beta\right], \tag{17.1c}$$

where n_α and n_β are numbers of sequences sampled from populations α and β, respectively, and d_{ij} and d_{kl} are the number of nucleotide substitutions between the i-th and the j-th sequences in population α and that between the k-th and the l-th sequences in population β, respectively. The summation is for all possible pairs. By subtracting $D\gamma$ from the overall DNA diversity ($D_{\alpha\beta}$) between populations α and β, we can obtain the net DNA divergence ($D_{net-\alpha\beta}$) between populations α and β, as first shown by Nei and Li (1979; [1]):

$$D_{\alpha\beta} = \frac{\sum d_{ik}}{\left[n_\alpha n_\beta\right]}, \tag{17.2a}$$

$$D_{net-\alpha\beta} = D_{\alpha\beta} - D_\gamma, \tag{17.2b}$$

where d_{ij} is the number of nucleotide substitutions between the i-th sequence of population α and the k-th sequence of population β. The summation is for all possible pairs.

$D_{\alpha\beta}$, D_γ, and $D_{net-\alpha\beta}$ correspond to $2\lambda(T_{within} + T_{between})$, $2\lambda T_{within}$, and $2\lambda T_{between}$, respectively. λ is the rate of nucleotide substitution per site per year. If we have only one DNA locus, such as mitochondrial DNA, Eq. 17.2b gives the estimate of DNA divergence after the two populations started to differentiate. If we have many independently evolving loci as in nuclear DNA, each estimate for one locus is averaged to obtain one genome-wide estimate, for the population differentiation should affect all the loci.

Tajima et al. (2003; [2]) determined mitochondrial DNA partial sequences for 180 human individuals from nine aboriginal populations of Taiwan. They presented DNA diversity values of all populations as well as $D_{\alpha\beta}$ and $D_{net-\alpha\beta}$ values of all pairwise populations. DNA diversity values ranged from 1.1 % to 1.5 %, and $D_{\alpha\beta}$ values were within the range of 1.2–1.5 %. Because $D_{net-\alpha\beta}$ values are obtained by subtracting the estimated DNA diversity of the common ancestor (D_γ) from $D_{\alpha\beta}$ values, $D_{net-\alpha\beta}$'s (0.07–0.23 %) are much smaller than $D_{\alpha\beta}$'s.

There is one caveat on the assumption of the constancy of the population size. As long as we consider only one population, there is no problem. When one population splits into two, the sum of population sizes for the two daughter populations should be the same as that of the mother population. We therefore have to assume the population expansion after the population split to set the sizes of three populations to be eventually constant. Therefore, in reality, there is no exact constancy of population sizes when we consider the evolutionary history of multiple populations.

Fig. 17.3 Overlayed gene
genealogies of genes 1–4
sampled from two species
A and B

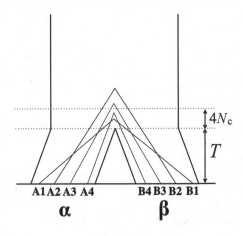

17.1.2   Distance Between Species Based on Gene Genealogy

Let us reexamine Fig. 17.1. If the divergence time ($T_{between}$) between populations α
and β is very large, the two "populations" may become different species. It is no
longer appropriate to assume the constant population size for the long time. Then
how can we estimate $T_{between}$ and T_{within} in this case? To solve this question, Takahata
et al. (1995; [3]) developed a maximum likelihood method which simultaneously
estimates the species divergence time and the effective size of the ancestral
population. While $T_{between}$ is the same for all nuclear DNA regions, T_{within} varies from
region to region. We already discussed this coalescent process in Chap. 4. For an
autosomal DNA locus in diploid organisms, the probability that two genes which
are destined to be the common ancestor for species α and β at the speciation time
coalesce at time T_{within} is

$$\text{Prob}\big[2 \to 1, T_{within}\big] \sim \big[1/2N_c\big]\exp\big[-T_{within}/2N_c\big], \qquad (17.3)$$

where N_c is the effective population size of the common ancestral species for α and
β. The problem is now to jointly estimate N_c and T (=$T_{between}$), the divergence time
between species α and β, from nucleotide sequence data of many loci. This situation is depicted in Fig. 17.3. Four genes sampled from both populations are no longer sequences from the same locus, but sampled from different loci, and Ai and Bi
($i = 1$–4) are orthologous genes. The heterogeneity of the pairwise sequence divergences is the clue for estimating N_c. A pairwise sequence divergence is expected to
follow the Poisson distribution (see Chap. 15); thus, the probability of having k
($k = 0, 1, 2, \ldots$) substitutions is

$$\text{Prob}\big[k; N_c, T\big] = \big\{2n_i\mu\left(T + T_{within}\right)\big\}^k / k! \bullet \exp\big\{2n_i\mu\left(T + T_{within}\right)\big\}, \qquad (17.4)$$

where n_i is the number of neutrally evolving nucleotide sites for the i-th DNA region
(locus) and μ is the mutation rate, which is assumed to be uniform in the genome ([3]).
Using the generating function of the above equation, Takahata et al. ([3]) introduced

a maximum likelihood function and applied this method to 13 gene sequence data available at that time for human and chimpanzee. Assuming μ as 1.5×10^{-8}/site/generation and one generation to be 15 years, they estimated $N_c = 87,000$ and $T = 4.6$ million years ago. The estimate for N_c is more than eight times higher than that for modern human [2], and the estimate for T seems to be too small if we consider a series of fossils on the line of the human lineages (e.g., [4]). One possibility of the overestimation of the N_c value was the assumption of the constant mutation rate over the genome. If the mutation rate varies from DNA region to region, this gives another source of heterogeneity. Another possibility is the estimation of the mutation rate itself. As we discussed in Chap. 2, recent estimates for the modern human is about half of the one based on the long-term evolution. If so, the estimate of T (the divergence time) may become much larger. Recently, Hara et al. (2012; [5]) assumed an evolutionary model in which mutation rates vary across lineages and chromosomes and estimated speciation times of the human lineage from chimpanzee, gorilla, and orangutan to be 5.9–7.6, 7.6–9.7, and 15–19 million years ago, respectively, from genome-wide sequence data. They also estimated the population size of the common ancestor of human and chimpanzee, that of human–chimpanzee and gorilla and that of human–chimpanzee–gorilla and orangutan to be 59,300–75,600, 51,400–66,000, and 159,000–203,000, respectively. These new estimates may be more compatible with those envisaged from paleoanthropological studies.

17.1.3 Distance Between Populations Based on Allele Frequency Differences

Let us consider two populations, α and β, which started to differentiate T years ago as in the former section. Instead of the gene genealogy, we now consider the temporal changes of allele frequencies (Fig. 17.4; from Fig. 8 of Saitou (1995; [6])). As we saw in Chap. 4, the random genetic drift always occurs at any time for any population. Frequencies of a particular allele for the two populations will gradually vary as the time goes on from the common ancestral population. Although the data for one allele is not appropriate as the measure of differentiation of the two populations, we can estimate the genetic distance between the two populations by using allele frequency data for many alleles at many loci.

Nei (1972; [7]) proposed the standard genetic distance (D_{std}) between populations based on allele frequency differences, as follows:

$$D_{std} = -\ln \left[\frac{J_{\alpha\beta}}{\sqrt{J_\alpha J_\beta}} \right]. \tag{17.5}$$

$J_{\alpha\beta}$, J_α, and J_β are the homozygosities (1 – heterozygosities or gene diversities) between populations α and β, population α, and population β, respectively, and they are defined as

$$J_{\alpha\beta} = \frac{\sum\sum \alpha_{ik}\beta_{ik}}{n}, \tag{17.6a}$$

Fig. 17.4 Dynamics
of allele frequency changes
during the population
differentiation (From
Saitou 1995; [6])

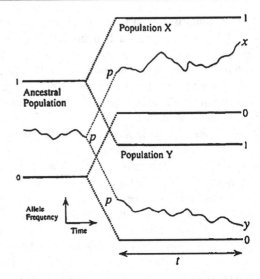

$$J_\alpha = \frac{\sum\sum \alpha_{ik}{}^2}{n},$$

(17.6b)

$$J_\beta = \frac{\sum\sum \beta_{ik}{}^2}{n},$$

(17.6c)

where α_{ik} and β_{ik} are frequencies of allele i of locus k for populations α and β, respectively, and the two summations are for all alleles at one locus and for all n loci [7].

Although Eq. 17.5 looks quite different from Eq. 17.2b, they are closely related. This is because $D_{\alpha\beta}$ and D_γ in Eq. 17.2b averaged for n DNA regions (loci) can be written as

$$D_{\alpha\beta} = \frac{\sum\sum \alpha_{ik}\beta_{ik}d_{ij}}{n},$$

(17.7a)

$$D_\gamma = \frac{1}{2}\left\{ \frac{\sum\sum \alpha_{ik}{}^2}{n} + \frac{\sum\sum \beta_{ik}{}^2}{n} \right\},$$

(17.7b)

where α_{ik} and β_{ik} are frequencies of allele i of locus (DNA region) k for populations α and β, respectively, and the two summations are for all alleles at one locus and for all n loci. Therefore, mathematically $J_{\alpha\beta}$ and $D_{\alpha\beta}$ are similar except that the former is identity and the latter is difference. Equation 17.5 can be written as

$$D_{std} = D*_{\alpha\beta} - \left(\frac{D*_\alpha + D*_\beta}{2} \right),$$

(17.8)

where

$$D^*_{\alpha\beta} = -\ln J_{\alpha\beta}, \tag{17.9a}$$

$$D^*_{\alpha} = -\ln J_{\alpha}, \tag{17.9b}$$

$$D^*_{\beta} = -\ln J_{\beta}. \tag{17.9c}$$

Under the neutral evolution, $J_{\alpha\beta}$ (the average homogeneity between populations α and β) is

$$J_{\alpha\beta} = J_0 (1-v)^{2T} \sim J_0 \exp(-2vT), \tag{17.10}$$

where J_0 is the initial J value when $T = 0$ and v is the mutation rate per year per locus [8]. Because we assume the constant population sizes for all three populations (α, β, and their common ancestral population γ), expected values of J_{α}, J_{β}, and J_0 are the same: $1/(1+4Nv)$. Therefore, putting these relationships to Eq. 17.8,

$$D_{std} = -\ln\left[\frac{1}{(1+4Nv)}\right] - \ln\left[\exp(-2vT)\right] + \ln\left[\frac{1}{(1+4Nv)}\right] \tag{17.11}$$

$$= 2vT.$$

Because of this expected linearity with the evolutionary time (T), D_{std}, or Nei's standard genetic distance [6, 7], has been widely used for many allele frequency data. If we consider the distance based on gene genealogy, $D_{net-\alpha\beta}$ in Eq. 17.2b corresponds to $2\lambda T_{between}$. Therefore, D_{std} and $D_{net-\alpha\beta}$ are essentially the same, if we compare neutrally evolving DNA regions in which the evolutionary rate λ is identical with the mutation rate.

Nei and Roychoudhury (1974; [9]) estimated the divergence of three human populations (African, East Eurasian, and West Eurasian, following Saitou's [6] nomenclatures of population names) as ~55,000 years for East Eurasian–West Eurasian divergence and ~120,000 years for the African and West and East Eurasian. Later, Nei and Ota (1991; [10]) obtained the essentially the same divergence time estimates from the allele frequency data of 181 loci. More recently, Gronau et al. (2011; [11]) obtained similar values from comparison of personal genomes: 38,000–64,000 years for East and West Eurasian divergence and 108,000–157,000 years for African and West and East Eurasian divergence.

Another genetic distance often used is the D_A distance [12]:

$$D_A = \sum_k \left(\frac{1-\sum_k \sqrt{\alpha_{ik}\beta_{ik}}}{n}\right). \tag{17.12}$$

D_A is closely related to the chord distance (D_{chord}) of Cavalli-Sforza and Edwards (1967; [13]):

$$D_{chord} = \frac{2\sqrt{2 D_A}}{\pi}. \tag{17.13}$$

Many other distance measures were proposed, and readers interested in this subject may refer to Nei (1987; [8]).

17.1.4 Evolutionary Distance Between Genomes

It is tempting to define a distance between homologous genomic sequences. When we consider haploid genome without recombination, such as animal mitochondrial DNAs, the genomic distance is simply the evolutionary divergence between the two genome sequences. If the recombination comes in as in the bacterial genomes, it is still possible to equate the sequence divergence as the genomic distance. However, the biological meaning of the genomic distance is now not clear. The situation becomes worse when we face diploid genomes. As we already discussed in Chap. 3, the so-called individual tree, which is based on genomic distance matrix, is biologically problematic.

17.2 Mitochondrial DNA Population Genomics

Horai et al. (1995; [14]) sequenced the mtDNA genome of one African individual, as well as determining those for chimpanzee, bonobo, gorilla, and orangutan. This was the second human mtDNA genome sequence determined after the first one, the so-called the Cambridge reference sequence [15]. The coalescent time of the two human mtDNA sequences was estimated to be $143,000 \pm 18,000$ years assuming 13 million years ago as the human–orangutan divergence time. Their paper was the starting point of the mitochondrial DNA population genomics. It is now becoming routine to determine complete mitochondrial DNA genomes in human population studies (e.g., Jinam et al. 2012; [16]). We will discuss various issues that have been examined using mitochondrial DNA sequences in this section.

17.2.1 Inference of Gene Genealogy

Ingman et al. (2000; [17]) determined mtDNA genome sequences of 53 human individuals distributed worldwide (see Fig. 3.8). Five years later, Tanaka et al. (2005; [18]) sequenced 672 Japanese mtDNA genomes. There are now close to 12,000 complete human mtDNA genomic sequences in the International Nucleotide Sequence Database as of December 2013. The primary interest on these mtDNA genome sequence comparison is inference of the human mtDNA gene genealogy. Historically, human mitochondrial DNA sequences have been classified alphabetically, named "haplogroups." The MITOMAP database (http://mitomap.org) provides a detailed haplogroup genealogy.

Jinam et al. (2012; [16]) determined 86 mtDNA genome sequences from four Malaysian populations. Figure 17.5 shows the phylogenetic tree of N haplogroup sequences. Haplogroups are designated as boxes, and numbers connecting

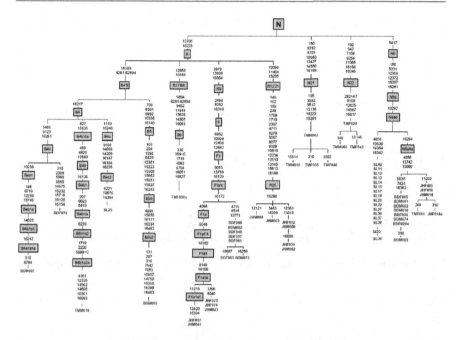

Fig. 17.5 A phylogenetic tree of *N* haplogroup mtDNA sequences of Malaysians (From Jinam et al. 2012; [16])

haplogroups are nucleotide positions in which mutations occurred on that branch. Sequences newly determined are shown starting with population ID such as TM. Through a comparison with published mtDNA genome sequences of human individuals distributed in Southeast Asia, they found that coalescent times of many Malaysian mtDNAs with those of people living in the Continental area were rather deep, and the "early train" hypothesis was proposed for peopling of Southeast Asia [16].

Behar et al. (2012; [19]) compared 18,843 complete human mtDNA genome sequences and constructed their phylogenetic tree, which is available at PhyloTree (http://www.phylotree.org/). Figure 17.6 shows the phylogenetic tree of major haplogroups.

17.2.2 Population Size Fluctuation

de Ririenzo and Wilson (1991; [20]) noticed a starlike gene genealogy for European mtDNA sequences. They also showed that the distributions of sequence differences between all possible pairs of individuals for non-African populations were approximately Poisson, while those for African populations were quite different from a Poisson distribution (Fig. 17.7). If the pattern of nucleotide substitutions follows a Poisson distribution at any branch of the gene genealogy, the distribution of pairwise distances for a starlike tree, in which the average length of branches is equal, is expected to show a Poisson distribution. This pattern is apparently caused by a rapid expansion

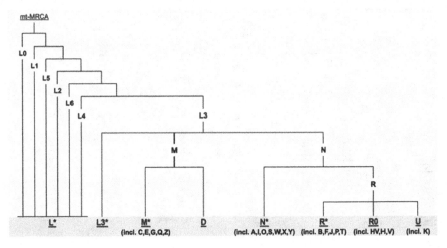

Fig. 17.6 Phylogenetic tree of major haplogroups of human mtDNAs (From http://www. phylotree.org/)

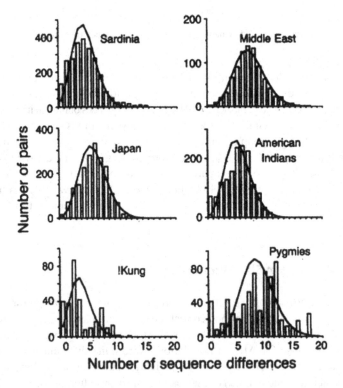

Fig. 17.7 Distributions of sequence differences between all possible pairs of individuals for non-African populations (From de Ririenzo and Wilson 1991; [20])

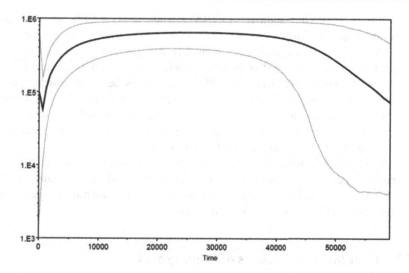

Fig. 17.8 An example of the Bayesian Skyline Plot (From Jinam et al. 2012; [16])

of mitochondrial DNA lineages after the Out-of-Africa. Harpending (1994; [21]) called this distribution as "mismatch distribution" and showed its utility as the mode of population expansion.

Drummond et al. (2005; [22]) introduced a method called "Bayesian Skyline Plot" for estimating past population dynamics through time from a sample of molecular sequences without dependence on a prespecified parametric model of demographic history. An example of the Bayesian Skyline Plot generated using mtDNA coding region sequences of 86 human individuals in Malysia is shown in Fig. 17.8 (from [16]). The estimated population dynamics can be classified into the four phases: (1) increase during 60,000–40,000 years before present (YBP), (2) stable during 40,000–10,000 YBP, (3) decline during 10,000–700 YBP, and (4) sharp increase from 700 YBP to present time. Because these estimates are based on a series of simple assumptions, we have to be careful to interpret them.

17.2.3 Estimation of Nucleotide Substitution Patterns

As we already introduced in Chap. 15, Tamura and Nei (1993; [23]) analyzed the pattern of mutation accumulations in human mitochondrial partial DNA sequences and showed that the distribution of nucleotide substitution numbers per site fits much better with the negative binomial distribution than the Poisson distribution. Kawai and Saitou (2011; [24]) and Kawai et al. (unpublished; [25]) estimated the nucleotide substitution patterns of complete mtDNA sequences of 145 cows (*Bos taurus*) and 7,264 humans (*Homo sapiens*), respectively. In both papers, nucleotide substitutions occurred only fourfold degenerate sites of protein coding regions were considered. A zebu (*Bos indicus*) mtDNA genome sequence (the International Nucleotide Sequence Database accession number = AY126697) was used as outgroup for cow sequences, and the 116 nucleotide substitutions on the fourfold degenerate sites were observed in the phylogenetic tree. Transitions were about 20 times higher than transversions in cow mtDNA sequences, but no clear heterogeneity was observed among the four transitions, except for slightly higher T=>C transitions [24]. It is a clear contrast to human mtDNA sequences, where G=>A transitions are more than two times, six times, and nine times higher than T=>C, A=G, and C=>T transitions, respectively (see Table 15.5; from [25]).

Estimation of evolutionary rates and mutation rates is another important problem in population genomics of mitochondrial DNA. Behar et al. ([19]) examined the possible violation from molecular clock assumption, but the violation was observed only in a restricted number of mtDNA lineages.

17.3 Population Genomics of Prokaryotes

There are now a huge number of complete bacterial genomes; however, most of them are remotely related. Therefore, comparison of these genome sequences is more inclined to phylogenetic interests rather than population studies. These genome comparisons were already discussed in Chap. 6. One exception may be *Helicobacter pylori*, which mainly live in human stomachs, and their genetic diversity is limited (e.g., [26, 27]). There are 56 complete *H. pylori* genome sequences (1.5~1.7 Mb size) available at the International Nucleotide Sequence Database as of December 2013. Figure 17.9 shows a partial multiple alignment result for 30 *H. pylori* genome sequences. The "first" option of MISHIMA [28] was used, and the computation took only 9 min in MacAir personal computer. This sort of highly conserved regions is scattered among the *H. pylori* genome sequences. Kawai et al. (2011; [29]) determined the complete genome sequences of four *H. pylori* strains isolated from different individuals in Fukui, Japan, and compared them with known 16 genome sequences. The phylogenetic tree of concatenated well-defined core genes showed divergence of the East Asian lineage from the European lineage ancestor, especially in genes for coding virulence factors, outer membrane proteins, and lipopolysaccharide synthesis enzymes.

Another bacterial population genomics study with multiple within-species genome sequence comparison can be seen in Shapiro et al. (2012; [30]). Using whole-genome sequences from two recently diverged *Vibrio cyclitrophicus* populations, they inferred recombination events. A gradual separation of bacterial gene pools was found as evidenced by increased habitat specificity of the most recent recombinations. Another bacterial species for the target of population genomics is *Pseudomonas aeruginosa*. There are seven complete genome sequences (all are more than 6-Mb lengths) of various strains of this species at the International Nucleotide Sequence Database as of December 2013. Pseudomonas genome database (http://www.pseudomonas.com) is available for population genomics studies (Winsor et al. 2011; [31]).

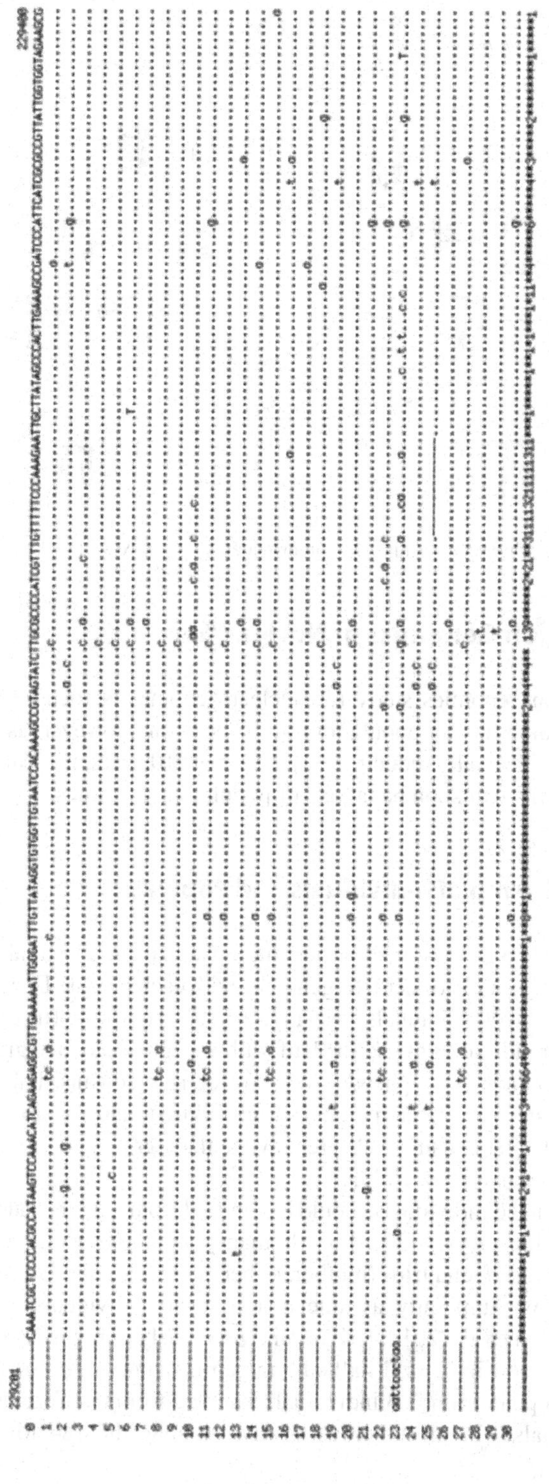

Fig. 17.9 Partial multiple alignment result for 30 *H. pylori* genome sequences using MISHIMA (Ref. [28])

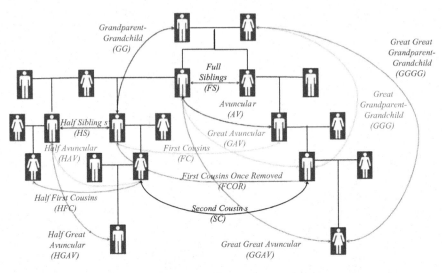

Fig. 17.10 Various kinship relationships (kindly provided by Sarah Voisin)

17.4 Population Genomics of Nuclear Genomes

Most of population genomics study for nuclear genomes is focused on humans. Therefore, we assume human populations at this section. However, as the genome sequencing cost is drastically reducing during the past 10 years, nuclear genomes of nonhuman organisms will soon be targets of population genomics.

17.4.1 Relationship of Individuals and Populations

Each human individual has two nuclear genomes, paternal and maternal. One person thus shares one or $\frac{1}{2}$ of his or her two genomes with each parent. There are various relationships of related individuals (see Fig. 17.10). The proportions of genome sharing are $\frac{1}{2}$ for full siblings; $\frac{1}{4}$ for half siblings, grandparents–grandchildren, and avunculars (uncle/aunt–nephew/niece); $\frac{1}{8}$ for first cousins, half-avunculars, great avunculars, and great grandparents–grandchildren; and $\frac{1}{16}$ for first cousins once removed, half first cousins, half great avunculars, great great avunculars, and great great grandparents–grandchildren.

These highly related individuals should be omitted from a large-scale SNP study, such as the HapMap project [32]. Four human populations were studied in the HapMap project phase I, and trios (mother, father, and their child) were included in two populations. We implicitly assume that non-trio individuals are unrelated. However, Pemberton et al. (2010; [33]) found previously unreported 2 parent–offspring pairs, 3 first cousin pairs, 1 half sibling pair, 3 avuncular pairs, and 2 grandparent–grandchild pairs from 203 individuals sampled from the Yoruba (YRI) tribe of Nigeria. They also found numerous relatives among the 184 Maasai (MKK) samples. Even the West Eurasian (CEU) samples contained two previously unreported avuncular pairs [33]. In contrast, two East Asian populations from Tokyo (JPT) and from Beijing (CHB) did not contain close relatives. Japanese collaborators carefully avoided close relatives by prescreening for the HapMap project (Dr. Katsushi Tokunaga, personal communication). We therefore should be careful for simply assuming unrelatedness for some HapMap samples.

Genetic relationships of individuals who are more remotely related than the genome sharing of $\frac{1}{16}$ can be estimated by examining the identity by descent (IBD), as shown by Browning and Browning (2010; [34]). They used linkage disequilibrium information between SNP loci and could detect ~2-cM lengths of IBD from genome-wide SNP data through massive computations using a hidden Markov model. Browning and Browning (2011; [35]) further developed a software called fastIBD and showed that this software can give the IBD values which are very close to the true values up to cases for the fifth cousins.

The genetic relationship of individuals can be analyzed in various ways. One is the principal component analysis (PCA). PCA is a standard statistical method for multivariate analysis, and the spatial distribution of individual samples is shown according to mutually independent (orthogonal in linear algebra) principal components (PCs). These PCs are defined by examining variances among samples, and all the samples are located along the first PC so as to maximize the variances among samples. The second PC can explain the second largest variances which are orthogonal to those explained by the first PC. These variance decompositions are obtained through matrix algebra (e.g., see [36]). Novembre et al. (2008; [37]) analyzed ~200,000 SNP loci data for 1,387 European individuals. Figure 17.11 shows their PCA result by using the smartpca program in the EIGENSOFT software package (Patterson et al., 2006; [38]). As the title of their paper indicates, individuals from the same geographic region cluster together and geographical features of Europe are well reproduced in this PCA plot. It is interesting to note that the proportions of total variances explained by the first and the second PC are only 0.30 % and 0.15 %, respectively [37]. One may question that a PCA plot with such a low proportion (0.45 % for the first and second PCs combined) does not deserve a convincing result for representing the genetic structure of human individuals. However, most of the allele frequency changes due to the random genetic drift are expected to have no correlations except for the population differentiation which should affect all the SNP loci. Under this logic, the top PC components (first PC, second PC, third PC, etc.) do show the patterns of population differentiation and admixture even if the proportion of variance explanation is very small. This argument should be tested by a series of computer simulations such as one conducted by Novembre and Stephens (2008; [39]).

If individuals belonging to one population focused at one particular sector of the multidimensional PCA space, we can consider the population as one unit. We can produce the phylogenetic tree of populations in this case. Japanese Archipelago Human Population Genetics Consortium (2012; [40]) compared various human populations in East Asia using ~15,000 genome-wide SNP data, and Fig. 17.12 shows a phylogenetic tree of nine populations constructed using a maximum likelihood method [41] implemented in the PHYLIP package [42]. Data for four ethnic

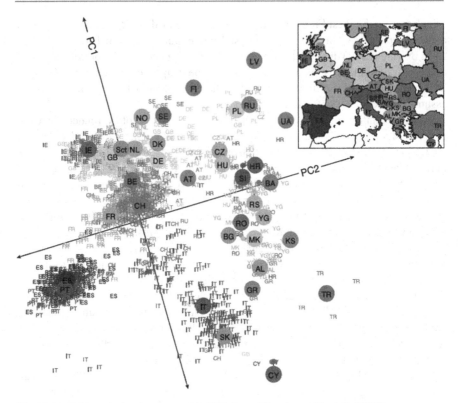

Fig. 17.11 PCA analysis of genome-wide SNP data of Europeans (From Ref. [37])

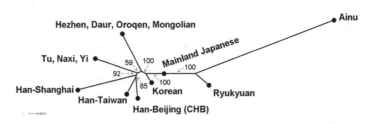

Fig. 17.12 Phylogenetic tree of nine human populations in East Asia (From Ref. [40])

minority populations (Hezhen, Daur, Oroqen, and Mongolian) in northern China and three populations (Tu, Naxi, and Yi) in southern China were combined because of small sample size for each population. The Ainu and the Ryukyuan were clearly clustered even though these two populations are geographically located at northern and southern regions of the Japanese Archipelago, in which the Mainland Japanese is sandwiched by the Ainu and the Ryukyuan. The three populations in the Japanese Archipelago were clustered with the Korean with the 100 % bootstrap probability,

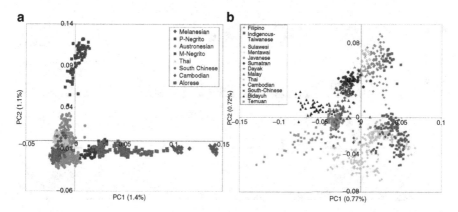

Fig. 17.13 PCA plots of genome-wide SNP data of human individuals in Southeast Asia (From Ref. [38]). (**a**) Eight populations. (**b**) Thirteen populations

and the four populations further form a tight cluster. Interestingly, the Ainu population is genetically different from the other East Asian populations in this figure with its long external branch. In contrast, the external branch length for Mainland Japanese is almost zero, and this population is located between the Ainu–Ryukyu cluster and the remaining populations. This suggests that the Mainland Japanese was formed by admixture between the first migrants (who probably came to the Japanese Archipelago during Paleolithic and Jomon periods) and the second migrants (probably Yayoi and later periods).

17.4.2   Admixture

One species often has some population structure, and long-term isolation may transform two closely related populations into two subspecies. If genetic isolation continues, these subspecies will become incipient species, and eventually reproductive isolation is established. If individuals of two generically differentiated populations happen to meet, however, admixture or gene flow can occur. This process is reducing effect of genetic differentiation between two populations. We show some examples of human admixture in this section.

Jinam et al. (2012; [16]) analyzed the ~50,000 SNP loci data of Southeast Asian populations [43]. Figure 17.13a, b present two PCA results using the smartpca program [38]. The first principal component (PC1) in Fig. 17.13a distinguishes between the Melanesians and Southeast Asians. Alorese (individuals living on the Alor Island and surrounding islands in Indonesia) is located between the Melanesians and Philippine Negritos. The second principal component (PC2) separates the Malaysian Negritos from other populations. Individuals belonging to Malaysian Negritos and Alorese are spread apart in a comet-like pattern, suggesting recent admixture between these groups with dominant populations in Southeast Asia who

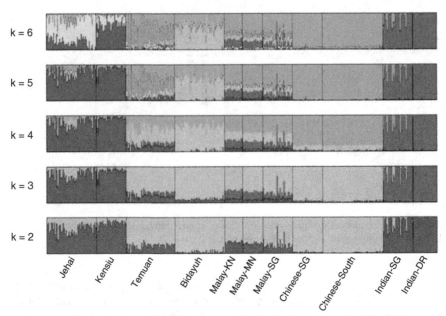

Fig. 17.14 STRUCTURE result of 13 human populations (From Jinam et al. 2013; [46])

form a tight cluster in the PCA plot. This comet-like pattern was also observed in PCA analyses of Bryc et al. (2010; [44]) and McEvoy et al. (2010; [45]). To explore this tight clustering of groups, another PCA plot (Fig. 17.11b) was drawn after omitting Melanesian, Alorese, Malaysian Negrito, and Philippine Negrito individuals. The southern Chinese and Thai individuals tend to cluster together, whereas the other Austronesian populations spread out more or less according to their geographic distribution.

Admixture as recent as one generation ago between Malaysian Negritos and Malays was studied by Jinam et al. (2013; [46]). They used PLINK software [47] for filtering SNP data in terms of SNP call rates and minor allele frequencies. The smart-pca program [38] for the PCA analysis and the STRUCTURE program developed by Pritchard et al. (2000; [48]) were used. STRUCTURE assigns individuals based on their genotypes into a user-defined number of ancestral populations, denoted as k. Under the admixture model, individuals who are jointly assigned to two or more ancestry components are considered to be admixed. Burn-in length and number of repeats were both set to 10,000. A "comet-like" pattern for the Malaysian Negritos is confirmed in the PCA plot, similar to what was observed in Jinam et al. (2012; [15]). Figure 17.14 is the STRUCTURE result. Each individual is represented by a vertical bar and their respective ancestry components are indicated by different colors. Multiple ancestry clusters within an individual (multiple colors in a single vertical bar) signify an admixed individual. At k=2, the two-population subdivision corresponds to the Negritos (Jehai and Kensiu) and Indian in blue component against the rest of the populations in green component. As k is increased to 3, the population clusters observed corresponded to the Indians, Negritos, and the other populations. At

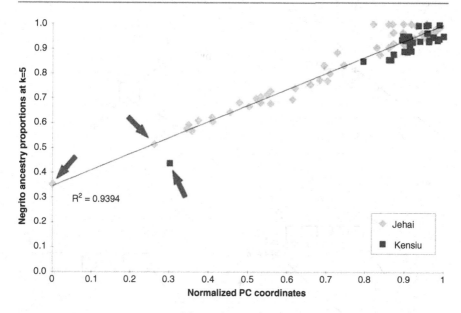

Fig. 17.15 Correlation of STRUCTURE and PCA results for two Malaysian Negrito populations. Three individuals with arrows are possible recent hybrids with Malays (From Jinam et al. 2013; [46])

k=4, there appeared to be a component shared mostly among the Chinese and indigenous Taiwanese (orange component). At this point, the population components observed correspond to the Negrito (blue), Indian (brown), Southeast Asian (light green), and Chinese (orange). As the value of k is increased, further differentiation of subgroups was observed which corresponds to the Temuan (dark green component) at k=5 and at k=6, whereby the previous Negrito component (dark blue) was further split into the Jehai and Kensiu. At k=6, the six ancestry components correspond to the Jehai, Kensiu, Indian, Temuan, Southeast Asian, and Chinese components. Both STRUCTURE and PCA results have high correlation, as shown in Fig. 17.15. Most of Kenshiu Negrito individuals are located at upper-right corners, suggesting no recent admixture, while one Kensiu individual (inside circle) and dozens of Jehai Negrito individuals are located in the middle of this plot, indicating recent admixture. In fact, one unique Kensiu individual inside circle was hybrid between Kensiu and Malay.

17.4.3 Introgression

When admixture occurs between different subspecies or species, it is usually called "introgression." We already discussed a possibility of introgression between modern humans and Neanderthals in Chap. 10. Before the full reproductive isolation is achieved, there is always opportunity for individuals from genetically differentiated populations to mate.

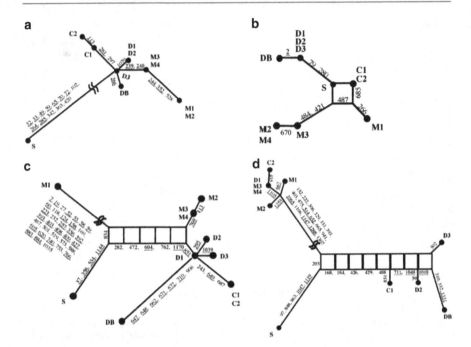

Fig. 17.16 Phylogenetic networks for four genes (**a**) b3GT2 gene, (**b**) b3GT1 gene, (**c**) Fut4 gene, (**d**) Dfy gene. Sequence symbols starting with C, D, and M are castaneus, domesticus, and musculus subspecies, S is Mus spicilegus, and DB is mouse genome database sequence (From Liu et al. 2008; [53])

Ferris et al. (1983; [49]) found a signature of introgression between *Mus musculus domesticus* and *Mus musculus musculus* at their distribution boundary in East Europe through comparison of mitochondrial DNA haplotype analysis using restriction enzymes. Wade et al. (2002; [50]) and Wiltshire et al. (2003; [51]) found evidence of introgression among mouse subspecies through nuclear genome-wide analysis. Abe et al. (2004; [52]) determined 176,000 BAC-end sequences of *Mus musculus molossinus* and compared them with the genome sequence of *Mus musculus domesticus*. They found the nucleotide difference of 0.96 % between the two subspecies in average, and 5 % of the *Mus musculus domesticus* genome was estimated to be derived from *Mus musculus molossinus* through introgression.

Introgression may occur at any time, and ancient introgression can be found through comparison of congeneric species. Liu et al. (2008; [53]) sequenced 21 nuclear protein coding genes for three subspecies of *Mus musculus* (*M. m. musculus*, *M. m. domesticus*, and *M. m. castaneus*) as well as *Mus spicilegus*. Phylogenetic networks for each gene revealed a mosaic structure of their genomes. While 14 out of 21 genes showed the expected pattern in which the *Mus spicilegus* sequence is most divergent from all *Mus musculus* sequences (e.g., Fig. 17.16a), *Mus spicilegus* sequence is within the *Mus musculus* gene cluster in five genes (e.g., Fig. 17.16b). Phylogenetic networks of the remaining two genes (Fig. 17.16c, d) showed that the within diversity in *Mus musculus* is larger than the divergence between *Mus musculus* and *Mus spicilegus*. There may be various possible explanations for the unique patterns of these two genes. First is the acceleration of the evolutionary rate due to directional natural selection or loss of function. The second possibility is the gene conversion from other paralogous genes. The third one is long-term coexistence of two distinct lineages through balancing selection. The fourth possibility is lineage sorting, or transmission of different allelic lineages (ancestral polymorphism) existed in the common ancestral population to different populations. The fifth one is introgression of a distant species, far apart from *M. spicilegus*, used in this study as an out-group species. The first and second possibilities were rejected by data analysis, and the third possibility is unlikely. The fourth and fifth possibilities are not easy to distinguish, for they are not exclusive [53].

Introgression is expected to occur rather frequently immediately after speciation, for the reproductive barrier may not be fully established and the two new species may be geographically closely located. In fact, Patterson et al. (2006; [54]) proposed that the ancestral lineages of humans and chimpanzees experienced temporal isolation followed by a hybridization event. However, Innan and his colleagues [55, 56] questioned this hypothesis by careful analysis of genome data, and they argued that the simplest speciation model with instantaneous split adequately described the human–chimpanzee speciation event. Hara et al. ([5]) also reached the same conclusion. If this is the case, two new populations which became the chimpanzee–bonobo common ancestral species and the bipedal human common ancestral species were geographically isolated immediately after their speciation.

17.4.4 Population Size Fluctuation

We discussed analyses of human population size fluctuation using mitochondrial DNA sequence data in Sect. 17.2.2. We now discuss the same problem using nuclear DNA data. Theunert et al. (2012; [57]) used genome-wide SNP data to reconstruct the demographic history of human populations. They introduced allele frequency-based identity by descent and identity by state statistics, and using approximate Bayesian computation, the population size fluctuation was estimated for two populations; the Yoruba experienced a rather stable ancestral population size with a mild recent expansion, and the French experienced a long-lasting severe bottleneck followed by a drastic population growth.

Human populations have experienced an explosive growth over the past 400 generations, expanding by at least three orders of magnitude. Keinan and Clark (2012; [58]) characterized the signatures of this explosive growth on the SNP allele frequency spectrum. Rapid population growth increases the frequency of rare variants, and they may become genetic burdens of complex disease risk.

Li and Durbin (2011; [59]) developed a new method for inferring past population size fluctuation from single diploid genome sequences by applying the coalescent theory (see Chap. 4), and named it the PSMC (pairwise sequentially Markovian coalescent) model. The PSMC infers the local time to the most recent common

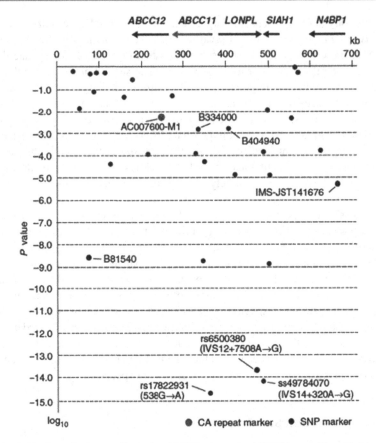

Fig. 17.17 P values of microsatellite loci and SNP loci near the ABCC11 gene (From Yoshiura et al. 2006; [61])

ancestor on the basis of the local density of heterozygotes, using a hidden Markov model. They applied the PSMC method to seven human individual genome data, and found an increase of the effective population size during 60,000–250,000 years ago. Although it is a great achievement, usage of multiple individual genome data is better for more accurate estimation of the temporal change of past population size than using single individual genome data. The PSMC software package is freely available at http://github.com/lh3/psmc [59].

17.4.5 Genome-Wide Association Study

After the determination of the human genome sequences, it became possible to expand a case–control association study to the genome-wide level for the search of disease-causing genes. Here "case" means patients and "control" means healthy people. Ozaki et al. (2002; [60]) examined ~93,000 SNPs for 1,133 myocardial infarction patients and 832 control individuals and found that a DNA region including lymphotoxin-α gene was closely associated with increased risk of that disease. This was the first study on genome-wide association study (GWAS). This GWAS strategy can of course be applied to non-disease phenotypes, and Yoshiura et al. (2006; [61]) identified a nonsynonymous SNP at the ABCC11 gene to be responsible for the wet/dry earwax polymorphism. Figure 17.17 shows the distribution and P values of microsatellite loci and SNP loci near the ABCC11 gene. The nonsynonymous SNP (rs17822931) with a probability of 10^{-15} turned out to be the causative polymorphism by experimental validation. Tamiya et al. (2005; [62]) used ~27,000 microsatellite (STR) DNA loci for a GWAS of rheumatoid arthritis. Because the average number of microsatellite DNA alleles is much larger than 2 for a typical SNP locus, microsatellite DNA loci are effective for the first screening to detect the region containing the gene responsible for a certain phenotype, and then SNP loci are used for fine mapping, for SNPs are more dense than microsatellite loci.

GWAS turned out to be quite effective to identify a gene which is mostly responsible for a particular trait, such as the earwax which was known to follow Mendelian inheritance (Matsunaga, 1962; [63]). However, a phenotype which seems to be linked to many loci is difficult to study. For example, Gudbjartsson et al. (2008; [64]) tried to identify genes that affect the human height through examination of a total of more than 20,000 individuals. They found 27 regions of the human genome showing statistically significant association with height. However, they explain only 3.7 % of the population variation in height. More recently, Liu et al. (2012; [65]) identified five candidate genes (PRDM16, PAX3, TP63, C5orf50, and COL17A1) which may be involved in determination of facial morphology in Europeans. Three of the five loci identified in their study have been shown to play an essential role in craniofacial development, and this time-consuming study turned out to show a more or less confirmatory conclusion.

It was almost 100 years ago when Fisher (1918; [66]) tried to explain quantitative traits by Mendelian genetics through inventing ANOVA (analysis of variance). Even now, however, it is still difficult to connect quantitatively defined morphological traits with DNA. This problem is left for future studies.

References

1. Nei, M., & Li, W.-H. (1979). Mathematical model for studying genetic variation in terms of restriction endonucleases. *Proceedings of the National Academy of Sciences of the United States of America, 76*, 5269–5273.
2. Tajima, A., Cheih-Shan Sun, C.-S., Pan, I.-H., Ishida, T., Saitou, N., & Horai, S. (2003). Mitochondrial DNA polymorphisms in nine aboriginal groups of Taiwan: Implications for the population history of aboriginal Taiwanese. *Human Genetics, 113*, 24–33.
3. Takahata, N., Satta, Y., & Klein, J. (1995). Divergence time and population size in the lineage leading to modern humans. *Theoretical Population Biology, 48*, 198–221.
4. White, T., WolderGabriel, G., Louchart, A., Suwa, G., Lovejoy, C. O., et al. (2009). *Ardipithecus ramidus* special issue. *Science, 326*(5949).

5. Hara, Y., Imanishi, T., & Satta, Y. (2012). Reconstructing the demographic history of the human lineage using whole-genome sequences from human and three great apes. *Genome Biology and Evolution 4*, 1133–1145.

6. Saitou, N. (1995). A genetic affinity analysis of human populations. *Human Evolution, 10*, 17–33.

7. Nei, M. (1972). Genetic distances between populations. *American Naturalist, 106*, 283–292.

8. Nei, M. (1987). *Molecular evolutionary genetics*. New York: Columbia University Press.

9. Nei, M., & Roychoudhury, A. (1974). Genetic variation within and between the three major races of man, caucasoids, negroids, and mongoloids. *American Journal of Human Genetics, 26*, 421–443.

10. Nei, M., & Ota, T. (1991). Evolutionary relationships of human populations at the molecular level. In T. Honjo & S. Osawa (Eds.), *Evolution of life* (pp. 415–428). New York: Springer.

11. Glonau, I., Hubisz, M. J., Gulko, B., Danko, C. G., & Siepel, A. (2011). Bayesian inference of ancient human demography from individual genome sequences. *Nature Genetics, 43*, 1031–1035.

12. Nei, M., Tajima, F., & Tateno, Y. (1983). Accuracy of estimated phylogenetic trees from molecular data. II. Gene frequency data. *Journal of Molecular Evolution, 19*, 153–170.

13. Cavalli-Sforza, L. L., & Edwards, A. W. F. (1967). Phylogenetic analysis: Models and estimation procedures. *American Journal of Human Genetics, 19*, 233–257.

14. Horai, S., Hayasaka, K., Kondo, R., Tsugane, K., & Takahata, N. (1995). Recent African origin of modern humans revealed by complete sequences of hominoid mitochondrial DNAs. *Proceedings of the National Academy of Sciences of the United States of America, 92*, 532–536.

15. Anderson, S., Bankier, A. T., Barrell, B. G., de Bruijn, M. H., Coulson, A. R., Drouin, J., Eperon, I. C., Nierlich, D. P., Roe, B. A., Sanger, F., Schreier, P. H., Smith, A. J., Staden, R., & Young, I. G. (1981). Sequence and organisation of the human mitochondrial genome. *Nature, 290*, 457–465.

16. Jinam, T. A., Hong, L.-C., Phipps, M. A., Stoneking, M., Ameen, M., Edo, J., Pan-Asian SNP Consortium, & Saitou, N. (2012). Evolutionary history of continental South East Asians: "Early train" hypothesis based on genetic analysis of mitochondrial and autosomal DNA data. *Molecular Biology and Evolution, 29* (Advance online access).

17. Ingman, M., Kaessman, H., Paabo, S., & Gyllensten, U. (2000). Mitochondrial genome variation and the origin of modern humans. *Nature, 408*, 708–713.

18. Tanaka, M., et al. (2005). Mitochondrial genome variation in Eastern Asia and the peopling of Japan. *Genome Research, 14*, 1832–1850.

19. Behar, D. M., van Oven, M., Rosset, S., Metspalu, M., Loogväli, E.-L., Silva, N. M., Kivisild, T., Torroni, A., & Villems, R. (2012). A "Copernican" reassessment of the human mitochondrial DNA tree from its root. *American Journal of Human Genetics, 90*, 675–684.

20. de Rienzo, A., & Wilson, A. C. (1991). Branching pattern in the evolutionary tree for human mitochondrial DNA. *Proceedings of the National Academy of Sciences of the United States of America, 88*, 1597–1601.

21. Harpending, H. C. (1994). Signature of ancient population growth in a low-resolution mitochondrial DNA mismatch distribution. *Human Biology, 66*, 591–600.

22. Drummond, A. J., Rambaut, A., Shapiro, B., & Pybus, O. G. (2005). Bayesian coalescent inference of past population dynamics from molecular sequences. *Molecular and Biological Evolution, 22*, 1185–1192.

23. Tamura, K., & Nei, M. (1993). Estimation of the number of nucleotide substitutions in the control region of mitochondrial DNA in humans and chimpanzees. *Molecular Biology and Evolution, 10*, 512–526.

24. Kawai, Y., & Saitou, N. (2011). Analysis of nucleotide substitution patterns through mitochondrial DNA SNP (in Japanese). *DNA Takei, 19*, 28–31.

25. Kawai, Y., Kikuchi, T., & Saitou, N. (unpublished) Evolutionary dynamics of nucleotide composition of primate mitochondrial DNA inferred from human SNP data and nuclear pseudogenes.

26. Falush, D., et al. (2003). Traces of human migrations in *Helicobacter pylori* populations. *Science, 299*, 1582–1585.
27. Suzuki, R., Shiota, S., & Yamaoka, Y. (2012). Molecular epidemiology, population genetics, and pathogenic role of *Helicobacter pylori*. *Infection, Genetics and Evolution, 12*, 203–213.
28. Kryukov, K., & Saitou, N. (2010). MISHIMA – A new method for high speed multiple alignment of nucleotide sequences of bacterial genome scale data. *BMC Bioinformatics, 11*, 142.
29. Kawai, M., Furuta, Y., Yahara, K., Tsuru, T., Oshima, K., Handa, N., Takahashi, N., Yoshida, M., Azuma, T., Hattori, M., Uchiyama, I., & Kobayashi, I. (2011). Evolution in an oncogenic bacterial species with extreme genome plasticity: *Helicobacter pylori* East Asian genomes. *BMC Microbiology, 11*, 104.
30. Shapiro, B. J., Friedman, J., Cordero, O. X., Preheim, S. P., Timberlake, S. C., Szabo, G., Polz, M. F., & Alm, E. J. (2012). Population genomics of early events in the ecological differentiation of bacteria. *Science, 336*, 48–51
31. Windor, G. L., Lam, D. K. W., Fleming, L., Lo, R., Whiteside, M. D., Yu, N. Y., Hancock, R. E. W., & Brinkman, F. S. L. (2011). Pseudomonas genome database: Improved comparative analysis and population genomics capability for *Pseudomonas* genomes. *Nucleic Acids Research, 39*, D596–D600.
32. International HapMap Consortium. (2005). The haplotype map of the human genome. *Nature, 437*, 1299–1320.
33. Pemberton, T. J., Wang, C., Li, J. Z., & Rosenberg, N. A. (2010). Inference of unexpected genetic relatedness among individuals in HapMap phase III. *American Journal of Human Genetics, 87*, 457–464.
34. Browning, S. R., & Browning, B. L. (2010). High-resolution detection of identity by descent in unrelated individuals. *American Journal of Human Genetics, 86*, 526–539.
35. Browning, B. L., & Browning, S. R. (2011). A fast, powerful method for detecting identity by descent. *American Journal of Human Genetics, 88*, 173–182.
36. Morrison, D. F. (1978). *Multivariate statistical methods*. Auckland: McGraw-Hill.
37. Novembre, J., et al. (2008). Genes mirror geography within Europe. *Nature, 456*, 98–101.
38. Patterson, N., Price, A. L., & Reich, D. (2006). Population structure and eigenanalysis. *PLoS Genetics, 2*, e190.
39. Novembre, J., & Stephens, M. (2008). Interpreting principal component analyses of spatial population genetic variation. *Nature Genetics, 40*, 646–649.
40. Japanese Archipelago Human Population Genetics Consortium. (2012). The history of human populations in the Japanese Archipelago inferred from genome-wide SNP data with a special reference to the Ainu and the Ryukyuan populations. *Journal of Human Genetics* (in press).
41. Felsenstein, J. (1981). Maximum likelihood estimation of evolutionary trees from continuous characters. *American Journal of Human Genetics, 25*, 471–492.
42. http://evolution.genetics.washington.edu/phylip/
43. The HUGO Pan-Asian SNP Consortium. (2009). Mapping human genetic diversity in Asia. *Science, 326*, 1541–1545.
44. Bryc, K., Velez, C., Karafet, T., Moreno-Estrada, A., Reynolds, A., Auton, A., Hammer, M., Bustamante, C. D., & Ostrer, H. (2010). Genome-wide patterns of population structure and admixture among Hispanic/Latino populations. *Proceedings of the National Academy of Sciences of the United States of America, 107*, 8954–8961.
45. McEvoy, B. P., Lind, J. M., Wang, E. T., Moyzis, R. K., Visscher, P. M., van Hoslt, P. S. M., & Wilton, A. N. (2010). Whole-genome genetic diversity in a sample of Australians with deep aboriginal ancestry. *American Journal of Human Genetics, 87*, 297–305.
46. Jinam, T. A., Phipps, M. A., & Saitou, N. (2012). Admixture patterns and genetic differentiation in Negrito groups from West Malaysia estimated from genome-wide SNP data. *Human Biology, 85*, Iss. 1, Article 8.
47. Purcell, S., et al. (2007). PLINK: A tool set for whole-genome association and population-based linkage analyses. *American Journal of Human Genetics, 81*, 559–575.

48. Pritchard, J. K., Stephens, M., & Donnelly, P. (2000). Inference of population structure using multilocus genotype data. *Genetics, 155,* 945–959.
49. Ferris, S. D., Sage, R. D., Huang, C. M., Nielsen, J. T., Ritte, U., & Wilson, A. C. (1983). Flow of mitochondrial DNA across a species boundary. *Proceedings of the National Academy of Sciences of the United States of America, 80,* 2290–2294.
50. Wade, C. M., Kulbokas, E. J., III, Kirby, A. W., Zody, M. C., Mullikin, J. C., Lander, E. S., Lindblad-Toh, K., & Daly, M. J. (2002). The mosaic structure of variation in the laboratory mouse genome. *Nature, 420,* 574–578.
51. Wiltshire, T., et al. (2003). Genome-wide single-nucleotide polymorphism analysis defines haplotype patterns in mouse. *Proceedings of the National Academy of Sciences of the United States of America, 100,* 3380–3385.
52. Abe, K., Noguchi, H., Tagawa, K., Yuzuriha, M., Toyoda, A., Kojima, T., Ezawa, K., Saitou, N., Hattori, M., Sakaki, Y., Moriwaki, K., & Shiroishi, T. (2004). Contribution of Asian mouse subspecies *Mus musculus* molossinus to genomic constitution of strain C57BL/6J, as defined by BAC end sequence-SNP analysis. *Genome Research, 14,* 2239–2247.
53. Liu, Y.-H., Takahashi, A., Kitano, T., Koide, T., Shiroishi, T., Moriwaki, K., & Saitou, N. (2008). Mosaic genealogy of the *Mus musculus* genome revealed by 21 nuclear genes from its three subspecies. *Genes and Genetic Systems, 83,* 77–88.
54. Patterson, N., Richter, D. J., Gnerre, S., Lander, E. S., & Reich, D. (2006). Genetic evidence for complex speciation of humans and chimpanzees. *Nature, 441,* 1103–1108.
55. Innan, H., & Watanabe, H. (2006). The effect of gene flow on the coalescent time in the human-chimpanzee ancestral population. *Molecular and Biological Evolution, 23,* 1040–1047.
56. Yamamichi, M., Gojobori, J., & Innan, H. (2012). An autosomal analysis gives no genetic evidence for complex speciation of humans and chimpanzees. *Molecular and Biological Evolution, 29,* 145–156.
57. Theunert, C., Tang, K., Lackmann, M., Hu, S., & Stoneking, M. (2012). Inferring the history of population size change from genome-wide SNP data. *Molecular Biology and Evolution,* 9 Aug 2012 (epub ahead of print).
58. Keinan, A., & Clark, A. G. (2012). Recent explosive human population growth has resulted in an excess of rare genetic variants. *Science, 336,* 740–743.
59. Li, H., & Durbin, R. (2011). Inference of human population history from individual whole-genome sequences. *Nature, 475,* 493–498.
60. Ozaki, K., Ohnishi, Y., Iida, A., Sekine, A., Yamada, R., Tsunoda, T., Sato, H., Sato, H., Hori, M., Nakamura, Y., & Tanaka, T. (2002). Functional SNPs in the lymphotoxin- gene that are associated with susceptibility to myocardial infarction. *Nature Genetics, 32,* 650–654.
61. Yoshiura, K., et al. (2006). A SNP in the ABCC11 gene is the determinant of human earwax type. *Nature Genetics, 38,* 324–330.
62. Tamiya, G., et al. (2005). Whole genome association study of rheumatoid arthritis using 27039 microsatellites. *Human Molecular Genetics, 14,* 2305–2321.
63. Matsunaga, E. (1962). The dimorphism in human normal cerumen. *Annals of Human Genetics, 25,* 273–286.
64. Gudbjartsson, D. F., et al. (2008). Many sequence variants affecting diversity of adult human height. *Nature Genetics, 40,* 609–615.
65. Liu, F., et al. (2012). A genome-wide association study identifies five loci influencing facial morphology in Europeans. *PLoS Biology, 8,* e1002932.
66. Fisher, R. A. (1918). The correlation between relatives on the superposition of Mendelian inheritance. *Transactions of the Royal Society of Edinburgh, 42,* 321–341.

Index

A

ABA, 329
ABCC11, 204
ABCC11 gene, 22, 23, 49, 50, 438, 439
ABCC11 transporter protein, 204
ABO blood group, 51, 64, 141, 277, 291–296, 390, 397, 399, 403
ABO blood group gene, 51, 64, 277, 291–295, 390
Absolute fitness, 126, 130, 132
Accepted point mutations, 361
Accumulated numbers of mutations, 404
Achondroplasia, 45
Acquired immune system, 28, 129, 223, 228
Actinopterygii, 168
Acute myeloid leukaemia, 254
Adaptation, 122, 126, 147
Adaptive radiation, 126
Additive tree method, 374
Adélie penguins, 48
Adenine, 4, 5, 13, 30, 31, 161, 340
Admixture, 84, 259, 431, 433–435
Advantageous mutation, 111, 125, 127, 136, 139, 140, 257
Aegilops tauschii, 211
African, 171, 256, 259, 423–426
  clawed frog, 239
  elephant, 242
Agaricomycotina, 211
Agarose,
Aggrecan, 228, 229
Agnatha, 168, 226, 227
Agrobacterium tumefaciens, 180
AIDS, 141
Ainu, 432, 433
Airborne odorant, 230
Ajellomyces dermatitidis, 211
Algae, 195, 198
Algorithm of the neighbor-joining method, 380–384

Alignment situation, 312
Align seeds only, 328
All biology aspire to evolution, xii
Allele, 40, 68, 91, 126, 277, 375, 421
Allele frequency(ies), 91–94, 101–107, 109, 127–129, 131–133, 135, 137, 139, 141, 143, 147, 375, 398, 399, 403, 407, 421–424, 431, 434, 437
Allo-duplog, 67
Alorese, 433, 434
Alpes, 259
Alphoid DNA, 203
Alu element, 201, 202, 234, 282
Alu sequence, 35, 37, 201, 242
American mastodon, 243
Amino acid molecule, 16
Amino acid replacing substitution, 118
Amino acid substitution, 16, 66, 113–117, 139, 145, 148, 205, 335, 336, 357, 360–362
Aminoacyl synthetase, 160, 161
Ammonium/methylammonium permease, 308
Ammonium transporter, 307, 309
Amniotes, 168, 223, 229, 230, 233, 237, 239–240
Amoebozoa, 163
Amount of divergence, 343, 356, 357, 361, 387
Amphibian, 168, 223, 237, 239–240
Amphioxus, 167, 168, 223, 225, 227, 228
Analogy, 301, 398
Analysis of variance (ANOVA), 439
Ancestral node, 75, 386
Ancestral polymorphism, 437
Anchor, 323–325, 328
Ancient DNA, 48, 243, 259, 266
Ancient genome, 204, 243
Ancient human genome, 243, 249, 259
And then there were none, 395
Angiosperms, 61, 163, 207, 209, 210

Printed in the United States
By Bookmasters